Ab Initio
Molecular
Orbital Theory

AB INITIO MOLECULAR ORBITAL THEORY

WARREN J. HEHRE
University of California, Irvine

LEO RADOM
Australian National University, Canberra

PAUL v.R. SCHLEYER
Universität Erlangen-Nurnberg, Erlangen, West Germany

JOHN A. POPLE
Carnegie-Mellon University, Pittsburgh

A Wiley-Interscience Publication

JOHN WILEY & SONS

New York   Chichester   Brisbane   Toronto   Singapore

Library of Congress Cataloging in Publication Data:

Main entry under title:

Ab initio molecular orbital theory.

"A Wiley-Interscience publication."
Includes bibliographies and index.
1. Molecular orbitals.   2. Quantum chemistry.
I.   Hehre, Warren J.

QD461.A185   1985   541.2'8   84-19524
ISBN 0-471-81241-2

Printed in the United States of America

10   9   8   7   6   5   4   3   2   1

To

NOKO, FAYE, INGE and JOY

PREFACE

The subject of *ab initio* quantum chemistry has long promised to become a major tool for the study of molecular problems of structure, stability, and reaction mechanism. The fundamental techniques, such as molecular orbital theory, date back to the earliest days of quantum mechanics more than half a century ago. However, widespread quantitative applications have only become practically possible in recent times, primarily because of explosive developments in computer hardware and associated achievements in the design of efficient mathematical algorithms. At the present time, the subject has become sufficiently mature and the technology sufficiently stable for practicing experimental chemists to use quantum mechanical methods directly for their own particular applications, rather than to rely on collaboration with a theoretician. In view of the great potential of the field, the authors have long felt that many chemists would welcome a treatise that gave an introductory account of the methodology together with assessments of the reliability of results and examples of potential applications. This book is an attempt to meet such a need.

The book begins with an introduction (Chapter 1) and a simple account of the main features of molecular orbital theory (Chapter 2). No attempt is made to provide detailed derivations of any of the mathematical results; the general user can benefit from the theory without such knowledge. Chapters 3, 4, and 5 describe further features of the theory and give some account of the way in which it is organized as a computational technique and software program. Much of this material is appropriate to the GAUSSIAN series of programs which are now widely available. Chapter 6 is an attempt to evaluate the predictive success of various levels of theory by systematic comparison with experimental data. This aims to educate readers so that they can attach some level of confidence to computational results. Finally, Chapter 7 is an account of recent progress in some fields of chemistry where theory is making major contributions. The examples used are necessarily limited in scope and reflect particular interests of the authors but they should provide a reasonable overview of the current power of *ab initio* theory.

Certain omissions may be noted. No account is given of semiempirical, parameterized methods which parallel *ab initio* theories to some extent. Neither have we attempted to describe some recent more advanced quantum-mechanical methods such as multiconfigurational self-consistent-field theories. The emphasis is on those *ab initio* techniques which are well documented, easy to use, and readily available.

WARREN J. HEHRE
LEO RADOM
PAUL v.R. SCHLEYER
JOHN A. POPLE

Irvine, California
Canberra, Australia
Erlangen, West Germany
Pittsburgh, Pennsylvania
September 1985

ACKNOWLEDGMENTS

Many people have contributed their ideas, the results of their research, and their time to this book during the eight years since its inception. The authors wish to express their gratitude to Jon Baker, Steve Binkley, Willem Bouma, Jayaraman Chandrasekhar, Tim Clark, Doug DeFrees, Kerwin Dobbs, Les Farnell, Michelle Francl, Mike Frisch, Peter Gill, Alan Hinde, Ken Houk, Bob Hout, Eluvathingal Jemmis, Scott Kahn, Elmar Kaufman, Michael Kausch, Alexander Kos, Bev Levi, Patty Pau, Bill Pietro, Steve Pollack, Krishnan Raghavachari, Noel Riggs, Cornelia Rohde, Svein Saezo, Berny Schlegel, Rolf Seeger, Guenther Spitznagel, Bob Whiteside, and Brian Yates for these contributions.

The drafts from which this book finally emerged were prepared by Meg Kessel, Sabrina Mullins, Denise Russell, Kathryn Severn, and Stacie Tibbetts. The drawings were skillfully executed by Arlene Saunders and the photographs prepared with the expert assistance of the staff of Irvine Photographics. Tom Hehre assisted greatly in the preparation of the index. To all of these people we express our sincere thanks. Finally, we acknowledge a large debt to Ross Nobes for his careful and thorough reading of the entire manuscript.

W. J. H.
L. R.
P. v. R. S.
J. A. P.

CONTENTS

Ab Initio
Molecular
Orbital Theory

PROLOGUE

The more progress physical sciences make, the more they tend to enter the domain of mathematics, which is a kind of centre to which they all converge. We may even judge the degree of perfection to which a science has arrived by the facility with which it may be submitted to calculation [1].

Adolphe Quetelet 1796–1874

This book helps to document the extent to which chemistry may now "be submitted to calculation."

The key to theoretical chemistry is molecular quantum mechanics. This is the science relating molecular properties to the motion and interactions of electrons and nuclei. Soon after its formulation in 1925 [2], it became clear that solution of the Schrödinger differential equation could, in principle, lead to direct quantitative prediction of most, if not all, chemical phenomena using only the values of a small number of physical constants (Planck's constant, the velocity of light, and the masses and charges of electrons and nuclei). Such a procedure constitutes an *ab initio* approach to chemistry, independent of any experiment other than determination of these constants. It was also early recognized that solution of the Schrödinger equation was a formidable if not completely impossible mathematical problem for any but the very simplest of systems.

The underlying physical laws necessary for the mathematical theory of a large part of physics and the whole of chemistry are thus completely known, and the difficulty is only that the exact application of these laws leads to equations much too complicated to be soluble [3].

P.A.M. Dirac 1902–1984

1

In practice, the Schrödinger equation has to be replaced by approximate mathematical models for which the possibility of solution exists. The advent of powerful digital computers and of increasingly efficient computer programs has led to significant progress in recent years, both in the development of ever more sophisticated approximate quantum mechanical models and in the application of these models to problems of chemical significance. It is fair to say that theory has now advanced sufficiently far as to provide the chemist with an alternative independent approach to his subject.

It is unlikely that chemistry will change overnight from an experimental to a theoretical science. Much of what now forms its basis may not be subjected to calculation, at least not at present. Nevertheless, it is reasonable to anticipate that an increasing number of chemical investigations which might previously have been performed experimentally will instead be carried out on a digital computer. A number of factors will be responsible for this change in strategy. In many respects theoretical calculations already are more powerful than experiment. They are not bound by practical considerations. Any chemical species may be scrutinized theoretically; calculations on cations, anions, and other reactive intermediates, which might be difficult to investigate experimentally, pose, in principle, no greater problem than calculations carried out on more stable and easily observed molecules. Detailed information about reaction transition structures, excited states, as well as on hypothetical molecular arrangements, deformed molecules, for example, may *only* be obtained by computation. Calculations are easy to perform and involve little human time and effort relative to the large amount of information obtained. They are becoming less and less costly, whereas experimental work is ever more expensive. Minicomputers, available at a price less than that of a double-focusing mass spectrometer or a research nuclear magnetic resonance instrument, are capable of performing all the calculations described in this book. Expense may well be the decisive factor in determining how investigations are to be carried out in the future.

The quantification of chemistry may not be welcomed by all. Thus,

> *Every attempt to employ mathematical methods in the study of chemical questions must be considered profoundly irrational and contrary to the spirit of chemistry. If mathematical analysis should ever hold a prominent place in chemistry—an aberration which is happily almost impossible—it would occasion a rapid and widespread degeneration of that science* [4].
>
> A. Compte 1798–1857

Even today many chemists are uncomfortable with the thought of using a digital computer as an investigative tool. Many are skeptical. Some are uninformed or even prejudiced. They do not believe that theory is capable of making accurate predictions of chemical phenomena. Some may feel that the theoretical methods are too difficult to learn, let alone apply. One of the primary goals of this book is to help overcome these reservations, and show the extent to which theoretical calculations, in their present stage of development, can be employed as a practical means of doing chemistry.

Although theory will continue to help understand in more detail systems that have already been investigated experimentally, an exciting prospect is the exploration of new areas of chemistry. Such areas are remarkably large; the vast majority of chemistry remains to be discovered if one considers the whole of the periodic table! Computational work enables the investigator to scrutinize quickly large numbers of seemingly reasonable molecules about which little or nothing may be known. Calculations can help to indicate whether such species are of sufficient interest to justify their preparation and examination, and whether they will be sufficiently stable to make experimental investigation practical. In many instances, the information available theoretically may suffice to answer questions of interest; there may be no need to carry out confirmatory experimental studies. This book describes some of the progress made by theory in "discovering" new chemistry.

REFERENCES

1. A. Quetelet, *Instructions Populaires sur le Calcul des Probabilities,* Tarlier, Brussels, 1828, p. 230.
2. E. Schrödinger, *Ann. Phys.,* **79**, 361 (1926).
3. P. A. M. Dirac, *Proc. Roy. Soc. (London),* **123**, 714 (1929).
4. A. Compte, *Philosophie Positive,* 1830.

1

INTRODUCTION

1.1. THEORETICAL MODELS

Two broadly different conceptual approaches to the approximate solution of the Schrödinger equation are possible. In the first, each problem is examined at the highest level of theory currently feasible for a system of its size. Very small systems, such as the helium atom or the hydrogen molecule, can clearly be handled at much higher levels of precision than are feasible for systems containing, say, a hundred or more electrons. In the second approach, with which this book is primarily concerned, a level of theory is first clearly defined after which it is applied *uniformly* to molecular systems of *all* sizes up to a maximum determined by available computational resources. Such a theory, if prescribed uniquely for any configuration of nuclei and any number of electrons, may be termed a *theoretical model*, within which all structures, energies, and other physical properties can be explored once the mathematical procedure has been implemented through a computer program. A *theoretical-model chemistry* results. The model may be *tested* by *systematic* comparison of its findings with known experimental results. If comparisons prove favorable, the model acquires some predictive value in situations where experimental data are unavailable. A number of theoretical models are described in this book, and their range of applicability, their successes, and their failures are thoroughly documented.

A theoretical model should possess a number of important characteristics. First, it should be both *unique* and *well defined*. The procedure for obtaining an energy and a wavefunction as an approximate solution of the Schrödinger equation should be completely specified in terms of nuclear positions and the number and spins of the electrons in the molecule. A second desirable feature is *continuity*: all *potential surfaces* should be continuous with respect to nuclear displacements. Special procedures must not be used for symmetrical molecules which might lead to results which are discontinuous with those for structures in which the nuclei are slightly displaced to nonsymmetrical positions. A theoretical model should also be *unbiased*. No appeal to "chemical intuition" should be made in setting up the details of the calculation. For example, while calculations in which electrons are assigned to certain "bond orbitals" might be satisfactory for many molecules, they are not suitable for those nuclear configurations where the locations of "bonds" are apt to be ambiguous. A theory can only be used for the analysis of such concepts as bonding if presuppositions have not been built into its formulation.

Another important requirement for a satisfactory theoretical model is *size-consistency*: relative errors involved in a calculation should increase more or less in proportion to the size of the molecule. This is particularly important if the model is to be used in a comparative manner, relating properties of molecules of different sizes. While it is generally not possible to satisfy this condition fully, it is often possible to construct models that are *size-consistent for infinitely separated systems*. This means that application of the model to a system of several molecules at infinite separation will yield properties that equal the sum of these same properties for the individual molecules.

It is also desirable that a theoretical model be *variational*, that is, yield a total energy that is an *upper bound* to that which would result from exact solution of the full Schrödinger equation.

Finally, a practical theoretical model should be capable of implementation on a computer and be usable with minimal human and computational effort. This enables application of the model to extensive exploration of the properties of a large number of molecules. The development of efficient programs for such models is a major area of research in present-day theoretical chemistry.

1.2. MOLECULAR ORBITAL MODELS

The theoretical models discussed in this book are all based on *molecular orbital (MO) theory*. This approximate treatment of electron distribution and motion assigns individual electrons to one-electron functions termed *spin orbitals*. These comprise a product of spatial functions, termed *molecular orbitals*, $\psi_1(x, y, z)$, $\psi_2(x, y, z)$, $\psi_3(x, y, z)$, . . . , and either α *or* β *spin components*. The spin orbitals are allowed complete freedom to spread throughout the molecule, their exact form being determined variationally to minimize the total energy. In the simplest version of the theory, a single assignment of electrons to orbitals (sometimes called an *electron configuration*) is made. These orbitals are then brought together to form a suitable *many-electron wavefunction* Ψ which is the simplest MO approximation to the solution of the Schrödinger equation.

In practical calculations, the molecular orbitals ψ_1, ψ_2, . . . are further restricted to be linear combinations of a set of N known one-electron functions $\phi_1(x, y, z)$, $\phi_2(x, y, z), \ldots, \phi_N(x, y, z)$:

$$\psi_i = \sum_{\mu=1}^{N} c_{\mu i} \phi_{\mu}. \tag{1.1}$$

The functions $\phi_1, \phi_2, \ldots, \phi_N$ (which are defined in the specification of the model) are known as *one-electron basis functions*, or simply as *basis functions*. They constitute the *basis set*. If the basis functions are the *atomic orbitals* for the atoms making up the molecule, Eq. (1.1) is often described as the *linear combination of atomic orbitals (LCAO) approximation*, and is frequently used in qualitative descriptions of electronic structure.

Given the basis set, the unknown coefficients $c_{\mu i}$ are determined so that the total electronic energy calculated from the many-electron wavefunction is minimized and, according to the *variational theorem*, is as close as possible to the energy corresponding to exact solution of the Schrödinger equation. This energy and the corresponding wavefunction represent the best that can be obtained within the *Hartree–Fock approximation*, that is, the best given the constraints imposed by: (a) the use of a limited basis set in the orbital expansion, and (b) the use of a single assignment of electrons to orbitals.

Hartree–Fock models are the simplest to use for chemical applications and have been employed in many of the studies carried out to date. To specify the model in full, it is only necessary to define a unique basis set $\phi_1, \phi_2, \ldots, \phi_N$ for any nuclear configuration. This is conveniently done by having a standard set of basis functions for each nucleus, centered at the nuclear position, which depend only on the corresponding atomic number. Thus, there would be a set of functions for each hydrogen atom and other sets for each carbon and so forth. In the simplest Hartree–Fock models, the number of basis functions on each atom will be as small as possible, that is, only large enough to accommodate all the electrons and still maintain spherical symmetry. As a consequence, the molecular orbitals (1.1) will have only limited flexibility. If larger basis sets are used, the number of adjustable coefficients in the variational procedure increases, and an improved description of the molecular orbitals is obtained. Very large basis sets will result in nearly complete flexibility. The limit of such an approach, termed the *Hartree–Fock limit*, represents the best that can be done with a single electron configuration. As is shown in later chapters, theoretical-model chemistry at the Hartree–Fock limit has become fairly well characterized, and many of its successes and limitations are now well documented.

The main deficiency of Hartree–Fock theory is its incomplete description of the *correlation* between motions of the electrons. Even with a large and completely flexible basis set, the full solution of the Schrödinger equation cannot be expressed in terms of a *single electron configuration*, that is, a unique assignment of electrons to orbitals. To correct for such a deficiency, it is necessary to use wavefunctions that go beyond the Hartree–Fock level, that is, that represent more than a single electron configuration. If Ψ_0 is the full Hartree–Fock many-electron wavefunction,

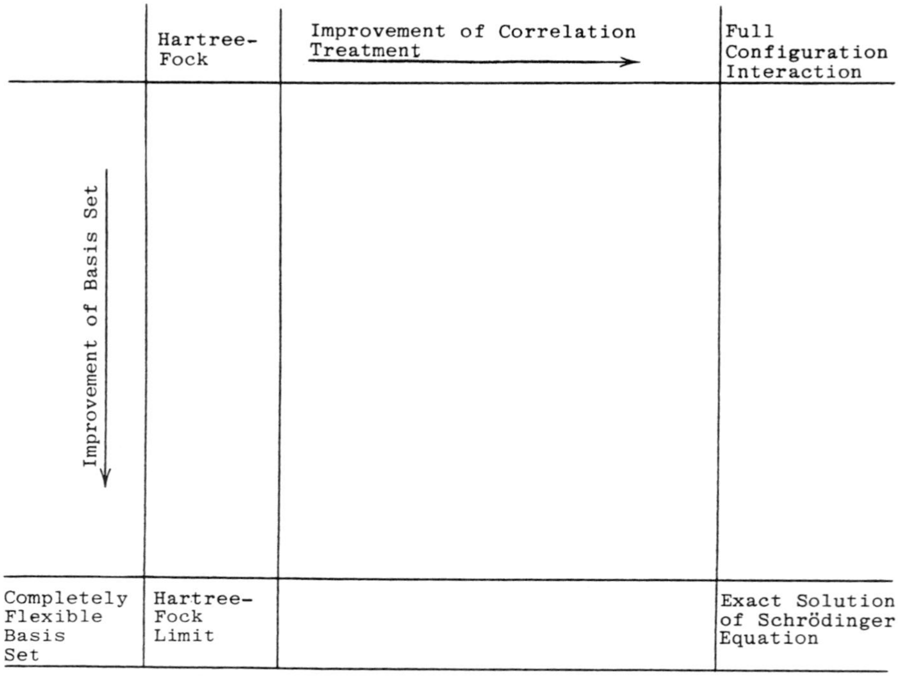

	Hartree-Fock	Improvement of Correlation Treatment $\longrightarrow$	Full Configuration Interaction
Improvement of Basis Set $\downarrow$			
Completely Flexible Basis Set	Hartree-Fock Limit		Exact Solution of Schrödinger Equation

FIGURE 1.1.   Schematic representation of theoretical models showing basis set improvement vertically and correlation improvement horizontally.

the extended approximate form for the more accurate wavefunction Ψ is

$$\Psi = a_0 \Psi_0 + a_1 \Psi_1 + a_2 \Psi_2 + \cdots. \tag{1.2}$$

Here Ψ_1, Ψ_2, . . . are wavefunctions for other configurations, and the linear coefficients a_0, a_1, . . . are to be determined. Inclusion of wavefunctions for all possible alternative electron configurations (within the framework of a given basis set) is termed *full configuration interaction*. It represents the best that can be done using that basis set. Practical methods, which may be sequenced in order of increasing sophistication and accuracy, seek either to limit the number of configurations or to approximate the effect which their inclusion has on the total wavefunction.

The two directions in which theoretical models may be improved can be shown with a two-dimensional chart as in Figure 1.1. The simplest type of model is a Hartree-Fock treatment using a small basis set. This would be placed at the top left of the diagram. As more sophisticated models are applied, an investigation may move downwards (improvement of the basis set) or from left to right (improvement of correlation technique). The bottom row, which may not be realizable in practice, represents various methods using a completely flexible basis. The right-hand column, which may also be impractical, represents full configuration interaction with a given basis. The bottom right-hand corner corresponds to full configuration interaction with a completely flexible basis set. Notice that the two directions on this chart correspond precisely to the two approximations which have been made in order to replace the full Schrödinger equation by practical molecular orbital schemes capable of application to diverse systems. Progression in the vertical direction ("Improvement of Basis Set") corresponds to increasing flexibility of the one-electron spin orbitals [Eq. (1.1)]. Progression in the horizontal direction ("Improvement of Correlation Treatment") corresponds to improved flexibility arising from taking the sum of an increasing number of many-electron functions [Eq. (1.2)]. It follows, therefore, that the bottom right-hand corner of the diagram constitutes the exact solution of the nonrelativistic Schrödinger equation.

The main objective of this book is to describe and to document the performance of a number of theoretical models which comprise such an investigational chart. Many of the applications to date have been at the Hartree–Fock level (left-hand column in Figure 1.1); this is reflected in the coverage of subsequent chapters. However, sufficient investigations have now been carried out beyond Hartree–Fock to enable some assessment of the performance of such theoretical models.

2

THEORETICAL
BACKGROUND

2.1. THE SCHRÖDINGER EQUATION

According to quantum mechanics [1], the energy and many properties of a stationary state of a molecule can be obtained by solution of the Schrödinger partial differential equation,

$$\hat{H}\Psi = E\Psi. \tag{2.1}$$

Here $\hat{H}$ is the *Hamiltonian*, a differential operator representing the total energy. E is the numerical value of the energy of the state, that is, the energy relative to a state in which the constituent particles (nuclei and electrons) are infinitely separated and at rest. Ψ is the *wavefunction*. It depends on the cartesian coordinates of all particles (which may take any value from $-\infty$ to $+\infty$) and also on the spin coordinates (which may take only a finite number of values corresponding to spin angular momentum components in a particular direction). The square of the wavefunction, Ψ^2 (or $|\Psi|^2$ if Ψ is complex), is interpreted as a measure of the probability distribution of the particles within the molecule.

The Hamiltonian $\hat{H}$, like the energy in classical mechanics, is the sum of kinetic and potential parts,

$$\hat{H} = \hat{T} + \hat{V}. \tag{2.2}$$

The kinetic energy operator $\hat{T}$ is a sum of differential operators,

$$\hat{T} = -\frac{h^2}{8\pi^2} \sum_i \frac{1}{m_i} \left(\frac{\partial^2}{\partial x_i^2} + \frac{\partial^2}{\partial y_i^2} + \frac{\partial^2}{\partial z_i^2} \right). \qquad (2.3)$$

The sum is over all particles i (nuclei + electrons) and m_i is the mass of particle i. h is *Planck's constant*. The potential energy operator is the coulomb interaction,

$$\hat{V} = \sum_{i<j} \sum \left(\frac{e_i e_j}{r_{ij}} \right), \qquad (2.4)$$

where the sum is over distinct pairs of particles (i, j) with electric charges e_i, e_j separated by a distance r_{ij}. For electrons, $e_i = -e$, while for a nucleus with atomic number Z_i, $e_i = +Z_i e$.

The Hamiltonian described above is *nonrelativistic*. It ceases to be appropriate as the velocities of the particles, particularly electrons, approach the velocity of light. (Although this is a significant effect for the inner-shell electrons of heavy atoms, the topic is not developed in this book [2].) Certain small magnetic effects, for example, spin–orbit coupling, spin–spin interactions, and so forth, are also omitted in this Hamiltonian; these are usually of minor significance in discussions of chemical energies.

One other restriction has to be imposed on the wavefunctions. The only solutions of (2.1) that are physically acceptable are those with appropriate symmetry under interchange of identical particles. For *boson* particles, the wavefunction is unchanged, that is, *symmetric*, under such interchange. For *fermion* particles, the wavefunction must be multiplied by -1, that is, *antisymmetric*. Electrons are fermions, so that Ψ must be antisymmetric with respect to interchange of the coordinates of any pair of electrons. This is termed the *antisymmetry principle*.

The Schrödinger equation for any molecule will have many solutions, corresponding to different *stationary states*. The state with lowest energy is the *ground state*. Most of the techniques and applications described in this book are concerned with the ground states of molecules.

2.2. SEPARATION OF NUCLEAR MOTION: POTENTIAL SURFACES

The first major step in simplifying the general molecular problem in quantum mechanics is the separation of the nuclear and electronic motions. This is possible because the nuclear masses are much greater than those of the electrons, and, therefore, nuclei move much more slowly. As a consequence, the electrons in a molecule adjust their distribution to changing nuclear positions rapidly. This makes it a reasonable approximation to suppose that the electron distribution depends only on the instantaneous *positions* of the nuclei and not on their *velocities*. In other words, the quantum-mechanical problem of electron motion in the field of *fixed* nuclei may first be solved, leading to an *effective* electronic energy $E^{eff}(\mathbf{R})$ which depends on the *relative nuclear coordinates*, denoted by $\mathbf{R}$. This effective energy is then used as a potential energy for a subsequent study of the nuclear motion. $E^{eff}(\mathbf{R})$ will depend on all of the relative nuclear coordinates. For a diatomic molecule, only the internuclear distance, R, is required and $E^{eff}(\mathbf{R})$ is the potential curve for the molecule. For a polyatomic system, more relative coordinates are needed, and $E^{eff}(\mathbf{R})$ is termed the *potential surface* for the molecule. This separation of the general problem into two parts is frequently called the *adiabatic or Born–Oppenheimer approximation*. It was first examined quantitatively by Born and Oppenheimer [3], who showed that it was valid, provided that the ratio of electron to nuclear mass was sufficiently small.

Quantitatively, the Born–Oppenheimer approximation may be formulated by writing down the Schrödinger equation for electrons in the field of fixed nuclei,

$$\hat{H}^{elec}\Psi^{elec}(\mathbf{r}, \mathbf{R}) = E^{eff}(\mathbf{R})\Psi^{elec}(\mathbf{r}, \mathbf{R}). \tag{2.5}$$

Here, Ψ^{elec} is the electronic wavefunction which depends on the electronic coordinates, $\mathbf{r}$, as well as on the nuclear coordinates, $\mathbf{R}$. The electronic Hamiltonian, $\hat{H}^{elec}$, corresponds to motion of electrons only in the field of fixed nuclei and is

$$\hat{H}^{elec} = \hat{T}^{elec} + \hat{V}, \tag{2.6}$$

where $\hat{T}^{elec}$ is the electronic kinetic energy,

$$\hat{T}^{elec} = -\left(\frac{h^2}{8\pi^2 m}\right) \sum_i^{electrons} \left(\frac{\partial^2}{\partial x_i^2} + \frac{\partial^2}{\partial y_i^2} + \frac{\partial^2}{\partial z_i^2}\right), \tag{2.7}$$

and $\hat{V}$ is the coulomb potential energy,

$$\hat{V} = -\sum_i^{electrons} \sum_s^{nuclei} \frac{Z_s e^2}{r_{is}} + \sum_{i<j}^{electrons} \frac{e^2}{r_{ij}}$$

$$+ \sum_{s<t}^{nuclei} \frac{Z_s Z_t e^2}{R_{st}}. \tag{2.8}$$

The first part of (2.8) corresponds to electron–nuclear attraction, the second to electron–electron repulsion, and the third to nuclear–nuclear repulsion. The last is independent of the electronic coordinates and is a constant contribution to the energy for any particular nuclear configuration.

The main task of theoretical studies of electronic structure is to solve, at least approximately, the electronic Schrödinger equation (2.5), and hence find the effective nuclear potential function $E^{eff}(\mathbf{R})$. From this point, we omit the superscripts in (2.5); it is assumed that the Hamiltonian, $\hat{H}$, wavefunction, Ψ, and energy, E, refer to electronic motion only, each quantity being implicitly a function of the relative nuclear coordinates, $\mathbf{R}$.

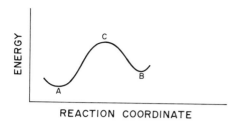

REACTION COORDINATE

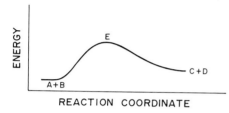

REACTION COORDINATE

FIGURE 2.1. Schematic sections of potential surfaces. In the top section, A and B are distinct minima, corresponding to isomers with different energies. C is the transition structure connecting them. In the bottom section, two species A and B may react (endothermically) to give C + D, E being the transition structure for the reaction.

The potential surface, $E(\mathbf{R})$, is fundamental to the quantitative description of chemical structures and reaction processes. If we deal with the lowest-energy solution of the electronic Schrödinger equation, $E(\mathbf{R})$ is the ground-state potential energy surface. When explored as a function of $\mathbf{R}$, it will generally have a number of local minima, as illustrated schematically in Figure 2.1 (top). These are *equilibrium structures*. The geometry corresponding to a minimum of $E(\mathbf{R})$ would be the geometry a molecule would have if the nuclei were in fact stationary. In practice, finite nuclear motion occurs because of *zero-point vibration*, even at low temperatures. Nevertheless, the potential minimum is usually a good approximation to the averaged structure.

If there are several distinct potential minima, the molecule has a number of *isomeric forms*, and the theory can be used to explore both their structures and their relative energies. In addition, the potential surface may also contain *saddle points*, that is, *stationary points* where there are one or more orthogonal directions in which the energy is at a maximum. In mathematical terms, the second derivative matrix of E with respect to nuclear coordinates has one or more negative *eigenvalues* at such a point. A saddle point with one negative eigenvalue frequently corresponds to a *transition structure* for a chemical reaction. This is defined as the point of lowest maximum energy on a valley connecting two minima on the potential surface. Transition structures also exist for reactions involving separated species. For example, Figure 2.1 (bottom) illustrates a section for a bimolecular reaction A + B → C + D with an intermediate transition structure (E).

2.3. ATOMIC UNITS

Before discussing approximate electronic wavefunctions, it is useful to adopt new units which eliminate the fundamental physical constants from the electronic

Schrödinger equation (2.5). This involves introduction of the *Bohr radius*, a_0, defined by (2.9).

$$a_0 = \frac{h^2}{(4\pi^2 m e^2)} \qquad (2.9)$$

This is the atomic unit of length (the *bohr*). New coordinates (x', y', z') may now be introduced:

$$x' = \frac{x}{a_0} . \qquad (2.10)$$

In a similar way, we introduce a new atomic unit of energy, E_H, which is the coulomb repulsion between two electrons separated by 1 bohr:

$$E_H = \frac{e^2}{a_0} . \qquad (2.11)$$

This unit is termed the *hartree*. New energies (E') are given by

$$E' = \frac{E}{E_H} . \qquad (2.12)$$

If (2.10) and (2.12) are substituted into the Schrödinger equation (2.5), we have

$$\hat{H}'\Psi' = E'\Psi' \qquad (2.13)$$

where the Hamiltonian, $\hat{H}'$, in atomic units, is

$$\hat{H}' = -\frac{1}{2} \sum_{i}^{\text{electrons}} \left(\frac{\partial^2}{\partial x_i'^2} + \frac{\partial^2}{\partial y_i'^2} + \frac{\partial^2}{\partial z_i'^2} \right)$$

$$- \sum_{i}^{\text{electrons}} \sum_{s}^{\text{nuclei}} \left(\frac{Z_s}{r_{is}'} \right) + \sum_{i<j}^{\text{electrons}} \left(\frac{1}{r_{ij}'} \right)$$

$$+ \sum_{s<t}^{\text{nuclei}} \left(\frac{Z_s Z_t}{R_{st}'} \right) . \qquad (2.14)$$

Throughout the rest of this book, we assume atomic units. The primes in all equations have been dropped.

2.4.   MOLECULAR ORBITAL THEORY

Molecular orbital theory is an approach to molecular quantum mechanics which uses one-electron functions or *orbitals* to approximate the full wavefunction. A molecular orbital, $\psi(x, y, z)$, is a function of the cartesian coordinates x, y, z of a single electron. Its square, ψ^2 (or square modulus $|\psi|^2$ if ψ is complex), is interpreted as the probability distribution of the electron in space. To describe the distribution of an electron completely, the dependence on the spin coordinates, ξ, also has to be included. This coordinate takes on one of two possible values $(\pm\frac{1}{2})$, and measures the spin angular momentum component along the z axis in units of $h/2\pi$. For spin aligned along the positive z axis, the spin wavefunction is written $\alpha(\xi)$. Thus,

$$\alpha(+\tfrac{1}{2}) = 1 \qquad \alpha(-\tfrac{1}{2}) = 0. \tag{2.15}$$

Similarly, for spin along the negative z axis, the spin wavefunction is $\beta(\xi)$, so that

$$\beta(+\tfrac{1}{2}) = 0 \qquad \beta(-\tfrac{1}{2}) = 1. \tag{2.16}$$

The complete wavefunction for a single electron is the product of a molecular orbital and a spin function, $\psi(x, y, z)\alpha(\xi)$ or $\psi(x, y, z)\beta(\xi)$. It is termed a *spin orbital*, $\chi(x, y, z, \xi)$.

It might appear that the simplest type of wavefunction appropriate for the description of an n-electron system would be in the form of a product of spin orbitals,

$$\Psi_{\text{product}} = \chi_1(1)\,\chi_2(2) \cdots \chi_n(n), \tag{2.17}$$

where $\chi_i(i)$ is written for $\chi_i(x_i, y_i, z_i, \xi_i)$, the spin orbital of electron i. However, such a wavefunction is not acceptable, as it does not have the property of antisymmetry. If the coordinates of electrons i and j are interchanged in this wavefunction, the product $\cdots \chi_i(i) \cdots \chi_j(j) \cdots$ becomes $\cdots \chi_i(j) \cdots \chi_j(i) \cdots$ which is not equivalent to multiplication by -1. To ensure antisymmetry, the spin orbitals may be arranged in a *determinantal wavefunction*.

$$\Psi_{\text{determinant}} = \begin{vmatrix} \chi_1(1)\,\chi_2(1) \cdots \chi_n(1) \\ \chi_1(2)\,\chi_2(2) \cdots \chi_n(2) \\ \vdots \\ \chi_1(n)\,\chi_2(n) \cdots \chi_n(n) \end{vmatrix} \tag{2.18}$$

Here the elements of the first row of the determinant contain assignations of electron 1 to all the spin orbitals $\chi_1, \chi_2, \ldots, \chi_n$, the second row all possible assignations of electron 2, and so forth.

The determinantal wavefunction (2.18) does have the property of antisymmetry. This is guaranteed because interchange of the coordinates of electrons i and j is equivalent to interchange of rows i and j in the determinant, which does have the effect of changing the sign [4]. On expansion, the determinant becomes a sum of products of spin orbitals,

$$\Psi_{\text{determinant}} = \sum_P (-1)^P \hat{P} [\chi_1(1)\,\chi_2(2)\cdots\chi_n(n)], \qquad (2.19)$$

where $\hat{P}$ is a permutation operator, changing the coordinates $1, 2, \ldots, n$ according to any of the $n!$ possible permutations among the n electrons. $(-1)^P$ is $+1$ or -1 for even and odd permutations, respectively. The wavefunction (2.19) is sometimes called an *antisymmetrized product function*.

In building up a determinantal wavefunction, the usual practice is to choose a set of molecular orbitals, $\psi_1, \psi_2, \psi_3, \ldots$, and then to assign electrons of α or β spin to these orbitals. Since each orbital is later associated with an energy, this assignation of electrons is often represented by an *electron configuration diagram* such as shown in Figure 2.2. The electrons are represented by arrows ($\uparrow$ for α, $\downarrow$ for β), orbitals of lowest energy being at the bottom of the diagram.

It is not possible for a molecular orbital to be occupied by two electrons of the same spin. This is the *Pauli exclusion principle* [5], which follows because the determinantal wavefunction (2.18) vanishes if two columns are identical [4]. Hence orbitals may be classified as doubly occupied (ψ_1, ψ_2 in Figure 2.2), singly occupied (ψ_3) or empty (ψ_4). Most molecules have an even number of electrons in their ground (lowest-energy) states and may be represented by *closed-shell wavefunctions* with orbitals either doubly occupied or empty.

Some further properties of molecular orbital wavefunctions are worth noting. It is possible to force the orbitals to be *orthogonal* to each other, that is,

$$S_{ij} = \int \psi_i^* \psi_j \, dx \, dy \, dz = 0 \quad \text{for } i \neq j. \qquad (2.20)$$

(The asterisk denotes complex conjugation.) This can be accomplished without changing the value of the whole wavefunction by mixing columns of the determinant [4]. We use orthogonal orbitals throughout. The spin functions, α and β, are orthogonal by integration over spin space (actually summation over the two possible values of ξ):

$$\sum_\xi \alpha(\xi)\beta(\xi) = \alpha(+\tfrac{1}{2})\beta(+\tfrac{1}{2}) + \alpha(-\tfrac{1}{2})\beta(-\tfrac{1}{2}) = 0. \qquad (2.21)$$

Molecular orbitals may be *normalized*, that is,

$$S_{ii} = \int \psi_i^* \psi_i \, dx \, dy \, dz = 1, \qquad (2.22)$$

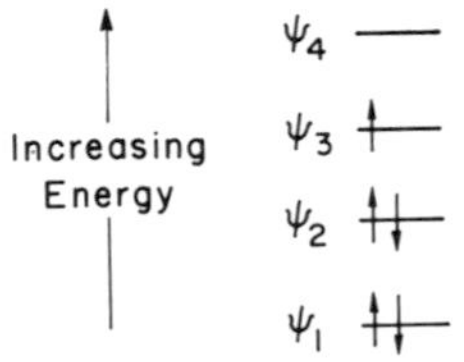

FIGURE 2.2. Electron configuration diagram for $(\psi_1\alpha)(\psi_1\beta)$ $(\psi_2\alpha)(\psi_2\beta)(\psi_3\alpha)$.

by multiplication of the individual ψ_i by a constant. Normalization corresponds to the requirement that the probability of finding the electron anywhere in space is unity. Given (2.22), the determinantal wavefunction (2.18 or 2.19) may be normalized by multiplication by a factor of $(n!)^{-1/2}$, that is,

$$\int \cdots \int \Psi^* \Psi \, d\tau_1 \, d\tau_2 \cdots d\tau_n = 1. \qquad (2.23)$$

Integration in (2.23) is over all coordinates (cartesian and spin) of all electrons.

With these features, we can write down a full many-electron molecular orbital wavefunction for the closed-shell ground state of a molecule with n(even) electrons, doubly occupying $n/2$ orbitals:

$$\Psi = (n!)^{-1/2} \begin{vmatrix} \psi_1(1)\,\alpha(1) & \psi_1(1)\,\beta(1) & \psi_2(1)\,\alpha(1) \cdots \psi_{n/2}(1)\,\beta(1) \\ \psi_1(2)\,\alpha(2) & \psi_1(2)\,\beta(2) & \psi_2(2)\,\alpha(2) \cdots \psi_{n/2}(2)\,\beta(2) \\ \vdots \\ \psi_1(n)\,\alpha(n) & \psi_1(n)\,\beta(n) & \psi_2(n)\,\alpha(n) \cdots \psi_{n/2}(n)\,\beta(n) \end{vmatrix} \qquad (2.24)$$

The determinant (2.24) is often referred to as a *Slater determinant* [6].

2.5. BASIS SET EXPANSIONS

In the previous section, we described how a many-electron wavefunction is constructed from molecular orbitals in the form of a single determinant. In practical applications of the theory, a further restriction is imposed, requiring that the individual molecular orbitals be expressed as linear combinations of a finite set of N prescribed one-electron functions known as *basis functions*. If the basis functions are $\phi_1, \phi_2, \ldots, \phi_N$, then an individual orbital ψ_i can be written

$$\psi_i = \sum_{\mu=1}^{N} c_{\mu i} \phi_\mu \qquad (2.25)$$

where $c_{\mu i}$ are the *molecular orbital expansion coefficients*. (We follow a convention of using Roman subscripts for molecular orbitals and Greek subscripts for basis functions.) These coefficients provide the orbital description with some flexibility, but clearly do not allow for complete freedom unless the ϕ_μ define a complete set. However, the problem of finding the orbitals is reduced from finding complete descriptions of the three-dimensional function ψ_i to finding only a finite set of linear coefficients for each orbital.

In simple qualitative versions of molecular orbital theory, atomic orbitals of constituent atoms are used as basis functions. Such treatments are often described as *linear combination of atomic orbital (LCAO)* theories. However, the mathematical treatment is more general, and any set of appropriately defined functions may be used for a basis expansion.

To provide a basis set that is well defined for any nuclear configuration and therefore useful for a theoretical model (Section 1.2), it is convenient to define a particular set of basis functions associated with each nucleus, depending only on the charge on that nucleus. Such functions may have the symmetry properties of atomic orbitals, and may be classified as $s, p, d, f, \ldots$ types according to their angular properties.

Two types of atomic basis functions have received widespread use. *Slater-type atomic orbitals* (STOs) have exponential radial parts. They are labeled like hydrogen atomic orbitals, $1s, 2s, 2p_x, \ldots$ and have the normalized form,

$$\phi_{1s} = \left(\frac{\zeta_1^3}{\pi}\right)^{1/2} \exp\left(-\zeta_1 r\right)$$

$$\phi_{2s} = \left(\frac{\zeta_2^5}{96\pi}\right)^{1/2} r \exp\left(\frac{-\zeta_2 r}{2}\right)$$

$$\phi_{2p_x} = \left(\frac{\zeta_2^5}{32\pi}\right)^{1/2} x \exp\left(\frac{-\zeta_2 r}{2}\right)$$
$$\vdots$$

$$(2.26)$$

where ζ_1 and ζ_2 are constants determining the size of the orbitals. STOs provide reasonable representations of atomic orbitals with standard ζ-values recommended by Slater [7]. They are, however, not well suited to numerical work, and their use in practical molecular orbital calculations has been limited.

The second type of basis consists of *gaussian-type atomic functions*. These are powers of x, y, z multiplied by $\exp(-\alpha r^2)$, α being a constant determining the size, that is, radial extent, of the function. In normalized form, the first ten such functions are

$$g_s(\alpha, \mathbf{r}) = \left(\frac{2\alpha}{\pi}\right)^{3/4} \exp\left(-\alpha r^2\right)$$

$$g_x(\alpha, \mathbf{r}) = \left(\frac{128\alpha^5}{\pi^3}\right)^{1/4} x \exp\left(-\alpha r^2\right)$$

$$g_y(\alpha, \mathbf{r}) = \left(\frac{128\alpha^5}{\pi^3}\right)^{1/4} y \exp\left(-\alpha r^2\right)$$

$$g_z(\alpha, \mathbf{r}) = \left(\frac{128\alpha^5}{\pi^3}\right)^{1/4} z \exp\left(-\alpha r^2\right)$$

$$g_{xx}(\alpha, \mathbf{r}) = \left(\frac{2048\alpha^7}{9\pi^3}\right)^{1/4} x^2 \exp\left(-\alpha r^2\right)$$

$$g_{yy}(\alpha, \mathbf{r}) = \left(\frac{2048\alpha^7}{9\pi^3}\right)^{1/4} y^2 \exp(-\alpha r^2)$$

$$g_{zz}(\alpha, \mathbf{r}) = \left(\frac{2048\alpha^7}{9\pi^3}\right)^{1/4} z^2 \exp(-\alpha r^2)$$

$$g_{xy}(\alpha, \mathbf{r}) = \left(\frac{2048\alpha^7}{\pi^3}\right)^{1/4} xy \exp(-\alpha r^2)$$

$$g_{xz}(\alpha, \mathbf{r}) = \left(\frac{2048\alpha^7}{\pi^3}\right)^{1/4} xz \exp(-\alpha r^2)$$

$$g_{yz}(\alpha, \mathbf{r}) = \left(\frac{2048\alpha^7}{\pi^3}\right)^{1/4} yz \exp(-\alpha r^2). \tag{2.27}$$

The gaussian functions g_s, g_x, g_y, and g_z have the angular symmetries of the s- and the three p-type atomic orbitals. The six second-order functions g_{xx}, g_{yy}, g_{zz}, g_{xy}, g_{xz}, and g_{yz} do not all have the angular symmetry of atomic orbitals. However, they may be combined to give a set of five d-type atomic functions, that is, g_{xy}, g_{xz}, g_{yz} and the two further functions

$$g_{3zz-rr} = \tfrac{1}{2}(2g_{zz} - g_{xx} - g_{yy})$$

$$g_{xx-yy} = (\tfrac{3}{4})^{1/2}(g_{xx} - g_{yy}). \tag{2.28}$$

A sixth linear combination yields an s-type function,

$$g_{rr} = 5^{-1/2}(g_{xx} + g_{yy} + g_{zz}). \tag{2.29}$$

In a similar manner, the ten third-order gaussian functions may be recombined into a set of seven f-type atomic functions and an additional set of three p functions.

Gaussian-type functions were introduced into molecular orbital computations by Boys [8]. They are less satisfactory than STOs as representations of atomic orbitals, particularly because they do not have a *cusp* at the origin. Nevertheless, they have the important advantage that all integrals in the computations can be evaluated explicitly without recourse to numerical integration.

A third possibility is to use linear combinations of gaussian functions as basis functions. For example, an s-type basis function ϕ_μ may be expanded in terms of s-type gaussians,

$$\phi_\mu = \sum_s d_{\mu s} g_s. \tag{2.30}$$

Here the coefficients $d_{\mu s}$ are fixed. Basis functions of this type are called *contracted gaussians*, the individual g_s being termed *primitive gaussians*.

2.6. VARIATIONAL METHODS AND HARTREE-FOCK THEORY

Up to this point, we have described how a determinantal wavefunction may be constructed from molecular orbitals, and how the orbitals may, in turn, be expanded in terms of a set of basis functions. It remains to specify a method for fixing the expansion coefficients. This is the realm of *Hartree-Fock theory*.

Hartree-Fock theory is based on the *variational method* in quantum mechanics [9]. If Φ is *any* antisymmetric normalized function of the electronic coordinates, then an expectation value of the energy corresponding to this function can be obtained from the integral

$$E' = \int \Phi^* \hat{H} \Phi \, d\tau, \tag{2.31}$$

where integration is over the coordinates of all electrons. The asterisk again denotes complex conjugation. If Φ happens to be the exact wavefunction, Ψ, for the electronic ground state, it will satisfy the Schrödinger equation (2.1). Since Ψ is normalized, E' will therefore be the exact energy E,

$$E' = E \int \Psi^* \Psi \, d\tau = E. \tag{2.32}$$

However, if Φ is *any other* normalized antisymmetric function, it can be shown that E' is greater than E,

$$E' = \int \Phi^* \hat{H} \Phi \, d\tau > E. \tag{2.33}$$

It follows then that if Φ is the antisymmetric molecular orbital function (Eq. 2.24), the energy E' calculated from (2.31) will be too high.

The variational method may be applied to determine optimum orbitals in single-determinant wavefunctions. We select a basis set for orbital expansion, and the coefficients $c_{\mu i}$ (as in Eq. 2.25) may then be adjusted to minimize the expectation value of the energy E'. The resulting value of E' will then be as close to the exact energy E as is possible within the limitations imposed by: (a) the single-determinant wavefunction, and (b) the particular basis set employed. Hence, the *best* single-determinant wavefunction, in an energy sense, is found by minimizing E' with respect to the coefficients $c_{\mu i}$. This implies the variational equations

$$\frac{\partial E'}{\partial c_{\mu i}} = 0 \quad (\text{all } \mu, i). \tag{2.34}$$

We first deal with these equations for closed-shell systems.

2.6.1. Closed-Shell Systems

The variational condition (2.34) leads to a set of algebraic equations for $c_{\mu i}$. They were derived independently for the closed-shell wavefunction (2.24) by Roothaan [10] and by Hall [11]. The *Roothaan–Hall equations* are

$$\sum_{\nu=1}^{N} (F_{\mu\nu} - \epsilon_i S_{\mu\nu})\, c_{\nu i} = 0 \qquad \mu = 1, 2, \ldots, N \tag{2.35}$$

with the normalization conditions

$$\sum_{\mu=1}^{N} \sum_{\nu=1}^{N} c_{\mu i}^{*} S_{\mu\nu} c_{\nu i} = 1. \tag{2.36}$$

Here, ϵ_i is the *one-electron energy* of molecular orbital ψ_i, $S_{\mu\nu}$ are the elements of an $N \times N$ matrix termed the *overlap matrix*,

$$S_{\mu\nu} = \int \phi_{\mu}^{*}(1)\, \phi_{\nu}(1)\, dx_1\, dy_1\, dz_1, \tag{2.37}$$

and $F_{\mu\nu}$ are the elements of another $N \times N$ matrix, the *Fock matrix*,

$$F_{\mu\nu} = H_{\mu\nu}^{\text{core}} + \sum_{\lambda=1}^{N} \sum_{\sigma=1}^{N} P_{\lambda\sigma}\, [(\mu\nu|\lambda\sigma) - \tfrac{1}{2}\,(\mu\lambda|\nu\sigma)]. \tag{2.38}$$

In this expression, $H_{\mu\nu}^{\text{core}}$ is a matrix representing the energy of a single electron in a field of "bare" nuclei. Its elements are

$$H_{\mu\nu}^{\text{core}} = \int \phi_{\mu}^{*}(1)\, \hat{H}^{\text{core}}(1)\, \phi_{\nu}(1)\, dx_1\, dy_1\, dz_1,$$

$$\hat{H}^{\text{core}}(1) = -\frac{1}{2}\left(\frac{\partial^2}{\partial x_1^2} + \frac{\partial^2}{\partial y_1^2} + \frac{\partial^2}{\partial z_1^2}\right) - \sum_{A=1}^{M} \frac{Z_A}{r_{1A}}. \tag{2.39}$$

Here Z_A is the atomic number of atom A, and summation is carried out over all atoms. The quantities $(\mu\nu|\lambda\sigma)$ appearing in (2.38) are *two-electron repulsion integrals*:

$$(\mu\nu|\lambda\sigma) = \int\int \phi_{\mu}^{*}(1)\, \phi_{\nu}(1)\left(\frac{1}{r_{12}}\right)\phi_{\lambda}^{*}(2)\, \phi_{\sigma}(2)\, dx_1\, dy_1\, dz_1\, dx_2\, dy_2\, dz_2.$$

$$\tag{2.40}$$

They are multiplied by the elements of the one-electron *density matrix*, $P_{\lambda\sigma}$,

$$P_{\lambda\sigma} = 2 \sum_{i=1}^{\text{occ}} c_{\lambda i}^{*} c_{\sigma i}. \qquad (2.41)$$

The summation is over *occupied* molecular orbitals only. The factor of two indicates that *two electrons* occupy each molecular orbital, and the asterisk denotes complex conjugation (required if the molecular orbitals are not real functions).

The electronic energy, E^{ee}, is now given by (2.42),

$$E^{\text{ee}} = \tfrac{1}{2} \sum_{\mu=1}^{N} \sum_{\nu=1}^{N} P_{\mu\nu}(F_{\mu\nu} + H_{\mu\nu}^{\text{core}}) \qquad (2.42)$$

which, when added to (2.43), accounting for the internuclear repulsion,

$$E^{\text{nr}} = \sum_{A<B}^{M} \frac{Z_A Z_B}{R_{AB}} \qquad (2.43)$$

(where Z_A and Z_B are the atomic numbers of atoms A and B, and R_{AB} is their separation) yields an expression for the total energy.

The Roothaan–Hall equations (2.35) are not linear since the Fock matrix $F_{\mu\nu}$ itself depends on the molecular orbital coefficients, $c_{\mu i}$, through the density matrix expression (2.41). Solution necessarily involves an iterative process. Since the resulting molecular orbitals are derived from their own effective potential, the technique is frequently called *self-consistent-field (SCF) theory*.

2.6.2. Open-Shell Systems

For open-shell systems, in which electrons are not completely assigned to orbitals in pairs, the Roothaan–Hall equations need modification. This applies, for example, to doublet free radicals or triplet states, for which one component will have an excess of α electrons. For doublets, there will be one extra α electron, for triplets, two extra α electrons, and so forth.

Simple molecular orbital theory can be extended to open-shell systems in two possible ways. The first is described as *spin-restricted Hartree-Fock (RHF) theory* [12]. In this approach, a single set of molecular orbitals is used, some being doubly occupied and some being singly occupied with an electron of α spin. This is the case illustrated for a five-electron doublet state in Figure 2.2. The spin orbitals used in the single determinant are then $(\psi_1 \alpha)\ (\psi_1 \beta)\ (\psi_2 \alpha)\ (\psi_2 \beta)\ (\psi_3 \alpha)$. The coefficients $c_{\mu i}$ are still defined by the expansion (2.25) and their optimum values are still obtained from the variational conditions (2.34). However, details are more complicated since different conditions apply to singly- and doubly-occupied orbitals [12].

The second type of molecular orbital theory in common use for open-shell systems is *spin-unrestricted Hartree-Fock (UHF) theory* [13]. In this approach, different

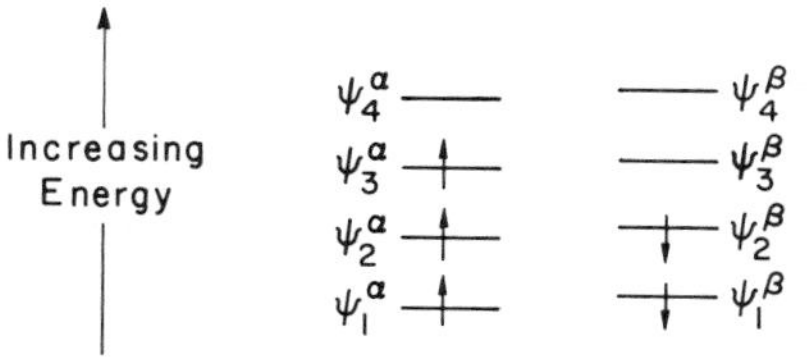

FIGURE 2.3. Electron configuration diagram for $(\psi_1^\alpha \alpha)(\psi_1^\beta \beta)(\psi_2^\alpha \alpha)(\psi_2^\beta \beta)(\psi_3^\alpha \alpha)$.

spatial orbitals are assigned to α and β electrons. Thus, there are two distinct sets of molecular orbitals ψ_i^α and ψ_i^β ($i = 1, \ldots, N$). The electron configuration for a five-electron doublet may be written as $(\psi_1^\alpha \alpha)(\psi_1^\beta \beta)(\psi_2^\alpha \alpha)(\psi_2^\beta \beta)(\psi_3^\alpha \alpha)$ and is illustrated in Figure 2.3. It is important to note that the previously doubly-occupied orbital ψ_1 is now replaced by two distinct orbitals, ψ_1^α and ψ_1^β. Since the RHF function is a special case of the UHF function, it follows from the variational principle that the optimized UHF energy must be below the optimized RHF value. On the other hand, UHF functions have the disadvantage that they are not true eigenfunctions of the total spin operator, unlike exact wavefunctions which necessarily are. Thus, UHF wavefunctions which are designed for doublet states (as in the example shown in Figure 2.3) are contaminated by functions corresponding to states of higher spin multiplicity, such as quartets.

In UHF theory, the two sets of molecular orbitals are defined by two sets of coefficients,

$$\psi_i^\alpha = \sum_{\mu=1}^N c_{\mu i}^\alpha \phi_\mu; \qquad \psi_i^\beta = \sum_{\mu=1}^N c_{\mu i}^\beta \phi_\mu. \tag{2.44}$$

These coefficients are varied independently, leading to the UHF generalizations of the Roothaan–Hall equations [13]. These are

$$\sum_{\nu=1}^N (F_{\mu\nu}^\alpha - \epsilon_i^\alpha S_{\mu\nu}) c_{\mu i}^\alpha = 0$$

$$\sum_{\nu=1}^N (F_{\mu\nu}^\beta - \epsilon_i^\beta S_{\mu\nu}) c_{\mu i}^\beta = 0 \qquad \mu = 1, 2, \cdots, N. \tag{2.45}$$

Here, the two Fock matrices are defined by

$$F_{\mu\nu}^\alpha = H_{\mu\nu}^{core} + \sum_{\lambda=1}^N \sum_{\sigma=1}^N [(P_{\lambda\sigma}^\alpha + P_{\lambda\sigma}^\beta)(\mu\nu|\lambda\sigma) - P_{\lambda\sigma}^\alpha(\mu\lambda|\nu\sigma)]$$

$$F_{\mu\nu}^\beta = H_{\mu\nu}^{core} + \sum_{\lambda=1}^N \sum_{\sigma=1}^N [(P_{\lambda\sigma}^\alpha + P_{\lambda\sigma}^\beta)(\mu\nu|\lambda\sigma) - P_{\lambda\sigma}^\beta(\mu\lambda|\nu\sigma)]. \tag{2.46}$$

The density matrix is also separated into two parts,

$$P_{\mu\nu}^{\alpha} = \sum_{i=1}^{\alpha \text{occ}} c_{\mu i}^{\alpha *} c_{\nu i}^{\alpha}; \qquad P_{\mu\nu}^{\beta} = \sum_{i=1}^{\beta \text{occ}} c_{\mu i}^{\beta *} c_{\nu i}^{\beta}. \qquad (2.47)$$

The integrals $S_{\mu\nu}$, $H_{\mu\nu}^{\text{core}}$ and $(\mu\nu|\lambda\sigma)$ appearing in the UHF equations are the same as those already defined in the Roothaan–Hall procedure for closed-shell calculations.

2.6.3. Koopmans' Theorem and Ionization Potentials

An important theorem due to Koopmans [14, 15] states that the eigenfunctions of the Fock operator for a closed-shell determinant, that is, $(\psi_1)^2(\psi_2)^2 \cdots (\psi_w)^2 \cdots (\psi_{n/2})^2$, are also appropriate for a determinant in which one electron has been removed from ψ_w, that is, $(\psi_1)^2(\psi_2)^2 \cdots (\psi_w) \cdots (\psi_{n/2})^2$. Furthermore, the energy difference between the original species and its ionized form is equal to the corresponding eigenvalue, ϵ_w. The significance of *Koopmans' theorem* is that of all the possible transformed sets of molecular orbitals, each giving the same total determinantal wavefunction, the original set, that is, those corresponding to the closed-shell determinant, is most appropriate for the description of the ionized species.

The Fock eigenvalues, ϵ_w, provide approximate values for the negatives of the ionization potentials of the closed-shell molecule if Koopmans' theorem is applied. However, it should be emphasized that the Koopmans' wavefunction for the ionized species, with orbitals identical to those in the parent, is *not* the full RHF wavefunction for the open-shell configuration $(\psi_1)^2(\psi_2)^2 \cdots (\psi_w) \cdots (\psi_{n/2})^2$. The energy of the ionized species can be further lowered if the RHF wavefunction is completely redetermined. This allows the remaining orbitals to "relax" following the ionization process. It follows that the Koopmans' ionization potentials, ϵ_w, are always numerically larger than those resulting from application of RHF theory separately to the parent and ionized species. On the other hand, the correlation energy correction is normally greater in the unionized molecule because there are more electrons. This means that relaxation and correlation corrections go in opposite directions, sometimes leading to good agreement between the ϵ_w and experimentally observed ionization potentials.

Koopmans' theorem also applies to spin-unrestricted (UHF) Hartree–Fock wavefunctions of the type considered in Section 2.6.2. Thus, for a radical with one extra α electron, the highest eigenvalue of the UHF α Fock matrix is an approximation to the radical ionization potential to give a singlet cation. The theorem may not be generally used, however, for the spin-restricted (RHF) open-shell case.

2.7. SYMMETRY PROPERTIES

In molecules for which the nuclear framework has elements of symmetry, the molecular orbitals belong to *irreducible representations* of the corresponding *point group*, and can be classified accordingly. For example, in an LCAO treatment of the hydrogen molecule, the lowest (unnormalized) molecular orbital is $1s_A + 1s_B$, where $1s_A$ and $1s_B$ are the atomic orbitals. This orbital may be labeled as σ_g, where σ indicates axial symmetry and the g subscript indicates that the orbital is unchanged under

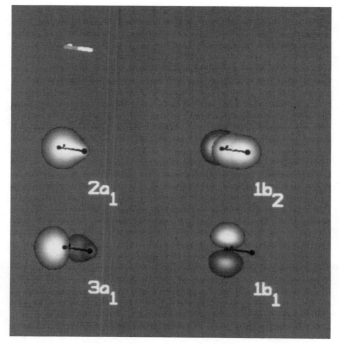

FIGURE 2.4. Valence molecular orbitals of water from lowest energy to highest energy (left to right, top to bottom). 3-21G//3-21G.

inversion. Similarly, a higher-energy orbital (unoccupied in the molecular ground state) is $1s_A - 1s_B$ and may be labeled σ_u, the u subscript indicating change of sign under inversion. For full details of the notation and possible symmetries of molecular orbitals, the reader is directed to other literature [16].

A common practice is to label molecular orbitals by the *irreducible representation* [16], preceded by an integer giving the energy ordering within that symmetry. Thus, the electronic configuration of the water molecule (which belongs to the C_{2v} point group) is found to be

$$(1a_1)^2(2a_1)^2(3a_1)^2(1b_2)^2(1b_1)^2. \tag{2.48}$$

The notation here indicates that the occupied orbitals include three totally symmetric orbitals ($1a_1$, $2a_1$, $3a_1$), one orbital with a node in the molecular plane ($1b_1$), and one with a node perpendicular to the molecular plane ($1b_2$). The valence molecular orbitals for water are shown in Figure 2.4.

2.8. MULLIKEN POPULATION ANALYSIS

The *electron density function* or *electron probability distribution function*, $\rho(\mathbf{r})$, is a three-dimensional function defined such that $\rho(\mathbf{r})\,d\mathbf{r}$ is the probability of finding an electron in a small volume element, $d\mathbf{r}$, at some point in space, $\mathbf{r}$. Normalization requires that

$$\int \rho(\mathbf{r}) \, d\mathbf{r} = n \tag{2.49}$$

where n is the total number of electrons. For a single-determinant wavefunction in which the orbitals are expanded in terms of a set of N basis functions, ϕ_μ, $\rho(\mathbf{r})$ is given by expression (2.50),

$$\rho(\mathbf{r}) = \sum_\mu^N \sum_\nu^N P_{\mu\nu} \phi_\mu \phi_\nu \tag{2.50}$$

where $P_{\mu\nu}$ are elements of the density matrix (2.41). Electron density surfaces for first-row, one-heavy-atom hydrides obtained from (2.50) are shown in Figure 2.5.

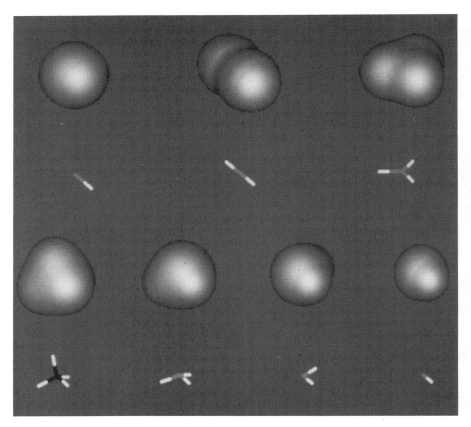

FIGURE 2.5. Electron density surfaces for lithium hydride (top, left), beryllium hydride (top, middle), borane (top, right), methane (bottom, left), ammonia (bottom, middle left), water (bottom, middle right) and hydrogen fluoride (bottom, right). 3-21G//3-21G.

It is desirable to allocate the electrons in some fractional manner among the various parts of a molecule (atoms, bonds, etc.). It may be useful, for example, to define a total electronic charge on a particular atom in a molecule in order that quantitative meaning may be given to such concepts as electron withdrawing or donating ability. Suggestions about how to do this, starting from the density matrix, were made by Mulliken [17]. Collectively they now constitute the topic of *Mulliken population analysis*.

Integration of (2.50) leads to

$$\int \rho(\mathbf{r})\, d\mathbf{r} = \sum_{\mu}^{N} \sum_{\nu}^{N} P_{\mu\nu} S_{\mu\nu} = n, \tag{2.51}$$

where $S_{\mu\nu}$ is the overlap matrix over basis functions (2.37). The total electron count n is thus composed of individual terms $P_{\mu\nu} S_{\mu\nu}$. Given that the basis functions ϕ_{μ} are normalized, that is, $S_{\mu\mu} = 1$, the diagonal terms in (2.51) are just $P_{\mu\mu}$, each of which represents the number of electrons directly associated with a particular basis function ϕ_{μ}. This is termed the *net population* of ϕ_{μ}. The off-diagonal components of (2.51) occur in pairs, $P_{\mu\nu} S_{\mu\nu}$ and $P_{\nu\mu} S_{\nu\mu}$, of equal magnitude. Their sum,

$$Q_{\mu\nu} = 2 P_{\mu\nu} S_{\mu\nu} \qquad (\mu \neq \nu), \tag{2.52}$$

is referred to as an *overlap population*. Note that it is associated with two basis functions, ϕ_{μ} and ϕ_{ν}, which may be on the same atom or on two different atoms.

The total electronic charge is now apportioned into two parts, the first associated with individual basis functions, the second with pairs of basis functions:

$$\sum_{\mu}^{N} P_{\mu\mu} + \sum_{\mu<\nu}^{N} \sum^{N} Q_{\mu\nu} = n. \tag{2.53}$$

Such a representation of the electron distribution is not always convenient, and it is sometimes desirable to partition the total charge among *only* the individual basis functions. One way this may be accomplished is to divide the overlap populations, $Q_{\mu\nu}$, equally between the basis functions ϕ_{μ} and ϕ_{ν}, adding half to each of the net populations $P_{\mu\mu}$ and $P_{\nu\nu}$. This gives a *gross population* for ϕ_{μ}, defined according to

$$q_{\mu} = P_{\mu\mu} + \sum_{\nu \neq \mu} P_{\mu\nu} S_{\mu\nu}. \tag{2.54}$$

The sum of gross populations for all N basis functions, ϕ_{μ}, is, of course, equal to the total electron count,

$$\sum_{\mu}^{N} q_{\mu} = n. \tag{2.55}$$

This particular partitioning scheme is not unique. Nor is any other! The choice of division of the overlap populations $Q_{\mu\nu}$ into equal contributions from ϕ_μ and ϕ_ν is arbitrary.

The gross basis function populations may be used to define *gross atomic populations* according to (2.56),

$$q_A = \sum_\mu^A q_\mu. \tag{2.56}$$

Here the summation is carried out for all functions ϕ_μ on a particular atom A. Assuming all basis functions to be atom centered, it follows that the sum of gross atomic populations is equal to the total electron count. Finally, a *total atomic charge* on A may be defined as $Z_A - q_A$, where Z_A is the atomic number of A. The sum of charges must be 0 for a neutral molecule, +1 for a singly charged cation and so forth.

A *total overlap population*, q_{AB}, between two atoms A and B may be defined in a similar manner,

$$q_{AB} = \sum_\mu^A \sum_\nu^B Q_{\mu\nu}. \tag{2.57}$$

Here summation is carried out for all μ on atom A and all ν on atom B. Total overlap populations provide quantitative information about the binding between atoms. A large positive value for q_{AB} indicates a significant electron population in the region between A and B, at the expense of density in the immediate vicinity of the individual atomic centers. It is a feature generally associated with strong bonding. Conversely, a significant negative value of q_{AB} implies that electrons have been displaced away from the interatomic region, indicating an antibonding interaction.

Although Mulliken populations often provide valuable information regarding electron distributions in molecules, they must be used with some caution for a number of reasons. As mentioned previously, the partitioning of $Q_{\mu\nu}$ into individual orbital contributions is arbitrary. Also, quantities such as gross atomic and total overlap populations are strongly dependent on the particular basis set employed. Finally, certain important features of electronic distributions are not addressed in the analysis. As an example, consider the electron distribution in a tetrahedral sp^3 hybrid orbital,

$$\psi = \tfrac{1}{2}\,\phi_s + \tfrac{1}{2}\,\sqrt{3}\,\phi_p, \tag{2.58}$$

the corresponding density (Eq. 2.50) for which is given by

$$\rho = \tfrac{1}{4}\,\phi_s^2 + \tfrac{1}{2}\,\sqrt{3}\,\phi_s\phi_p + \tfrac{3}{4}\,\phi_p^2. \tag{2.59}$$

The real hybrid orbital is directed away from the nucleus, and the center of electron distribution is some distance from the nuclear position. This feature is primarily due to the cross term, $\frac{1}{2}\sqrt{3}\,\phi_s\phi_p$, which within the framework of the Mulliken population analysis is ignored, because the corresponding one-center overlap integral, $\int\phi_s\phi_p\,d\tau$, is zero. Thus, in this instance the model fails to describe properly the effect.

2.9. MULTIPLE-DETERMINANT WAVEFUNCTIONS

Up to this point, the theory has been developed in terms of single-determinant wavefunctions. Much effort has been put into this level of theory and into the related program development. Indeed, the greater part of the chemical applications described later in this book have been carried out within the framework of Hartree–Fock theory. This effort has largely been justified by the adequacy of the Hartree–Fock descriptions of the ground states of most molecules. Nevertheless, it must be recognized that exact wavefunctions cannot generally be expressed as single determinants. There is a need for further refinement of the theory to eliminate errors implicit in the Hartree–Fock approximation.

In this section, we briefly outline two of the methods that address this problem. Both involve use of a linear combination of Slater determinants, each of which represents an individual electron configuration. Such a wavefunction is said to invoke *configuration interaction*.

The primary deficiency of Hartree–Fock theory is the inadequate treatment of the *correlation between motions of electrons*. In particular, single-determinant wavefunctions take no account of correlation between electrons with opposite spin. Correlation of the motions of electrons with the same spin is partially, but not completely, accounted for by virtue of the determinantal form of the wavefunction. These limitations lead to calculated (Hartree–Fock) energies that are above the exact values. By convention, the difference between the Hartree–Fock and exact (nonrelativistic) energies is the *correlation energy*,

$$E(\text{exact}) = E(\text{Hartree–Fock}) + E(\text{correlation}). \qquad (2.60)$$

The neglect of correlation between electrons of opposite spin leads to a number of qualitative deficiencies in the description of electronic structure. One very important consequence is that the closed-shell Hartree–Fock function often does not dissociate correctly when nuclei are moved to infinite separation. For example, Hartree–Fock theory assigns both electrons in the hydrogen molecule to a symmetric σ_g-type bonding orbital. The molecular orbital wavefunction corresponding to this electron configuration is given by (2.61).

$$2^{-1/2}\,\sigma_g(1)\,\sigma_g(2)\,[\alpha(1)\,\beta(2) - \beta(1)\,\alpha(2)] \qquad (2.61)$$

As the two hydrogen nuclei are moved apart, this description continues to treat the motion of the electrons in the σ_g molecular orbital as uncorrelated. Within a minimal basis set framework, the form of this orbital at large separations is

$$\sigma_g = 2^{-1/2}[1s_A + 1s_B], \tag{2.62}$$

where $1s_A$ and $1s_B$ are the $1s$ atomic orbitals for the constituent hydrogen atoms. Consequently, the cartesian part of the wavefunction is

$$\sigma_g(1)\,\sigma_g(2) = \tfrac{1}{2}\,[1s_A(1)\,1s_A(2) + 1s_A(1)\,1s_B(2) + 1s_B(1)\,1s_A(2)$$
$$+ 1s_B(1)\,1s_B(2)]. \tag{2.63}$$

According to this wavefunction, the two electrons spend half of the time on the same atom (both on A or both on B) and half of the time on different atoms (one on A and one on B), even when the centers are infinitely separated. This is clearly incorrect, since the molecule should dissociate into two neutral hydrogen atoms. In fact, the correct wavefunction for the singlet state of hydrogen at large separation, (2.64),

$$\tfrac{1}{2}\,[1s_A(1)\,1s_B(2) + 1s_B(1)\,1s_A(2)]\,[\alpha(1)\,\beta(2) - \beta(1)\,\alpha(2)] \tag{2.64}$$

cannot be expressed in terms of a single determinant.

The best single-determinant wavefunction may not always have the full symmetry of the exact wavefunction. The benzene radical anion, with the nuclear framework held in a fixed hexagonal geometry (point group D_{6h}), provides an example. Here the odd electron will occupy one of a degenerate pair of antibonding π orbitals with symmetry e_{2u}, leading to a degenerate $^2E_{2u}$ electronic state. The two components have nodal properties as follows:

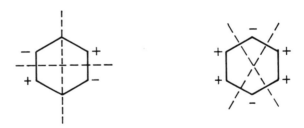

Independent Hartree–Fock calculations, carried out starting with these functions as initial guesses, will yield *two different energies.* Even at the level of the best single determinant, the degeneracy of the state is incorrectly handled, and it is not until configuration interaction is introduced that a proper description results.

Many other deficiencies of Hartree–Fock theory have come to light in the course of carrying out its application. Some of these are documented in Chapter 6 of this book. For example, it has long been recognized that bond dissociation energies are seriously underestimated if correlation between the bonding pair of electrons is not adequately taken into account. Thus, applications to the energies of transition structures, which often involve the partial breaking of bonds, may be suspect if single-determinant wavefunctions are used. Even the properties of "normal" molecules are sometimes subject to considerable errors because of restrictions inherent in the Hartree–Fock approximation. While equilibrium geometries are *usually* given well at the single-determinant level, many examples of significant deviations between Hartree–Fock and experimental bond lengths are now well documented. Calculated molecular vibration frequencies, providing a measure of the shape of the potential surface in the vicinity of the equilibrium structure, have also been shown to be sensitive to the level of treatment given to correlation effects.

At this point, it is useful to review some of the requirements for a satisfactory model chemistry that were introduced in Chapter 1. They play an important role in the selection of methods used in the study of electron correlation. The first requirement is that the method should be *well defined* and *applicable in a continuous manner to any arrangement of nuclei and any number of electrons*. This implies that the choice of determinants or electron configurations must be made *without appeal to special symmetry features*; calculated properties for unsymmetrical structures must be continuous with those for symmetrical arrangements to which they are closely related. The restriction also requires care in the elimination of electron configurations that are presumed to make only small contributions to the total wavefunction, that is, they can only be neglected if their contribution to the total is so small as to fall below an acceptable round-off level. A second requirement is that, whatever method of configuration selection is employed, it must not lead to such a rapid increase in required computation with molecular size as to preclude its use in systems of chemical interest.

A third requirement for a satisfactory model is *size-consistency*. This means that any method must give additive results when applied to an assembly of isolated molecules. Unless this is true, comparison of properties of molecules of different size will not lead to quantitatively meaningful results. Size-consistency is not easy to achieve and plays a major role in the selection of appropriate methods for the calculation of the correlation energy.

A fourth desirable model feature is that the calculated electronic energy be *variational*, that is, it should correspond to an *upper bound* to the energy that would derive from exact solution of the Schrödinger equation. This will be true if the energy is calculated as an expectation value of the Hamiltonian according to the variational theorem (Section 2.6). The advantage of variational methods is that they provide a criterion by which to judge the quality of the theoretical model.

Most Hartree–Fock methods satisfy these four model requirements. Thus, if the basis functions are specified for each atom according only to its atomic number, and if they are centered only at the nuclear positions, the resulting energy surface is usually well defined and variational, at least for electronic ground states. Hartree–Fock models are also generally size-consistent and, given present computer technology, may be widely applied to a variety of chemical systems.

Practical models incorporating electron correlation do not usually satisfy all the requirements! Calculations beyond the Hartree–Fock level first require the selection of the determinants (or electron configurations) which are allowed to participate in the multiple-determinant wavefunction, and then determination of the appropriate linear coefficients. These tasks must be carried out in ways that satisfy the model requirements as closely as possible. In the remainder of this section, we outline a number of methods, beginning with full configuration interaction, which is perfect in principle but usually unachievable in practice, that is, it fails the second requirement, and then continue with other schemes that are practical, but which only satisfy the other requirements to a limited degree.

2.9.1. Full Configuration Interaction

Consider a system comprising n electrons described by a basis set of N functions, ϕ_μ. There will then be $2N$ spin orbital basis functions of the type $\phi_\mu \alpha$ and $\phi_\mu \beta$, which in turn may be linearly combined into $2N$ spin orbitals χ_i. Suppose that we have already solved the Hartree–Fock problem using these basis functions and have obtained the single-determinant wavefunction Ψ_0,

$$\Psi_0 = (n!)^{-1/2} \left| \chi_1 \chi_2 \cdots \chi_n \right|. \tag{2.65}$$

(Here we have specified the determinant (2.24) in abbreviated form.) Note that the spin orbitals utilized in this determinant, $\chi_1, \chi_2, \ldots, \chi_n$, are a subset of the total set which have been determined in the variational procedure. The unused spin orbitals correspond to unoccupied or *virtual* spin orbitals $\chi_a (a = n + 1, n + 2, \ldots, 2N)$. We shall find it useful to denote occupied spin orbitals by subscripts $i, j, k, \ldots$ and virtual ones by $a, b, c, \ldots$.

Determinantal wavefunctions, other than the Hartree–Fock function Ψ_0, may now be constructed by replacing one or more of the occupied spin orbitals χ_i, $\chi_j, \ldots$ in (2.65) by virtual spin orbitals $\chi_a, \chi_b, \ldots$. We shall denote these determinants as Ψ_s with $s > 0$. They may be further classified into single-substitution functions, Ψ_i^a in which χ_i is replaced by χ_a, double-substitution functions, Ψ_{ij}^{ab} in which χ_i is replaced by χ_a and χ_j by χ_b, triple-substitution functions, and so forth. We may write the general substitution determinant, $\Psi_{ijk}^{abc} \cdots$, with the restrictions $i < j < k < \cdots$ and $a < b < c < \cdots$ to avoid repetition of the same configuration. This series of substituted determinants goes all the way to n-substituted terms in which all occupied spin orbitals are replaced by virtual spin orbitals.

In the full configuration interaction method, a trial wavefunction

$$\Psi = a_0 \Psi_0 + \sum_{s > 0} a_s \Psi_s \tag{2.66}$$

is used, where the summation $\sum_{s > 0}$ is over all substituted determinants. The unknown coefficients, a_s, are then determined by the linear variational method, leading to Eq. (2.67),

$$\sum_s (H_{st} - E_i \delta_{st}) a_{si} = 0 \qquad t = 0, 1, 2, \ldots . \tag{2.67}$$

Here, H_{st} is a *configurational matrix element*,

$$H_{st} = \int \cdots \int \Psi_s H \Psi_t \, d\tau_1 \, d\tau_2 \cdots d\tau_n \tag{2.68}$$

and E_i is an energy. The lowest root E of Eq. (2.67) leads to the energy of the electronic ground state. Note the similarity of Eq. (2.67) to the Roothaan–Hall equations (2.35). Also note that, because the determinantal wavefunctions Ψ_s are mutually orthogonal, the overlap matrix S in Eq. (2.35) is replaced by a simple delta function.

The full configuration interaction method represents the most complete treatment possible within the limitations imposed by the basis set. In fact, full CI represents the right-hand column of the two-dimensional model chart (Figure 1.1) discussed in Chapter 1. The difference between the Hartree–Fock energy with a given basis set and the full CI energy is the *correlation energy within the basis*. As the basis set becomes more complete, that is, as $N \rightarrow \infty$, the result of a full configuration interaction treatment will approach the exact solution of the nonrelativistic Schrödinger equation. The full CI method is well-defined, size-consistent, and variational. It is not practical except for very small systems, however, because of the very large number of substituted determinants, the total number of which in Eq. (2.66) is $(2N!)/[n!(2N-n)!]$.

Before proceeding to practical correlation methods, it is useful to examine the general form of the full Hamiltonian matrix, H_{st}. This is illustrated in Figure 2.6, where rows and columns have been arranged starting with Ψ_0, the Hartree–Fock function, and proceeding through single, double, triple substitutions and so forth. It follows from Eq. (2.68) that the element in the upper left-hand corner, designated H_{00}, is the Hartree–Fock energy. Certain blocks of elements in the first row (H_{0s}) or first column (H_{s0}) vanish. If s is a single substitution, H_{0s} vanishes by Brillouin's theorem [18]. If s is a substitution which is triple or higher, H_{0s} again vanishes, due to the fact that the Hamiltonian contains only one- and two-electron

		Single Substitutions	Double Substitutions	Triple and Higher Substitutions
	H_{00}	0	H_{0s}	0
Single Substitutions	0			
Double Substitutions	H_{s0}			
Triple and Higher Substitutions	0			

FIGURE 2.6. Partitioning of full configuration interaction Hamiltonian showing zero sections.

terms. It is only the double substitutions which lead to nonvanishing H_{0s}. As a result, the simplest correlation models account only for determinants formed from Ψ_0 by double substitutions.

2.9.2. Limited Configuration Interaction

The most straightforward way of limiting the length of the CI expansion (2.66) is to truncate the series at a given level of substitution. If no substitutions are permitted, $\Psi = \Psi_0$, corresponding to the Hartree–Fock solution. Inclusion of single substitution functions only, termed *Configuration Interaction, Singles* or CIS,

$$\Psi_{\text{CIS}} = a_0 \Psi_0 + \sum_i^{\text{occ}} \sum_a^{\text{virt}} a_i^a \Psi_i^a, \qquad (2.69)$$

normally leads to no improvement relative to the Hartree–Fock wavefunction or energy. In general, the simplest procedure to have any effect on the calculated energy is limited to double substitutions only, and is termed *Configuration Interaction, Doubles* or CID,

$$\Psi_{\text{CID}} = a_0 \Psi_0 + \sum_{i<j}^{\text{occ}} \sum_{a<b}^{\text{virt}} a_{ij}^{ab} \, \Psi_{ij}^{ab}. \qquad (2.70)$$

This is an important practical procedure. Its execution requires evaluation of the matrix elements H_{0s} and H_{st}, Eq. (2.68), for double substitutions, and solution of Eq. (2.67). Two major computational tasks are involved. The first is a transformation of two-electron integrals $(\mu\nu|\lambda\sigma)$ over basis functions, Eq. (2.40), into corresponding integrals with the Hartree–Fock spin orbitals χ_i replacing the basis functions ϕ_μ. The second is the determination of the lowest (or lowest few) energy solutions of Eq. (2.67) and the associated wavefunction coefficients. As will be shown in Section 3.3.4, both tasks are significant computationally, and considerable effort has been expended toward the development of efficient algorithms.

At a slightly higher level of theory, both single and double substitutions can be included in the configuration interaction treatment. The model is termed *Configuration Interaction, Singles and Doubles*, or CISD. The trial wavefunction is given by

$$\Psi_{\text{CISD}} = a_0 \Psi_0 + \sum_i^{\text{occ}} \sum_a^{\text{virt}} a_i^a \Psi_i^a + \sum_{i<j}^{\text{occ}} \sum_{a<b}^{\text{virt}} a_{ij}^{ab} \Psi_{ij}^{ab}. \qquad (2.71)$$

Here, all coefficients are varied to minimize the expectation value of the energy. Although the single substitutions do not contribute by themselves (in CIS), they do contribute to the wavefunction, Eq. (2.71), since there are nonzero matrix elements of the Hamiltonian between singly- and doubly-substituted determinants. However, since the participation is indirect, the energy lowering due to inclusion of single substitutions is considerably less than that due to doubles.

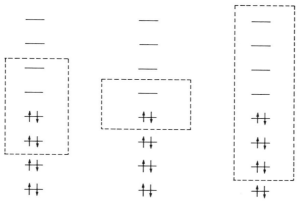

FIGURE 2.7. Use of a window for designating substitutions to be considered in limited configuration interaction. Left: General window. Middle: Minimum sized window allowing only excitations from highest-occupied to lowest-occupied molecular orbitals. 3 × 3 CI. Right: Window excluding participation only of inner-shell orbitals. Frozen-core approximation.

The computational task in CID or CISD calculations can be reduced by limiting the set of spin orbitals that are involved in the single or double substitutions. This is most conveniently done in terms of a *window* (Figure 2.7, left) in which only a set of high-energy occupied and low-energy virtual spin orbitals is used. Two special cases of the use of windows in CI studies are worth noting. The first is the use of the smallest possible size of window for a closed-shell calculation (Figure 2.7, middle). This consists of the *highest-occupied molecular orbital (HOMO)* and the *lowest-unoccupied molecular orbital (LUMO)*. Within this window, there is the Hartree-Fock configuration, $\ldots(\psi_i\alpha)(\psi_i\beta)$, four singly-substituted configurations, $\ldots(\psi_i\alpha)(\psi_a\alpha)$, $\ldots(\psi_i\alpha)(\psi_a\beta)$, $\ldots(\psi_a\alpha)(\psi_i\beta)$, $\ldots(\psi_i\beta)(\psi_a\beta)$, and one doubly-substituted configuration, $\ldots(\psi_a\alpha)(\psi_a\beta)$. Of these six determinants, the two with both α or both β electrons do not mix with Ψ_0 and represent two of the components of the triplet state wavefunctions for the configuration $\ldots(\psi_i)(\psi_a)$. Solution of Eq. (2.67) for the remaining four determinants leads to a third component of the triplet wavefunction, and three solutions corresponding to the ground and two low-lying singlet states. The triplet wavefunction is

$$(2n)^{-1/2}\left\{\left|\ldots(\psi_i\alpha)(\psi_a\beta)\right| - \left|\ldots(\psi_a\alpha)(\psi_i\beta)\right|\right\}, \qquad (2.72)$$

and the three singlet wavefunctions are appropriate linear combinations of the functions (2.73):

$$n^{-1/2}\left|\ldots(\psi_i\alpha)(\psi_i\beta)\right|$$
$$(2n)^{-1/2}\left\{\left|\ldots(\psi_i\alpha)(\psi_a\beta)\right| + \left|\ldots(\psi_a\alpha)(\psi_i\beta)\right|\right\}$$
$$n^{-1/2}\left|\ldots(\psi_a\alpha)(\psi_a\beta)\right|. \qquad (2.73)$$

This technique, which because of its simplicity has received considerable application in studies involving the elucidation of reaction transition structures [19], is sometimes referred to as *3 × 3 configuration interaction*.

For the two-electron problem, the wavefunctions in a 3 × 3 configuration interaction treatment are particularly simple. The triplet functions become

$$2^{-1/2} [\psi_i(1) \, \psi_a(2) - \psi_a(1) \, \psi_i(2)] \, \alpha(1) \, \alpha(2)$$
$$2^{-1} [\psi_i(1) \, \psi_a(2) - \psi_a(1) \, \psi_i(2)] \, [\alpha(1) \, \beta(2) + \beta(1) \, \alpha(2)]$$
$$2^{-1/2} [\psi_i(1) \, \psi_a(2) - \psi_a(1) \, \psi_i(2)] \, \beta(1) \, \beta(2) \tag{2.74}$$

and the singlet functions become linear combinations of

$$2^{-1/2} \psi_i(1) \, \psi_i(2) \, [\alpha(1) \, \beta(2) - \beta(1) \, \alpha(2)]$$
$$2^{-1} [\psi_i(1) \, \psi_a(2) + \psi_a(1) \, \psi_i(2)] \, [\alpha(1) \, \beta(2) - \beta(1) \, \alpha(2)]$$
$$2^{-1/2} \psi_a(1) \, \psi_a(2) \, [\alpha(1) \, \beta(2) - \beta(1) \, \alpha(2)] . \tag{2.75}$$

They are represented pictorially in Figure 2.8.

A second useful type of window constraint is the *frozen-core approximation* (Figure 2.7, right). In this method, all virtual spin orbitals are included in the window, but those occupied spin orbitals which correspond principally to inner-shell electrons are omitted. For each first-row element, that is, lithium to neon, in a molecule, a single molecular orbital is eliminated; for each second-row atom, that is, sodium to argon, four orbitals are eliminated, and so forth. The contributions to the total correlation energy due to inner-shell electrons are not particularly small, but appear to be relatively constant from one molecular environment to another. As a result, the

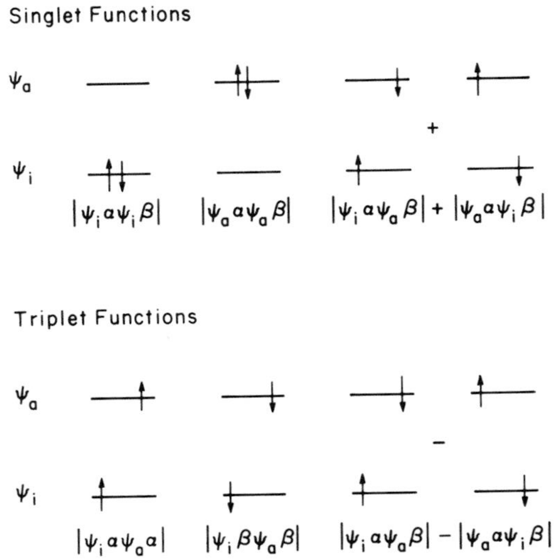

FIGURE 2.8. Wavefunctions resulting from HOMO–LUMO excitations only.

shapes of potential surfaces are little affected by omission of these contributions. The frozen-core approximation has been employed for all correlation energy calculations provided in this book.

These limited configuration interaction methods satisfy some of the general conditions discussed earlier in this section. They are well defined in the general form, although use of a window may lead to difficulties. With the HOMO–LUMO window, for example, the method is ill defined if either the highest-occupied or lowest-unoccupied molecular orbitals belong to a degenerate set. The solution of Eq. (2.67) is practical for reasonably large n and N. The methods are variational, since the energy is calculated as an expectation value. The most serious deficiency of the CID and CISD limited configuration interaction methods is that they fail to satisfy the size-consistency condition. This may be easily seen in terms of a simple example. Suppose that we treat a two-electron system, such as a helium atom, using just two basis functions. With a closed-shell Hartree–Fock treatment there will be one occupied molecular orbital, ψ_0, and one virtual orbital, ψ_1. The CID wavefunction will have the form (considering only the space part)

$$\Psi_{CID} = 2^{-1/2} [a_0 \psi_0(1) \psi_0(2) + a_1 \psi_1(1) \psi_1(2)], \qquad (2.76)$$

where the coefficients a_0 and a_1 are determined variationally. If the theory is now applied to two helium atoms A and B, infinitely separated from each other, the wavefunction in which both atoms are described independently by functions of the form (2.76) will have the product form (2.77),

$$[a_0 \psi_0^A(1) \psi_0^A(2) + a_1 \psi_1^A(1) \psi_1^A(2)] [a_0 \psi_0^B(3) \psi_0^B(4) + a_1 \psi_1^B(3) \psi_1^B(4)]$$
$$= a_0^2 \psi_0^A(1) \psi_0^A(2) \psi_0^B(3) \psi_0^B(4) + a_0 a_1 \psi_0^A(1) \psi_0^A(2) \psi_1^B(3) \psi_1^B(4)$$
$$+ a_1 a_0 \psi_1^A(1) \psi_1^A(2) \psi_0^B(3) \psi_0^B(4) + a_1^2 \psi_1^A(1) \psi_1^A(2) \psi_1^B(3) \psi_1^B(4). \qquad (2.77)$$

The correct Hartree–Fock wavefunction for the whole four-electron system is $\psi_0^A(1) \psi_0^A(2) \psi_0^B(3) \psi_0^B(4)$ (space part only). The corresponding CID wavefunction for the whole system contains terms like the first three appearing in Eq. (2.77), but does *not* include any terms analogous to the fourth part, which represents a *quadruple substitution*. It follows that application of the CID method to the composite four-electron system does *not* give the same wavefunction or energy as independent treatments of the two two-electron systems, and therefore the method is not size consistent. Serious attempts have been made to obtain corrections to the CID and CISD methods. The most commonly used of these is due to Langhoff and Davidson [20], who proposed the approximate formula

$$\Delta E_{correction} = (1 - a_0^2) \Delta E_{CISD} \qquad (2.78)$$

where ΔE_{CISD} is the correlation energy at the CISD level and a_0 is the coefficient of the Hartree–Fock function in the CISD expansion. This corrects a major part of the discrepancy. However, difficulties about this formula, which should be kept in mind, include: (a) the total energy is still not precisely size consistent and (b) a correction is incorrectly applied to a two-electron system, when CISD is equivalent to full configuration interaction.

2.9.3. Møller–Plesset Perturbation Theory

The perturbation theory of *Møller* and *Plesset* [21], closely related to *many-body perturbation theory*, is an alternative approach to the correlation problem. Within a given basis set, its aim is still to find the lowest eigenvalue and corresponding eigenvector of the full Hamiltonian matrix illustrated in Figure 2.6. However, the approach is not to truncate the matrix as in limited CI, but rather to treat it as the sum of two parts, the second being a perturbation on the first.

Møller–Plesset models are formulated by first introducing a generalized electronic Hamiltonian, $\hat{H}_\lambda$, according to

$$\hat{H}_\lambda = \hat{H}_0 + \lambda \hat{V}. \tag{2.79}$$

Here, $\hat{H}_0$ is an operator such that the matrix with elements

$$\int \cdots \int \Psi_s \hat{H}_0 \Psi_t \, d\tau_1 \, d\tau_2 \cdots d\tau_n \tag{2.80}$$

is diagonal. The *perturbation*, $\lambda \hat{V}$, is defined by

$$\lambda \hat{V} = \lambda (\hat{H} - \hat{H}_0), \tag{2.81}$$

where $\hat{H}$ is the correct Hamiltonian and λ is a dimensionless parameter. Clearly $\hat{H}_\lambda$ coincides with $\hat{H}_0$ if $\lambda = 0$, and with $\hat{H}$ if $\lambda = 1$. In Møller–Plesset theory, the zero-order Hamiltonian, $\hat{H}_0$, is taken to be the sum of the one-electron Fock operators. The eigenvalue, E_s, corresponding to a particular determinant, Ψ_s, is the sum of the one-electron energies, ϵ_i, for the spin orbitals which are occupied in Ψ_s.

Ψ_λ and E_λ, the exact or full CI (within a given basis set) ground-state wavefunction and energy for a system described by the Hamiltonian $\hat{H}_\lambda$, may now be expanded in powers of λ according to Rayleigh–Schrödinger perturbation theory [22],

$$\Psi_\lambda = \Psi^{(0)} + \lambda \Psi^{(1)} + \lambda^2 \Psi^{(2)} + \cdots$$
$$E_\lambda = E^{(0)} + \lambda E^{(1)} + \lambda^2 E^{(2)} + \cdots. \tag{2.82}$$

Practical correlation methods may now be formulated by setting the parameter $\lambda = 1$, and by truncation of the series in Eq. (2.82) to various orders. We refer to the methods by the highest-order energy term allowed, that is, truncation after second-order as MP2, after third-order as MP3 and so forth.

The leading terms in expansions (2.82) are

$$\Psi^{(0)} = \Psi_0 \tag{2.83}$$

$$E^{(0)} = \sum_{i}^{occ} \epsilon_i \tag{2.84}$$

$$E^{(0)} + E^{(1)} = \int \cdots \int \Psi_0 \hat{H} \Psi_0 \, d\tau_1 \, d\tau_2 \cdots d\tau_n \tag{2.85}$$

where Ψ_0 is the Hartree–Fock wavefunction and ϵ_i are the one-electron energies defined by Eq. (2.35). The Møller–Plesset energy to first-order is thus the Hartree–Fock energy. Higher terms in the expansion involve other matrix elements of the operator $\hat{V}$. The first-order contribution to the wavefunction is

$$\Psi^{(1)} = \sum_{s>0} (E_0 - E_s)^{-1} V_{s0} \Psi_s, \tag{2.86}$$

where V_{s0} are matrix elements involving the perturbation operator, $\hat{V}$,

$$\int \cdots \int \Psi_s \hat{V} \Psi_0 \, d\tau_1 \, d\tau_2 \cdots d\tau_n. \tag{2.87}$$

It follows that the first-order contribution to the coefficient a_s in Eq. (2.67) is given by

$$a_s^{(1)} = (E_0 - E_s)^{-1} V_{s0}. \tag{2.88}$$

As noted previously (Figure 2.6), V_{s0} vanishes unless s corresponds to a double substitution, so that only such substitutions contribute to the first-order wavefunction.

The second-order contribution to the Møller–Plesset energy is

$$E^{(2)} = -\sum_{s}^{D} (E_0 - E_s)^{-1} \left| V_{s0} \right|^2 \tag{2.89}$$

where $\sum^{D}$ indicates that summation is to be carried out over all double substitutions. *This probably represents the simplest approximate expression for the correlation energy.* If Ψ_s is the double substitution $ij \to ab$, the explicit expression for V_{s0} is

$$V_{s0} = (ij\|ab), \tag{2.90}$$

where $(ij\|ab)$ is a two-electron integral over spin orbitals, defined by

$$(ij\|ab) = \iint \chi_i^*(1) \chi_j^*(2) \left(\frac{1}{r_{12}}\right) [\chi_a(1) \chi_b(2) - \chi_b(1) \chi_a(2)] \, d\tau_1 \, d\tau_2. \tag{2.91}$$

Here integration is over all coordinates (cartesian and spin) for both electrons. The final formula for the second-order contribution to the energy then becomes

$$E^{(2)} = - \sum_{i<j}^{occ} \sum_{a<b}^{virt} (\epsilon_a + \epsilon_b - \epsilon_i - \epsilon_j)^{-1} |(ij||ab)|^2. \tag{2.92}$$

An important point to note is that, unlike the simple CID and CISD configuration interaction schemes, MP2 requires only a partial transformation of the two-electron integrals of Eq. (2.40) into a spin orbital basis.

The third-order contribution to the Møller–Plesset energy also follows directly from Rayleigh–Schrödinger theory. It is

$$E^{(3)} = \sum_s^D \sum_t^D (E_0 - E_s)^{-1} (E_0 - E_t)^{-1} V_{0s}(V_{st} - V_{00}\delta_{st}) V_{t0}, \tag{2.93}$$

where the summations are again carried out over double substitutions only. The matrix elements V_{st} between different double substitutions require a full integral transformation or other techniques of comparable complexity [23]. At the fourth-order level of theory, single, triple, and quadruple substitutions also contribute, since they have nonzero Hamiltonian matrix elements with the double substitutions. The triple substitutions are the most difficult computationally, and some computations have been carried out using only singles, doubles, and quadruples. This partial fourth-order level of theory is termed MP4SDQ.

MP2, MP3, and MP4 energy expressions again satisfy some, but not all, of the model conditions discussed earlier. They are well defined. They can be applied quite widely. In this respect the computational labor for MP2 is dominated by the partial transformation of two-electron integrals. This can be accomplished in $O(nN^4)$ steps. For the MP3 energy, evaluation of Eq. (2.93) requires $O(n^2N^4)$ steps, comparable to the labor involved in one iteration of a CID calculation. At the MP4 level, evaluation of the triple contribution requires $O(n^3N^4)$ steps [24]. MP2, MP3, and MP4 energies do satisfy the size-consistency requirement, as do Møller–Plesset energy expansions terminated at any order. This follows since full CI is size consistent with the Hamiltonian $\hat{H}_\lambda$ for any value of λ; hence, individual terms in Eq. (2.80) must be size consistent. In this respect, the perturbation expressions are more satisfactory than the CID or CISD methods for determining correlation energies. On the other hand, perturbation theory results, terminated at any order, are no longer variational, since they are not derived as expectation values of the Hamiltonian.

2.10. ONE-ELECTRON PROPERTIES: ELECTRIC DIPOLE MOMENTS

We have seen that a general theoretical model for investigating electronic structure leads to an energy, E, and a wavefunction, Ψ. The same method may be applied in the presence of an external perturbation, using the Hamiltonian

$$H(\lambda) = H(0) + \lambda\hat{M}. \tag{2.94}$$

Here $H(0)$ is the full many-electron Hamiltonian for the unperturbed molecule, $\hat{M}$ is an operator describing some physical property of the molecule, and λ is a parameter measuring the strength of the interaction with the external perturbation. In many interactions, $\hat{M}$ is a *one-electron operator*, being the sum of independent contributions from each electron. The *electric dipole moment operator* is an example of such an operator; the corresponding perturbation parameter λ would then be the negative of an applied uniform electric field. In the presence of the perturbation, the theoretical method will lead to an energy $E(\lambda)$ and a wavefunction $\Psi(\lambda)$, depending on the value of λ.

For small perturbations, the value of the property corresponding to the operator $\hat{M}$ is the derivative of the interaction energy with respect to λ at $\lambda = 0$. This is $[dE(\lambda)/d\lambda]_{\lambda=0}$. Alternatively, the value of $\hat{M}$ may be evaluated as an expectation value using the unperturbed wavefunction. This is $\int \Psi^*(0)\hat{M}\Psi(0)\,d\tau$. For the exact wavefunction and energy (obtained by solution of the Schrödinger equation), these procedures are equivalent so that

$$\left[\frac{dE(\lambda)}{d\lambda}\right]_{\lambda=0} = \int \Psi^*(0)\hat{M}\Psi(0)\,d\tau. \tag{2.95}$$

This is a general form of the Hellmann–Feynman theorem [25, 26].

Equation (2.95) also holds for certain approximate theoretical models. In particular, it is true in Hartree–Fock (HF) theory, even if a finite basis is used (provided that the basis is independent of λ) [27, 28]. Hartree–Fock calculations of molecular properties are usually carried out by finding expectation values (right-hand side of Eq. 2.95). For example, evaluation of the electric dipole moment corresponding to a closed-shell single-determinant wavefunction (see Section 2.6.1) is given by

$$\mu(\text{debyes}) = 2.5416\left[\sum_A Z_A \mathbf{r}_A - \sum_\mu\sum_\nu P_{\mu\nu}\mathbf{r}_{\mu\nu}\right]. \tag{2.96}$$

Here the first summation is over atoms A, and the second pair of summations is over basis functions ϕ_μ and ϕ_ν. Z_A is the atomic number of atom A, $\mathbf{r}_A$ the position of atom A relative to the origin, $P_{\mu\nu}$ is an element of the one-electron density matrix defined according to (2.41), and $\mathbf{r}_{\mu\nu}$ is given by

$$\mathbf{r}_{\mu\nu} = \int \phi_\mu(1)\,\mathbf{r}(1)\,\phi_\nu(1)\,d\tau. \tag{2.97}$$

Here $\mathbf{r}$ is a position vector, and integration is carried out over the coordinates of a single electron.

For many theoretical models which go beyond the Hartree–Fock level, Eq. (2.95) does not hold, and the two sides give different values for the operator $\hat{M}$. In particular, Eq. (2.95) does not hold for configuration interaction (CI) with all single and double substitutions from the Hartree–Fock reference (CISD). The extent of failure of Eq. (2.95) will be explored briefly in Section 6.6.

REFERENCES

1. (a) E. Schrödinger, *Ann. Physik,* **79,** 361 (1926). Representative general texts include: (b) E. C. Kemble, *Fundamental Principles of Quantum Mechanics,* McGraw-Hill, New York, 1965; (c) I. N. Levine, *Quantum Chemistry,* 3rd ed., Allyn and Bacon, Boston, 1983; (d) F. L. Pilar, *Elementary Quantum Chemistry,* McGraw-Hill, New York, 1968.

2. For an interesting and readable account of relativistic effects on chemical properties, see: (a) K. S. Pitzer, *Accounts Chem. Res.,* **12,** 271 (1979); also (b) P. Pyykko and J. P. Desclaux, *ibid.,* **12,** 276 (1979).

3. M. Born and J. R. Oppenheimer, *Ann. Physik,* **84,** 457 (1927).

4. For a brief review of the properties of determinants, see reference 1c, pp. 178–183.

5. W. Pauli, *Z. Physik,* **31,** 765 (1925).

6. J. C. Slater, *Phys. Rev.,* **34,** 1293 (1929); **35,** 509 (1930).

7. J. C. Slater, *Phys. Rev.,* **36,** 57 (1930).

8. (a) S. F. Boys, *Proc. Roy. Soc. (London),* **A200,** 542 (1950). For a readable discussion of the properties and uses of gaussian functions in quantum mechanics, see: I. Shavitt in *Methods in Computational Physics,* vol. 2, Wiley, New York, 1962, p. 1.

9. For a discussion, see reference 1c, pp. 172–192.

10. C. C. J. Roothaan, *Rev. Mod. Phys.,* **23,** 69 (1951).

11. G. G. Hall, *Proc. Roy. Soc. (London),* **A205,** 541 (1951).

12. (a) C. C. J. Roothaan, *Rev. Mod. Phys.,* **32,** 179 (1960); (b) J. S. Binkley, J. A. Pople, and P. A. Dobosh, *Mol. Phys.,* **28,** 1423 (1974).

13. J. A. Pople and R. K. Nesbet, *J. Chem. Phys.,* **22,** 571 (1954).

14. T. A. Koopmans, *Physica,* **1,** 104 (1933).

15. See also reference 1d, p. 350.

16. F. A. Cotton, *Chemical Applications of Group Theory,* 2nd ed., Wiley-Interscience, New York, 1971.

17. R. S. Mulliken, *J. Chem. Phys.,* **23,** 1833, 1841, 2338, 2343 (1955).

18. L. Brillouin, *Actualities Sci. Ind.,* **71,** 159 (1934).

19. For a discussion of the use of 3 × 3 CI, see: (a) L. Salem and C. Rowland, *Angew. Chem., Int. Ed. Engl.,* **11,** 92 (1972); applications include: (b) W. J. Hehre, L. Salem, and M. R. Willcott, *J. Am. Chem. Soc.,* **96,** 4328 (1974); (c) R. E. Townshend, G. Ramunni, G. Segal, W. J. Hehre, and L. Salem, *ibid.,* **98,** 2190 (1976).

20. (a) S. R. Langhoff and E. R. Davidson, *Int. J. Quantum Chem.,* **8,** 61 (1974); (b) J. A. Pople, R. Seeger, and R. Krishnan, *ibid., Symp.* **11,** 149 (1977).

21. C. Møller and M. S. Plesset, *Phys. Rev.,* **46,** 618 (1934).

22. For a discussion, see reference 1c, p. 193 ff.

23. J. A. Pople, J. S. Binkley, and R. Seeger, *Int. J. Quantum Chem., Symp.* **10,** 1 (1976).

24. (a) J. A. Pople, R. Krishnan, H. B. Schlegel, and J. S. Binkley, *Int. J. Quantum Chem.,* **14,** 545 (1978); (b) R. Krishnan, M. J. Frisch and J. A. Pople, *J. Chem. Phys.,* **72,** 4244 (1980); (c) M. J. Frisch, R. Krishnan, and J. A. Pople, *Chem. Phys. Lett.,* **75,** 66 (1980).

25. (a) H. Hellmann, *Einführung in die Quantenchemie,* Franz Deuticke, Leipzig, Germany, 1937; (b) R. P. Feynman, *Phys. Rev.,* **41,** 721 (1939). For further discussion, see reference 1c, pp. 404–407.

26. R. Krishnan and J. A. Pople, *Int. J. Quantum Chem.,* **20,** 1067 (1981).

27. A. C. Hurley, *Proc. Roy. Soc. (London),* **A226,** 179 (1954).

28. J. A. Pople, J. W. McIver, Jr., and N. S. Ostlund, *J. Chem. Phys.,* **49,** 2960 (1968).

3

THE COMPUTATIONAL PROBLEM

3.1. INTRODUCTION

The application of quantum mechanical methods to the calculation of molecular properties requires a detailed algorithm or program to translate the mathematical formalism into "step-by-step" instructions suitable for execution on a digital computer. Several factors must be considered in the construction of such an algorithm. It should be as efficient as possible, both in terms of time required to complete a given task, and in its optimal usage of available computational resources. It should be convenient to use, a given task requiring as little human input and intervention as possible. Finally, it should be constructed in a flexible manner, enabling a variety of different tasks to be performed without the necessity of changes to the overall structure of the program. The construction of such an algorithm is a highly involved task; each of the currently available *ab initio* molecular orbital programs represents the combined efforts of several people over a period of years. Most current applications of molecular orbital theory stem from a relatively small number of available programs, and only a few scientists have actually contributed to the coding. Thus, the community of users of molecular orbital theory is at present, and will probably continue to be, mostly unaware of the overall architecture of the programs used and of the details of the numerical procedures involved. The purpose of this chapter is to provide the interested practitioner with a glimpse of the strategies involved in the construction of practical algorithms for molecular orbital calculations. In so doing, we hope to impart to the reader some sense of the relative ease or difficulty of the various tasks involved in an actual calculation, and of the practical constraints and limitations associated with each. This chapter is not an attempt at a broad or in-depth coverage of the highly technical subject of computer algorithm development. Neither is it a user's manual for any particular computer program. Rather, it is a statement of what the authors believe is fundamental to the intelligent and efficient application of available programs to the solution of chemical problems.

The *ab initio* molecular orbital programs that have been written to date, including those which have now been widely distributed (POLYATOM [1], IBMOL [2], HONDO 76 [3], GAUSSIAN 70 [4], GAUSSIAN 76 [5], and GAUSSIAN 80 [6]), differ substantially in many fundamental respects, including overall design or architecture, and in the details of the algorithms for purposes such as integral evaluation and solution of the closed- and open-shell self-consistent-field (SCF) equations. We limit our discussion of general and specific features to the closely related GAUSSIAN series of computer programs. GAUSSIAN 70 [4], GAUSSIAN 76 [5], and GAUSSIAN 80 [6] were released several years ago and are presently in widespread use. Later program systems (GAUSSIAN 82 [7] and GAUSSIAN 85 [8]) have capabilities in excess of previous releases, and are likely largely to replace these earlier versions in the coming years.

3.2. LOGICAL STRUCTURE OF THE COMPUTER PROGRAM

A schematic representation of the overall logical structure of a typical computer program capable of performing the types of molecular orbital calculations described in this book is presented in Figure 3.1. This structure follows closely that of GAUSSIAN 85, and is derived from its predecessors. Two types of logical elements are depicted in the figure, namely program elements (shown as rectangles) and a mass storage element (represented as a circle). The latter actually comprises several distinct logical units, one of random-access type for the storage of the various matrices that appear during the course of a calculation, and others of somewhat greater capacity and of sequential-access type for the storage of two-electron integrals. The individual and independent program elements each serve to perform a specific task, for example, calculation of integrals. They are interconnected to one another only via a master program, labeled *Control program*, and exchange data only via two-way access to the various mass storage units. This type of architecture, the whole being constructed from a number of self-contained and independent modules, is an essential feature. It enables program development and changes to be handled at a local level, that is, on a particular module, without affecting the remaining sections. Just as important, the modular design facilitates user application of the complete program system. Thus, by controlling the flow of logic through the various program segments, it is possible to automate completely tasks such as geometry optimization, and, therefore, to reduce the need for human intervention. Examples are provided in the following sections.

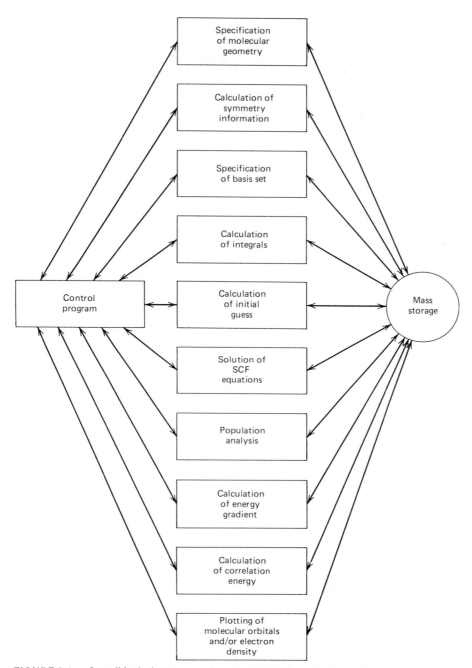

FIGURE 3.1. Overall logical structure of a typical *ab initio* molecular orbital program. Program modules are indicated by rectangles, mass storage elements by circles. The double-headed arrows depict two-way communication between the individual modules and the control program or the mass storage files. Sequence of execution of individual modules is dictated by user input and is controlled by a single master program labeled *Control program*.

Although some general molecular orbital programs may differ considerably from the arrangement shown in Figure 3.1, all contain most of the specified basic logical elements. Thus, our discussion, while directed toward one specific genus of computer programs, is also broadly applicable to many of the other molecular orbital programs.

A single *ab initio* molecular orbital calculation on a molecule with a given geometry might typically proceed as in Figure 3.2. Here the arrows connecting the various modules indicate the direction of logic flow. For example, the arrow linking the element labeled *Specification of basis set* to the element *Calculation of integrals* indicates that upon completion of the former, control is transferred (by way of the control program) to the latter. For clarity, the various communication links between the program modules and the mass storage units have been omitted from the figure.

The calculation starts by specifying the molecular geometry, either in terms of bond lengths and bond angles, or as cartesian coordinates. The former, generally more convenient and better suited for molecular structure optimizations, requires a program to transform the bond lengths and angles into cartesian coordinates. The point group of the molecule may now be determined by the program (using only the available cartesian coordinates and atomic charges) and information relating to molecular symmetry computed and transferred into mass storage. This information may be used to reduce the time required for integral evaluation and for solution of the SCF equations. Section 3.3.5 provides additional details regarding the use of symmetry.

A basis set must now be specified. If it is standard, it is prestored in a library, thereby eliminating the need for its input and reducing the possibility of error. Many of the standard basis sets to be discussed in Chapter 4 have been incorporated directly into the GAUSSIAN series of programs. If a nonstandard basis set is desired, it must be specified in detail for each atom. At this point the one-electron overlap, kinetic energy, and potential energy integrals, and the two-electron repulsion integrals, all of which will be required for the solution of the Roothaan–Hall equations (2.35) [9] (or related equations for systems with unpaired electrons [10]) are calculated. In addition, any one-electron integrals which, although not required in the wavefunction calculation, will be needed later for the evaluation of specific molecular properties, for example, electric dipole moment, are computed. Molecular symmetry may be employed to ensure that the minimum number of unique integrals is evaluated, thereby significantly reducing the computer time required. All one-electron integrals are stored as matrices on random-access files; only nonzero members of the much larger set of electron–electron repulsion integrals are kept; their values are stored sequentially, along with sufficient information to identify each particular element.

As a starting point for the solution of the self-consistent-field equations, a guess at the wavefunction or at the density matrix, Eq. (2.41), is required. The simplest general procedure to obtain such a guess is to diagonalize the one-electron Hamiltonian, Eq. (2.39). A generally more suitable approach is to use the wavefunction from a semiempirical molecular orbital procedure, preferably one which has been specially parameterized to reproduce the results of a given *ab initio* basis set. One possibility is to use an extended Hückel type of scheme [11]. It is computationally

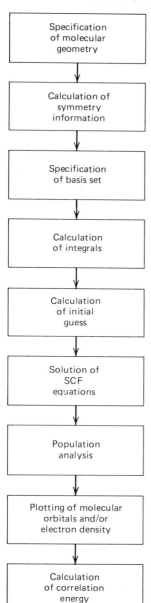

FIGURE 3.2. Sequence of execution of program modules resulting in a single calculation for a fixed basis set and geometry. For clarity, communication links involving the control program and the mass storage elements are omitted from the diagram.

inexpensive and lends itself to easy parameterization. A solution to the Roothaan–Hall equation (or analogous open-shell formalism) may now be attempted. As these equations are not directly soluble, and an iterative procedure must be used, the time required for their solution is directly proportional to the number of iterations taken. Any reduction in this number saves a corresponding fraction of the time.

Considerable effort has been expended in this direction, particularly in the specification of more precise initial wavefunction guesses, and on the development of numerical extrapolation procedures to speed convergence.

Once self-consistency has been achieved, the wavefunction may be printed, and a Mulliken population analysis [12] carried out. This yields atomic and overlap electron populations (see Section 2.8 for a discussion). One-electron properties, for example, electric dipole moments, may also be calculated, making use of integrals which have previously been evaluated and stored. Some or all of the molecular orbitals and/or the total electron density may be plotted. Finally, the Hartree–Fock study may be followed by a calculation of the correlation energy, using one of the methods described in this book (see Section 2.9). Computationally this will generally involve transformation of the two-electron integrals from an atomic orbital to a molecular orbital basis, and the use of these transformed integrals to determine the linear coefficients in a multiple-determinant wavefunction.

A different interconnection of basic program elements enables calculation of molecular equilibrium geometry. As detailed in Figure 3.3, the modular design of the overall program, and the ability to access individual elements *in any order*, makes possible the construction of the loop structures required for such tasks. The computation begins in exactly the same way as previously described for a fixed geometry. After an initial geometry and a basis set have been specified, the required integrals are evaluated and a guess at the wavefunction provided. Execution then proceeds into a programmed loop. First the SCF equations are solved for the total energy and wavefunction. Using this wavefunction, the gradient of the energy, that is, first derivatives of the energy with respect to displacements in the nuclear coordinates, is evaluated. If the gradient is below some preset limit, then the originally specified geometry represents, within that limit, a stationary point on the potential energy surface. The optimization procedure terminates and control transfers out of the loop, in this instance for final printing and population analysis. If, on the other hand, the calculated gradient is larger than the allowed tolerance, the original geometry is varied and a new calculation of integrals, SCF energy, and energy gradient follows. Note that at this point the basis set has already been specified, and may simply be retrieved from mass storage, entirely eliminating the need for further input. Also, it is no longer necessary to perform a semiempirical molecular orbital calculation to furnish an initial guess. The wavefunction resulting from the previous Hartree–Fock calculation conveniently provides an excellent starting point. How many passes through the optimization loop are required depends on several factors, such as the number of independent geometrical variables, the stringency of the convergence requirement, and the choice of starting structure. Experience indicates that for a system of n variables, between n and $2n$ gradient cycles need to be executed in order to ensure that bond lengths and bond angles have converged to within 0.001 Å and $0.1°$, respectively. Additional effort is generally required in the optimization of transition structure geometries using more sophisticated search algorithms [13] frequently involving computation of energy second derivatives.

Evaluation of the gradient by strictly analytical means for some of the multi-determinant schemes may be a formidable task, not yet always computationally feasible. An alternative in these instances is to utilize optimization procedures that

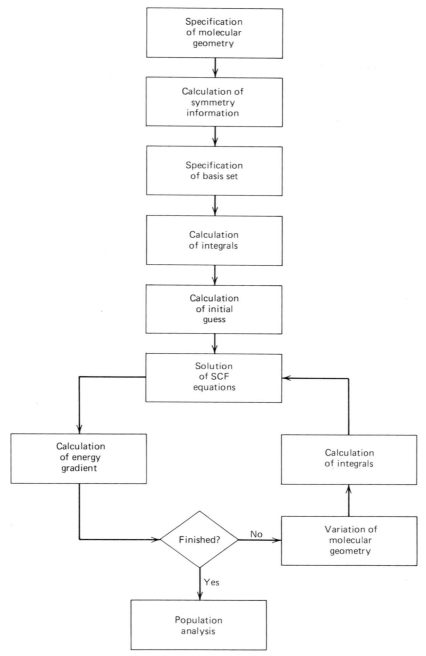

FIGURE 3.3. Sequence of execution of program modules resulting in the optimization of molecular geometry for a particular basis set. The programmed loop involving the elements labeled "Variation of molecular geometry," "Calculation of integrals," "Solution of SCF equations," and "Calculation of energy gradient" is terminated automatically when the calculated geometry has reached a required degree of convergence. For clarity, communication links involving the control program and the mass storage elements have been omitted from the diagram.

do not require the gradient, for example, sequential variation of geometrical parameters with the energy function for each fitted to a parabola, or to obtain energy gradients by numerical as opposed to analytical means. In the GAUSSIAN 82 and GAUSSIAN 85 computer programs, such cases are sometimes handled in the latter manner, using the Fletcher–Powell method [14], and approximate gradients (and diagonal second derivatives) obtained numerically by finite-difference methods. The use of such procedures, as opposed to analytical gradient methods, generally results in a significant (a factor of approximately $n/2$, where n is the number of degrees of freedom) increase in the computer time required for optimization of a given geometrical structure. The exact difference depends on the number of geometrical parameters being varied. Whereas the time for calculation of the energy gradient using analytical methods is largely independent of the number of parameters, numerical evaluation of the total gradient requires at least one energy calculation for each independent variable.

A common practice has been to employ the molecular geometry obtained by optimization at a low level of theory for a single computation at a higher level. This strategy is especially useful for studies using multi-determinant methods, where, as mentioned previously, evaluation of the energy gradient is far more difficult than it is using single-determinant (Hartree–Fock) wavefunctions. What is commonly done here is to utilize Hartree–Fock-level structures for single calculations that treat electron correlation. The overall task (structure optimization at one level of theory followed by a single calculation at another) may be completed either in two independent computational steps, that is, (a) obtain geometry at the Hartree–Fock level, and (b) perform a single calculation using the correlated method, or these steps may be combined. The latter mode not only eliminates all need for human intervention at the intermediate stage, but also allows for the possible sharing of information, for example, integrals, between the separate calculations.

It should by now be obvious to the reader that the inherent design of computer programs for practical molecular orbital calculations allows the user considerable flexibility in their application. In the following chapters we attempt to provide examples of the types of tasks to which these programs can and have been put. In so doing, it is our hope that the enormous potential of the theoretical methods, as presently implemented, wil become abundantly clear.

3.3. COMPUTATIONAL METHODS

This section provides a brief description of some of the more important elements incorporated in existing molecular orbital programs. It may not be of immediate concern to those most interested in the application of molecular orbital techniques, who really wish not to be "bothered" with details of the innermost workings of their tools. For others, it may shed a bit of light inside the "black box". Readers who wish to comprehend fully the detailed numerical algorithms will, on the other hand, probably not be satisfied with the present coverage. They should ultimately familiarize themselves both with the original literature and with the actual code of available programs.

Our treatment touches on four general areas of importance to practical molecular orbital theory: (a) calculation of integrals, (b) solution of the self-consistent equations, (c) evaluation of the energy gradient, and (d) the transformation of integrals to a molecular orbital basis. In addition, we provide brief descriptions of the use of molecular symmetry to simplify the overall calculation, and of procedures for the graphical presentation of molecular orbitals and electron density distributions.

3.3.1. Methods for Integral Evaluation

Before commencing the solution of the Roothaan–Hall equations (2.35), it is necessary to evaluate the elements of both the overlap and one-electron Hamiltonian matrices, Eqs. (2.37) and (2.39), respectively, as well as the set of two-electron integrals, $(\mu\nu|\lambda\sigma)$, defined by Eq. (2.40). These integrals involve basis functions ϕ_μ, which are themselves either individual gaussian functions, termed primitives,

$$g_X(\alpha, \mathbf{r}) = NX \exp(-\alpha r^2) \tag{3.1}$$

or finite linear combinations of gaussians,

$$\phi_\mu = \sum_{k=1}^{K} d_{k\mu} g_X(\alpha_k, \mathbf{r}). \tag{3.2}$$

Here N is a normalization constant (see Section 2.5 for detailed forms), α_k is the gaussian exponent, r is the distance from the center of the function, and X specifies the type (s, p, or d) of gaussian function. X is 1 for zero-order gaussians (s-type), x, y, or z for first-order gaussians (p-type), and x^2, y^2, z^2, xy, xz, or yz for second-order gaussians (d-type), where x, y, and z are cartesian coordinates relative to the center of the function. Note that integral evaluation involving d-type functions is carried out using the six second-order cartesian components, so that an additional s-type function, that is, $x^2 + y^2 + z^2$, is effectively included. Transformation to the more familiar set of five d-type gaussians ($3z^2 - r^2$, $x^2 - y^2$, xy, xz, and yz) may occur at a later stage if desired.

All individual integrals may be evaluated analytically in a straightforward manner following the original method of Boys [15] or the more recent techniques of Rys, Dupuis, and King [16]. It is the relative ease of integral evaluation over gaussians that prompted their use as alternatives to Slater-type (exponential) functions in practical molecular orbital calculations [17]. Early computational work involving basis sets made up of Slater (exponential) functions was plagued with problems of accuracy, resulting from the unavailability of direct analytical expressions for two-electron multi-center integrals, and the need to resort to numerical procedures for their evaluation.

Note that, even for moderate expansion lengths, K, in Eq. (3.2), the number of integrals involving gaussian primitives that need to be evaluated for a single two-electron integral over basis functions is considerable ($O(K^4)$). Consequently, the overall computation time for two-electron integral evaluation may easily become significant, and it is highly desirable to utilize as efficient algorithms as are available.

Computational efficiency of two-electron integral evaluation may be improved in a number of ways [18]. The fact that basis functions ϕ_μ may be grouped into sets termed *shells*, which share common information, may be used to advantage. For example, the set of p-type basis functions [Eq. (3.1), $X = x, y, z$] is generally specified such that all three components have the same contraction coefficients, $d_{k\mu}$, and gaussian exponents, α_k, and differ only in the choice of X. Because of this, it is significantly less costly to evaluate all 81 (3^4) two-electron integrals (2.40) involving four such sets of p functions in one step, rather than to perform the integration individually. In this way, computations involving common information are not repeated unnecessarily. Similarly, simultaneous evaluation of the 1296 (6^4) two-electron integrals involving the set of second-order (d-type) gaussians ($X = x^2$, y^2, z^2, xy, xz, yz) is much more efficient than calculation of the individual terms in an independent manner. Even greater savings may be achieved if the linear combinations defining s- and p-type basis functions are constrained to share a common set of gaussian exponents, α_k, even if different expansion coefficients, $d_{k\mu}$, are employed for different types of functions. Such a set of related basis functions is described as an *sp shell*. Here, efficiency gains are achieved because most of the information required for the evaluation of integrals involving s-type basis functions is also needed in the calculation of integrals involving the p-type functions. The same line of reasoning further suggests the potential benefits of constraints among all three of s-, p-, and d-type basis functions.

Other significant computational advantages stem from the association of related basis functions. For example, grouping into shells facilitates estimation, prior to actual computation, of the magnitudes of two-electron integrals. Many integrals that involve highly compacted inner-shell basis functions, or comprise functions centered on widely distant nuclei, may be too small to warrant keeping. It is often possible to eliminate entire sets of integrals with relatively little effort, far less than would be required if checks were needed on the individual terms.

The actual computer programs in use for calculation of two-electron integrals have been written in such a way as to simplify as much as possible those computational steps that need to be performed K^4 times, at the expense of more elaborate procedures elsewhere. For example, it is often worthwhile to use a different system of coordinate axes in the innermost (K^4) part of the computation, and then transform the integrals to the prescribed axes at a later stage [18]. While modifications of this sort will not be of value for uncontracted basis functions ($K = 1$), their utility increases rapidly with the degree of contraction K. The s and p function integral packages incorporated into all of the GAUSSIAN programs operate in this manner. In addition, the d integral program in GAUSSIAN 85 [8] optionally provides for axis transformation for high contraction levels.

3.3.2. Methods for Solution of the Self-Consistent Equations

The mathematical steps required for the solution of the Roothaan–Hall equations for a closed-shell system [9] are outlined in Figure 3.4. This diagram not only

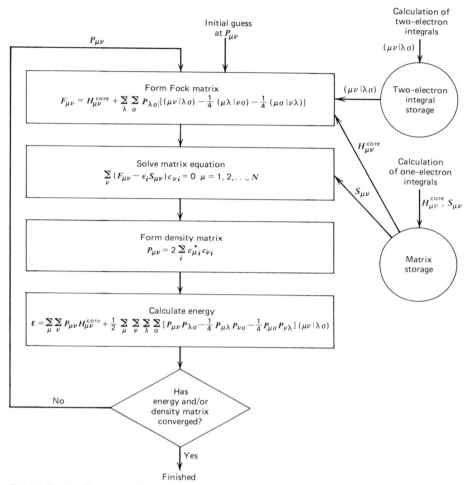

FIGURE 3.4. Sequence of program steps required for the solution of the Roothaan–Hall equations for a closed-shell system. The loop involving the steps labeled "Form Fock matrix," "Solve matrix equation," "Form density matrix," and "Calculate energy" is initiated by specification of a guess at the density matrix, and terminated upon satisfaction of convergence for either the density matrix or the energy. The one- and two-electron integrals ($S_{\mu\nu}$, $H_{\mu\nu}^{core}$, and $(\mu\nu|\lambda\sigma)$) have been calculated and placed in mass storage prior to the start of execution of the SCF procedure. They are accessed during each cycle of the SCF procedure.

represents the overall flow of logic, but also depicts the interaction between the various program elements and the mass storage files containing the matrices and two-electron integrals. At the outset of the calculation, the one-electron matrix elements, $S_{\mu\nu}$ and $H_{\mu\nu}^{core}$ (Eqs. 2.37 and 2.39), and two-electron integrals $(\mu\nu|\lambda\sigma)$ (Eq. 2.40) are available in mass storage. A guess at the density matrix has been provided. The Fock matrix, **F**, is now formed. This is done in the reverse of the order indicated by the formal expression in Eq. (2.38). Thus, each finite two-electron integral, $(\mu\nu|\lambda\sigma)$, is retrieved from sequential-access mass storage, together with its atomic orbital labels, μ, ν, λ, and σ, and its contribution to the appropriate elements of the Fock matrix computed according to Eq. (3.3).

$$
\begin{aligned}
&\text{to } F_{\mu\nu} && P_{\lambda\sigma}(\mu\nu|\lambda\sigma) \\
&\text{to } F_{\lambda\sigma} && P_{\mu\nu}(\mu\nu|\lambda\sigma) \\
&\text{to } F_{\mu\lambda} && -\tfrac{1}{2}P_{\nu\sigma}(\mu\nu|\lambda\sigma) \\
&\text{to } F_{\nu\sigma} && -\tfrac{1}{2}P_{\mu\lambda}(\mu\nu|\lambda\sigma) \\
&\text{to } F_{\mu\sigma} && -\tfrac{1}{2}P_{\nu\lambda}(\mu\nu|\lambda\sigma) \\
&\text{to } F_{\nu\lambda} && -\tfrac{1}{2}P_{\mu\sigma}(\mu\nu|\lambda\sigma).
\end{aligned}
\tag{3.3}
$$

Some modifications are required in the event of coincidences among the μ, ν, λ, and σ. An alternative, and computationally more efficient, scheme for the formation of the two-electron contribution to the Fock matrix was originally proposed by Raffenetti [19] and is now incorporated into the GAUSSIAN 82 and GAUSSIAN 85 computer programs. In this method, integrals over atomic orbitals are first linearly combined,

$$
\begin{aligned}
I_{\mu\nu\lambda\sigma} &= (\mu\nu|\lambda\sigma) - \tfrac{1}{4}(\mu\lambda|\nu\sigma) - \tfrac{1}{4}(\mu\sigma|\nu\lambda) \\
I_{\mu\lambda\nu\sigma} &= (\mu\lambda|\nu\sigma) - \tfrac{1}{4}(\mu\nu|\lambda\sigma) - \tfrac{1}{4}(\mu\sigma|\lambda\nu) \\
I_{\mu\sigma\nu\lambda} &= (\mu\sigma|\nu\lambda) - \tfrac{1}{4}(\mu\nu|\sigma\lambda) - \tfrac{1}{4}(\mu\lambda|\sigma\nu),
\end{aligned}
\tag{3.4}
$$

and the resulting combinations used in lieu of their components to form the Fock matrix according to Eq. (3.5):

$$
\begin{aligned}
&\text{to } F_{\mu\nu} && P_{\lambda\sigma}I_{\mu\nu\lambda\sigma} \\
&\text{to } F_{\lambda\sigma} && P_{\mu\nu}I_{\mu\nu\lambda\sigma}.
\end{aligned}
\tag{3.5}
$$

Again, modifications are required to take account of coincidences [19]. Note that now each integral combination contributes only to two (not six) Fock matrix locations. Although the precombination step (which needs to be performed only once) requires some added computation, it is modest when compared with the savings achieved in the subsequent formation of the Fock matrix (which must normally be repeated many times in order to achieve convergence of the SCF procedure). The Raffenetti procedure has recently been generalized to treat open-shell systems within the framework of the unrestricted Hartree-Fock (UHF) model [20].

Before the Roothaan–Hall equations (2.35) can be solved, they need to be put into the form of a standard eigenvalue problem,

$$
\mathbf{F'c'} = E\mathbf{c'}.
\tag{3.6}
$$

This may be accomplished by way of the matrix transformation

$$\mathbf{F}' = \mathbf{S}^{-1/2} \mathbf{F} \mathbf{S}^{-1/2} \tag{3.7}$$

and subsequent back transformation of the resulting eigenvectors, $\mathbf{c}'$, to their original basis, according to Eq. (3.8),

$$\mathbf{c} = \mathbf{S}^{-1/2} \mathbf{c}'. \tag{3.8}$$

Here $\mathbf{S}^{-1/2}$ is the inverse square root of the matrix of overlap integrals, $S_{\mu\nu}$. Readers may verify for themselves that substitution of Eqs. (3.7) and (3.8) into Eq. (3.6) yields the matrix equivalent of Eq. (2.35).

A new density matrix may now be computed, Eq. (2.41), an electronic energy calculated, Eq. (2.42), and either or both tested for convergence. If final values have been reached to within preset tolerance limits, the iteration cycle is terminated. If, on the other hand, convergence has not been achieved, the cycle is repeated, starting with the formation of a new Fock matrix.

In practice, the density matrix resulting from the termination of one cycle is usually, but not always, used as a starting point for the next. In some instances, convergence to a final solution may be accelerated by extrapolation of a density matrix from the past few SCF iterations [21].

3.3.3. Methods for Evaluation of the Energy Gradient

As noted in Section 3.2, efficient evaluation of the energy gradient (negative forces on the nuclei) plays a central role in searching potential surfaces for stationary points. For single-determinant (Hartree–Fock) theories, this can be handled by analytical differentiation of the corresponding energy expressions. Solution of the Roothaan–Hall equations (2.35) for closed-shell systems leads to a total energy E, which is given by the sum of Eqs. (2.42) and (2.43). Differentiation of this energy with respect to any nuclear coordinate R, leads, after some algebraic manipulation, to the expression [22]

$$\frac{\partial E}{\partial R} = \sum_{\mu} \sum_{\nu} P_{\mu\nu} \left(\frac{\partial H_{\mu\nu}^{\text{core}}}{\partial R} \right) + \frac{1}{2} \sum_{\mu} \sum_{\nu} \sum_{\lambda} \sum_{\sigma} P_{\mu\nu} P_{\lambda\sigma} \left(\frac{\partial}{\partial R} \right) (\mu\nu | \lambda\sigma)$$

$$+ \left(\frac{\partial E^{nr}}{\partial R} \right) - \sum_{\mu} \sum_{\nu} W_{\mu\nu} \left(\frac{\partial S_{\mu\nu}}{\partial R} \right). \tag{3.9}$$

Here $P_{\mu\nu}$ is the density matrix (2.41), $S_{\mu\nu}$ and $H_{\mu\nu}^{\text{core}}$ the overlap and core Hamiltonian matrices, Eqs. (2.37) and (2.39), respectively, and E^{nr} the nuclear repulsion energy (2.43). The two-electron integrals $(\mu\nu | \lambda\sigma)$ are defined according to Eq. (3.10),

$$(\mu\nu | \lambda\sigma) = \int\int \phi_{\mu}^{*}(1) \phi_{\lambda}^{*}(2) \left(\frac{1}{r_{12}} \right) [(\phi_{\nu}(1) \phi_{\sigma}(2) - \phi_{\sigma}(1) \phi_{\nu}(2)] \, d\tau_1 \, d\tau_2$$

$$\tag{3.10}$$

and the energy-weighted density matrix, W, by Eq. (3.11),

$$W_{\mu\nu} = \sum_i^{\text{occ}} \epsilon_i c_{\mu i}^* c_{\nu i}. \tag{3.11}$$

Here, the ϵ_i are the one-electron energies appearing in Eq. (2.35). It is important to note that explicit differentiation of the density matrix $P_{\mu\nu}$ is not required to evaluate Eq. (3.9). Only integral derivatives remain. For gaussian-type basis functions, these may be calculated analytically using the same numerical techniques [15, 16, 18] as employed for evaluation of the original one- and two-electron integrals, This is because the derivatives of gaussian-type functions may themselves be expressed in terms of gaussians [15, 17]. Thus, the cartesian derivative of a gaussian s-type function involves a p-type function. Similarly, differentiation of a gaussian p-type function leads to both s- and d-type gaussians.

Since the general two-electron integral $(\mu\nu|\lambda\sigma)$ involves gaussians on four centers, there will be a total of twelve cartesian derivatives for each. In practice, however, it is not actually necessary to evaluate all of these, since invariance to translation demands that the sum of the four derivatives for a given cartesian direction be zero. This reduces the number of independent derivatives to nine. By going to an intermediate (three-center) axis system in the inner-loop (see Section 3.3.1), it is possible to reduce the number of required integral derivatives to a maximum of six. The reader is referred elsewhere for a more thorough discussion of the computational methods for efficient gradient evaluation [22, 23].

An important point about Hartree–Fock gradient calculations is that the many integral derivatives do not need to be kept in mass storage. After evaluation, each such derivative is multiplied by an appropriate density matrix element to give contributions to the forces on the four nuclei involved and may then be discarded.

3.3.4. Methods for Integral Transformation

As noted in Section 2.9, theoretical treatments beyond the Hartree–Fock level generally commence with a full or partial transformation of two-electron integrals from the original localized atom-centered basis, ϕ_μ, into delocalized molecular orbitals, ψ_p:

$$(pq|rs) = \int \int \psi_p(1)\, \psi_q(1) \left(\frac{1}{r_{12}}\right) \psi_r(2)\, \psi_s(2)\, d\tau_1\, d\tau_2$$

$$= \sum_\mu^N \sum_\nu^N \sum_\lambda^N \sum_\sigma^N c_{\mu p} c_{\nu q} c_{\lambda r} c_{\sigma s} (\mu\nu|\lambda\sigma). \tag{3.12}$$

Here, the $(\mu\nu|\lambda\sigma)$ are the two-electron integrals over atomic orbitals (2.40), and molecular orbitals ψ_p are defined in the usual manner as linear combinations of N basis functions ϕ_μ according to Eq. (3.13):

$$\psi_p = \sum_{\mu}^{N} c_{\mu p} \phi_{\mu}. \tag{3.13}$$

At first sight, this transformation appears to require $O(N^8)$ steps, since there is an $O(N^4)$ sum for each of the $O(N^4)$ new integrals $(pq|rs)$. However, this requirement can be reduced substantially if the transformation is carried out in four separate steps, the first of these to the set of mixed integrals of Eq. (3.14),

$$(p\nu|\lambda\sigma) = \int\int \psi_p(1)\,\phi_\nu(1)\left(\frac{1}{r_{12}}\right)\phi_\lambda(2)\,\phi_\sigma(2)\,d\tau_1\,d\tau_2$$

$$= \sum_{\mu} c_{\mu p}(\mu\nu|\lambda\sigma). \tag{3.14}$$

Evaluation of the array $(p\nu|\lambda\sigma)$ requires only $O(N^5)$ operations since an N-fold sum is required for each of the $O(N^4)$ new integrals. We can proceed by successive transformations,

$$(\mu\nu|\lambda\sigma) \longrightarrow (p\nu|\lambda\sigma) \longrightarrow (pq|\lambda\sigma) \longrightarrow (pq|r\sigma) \longrightarrow (pq|rs), \tag{3.15}$$

each step also requiring $O(N^5)$ operations.

For correlation treatments using Møller–Plesset perturbation theory [24] terminated at second order (MP2), only integrals with two occupied and two virtual orbitals are required. Thus, the transformation, Eq. (3.12), needs to be carried out only for p occupied. This reduces the computational problem from $O(N^5)$ to $O(nN^4)$ steps, where n is the number of electrons. Higher-level (third- and fourth-order) Møller–Plesset expansions require more integrals, but a complete transformation can still be avoided.

In addition to concerns over the *computation time* required for integral transformation, attention must also be focused on the availability of random-access memory. The two-electron integral arrays are generally too large to be accommodated in memory at one time. An algorithm has been developed enabling the sequential transformation outlined above to be carried out more efficiently using one random-access array with $\frac{1}{2}N^3$ elements together with a few smaller $O(N^2)$ arrays. The scheme outlined below is incorporated into the GAUSSIAN 82 and GAUSSIAN 85 computer programs.

1. For a *single* molecular orbital ψ_p, read from mass storage the two-electron integrals $(\mu\nu|\lambda\sigma)$ in turn, multiply by $c_{\mu p}$, $c_{\nu p}$, $c_{\lambda p}$, and $c_{\sigma p}$ and accumulate the contributions to $(p\nu|\lambda\sigma)$ in the $\frac{1}{2}N^3$ array. Note that the original integral set $(\mu\nu|\lambda\sigma)$ needs to be read only once for each molecular orbital ψ_p. Given the integrals $(p\nu|\lambda\sigma)$ for one p and all ν, λ, and σ, steps 2–4 in the transformation can be carried out entirely in the random-access memory.

2. For a single value of q, obtain the $O(N^2)$ set $(pq|\lambda\sigma)$ according to Eq. (3.16),

$$(pq|\lambda\sigma) = \sum_{\nu} c_{\nu q}(p\nu|\lambda\sigma). \tag{3.16}$$

3. For a single value of r, obtain the $O(N)$ set $(pq|r\sigma)$ according to Eq. (3.17),

$$(pq|r\sigma) = \sum_\lambda c_{\lambda r}(pq|\lambda\sigma). \tag{3.17}$$

4. Finally, for a single value of s, obtain a single $(pq|rs)$,

$$(pq|rs) = \sum_\sigma c_{\sigma s}(pq|r\sigma). \tag{3.18}$$

If necessary, each integral $(pq|rs)$ may be written out to mass storage as soon as it is calculated. Note, however, that for the second-order (MP2) correlation treatment, it is not necessary to store the transformed integrals. The energy expression, Eq. (2.92), is a sum over individual double substitutions; it is possible to collect all the $(pq|rs)$ for one term in memory at the same time. Each term may then be added into the total as the integrals become available.

The storage requirement of a $\frac{1}{2}N^3$ array is still quite large; for 80 basis functions it is over a quarter of a million words. Larger cases may be handled by performing multiple passes over the original two-electron integral file.

3.3.5. Use of Molecular Symmetry

Molecular symmetry can be employed both to simplify integral evaluation and to aid in the solution of the Roothaan–Hall equations. For example, symmetry may be used to block the various matrices occurring in the calculation, thereby reducing the number of steps required for operations such as matrix multiplication and diagonalization. Perhaps the most important application of symmetry in Hartree–Fock theory is to reduce the number of two-electron repulsion integrals and integral derivatives to be evaluated, ideally to those that are *unique by symmetry*. For example, a calculation on the H_2 molecule, in which each hydrogen is represented by a single s-type basis function (S_A and S_B, respectively) requires evaluation of six two-electron integrals,

$$(S_A S_A | S_A S_A)$$
$$(S_B S_A | S_A S_A)$$
$$(S_B S_B | S_A S_A)$$
$$(S_B S_A | S_B S_A)$$
$$(S_B S_B | S_B S_A)$$
$$(S_B S_B | S_B S_B). \tag{3.19}$$

Here, we have already taken account of the trivial (permutational) symmetry of two-electron integrals, that is, $(\mu\nu|\lambda\sigma) = (\mu\nu|\sigma\lambda) = (\nu\mu|\lambda\sigma) = (\nu\mu|\sigma\lambda) = (\lambda\sigma|\mu\nu) = (\lambda\sigma|\nu\mu) = (\sigma\lambda|\mu\nu) = (\sigma\lambda|\nu\mu)$. Given that the two hydrogen atoms are identical by symmetry, that is, are interconverted by reflection in a σ plane, the first two of the integrals are numerically identical to the last two:

$$(S_A S_A | S_A S_A) \xrightarrow{\sigma} (S_B S_B | S_B S_B)$$
$$(S_B S_A | S_A S_A) \xrightarrow{\sigma} (S_A S_B | S_B S_B) \equiv (S_B S_B | S_B S_A). \qquad (3.20)$$

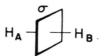

Therefore, only the first four (or the last four) members of the complete set of six two-electron integrals actually need to be calculated.

The same arguments apply to two-electron integrals involving p-type and higher-order atomic basis functions. Note that integrals that are related by symmetry will not necessarily have the same numerical values. Rather, they can be transformed into one another by symmetry-group operations, transformations which are effected by matrix multiplication. For example, a clockwise rotation through an angle θ about the z axis is described by the following matrix transformation:

$$\begin{bmatrix} \cos\theta & -\sin\theta & 0 \\ \sin\theta & \cos\theta & 0 \\ 0 & 0 & 1 \end{bmatrix}. \qquad (3.21)$$

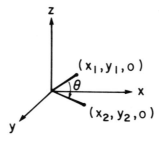

Thus, an integral not in the symmetry-unique set involving a p_x-type basis function would be related to integrals involving both p_x and p_y components.

In computer programs such as GAUSSIAN 85, which take full advantage of molecular symmetry, integrals which may be derived from some unique set by operations of the molecular point group are not evaluated. Their contribution to the Fock matrix may be obtained according to the method of Dacre [25] and Elder [26] as modified by Dupuis and King [27] and first incorporated into the HONDO 76 computer program [3].

3.3.6. Methods for Generation of Three-Dimensional Molecular Orbital and Total Electron Density Plots

In the simplest approach to the graphical presentation of molecular orbitals, contributions to a given molecular orbital from the various atomic functions are first independently sketched onto a picture of the skeleton of the molecule, and are then "smeared together" to indicate delocalization of the orbital away from the atomic centers. The resulting "smears" may be portrayed simply as ellipses, or the services of an artist may be employed to represent them as smooth half-tone ellipsoids, yielding some sense of three-dimensionality. This approach is clearly too qualitative to be of use for the examination of differences in orbital structure between closely related molecules. It requires significant human intervention in order to construct realistic three-dimensional representations, and because of this, the quality and character of the resulting portrayals will be nonuniform. Construction of accurate graphical representations of total electron densities (defined according to Eq. (2.50) for closed-shell systems) by inspection is likely to be even less straightforward; they are comprised of contributions from not one but all occupied molecular orbitals.

The orbital pictures for water (Figure 2.4), as well as those for the total electron densities for this and other first-row hydrides (Figure 2.5), have been obtained using an automated scheme [28] that seeks to view the function surface, that is, $|\psi(x, y, z)| = $ constant or $\rho(x, y, z) = $ constant, as a solid object. The best analogy is to a camera, the purpose of which is to construct an accurate two-dimensional portrayal (photograph) of an actual three-dimensional object.

Definition of the viewable surface of a particular molecular orbital or total electron density begins with the construction of a three-dimensional box that will contain the entire surface (or that portion of the surface which is of interest). The front panel, which acts as a viewing screen, is then grided, and for each point in the grid a one-dimensional search is carried out, perpendicular to the surface, and from the front to the back of the box. Each search determines whether or not the absolute value of the function reaches some preset level, that is, defining the surface of the function, and if so, at what depth this value is encountered. Details have been presented elsewhere [28].

What remains to be done is to assign a shading value (color and intensity) to each of the exposed points on the surface. The shading model used is due to Cook and Torrance [29] and takes into account *ambient* illumination as well as the *diffuse* and *specular reflection* which occurs with low- and high-intensity light sources. Specific parameters, determined experimentally, that characterize the material out of which the object being viewed is fabricated, are introduced into the scheme. Specific details have already been presented [28a, 29]. Implementation of the Cook–Torrance shading model requires calculation of the surface normal, that is, gradient, at every sampling point. This is greatly facilitated by separation of the atomic orbitals ϕ_μ, which linearly combine to make up the set of delocalized molecular orbitals, that is, Eq. (2.25), into radial and angular parts,

$$\phi_\mu(r, \theta, \phi) = R_\mu(r)\, \Theta_\mu(\theta, \phi). \qquad (3.22)$$

The individual cartesian derivatives, that is,

$$\left(\frac{\partial \psi_i}{\partial x}\right) = \sum_{\mu=1}^{N} c_{\mu i} \left(\frac{\partial R_\mu}{\partial x}\right) \Theta_\mu + R_\mu \left(\frac{\partial \Theta_\mu}{\partial x}\right) \tag{3.23}$$

are evaluated by interpolation of the radial component (which depends only on the *distance* between the contributing atomic center and the surface point) and calculation of the appropriate angular terms. The gradient of the total electron density, Eq. (2.50) for closed-shell systems, may also be expressed in terms of the individual molecular orbitals and their derivatives

$$\left(\frac{\partial \rho}{\partial x}\right) = 4 \sum_{i=1}^{occ} \psi_i \left(\frac{\partial \psi_i}{\partial x}\right). \tag{3.24}$$

REFERENCES

1. POLYATOM 2, D. B. Newmann, H. Basch, R. L. Korregay, L. C. Snyder, J. Moskowitz, C. Hornback, and P. Liebman, Program no. 199, Quantum Chemistry Program Exchange, Indiana University, Bloomington, Ind.

2. IBMOL, A. Veillard, available from IBM, San Jose, Ca.

3. HONDO 76, M. Dupuis, J. Rys, and H. F. King, Program no. 336, Quantum Chemistry Program Exchange, Indiana University, Bloomington, Ind.

4. GAUSSIAN 70, W. J. Hehre, W. A. Lathan, M. D. Newton, R. Ditchfield, and J. A. Pople, Program no. 236, Quantum Chemistry Program Exchange, Indiana University, Bloomington, Ind.

5. GAUSSIAN 76, J. S. Binkley, R. Whiteside, P. C. Hariharan, R. Seeger, W. J. Hehre, M. D. Newton, and J. A. Pople, Program no. 368, Quantum Chemistry Program Exchange, Indiana University, Bloomington, Ind.

6. GAUSSIAN 80, J. S. Binkley, R. A. Whiteside, R. Krishnan, R. Seeger, D. J. DeFrees, H. B. Schlegel, S. Topiol, L. R. Kahn, and J. A. Pople, Program no. 406, Quantum Chemistry Program Exchange, Indiana University, Bloomington, Ind.

7. GAUSSIAN 82, J. S. Binkley, M. J. Frisch, D. J. DeFrees, K. Rahgavachari, R. A. Whiteside, H. B. Schlegel, E. M. Fluder, and J. A. Pople, Department of Chemistry, Carnegie-Mellon University, Pittsburgh, PA.

8. GAUSSIAN 85, R. F. Hout, Jr., M. M. Francl, S. D. Kahn, K. D. Dobbs, E. S. Blurock, W. J. Pietro, D. J. DeFrees, S. K. Pollack, B. A. Levi, R. Steckler, and W. J. Hehre, program to be submitted to Quantum Chemistry Program Exchange, Indiana University, Bloomington, Ind.

9. (a) C. C. J. Roothaan, *Rev. Mod. Phys.,* **23,** 69 (1951); (b) G. G. Hall, *Proc. Roy. Soc. (London),* **A205,** 541 (1951).

10. (a) C. C. J. Roothaan, *Rev. Mod. Phys.,* **32,** 179 (1960); (b) J. S. Binkley, J. A. Pople, and P. A. Dobosh, *Mol. Phys.,* **28,** 1432 (1974); (c) J. A. Pople and R. K. Nesbet, *J. Chem. Phys.,* **22,** 571 (1954).

11. R. Hoffmann, *J. Chem. Phys.,* **39,** 1397 (1963).

12. R. S. Mulliken, *J. Chem. Phys.,* **23,** 1833, 1841, 2338, 2343 (1955).

13. See: H. B. Schlegel, *J. Comput. Chem.,* **3,** 214 (1982).

14. (a) R. Fletcher and M. J. D. Powell, *Comput. J.*, **6**, 163 (1963); see also (b) J. B. Collins, P. v. R. Schleyer, J. S. Binkley, and J. A. Pople, *J. Chem. Phys.*, **64**, 5142 (1976).

15. S. F. Boys, *Proc. Roy Soc. (London)*, **A200**, 542 (1950).

16. (a) M. Dupuis, J. Rys, and H. F. King, *J. Chem. Phys.*, **65**, 111 (1976); (b) H. F. King and M. Dupuis, *J. Comput. Phys.*, **21**, 144 (1976).

17. For a discussion of the properties of gaussian functions, see: I. Shavitt in *Methods in Computational Physics,* vol. 2, Wiley, New York, 1962, p. 1.

18. (a) J. A. Pople and W. J. Hehre, *J. Comput. Phys.*, **27**, 161 (1978); (b) E. S. Blurock and W. J. Hehre, manuscript in preparation.

19. R. C. Raffenetti, *Chem. Phys. Lett.*, **20**, 335 (1973).

20. J. S. Binkley, J. A. Pople, B. A. Levi, D. J. DeFrees, and W. J. Hehre, unpublished.

21. (a) For a discussion of numerical techniques for assisting convergence of the self-consistent equations, see: D. R. Hartree, *The Calculation of Atomic Structures,* Wiley, New York, 1957. See also: (b) V. R. Saunders and I. H. Hillier, *Int. J. Quantum Chem.*, **7**, 699 (1973).

22. For a review, see: (a) P. Pulay in *Modern Theoretical Chemistry,* vol. 4, H. F. Schaefer III, ed., Plenum Press, New York, 1977, p. 153; other discussions include (b) S. Bratoz, *Colloq. Intern. Centre Natl. Rech. Sci. (Paris),* **82**, 287 (1958); (c) P. Pulay, *Mol. Phys.,* **17**, 197 (1969); (d) H. B. Schlegel, S. Wolfe, and F. Bernardi, *J. Chem. Phys.,* **63**, 3632 (1975); (e) A. Komornicki, K. Ishida, K. Morokuma, R. Ditchfield, and M. Conrad, *Chem. Phys. Lett.,* **45**, 595 (1977); (f) J. A. Pople, R. Krishnan, H. B. Schlegel, and J. S. Binkley, *Int. J. Quantum Chem., Symp.* **13**, 225 (1979); (g) H. B. Schlegel in *Computational Theoretical Organic Chemistry,* I. G. Csizmadia and R. Daudel, eds., Reidel, Holland, 1981, p. 129.

23. E. S. Blurock and W. J. Hehre, manuscript in preparation.

24. (a) C. Møller and M. S. Plesset, *Phys. Rev.,* **46**, 618 (1934); more recently: (b) J. S. Binkley and J. A. Pople, *Int. J. Quantum Chem.,* **9**, 229 (1975); (c) J. A. Pople, J. S. Binkley, and R. Seeger, *ibid.,* **105**, (1976).

25. P. D. Dacre, *Chem. Phys. Lett.,* **7**, 47 (1970).

26. M. Elder, *Int. J. Quantum Chem.,* **7**, 75 (1973).

27. M. Dupuis and H. F. King, *Int. J. Quantum Chem.,* **10**, 613 (1976).

28. (a) R. F. Hout, Jr., W. J. Pietro, and W. J. Hehre, *J. Comput. Chem.,* **4**, 276 (1983); (b) R. F. Hout, Jr., W. J. Pietro, and W. J. Hehre, *A Pictorial Guide to Molecular Structure and Reactivity*, Wiley, New York, 1984; (c) R. F. Hout, Jr., W. J. Pietro, and W. J. Hehre, program to be submitted to Quantum Chemistry Program Exchange, Indiana University, Bloomington, Ind.

29. R. L. Cook and K. E. Torrance, *Comput. Graphics,* **15**, 307 (1981).

4

SELECTION OF A MODEL

4.1. INTRODUCTION

How does one select an appropriate model to link the formal molecular orbital theory with the actual task of calculating specific molecular properties? Does one always select the most sophisticated approach available? Certainly this will be expected to yield the most accurate results. On the other hand, will the usage of a particular model be restricted by the size of the system? Generally the more sophisticated the quantum mechanical treatment, the more limited the range of its application. Obviously, some compromise between sophistication and applicability must be sought, keeping in mind that certain features of particular quantum mechanical models will make them more or less suitable for a given application.

As was illustrated in Figure 1.1, quantum mechanical models may broadly be classified in two dimensions, the first referring to the size of the basis set expansion used and the second to the level of treatment of electron correlation. Our discussion of model selection follows from this classification. After a brief description of available single-determinant (Hartree–Fock) methods for closed- and open-shell systems (Section 4.2), we consider in Section 4.3 the various gaussian basis sets that have been established for routine use. This is followed in Section 4.4 by a discussion of the principal features of various electron correlation treatments. Finally, Section 4.5 addresses the ways of specifying molecular geometry for particular investigations. Taken together, these components comprise the theoretical model. Each needs to be specified before the model can be applied.

4.2. HARTREE-FOCK METHODS

As discussed in Section 2.6, Hartree–Fock (HF) theory approximates the true many-electron wavefunction as a single determinant of orthonormalized spin orbitals. Various levels of HF theory are possible, and some selection is required depending on the kind of system to be investigated. These levels will be described in turn.

4.2.1. Closed-Shell Determinantal Wavefunctions

Closed-shell single-determinant wavefunctions represent the most commonly used form of HF theory and are appropriate for the description of the ground states of most molecules with an even number of electrons (n). The occupied molecular orbitals ψ_1, ψ_2, ..., $\psi_{n/2}$ each contain an α electron and a β electron. The determinantal wavefunction, $\Psi(1, 2, ..., n/2)$, is then formed from the spin orbitals $\psi_1\alpha$, $\psi_1\beta$, $\psi_2\alpha$, $\psi_2\beta$, ..., $\psi_{n/2}\beta$. It is written explicitly in Eq. (2.24). This wavefunction is precisely of singlet type, that is, it is an eigenfunction of the spin-squared operator, $\hat{S}^2$, with eigenvalue zero, and may, therefore, be classified as spin restricted (RHF). It is undoubtedly the most widely applied type of wavefunction for large molecules. Its principal merit is ease of application. Its principal deficiency is that it does not always give the lowest possible energy. Thus, the constraint of assigned pairs of electrons (α and β) to each orbital ψ_i may lead to an unrealistic distribution of electrons. This is particularly true for partially broken bonds when the RHF function may fail to dissociate into the correct products (as discussed in Section 2.9).

One other aspect of closed-shell wavefunctions should be mentioned. The RHF equations, in their algebraic Roothaan–Hall [1] form (2.35), normally have *real* solutions. If the iterative procedure outlined in Section 3.3.2 is followed using a real initial guess, the unknown coefficients are real at every iteration and will converge to a real solution. However, it is possible that the Roothaan–Hall equations have *complex* solutions that lead to total energies lower than the best real solution [2]. Convergence to a complex solution would require an appropriate, complex initial guess. Such solutions are significant if two distinct real molecular orbital electron configurations give results close in energy, as in the vicinity of orbital symmetry crossings of the Woodward–Hoffmann type [3].

4.2.2. Open-Shell Determinantal Wavefunctions

Open-shell single-determinant wavefunctions, introduced in Section 2.6.2, are appropriate for the description of systems with an odd number of electrons, as well as for systems with an even number of electrons characterized by electronic states other than closed-shell singlets. The spin-restricted (RHF) wavefunctions are those with some molecular orbitals doubly occupied and some singly occupied, the latter containing all α or all β electrons [4]. Again, this type of determinantal function represents a pure doublet, triplet, and so on, spin state, that is, it is an eigenfunction of $\hat{S}^2$. Its use is appropriate if it is important to obtain such a pure spin state. Disadvantages are: (a) that it is computationally cumbersome, and (b) that it is somewhat unsatisfactory as a starting point for a perturbation treatment of electron correlation.

The alternative open-shell procedure is the spin-unrestricted (UHF) method [5], in which the orbitals associated with α and β electrons are treated completely independently. The advantages of this method are: (a) that it generally gives a lower energy than the corresponding RHF treatment, (b) that it is capable of providing a qualitatively correct description of bond dissociation, and (c) that it is generally computationally more efficient than the corresponding RHF procedure. The principal disadvantage is that the resulting wavefunction is no longer spin pure. Thus, a UHF calculation on a system with one extra α electron might lead to a wavefunction that is a mixture of a doublet and quartet rather than a pure doublet. The extent of such "spin contamination" can be assessed by examining the expectation value of the S^2 operator, $\langle S^2 \rangle$. For a pure doublet wavefunction, this should be 0.75. Quartet contamination increases this value. Hence, if values of $\langle S^2 \rangle$ much greater than 0.75 are found for studies that are supposed to be on doublet states, the results are suspect and should be compared carefully with those of the corresponding RHF model. $\langle S^2 \rangle$ for a pure triplet wavefunction should be 2.00; again contamination by states of higher spin multiplicity leads to larger values. Discussion of the consequences of spin contamination on other molecular properties, for example, equilibrium geometries, is provided in Chapter 6.

The UHF method can also be applied to supposed closed-shell systems. Usually such an application will lead back to the RHF solution, indicating that the best UHF orbitals, that is, those leading to lowest energy, are identical in α and β pairs. Sometimes, however, the use of an initial guess with *different* α and β orbitals will lead to energies *lower* than with RHF. This usually occurs at large internuclear separations, when the UHF method permits electrons of different spin to become detached from one another and become localized on different atoms. It is indeed possible to test the *stability* of RHF solutions relative to UHF [6]. For some nuclear configurations it is found that the RHF orbitals do *not* correspond to a minimum on the surface of energy versus UHF coefficients, but rather to a saddle point. The energy minimum, that is, the best UHF determinant in an energy sense, then has to be found by solving the UHF equations after making a suitable asymmetric initial guess.

4.3. ATOMIC BASIS SETS OF GAUSSIAN FUNCTIONS

As discussed in Section 2.5, the molecular orbitals ψ_i in a Hartree–Fock treatment are expressed as linear combinations of N nuclear-centered basis functions ϕ_μ ($\mu = 1, 2, \ldots, N$),

$$\psi_i = \sum_\mu^N c_{\mu i} \phi_\mu. \tag{4.1}$$

For UHF wavefunctions, two sets of coefficients are needed:

$$\psi_i^\alpha = \sum_\mu^N c_{\mu i}^\alpha \phi_\mu$$

$$\psi_i^\beta = \sum_\mu^N c_{\mu i}^\beta \phi_\mu, \tag{4.2}$$

but the same basis functions are used for α and β orbitals. In selecting a set of basis functions ϕ_μ, the size (N) of the expansion and the nature of the functions ϕ_μ need to be considered.

A limiting Hartree–Fock treatment would involve an infinite set of basis functions ϕ_μ. This is clearly impractical since the computational expense of Hartree–Fock molecular orbital calculations is formally proportional to the fourth power of the total number of basis functions (see Sections 3.3.1 and 3.3.2). Therefore, the ultimate choice of basis set size depends on a compromise between accuracy and efficiency. The size of the basis set expansion required to describe various properties satisfactorily differs from property to property and is discussed in detail in Chapter 6.

For computational facility, it is desirable to use gaussian basis functions $g(\mathbf{r})$ [7], where $\mathbf{r}$ is the position vector (x, y, z) (Sections 2.5 and 3.3.1). These may be chosen as single (*uncontracted* or *primitive*) gaussian functions (2.27) or as fixed, predefined linear combinations of such functions (2.30), termed *contracted*. Evaluation of one- and two-electron integrals involving d-type functions is most efficiently performed using the complete set of six second-order gaussians given in Eq. (2.27). Later in the calculation, these may be linearly combined to give integrals corresponding to the familiar set of five d functions. Correspondingly, f-integral evaluation is accomplished in practice by transforming the complete set of ten third-order gaussians into the seven more familiar components.

The discussion which follows largely concerns basis sets developed in conjunction with the GAUSSIAN series of computer programs and formulated to take maximum advantage of these programs. The reader is directed elsewhere for discussion of other comprehensive series of gaussian basis sets [8].

4.3.1. Minimal Basis Sets

The simplest level of *ab initio* molecular orbital theory involves the use of a *minimal basis set* of nuclear-centered functions. In the strictest sense, such a representation comprises exactly that number of functions required to accommodate all of the electrons of the atom, while maintaining overall spherical symmetry. Thus, within the framework of a minimal basis set, hydrogen and helium are represented by a single s-type function, lithium and beryllium by a pair of such functions, and the remaining first-row elements (boron to neon) by two s functions and a complete set of three p-type functions. Second-row atoms are treated similarly by adding to the $1s, 2s, 2p$ inner-shell description a single s-type basis function for sodium and magnesium, and an s- and three p-type functions for the elements aluminum to argon. For the third-row elements, $1s, 2s, 2p, 3s,$ and $3p$ basis functions comprise the inner shell. For potassium and calcium, these are supplemented by a single $4s$ function. The set of (five) $3d$ functions is required for the remaining elements in the row (scandium through krypton), the proper atomic description of the main-group elements (gallium through krypton) also necessitating addition of valence $4p$-type basis functions. Fourth-row elements are treated similarly. Description of rubidium and strontium requires a single $5s$ function (on top of $1s$, $2s$, $2p$, $3s$, $3p$, $3d$, $4s$, and $4p$ functions, which make up the core). For the remaining elements in the row, the set of $4d$ functions is (partially) occupied, as are $5p$ functions for the main-group elements indium through xenon.

Practical experience in dealing with molecules suggests that it is necessary to include in minimal basis set descriptions of lithium and beryllium the set of unoccupied (in the atom) but energetically low-lying $2p$ functions. Similarly, for sodium and magnesium, potassium and calcium, and for rubidium and strontium, $3p$, $4p$, and $5p$ functions, respectively, need to be added. $4p$ and $5p$ functions are also thought to be involved in the bonding of compounds containing first- and second-row transition metals, respectively, although again they are unoccupied in the ground-state atoms. Therefore, useful minimal basis set representations for the first 54 elements of the Periodic table comprise the following atomic functions:

H, He:	$1s$
Li to Ne:	$1s$
	$2s, 2p_x, 2p_y, 2p_z$
Na to Ar:	$1s$
	$2s, 2p_x, 2p_y, 2p_z$
	$3s, 3p_x, 3p_y, 3p_z$
K, Ca:	$1s$
	$2s, 2p_x, 2p_y, 2p_z$
	$3s, 3p_x, 3p_y, 3p_z$
	$4s, 4p_x, 4p_y, 4p_z$
Sc to Kr:	$1s$
	$2s, 2p_x, 2p_y, 2p_z$
	$3s, 3p_x, 3p_y, 3p_z, 3d_{3zz-rr}, 3d_{xx-yy}, 3d_{xy},$
	$3d_{xz}, 3d_{yz}$
	$4s, 4p_x, 4p_y, 4p_z$
Rb, Sr:	$1s$
	$2s, 2p_x, 2p_y, 2p_z$
	$3s, 3p_x, 3p_y, 3p_z, 3d_{3zz-rr}, 3d_{xx-yy}, 3d_{xy},$
	$3d_{xz}, 3d_{yz}$
	$4s, 4p_x, 4p_y, 4p_z$
	$5s, 5p_x, 5p_y, 5p_z$
Y to Xe:	$1s$
	$2s, 2p_x, 2p_y, 2p_z$
	$3s, 3p_x, 3p_y, 3p_z, 3d_{3zz-rr}, 3d_{xx-yy}, 3d_{xy},$
	$3d_{xz}, 3d_{yz}$
	$4s, 4p_x, 4p_y, 4p_z, 4d_{3zz-rr}, 4d_{xx-yy}, 4d_{xy},$
	$4d_{xz}, 4d_{yz}$
	$5s, 5p_x, 5p_y, 5p_z$

The number of basis functions per atom is now completely defined:

H, He:	1
Li to Ne:	5
Na to Ar:	9
K, Ca:	13
Sc to Kr:	18
Rb, Sr:	22
Y to Xe:	27

All that now remains is the detailed specification of the individual basis functions.

a. STO-KG Minimal Basis Sets: The STO-3G Basis Set. The series of minimal basis sets termed STO-KG [9] consists of expansions of Slater-type atomic orbitals (STOs) in terms of K gaussian functions:

$$\phi_{nl}(\zeta = 1, \mathbf{r}) = \sum_{k=1}^{K} d_{nl,k} g_l(\alpha_{n,k}, \mathbf{r}) \tag{4.3}$$

where the subscripts n and l define the specific principal and angular quantum numbers, for example, ϕ_{1s}, and g_l are normalized gaussian functions. The values of the gaussian exponents, α, and the linear expansion coefficients, d, have been determined by minimizing, in a least-squares sense, the error in the fit of the gaussian expansion to the exact Slater orbital,

$$\epsilon_{nl} = \int (\phi_{nl}^{\text{Slater}} - \phi_{nl}^{\text{gaussian expansion}})^2 \, d\tau. \tag{4.4}$$

Minimization is performed simultaneously for all expansions of given n quantum number. For example, $3s$, $3p$, and $3d$ gaussian expansions for use in calculations on third- and fourth-row elements are obtained by minimization of a sum of integrals,

$$\epsilon_{3s} + \epsilon_{3p} + \epsilon_{3d} = \int (\phi_{3s}^{\text{Slater}} - \phi_{3s}^{\text{gaussian expansion}})^2 \, d\tau$$

$$+ \int (\phi_{3p}^{\text{Slater}} - \phi_{3p}^{\text{gaussian expansion}})^2 \, d\tau$$

$$+ \int (\phi_{3d}^{\text{Slater}} - \phi_{3d}^{\text{gaussian expansion}})^2 \, d\tau. \tag{4.5}$$

Values for α and d for expansion lengths K between 2 and 6 along with the least-squares errors are tabulated in the original literature [9].

Two features of the gaussian expansions which comprise the STO-KG basis sets should be noted. First, the Slater $2s$, $3s$, $4s$, and $5s$ functions are expanded in terms of the simplest s-type gaussians, that is, zero-order s gaussians, rather than as combinations of higher-order gaussian forms. This is also the case for the Slater $3p$, $4p$, and $5p$ atomic orbitals, all of which are written in terms of first-order p gaussian functions, and for the $4d$ orbitals, which are expressed as linear combinations of second-order d gaussian primitives. The reason for expressing all Slater orbitals, ϕ_{nl}, in terms of the simplest gaussians of the same symmetry type is simply that integrals involving the higher-order gaussian functions are more difficult to evaluate.

A second feature to note is that the expansions for atomic functions of given principal quantum number n often share a common set of gaussian exponents, α. Thus, ϕ_{2s} and ϕ_{2p} utilize a single set of exponents, α_{2k}. ϕ_{3s} and ϕ_{3p} for use on second-row elements share common exponents, α_{3k}. Basis sets for the third- and fourth-row main-group elements have been constructed along similar lines. Here, ϕ_{3s}, ϕ_{3p}, and ϕ_{3d} share a common set of gaussian exponents. For third-row representations, ϕ_{4s} and ϕ_{4p} share exponents; for fourth-row elements, ϕ_{4s}, ϕ_{4p}, and ϕ_{4d} are restricted to a single set of gaussian exponents, as are ϕ_{5s} and ϕ_{5p}. Expansions for first- and second-row transition metals are formulated with fewer exponent constraints involving the valence functions. For first-row transition metals, ϕ_{3s} and ϕ_{3p} share common exponents as do ϕ_{4s} and ϕ_{4p}. ϕ_{3d} is, however, independently constructed. In an analogous manner, ϕ_{4s} and ϕ_{4p}, and ϕ_{5s} and ϕ_{5p}, for second-row transition metals are constrained to share common exponents while ϕ_{4d} is treated independently. The underlying $3s$, $3p$, $3d$ shell is constructed (as for the main-group elements) with a single set of gaussian exponents.

Exponent restrictions do reduce somewhat the overall flexibility of the basis representations. However, they give rise to the possibility of significant gains in the efficiency of evaluation of the various integrals that arise in the course of a Hartree-Fock calculation. A brief discussion has already been provided in Section 3.3.1.

Least-squares fits were originally carried out on Slater functions of exponent $\zeta = 1$. Best (least-squares) fits to orbitals of arbitrary ζ may be obtained from the tabulated functions by uniform scaling,

$$\phi(\zeta, \mathbf{r}) = \zeta^{3/2} \phi(1, \zeta \mathbf{r}). \qquad (4.6)$$

Numerical values for Slater exponents suitable for use in calculations on molecules are presented in Table 4.1 for the first 54 elements of the Periodic Table. Inner-shell exponents are those leading to the lowest UHF energy for the ground state in the free atom [10], while valence-shell exponents have been selected as appropriate for atoms in typical molecular environments. Details are given in the original papers [9]. Note that the valence-shell exponents listed here differ significantly from those originally proposed by Slater [11]. The differences are greatest for hydrogen and the elements to the left-hand side of the Periodic Table. In all cases, the Slater exponents (based on atoms) lead to basis functions that are too diffuse to be suitable for use in molecules.

While absolute energies of atoms and molecules calculated using the STO-KG minimal basis sets exhibit strong dependence on the value of K, other properties, such as energy differences, optimum geometries, charge distributions, and electric

TABLE 4.1. Standard Molecular Scaling Factors for STO-KG Minimal Basis Sets

Atom	ζ_{1s}	$\zeta_{2s} = \zeta_{2p}$	$\zeta_{3s} = \zeta_{3p}$	ζ_{3d}	$\zeta_{4s} = \zeta_{4p}$	ζ_{4d}	$\zeta_{5s} = \zeta_{5p}$
H	1.24						
He	1.69						
Li	2.69	0.80					
Be	3.68	1.15					
B	4.68	1.50					
C	5.67	1.72					
N	6.67	1.95					
O	7.66	2.25					
F	8.65	2.55					
Ne	9.64	2.88					
Na	10.61	3.48	1.75				
Mg	11.59	3.90	1.70				
Al	12.56	4.36	1.70				
Si	13.53	4.83	1.75				
P	14.50	5.31	1.90				
S	15.47	5.79	2.05				
Cl	16.43	6.26	2.10				
Ar	17.40	6.74	2.33				
K	18.61	7.26	2.75		1.43		
Ca	19.58	7.74	3.01		1.36		
Sc	20.56	8.22	3.21	1.10	1.60		
Ti	21.54	8.70	3.44	1.90	1.70		
V	22.53	9.18	3.67	2.55	1.70		
Cr	23.52	9.66	3.89	3.05	1.75		
Mn	24.50	10.13	4.11	3.45	1.65		
Fe	25.49	10.61	4.33	3.75	1.55		
Co	26.44	11.07	4.56	4.10	1.55		
Ni	27.46	11.56	4.76	4.35	1.60		
Cu	28.44	12.04	4.98	4.60	1.60		
Zn	29.43	12.52	5.19	4.90	1.90		
Ga	30.42	12.99	5.26	5.26	1.80		
Ge	31.40	13.47	5.58	5.58	2.00		
As	32.39	13.94	5.90	5.90	2.12		
Se	33.37	14.40	6.22	6.22	2.22		
Br	34.36	14.87	6.54	6.54	2.38		
Kr	35.34	15.34	6.86	6.86	2.54		
Rb	36.32	15.81	7.18	7.18	3.02		1.90
Sr	37.31	16.28	7.49	7.49	3.16		1.80
Y	38.29	16.72	7.97	7.97	3.29	1.40	1.80
Zr	39.27	17.19	8.21	8.21	3.48	1.95	1.90
Nb	40.26	17.66	8.51	8.51	3.67	2.40	1.90
Mo	41.24	18.12	8.82	8.82	3.87	2.70	1.95
Tc	42.22	18.59	9.14	9.14	4.05	3.00	1.85
Ru	43.21	19.05	9.45	9.45	4.24	3.20	1.75
Rh	44.19	19.51	9.77	9.77	4.41	3.45	1.75
Pd	45.17	19.97	10.09	10.09	4.59	3.60	1.80

TABLE 4.1.   (Continued)

Atom	ζ_{1s}	$\zeta_{2s} = \zeta_{2p}$	$\zeta_{3s} = \zeta_{3p}$	ζ_{3d}	$\zeta_{4s} = \zeta_{4p}$	ζ_{4d}	$\zeta_{5s} = \zeta_{5p}$
Ag	46.15	20.43	10.41	10.41	4.76	3.75	1.80
Cd	47.14	20.88	10.74	10.74	4.93	3.95	2.10
In	48.12	21.33	11.08	11.08	4.65	4.65	2.05
Sn	49.10	21.79	11.39	11.39	4.89	4.89	2.15
Sb	50.09	22.25	11.71	11.71	5.12	5.12	2.20
Te	51.07	22.71	12.03	12.03	5.36	5.36	2.28
I	52.05	23.17	12.35	12.35	5.59	5.59	2.42
Xe	53.03	23.63	12.67	12.67	5.82	5.82	2.57

dipole moments, are far less sensitive to the size of the expansion. The data in Table 4.2 illustrate this point for the case of formaldehyde. Note that, with the exception of the total energy, molecular properties calculated using the STO-4G and higher expansion levels have nearly converged to their respective STO limits. Even the STO-3G basis set yields properties that are reasonably close to limiting values, and, in view of the relative computational times of the various expansions, it is this level that has been selected as an optimum compromise for widespread application. One possible exception to this generalization is for the calculation of equilibrium geometries. Poppinger has demonstrated [12a] that the STO-2G basis set yields equilibrium structures for molecules containing first-row elements in good accord with those derived from STO-3G, even though, as is apparent in Table 4.2, other properties, for example, relative energies, charge distributions, and electric dipole moments, calculated using the two basis sets are not generally in good overall agreement. Differences between STO-2G and STO-3G equilibrium structures appear, however, to be larger for molecules containing heavier (second-, third-, and fourth-row) atoms [9c-e].

Another possible exception to the use of the three-gaussian expansion as the standard minimal basis level occurs in the consideration of the properties of weakly-bound complexes where long-range forces are important. Specifically, the STO-4G representation has been found to be more suitable than STO-3G in the description of hydrogen-bonded complexes [12b]. Presumably the additional gaussian in the STO-4G expansion contributes significantly to the description of the outermost regions of the atomic functions.

4.3.2.   Extended *sp* Basis Sets

Minimal basis sets, such as STO-3G, have several inherent inadequacies. Because the number of atomic basis functions is not apportioned according to electron count—for example, the lithium atom, which has only three electrons, is provided with the same number of functions (five) as fluorine with its nine electrons—it follows that minimal-basis-set descriptions of compounds containing elements such as oxygen and fluorine are likely to be poorer than those for molecules comprising elements with fewer electrons. A second problem arises because a minimal basis set using

TABLE 4.2. Properties of Formaldehyde Calculated with Various STO-KG Basis Sets

Property		Basis Set				
		STO-2G	STO-3G	STO-4G	STO-5G	STO-6G
Total energy (hartrees)		−109.0244	−112.3525	−113.1611	−113.3752	−113.4408
Atomization energy (hartrees)		0.4830	0.4168	0.4093	0.4096	0.4095
Equilibrium geometry	r_{CO}	1.220	1.217	1.216	1.216	1.216
(distances in angstroms,	r_{CH}	1.110	1.101	1.099	1.098	1.098
angles in degrees)	<HCH	113.1	114.5	114.8	114.8	114.8
Atomic charges (electrons)	C	+0.092	+0.059	+0.075	+0.078	+0.079
	O	−0.120	−0.186	−0.197	−0.198	−0.198
	H	+0.014	+0.063	+0.061	+0.060	+0.059
Dipole moment (debyes)		1.118	1.520	1.592	1.596	1.596
Relative computation times[a]		1	2	3	6	10

[a]Single-point energy calculation.

fixed gaussian exponents is unable to expand and contract in response to differing molecular environments. This is because a minimal basis set contains only a *single* valence function of each particular symmetry type, for example, $2s, 2p_x$. Thus, in the absence of radial exponent optimization for each atom in a molecule, there is no mechanism for the individual sets of functions to adjust their sizes. Finally, minimal representations lack the ability to describe adequately the nonspherical anisotropic aspects of molecular charge distributions. In this context it should be recognized that the only feature in the minimal basis representation of a first- or second-row atom that might allow for anisotropy is the set of p-type functions. The s functions are, of course, limited to handling only totally spherical environments. For third- and fourth-row elements, both p- and d-type functions may be employed to describe aspherical charge distributions. As minimal-basis-set representations incorporate only a single set of valence functions of each symmetry type, and as the radial parts of all components, for example, x, y, and z parts of valence p-type functions, are constrained to be the same, molecules in which the electron distribution is roughly spherical or *isotropic* about the individual atomic centers, for example, methane and ethane, should be more accurately described than those in which a high degree of *anisotropy* is present, for example, ethylene and acetylene. The assessment of the STO-3G minimal basis set (Chapter 6) shows considerable evidence for such nonuniformity. For example, calculated equilibrium geometries of highly *polar* and *anisotropic* molecules are often significantly in error, as are the calculated energies of chemical reactions involving reactants and products of markedly different structure.

In principle, the first two deficiencies may be alleviated simply by allowing for more than a single valence function of each symmetry type in the basis set description. In this way, the number of basis functions for all elements, not just those on the left-hand side of the Periodic Table, would be substantially in excess of the number actually required. Furthermore, the allocation of two or more valence basis functions of each given symmetry type would provide the needed flexibility for overall radial size to be determined simply by the adjustment of the relative weights of the individual components in the variational procedure. For example, the relative contributions of two s-type valence functions, one highly contracted, the other more diffuse, may be adjusted so as to yield an s function, the radial description of which may lie anywhere between the "contracted" and "diffuse" limits.

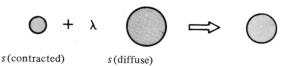

s(contracted) s(diffuse)

The same arguments apply to p-type and higher-order functions.

The third deficiency, the inability of a minimal basis set to describe properly *anisotropic* molecular environments, may be alleviated in one of two ways. The conceptually simpler way would be to allow each of the x, y, and z p components describing the valence region of a main-group element to have a different radial distribution, that is, to employ an *anisotropic* rather than an *isotropic* minimal basis set [13]. Similarly, the five d-function components describing the valence manifold of a first- or second-row transition metal could be independently specified. A limited number of minimal basis set studies of this kind have been carried out [13]. These have generally shown that optimum radial exponents do vary significantly with orientation, and that no single scaling factor provides a particularly appropriate overall description. For example, the optimum radial exponents for the p_y and p_z basis functions on carbon in hydrogen cyanide, aligned along the x axis, are considerably smaller than that of p_x. In simple terms, the molecular π system is more diffuse than the σ system [13b].

$$H - C \equiv N$$

$$\zeta_x = 2.08$$
$$\zeta_y = \zeta_z = 1.51.$$

Although this type of approach, requiring independent specification of exponents for individual cartesian directions, might be useful for molecules possessing a high degree of symmetry, for example, those in which separation of σ and π systems is possible, it is unsatisfactory in situations with little or no symmetry. Here, the only unbiased manner in which to choose the radial distributions of the individual atomic orbital components would require optimization with respect to total energy for each atom in every molecule considered. Such a procedure is clearly undesirable.

A more reasonable way to surmount the difficulties inherent in an *isotropic* minimal basis set is to include more than a single set of valence p- and/or d-type functions, exactly the remedy noted above for the problem of flexibility of size of basis functions. The allocation of two sets of isotropic valence functions in the basis, one more tightly held to the nucleus than the other, permits independent adjustment through the SCF procedure of the individual radial components between contracted and diffuse extremes. As an example, consider the acetylene molecule oriented along the x axis.

$$H - C \equiv C - H$$

Here, adjustment of the two components of the p_x function on carbon will produce the highly contracted function required in the description of the σ system of the molecule.

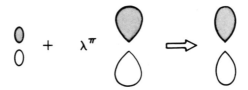

On the other hand, the p_y and p_z functions, each resulting from combination of two components, will be far more diffuse. These form the basis for the degenerate set of π-type functions.

In summary, while an isotropic minimal basis set constrains the valence functions to have the same fixed radial size, the extended representations allow these individual components to adjust themselves independently to be more suitable for a given molecular environment.

A basis set formed by doubling all functions of a minimal representation is usually termed a *double-zeta basis*. An even simpler extension of a minimal basis set is to double only the number of basis functions representing the valence region. While the inner-shell electrons are important with regard to total energy, their effect on molecular bonding is of little consequence. In a *split-valence-shell* or more simply *split-valence basis set*, hydrogen and helium are represented by two s-type functions and first- and second-row atoms by two complete sets of valence s and p functions.

H, He:	$1s'$
	$1s''$
Li to Ne:	$1s$
	$2s', 2p'_x, 2p'_y, 2p'_z$
	$2s'', 2p''_x, 2p''_y, 2p''_z$
Na to Ar:	$1s$
	$2s, 2p_x, 2p_y, 2p_z$
	$3s', 3p'_x, 3p'_y, 3p'_z$
	$3s'', 3p''_x, 3p''_y, 3p''_z$

Here, the basis functions comprising the two valence shells are denoted $'$ and $''$, respectively.

Development of split-valence basis sets for the elements beyond argon proceeds along similar lines. For the third-row main-group elements, $4s$ and $4p$ functions make up the valence shell. Each of these is described in terms of two components.

K, Ca: $1s$
               $2s, 2p_x, 2p_y, 2p_z$
               $3s, 3p_x, 3p_y, 3p_z$
               $4s', 4p'_x, 4p'_y, 4p'_z$
               $4s'', 4p''_x, 4p''_y, 4p''_z$

Ga to Kr: $1s$
               $2s, 2p_x, 2p_y, 2p_z$
               $3s, 3p_x, 3p_y, 3p_z, 3d_{3zz-rr}, 3d_{xx-yy}, 3d_{xy},$
               $3d_{xz}, 3d_{yz}$
               $4s', 4p'_x, 4p'_y, 4p'_z$
               $4s'', 4p''_x, 4p''_y, 4p''_z$

Both $3d$ and $4sp$ shells make up the valence manifold of a first-row transition metal. Thus, a proper split-valence representation for these elements requires splitting of all three sets of functions.

Sc to Zn: $1s$
               $2s, 2p_x, 2p_y, 2p_z$
               $3s, 3p_x, 3p_y, 3p_z$
               $3d'_{3zz-rr}, 3d'_{xx-yy}, 3d'_{xy}, 3d'_{xz}, 3d'_{yz}$
               $3d''_{3zz-rr}, 3d''_{xx-yy}, 3d''_{xy}, 3d''_{xz}, 3d''_{yz}$
               $4s', 4p'_x, 4p'_y, 4p'_z$
               $4s'', 4p''_x, 4p''_y, 4p''_z$

Similar procedures would need to be followed in the construction of split-valence basis sets for heavier elements. In particular, basis sets for fourth-row main-group elements would involve splitting of $5s$ and $5p$ functions while those for second-row transition metals would in addition allow for splitting of the $4d$ shell.

a.   The 6-21G and 3-21G Split-Valence Basis Sets.   The 6-21G and 3-21G basis sets [14], the former presently defined through the second row of the Periodic Table and the latter through the fourth row, typify representations in which two basis functions, instead of one, have been allocated to describe each valence atomic orbital. In the 6-21G basis set, each inner-shell atomic orbital is represented by a single function, which in turn is written in terms of six gaussian primitives, that is, Eq. (4.7) with $K = 6$.

$$\phi_{nl}(\mathbf{r}) = \sum_{k=1}^{K} d_{nl, k} g_l(\alpha_{n, k}, \mathbf{r}). \qquad (4.7)$$

The contracted and diffuse basis functions representing valence-shell orbitals are written in terms of Eqs. (4.8) and (4.9), with expansion lengths K' and K'' chosen as 2 and 1, respectively.

$$\phi'_{nl}(\mathbf{r}) = \sum_{k=1}^{K'} d'_{nl, k} g_l(\alpha'_{n, k}, \mathbf{r}) \qquad (4.8)$$

$$\phi''_{nl}(\mathbf{r}) = \sum_{k=1}^{K''} d''_{nl,k} g_l(\alpha''_{n,k}, \mathbf{r}). \tag{4.9}$$

The subscripts n and l in Eqs. (4.7) to (4.9) define the specific atomic functions, for example, ϕ_{2p}.

For first- and second-row elements, the linear expansion coefficients, d, and gaussian exponents, α, have been determined by minimization of the UHF atomic energies, subject to overall normalization. Except for lithium, beryllium, sodium, and magnesium, the atomic ground state has been used. For Be and Mg, the ^3P excited states $((1s)^2 (2s)(2p)$ and $(1s)^2 (2s)^2 (2p)^6 (3s)(3p)$, respectively) have been used in order to obtain a good simultaneous description of valence s- and p-type functions. For Li and Na, the valence s-type basis functions are determined using the atomic ground states $((1s)^2 (2s)$ and $(1s)^2 (2s)^2 (2p)^6 (3s)$, respectively) and the contraction coefficients for the valence p-functions are determined subsequently using the ^2P excited states $((1s)^2 (2p)$ and $(1s)^2 (2s)^2 (2p)^6 (3p)$, respectively), holding the other parameters fixed. The 21G basis functions for hydrogen and helium are due to van Duijneveldt [15].

Recent developments in computer program technology have encouraged the formulation of basis sets that comprise fewer and fewer gaussian primitives. In particular, programs have been developed to calculate analytically the first and second derivatives of the energy with respect to displacements in nuclear coordinates. These programs, which are of great value in investigations of equilibrium structures, transition structures, force constants, and molecular vibrational frequencies, are particularly efficient when the number of primitive gaussians is reduced. This is not necessarily true for single-point (nonderivative) calculations, where the computation is often dominated by the self-consistent-field (SCF) procedure, which depends only on the number of basis functions, and not on the number of primitives. From a purely computational standpoint, therefore, basis sets should be constructed so as to contain as few primitive gaussians per basis function as possible. In this regard, the 6-21G split-valence representations, which utilize six primitives for each inner-shell basis function, must be viewed as relatively inefficient.

The search for related, smaller basis sets, that is, containing fewer primitive gaussians, is not without its complications. In particular, it should be noted that the same procedure used to obtain the 6-21G basis sets, that is, optimization of all gaussian exponents and expansion coefficients to give the lowest UHF energy for the atom in its ground state, will present difficulties if followed using a small number of primitives, specifically if only a small number is used for the inner-shell basis functions. If there are few valence electrons, as in Li or Be, there is a tendency for the functions to "fall inward" toward the nucleus. This presumably occurs because the total energy minimization criterion prefers additional functions in the inner-shell region rather than a good description of the valence region. The "falling inward" of the valence part of split-valence basis sets is clearly undesirable since a good description of bonding interactions must involve the overlap of valence basis functions on neighboring atoms, which in turn depends on a good description of the outer part of the atomic electron distribution. Ideally, the valence part of a split-valence basis should be determined with very good inner-shell functions, so as to prevent such an unwanted collapse. Thus, for example, we might define a "best" 21G valence part of a split-valence K-21G basis as one in which all parameters are optimized with K large, approximating an ∞-21G basis. Such a basis would no longer be computationally efficient because of the large value of K. However, the inner-shell basis functions could then be replaced by ones with fewer primitives (smaller K) *without reoptimizing the valence functions.* This would yield a computationally efficient basis set in which the valence functions would not have collapsed to an appreciable extent.

In practice, the previously mentioned 6-21G basis set for first- and second-row elements appears to mimic closely the properties of the hypothetical ∞-21G representations; all valence-shell parameters (gaussian exponents and linear expansion coefficients) for an 8-21G basis set determined for the carbon atom lie within 1% of the values found optimum for the corresponding 6-21G set. The 6-21G representations seem suitable, therefore, as a foundation on which to construct smaller, more efficient split-valence basis sets.

The small split-valence basis sets formed in this manner, optimizing only inner-shell exponents and coefficients while leaving valence-shell parameters alone, comprise three gaussian functions per inner-shell orbital. Collectively they are termed 3-21G [14]. Specification for second-row atoms requires that intermediate 63-21G basis sets ($K_1 = 6$, $K_2 = 3$, $K' = 2$, $K'' = 1$) first be constructed from the 66-21G (6-21G) sets by reoptimization of the $1s$, $2s$, and $2p$ inner shells while keeping the valence $(3s, 3p)$ functions fixed. 3-21G (33-21G) basis sets may then be derived from these latter representations by reoptimization of only the ϕ_{1s} coefficients and exponents for a value of $K_1 = 3$. The situation is even more involved for third- and fourth-row elements, where two and three intermediate basis sets would need to be constructed. Thus far, this procedure has not led to a uniform series of basis sets, and an alternative has been adopted. This involves reformulation of a series of minimal basis sets proposed by Huzinaga [8b] in which each atomic orbital has been written in terms of three gaussian functions. Specifically, the original expansions for s- and p-type orbitals have been replaced by new combinations in which the two sets of orbitals (of the same n quantum number) share gaussian exponents. Valence functions have been split into two and one gaussian parts. Full details are provided in the original papers [14c, d].

Except for hydrogen, the 6-21G and 3-21G basis sets are employed as is, that is, without rescaling of the valence functions to account better for changes that might occur as the result of molecule formation. There are two main reasons behind this choice. In the first place, optimum scale factors both greater and less than unity have been found in molecules for all first-row atoms except boron. Although scale factor optimizations for compounds containing heavier elements have not been carried out, it is likely that the situation here would be similar. Therefore, choice of an average scale factor greater (or less) than unity is quite arbitrary. Secondly, if the basis sets are to be used for exploration of reaction potential surfaces, where molecules may be partly dissociated, a good description of the free atom is also important. The most serious possible inadequacy of using unit scale factors discovered thus far occurs for lithium, where an optimum scale value of 1.5 has been found for the outer valence shell in lithium fluoride. This reflects the contracted character of the

TABLE 4.3. Properties of Fluoromethane Derived with the 3-21G
and 6-21G Basis Sets

Property		Basis Set	
		3-21G	6-21G
Total energy (hartrees)		-138.2819	-138.8362
Atomization energy (hartrees)		0.4649	0.4560
Equilibrium geometry	r_{CF}	1.404	1.408
(distances in angstroms,	r_{CH}	1.080	1.080
angles in degrees)	$<HCH$	109.5	109.5
Atomic charges (electrons)	C	-0.178	-0.177
	F	-0.408	-0.401
	H	$+0.195$	$+0.193$
Dipole moment (debyes)		2.34	2.32
Relative computation times[a]		1	2

[a]Single-point energy calculation; the time differential for calculations involving energy derivatives would be significantly greater.

valence functions in this highly ionic compound. Other highly electropositive elements, for example, sodium and potassium, would probably show similar effects.

Hydrogen, the optimum molecular scale factors of which are almost always
greater than unity, is treated as an exception. Both ζ' and ζ'' are chosen, according
to Eq. (4.10),

$$\phi^{molecule}(\mathbf{r}) = \zeta^{3/2}\phi^{atom}(\zeta\mathbf{r}), \qquad (4.10)$$

to be 1.10, the same values as proposed by Pulay et al. [16] for use with their
4-21G split-valence basis set.

As seen by the example provided in Table 4.3, relative energies, equilibrium
geometries, charge distributions, and electric dipole moments calculated using the
3-21G split-valence basis sets are uniformly close to those obtained with the 6-21G
representation. It is likely that other properties that do not depend to a significant
extent on the description in the region of the atomic nuclei will likewise be handled
equally well by the two basis sets. The larger basis set, with its improved inner-shell
description, does, however, lead to significantly lower total energies.

b. Larger Split-Valence Representations: The 4-31G and 6-31G Basis Sets. An
early split-valence basis set which, while now largely replaced by 3-21G, has received
widespread use, is the 4-31G representation [17]. It is defined for all first-row elements, as well as for the second-row elements phosphorus, sulfur, and chlorine.
4-31G utilizes inner-shell expansions of four gaussian functions and two valence
shells comprising three and one gaussians, respectively. Closely related is the 6-31G
basis set [18], which has now been defined through the second row of the Periodic
table. It comprises inner-shell functions each written in terms of a linear combination of six gaussians, and two valence shells represented by three and one gaussian

primitives, respectively, that is, Eqs. (4.7) to (4.9), where $K_1 = K_2 = K_3 = 6$, $K' = 3$, $K'' = 1$. 4-31G and 6-31G basis sets for hydrogen, helium and all first-row elements, and 4-31G representations for phosphorus, sulfur, and chlorine have been determined by variation of all exponent and coefficient parameters so as to yield the lowest possible (UHF) energy for the free atom. 6-31G basis sets for second-row elements are constructed utilizing inner-shell descriptions taken directly from the previously defined 6-21G basis sets, and optimized only with respect to valence-shell parameters.

The valence parts of the 4-31G and 6-31G representations for hydrogen and first-row elements lithium through oxygen have been scaled according to Eq. (4.10) so as to be more appropriate for use in calculations on molecules. "Standard" valence-shell scale factors, determined by optimization on a number of simple molecules, may be found in the original paper [17a]. Basis sets for helium, fluorine, neon, and all second-row elements are employed unscaled, that is, $\zeta = 1$ in Eq. (4.10). The inner shells of all atoms are left unscaled.

For reasons of economy, most applications requiring the use of a split-valence basis set are best carried out at the 3-21G level rather than with the larger 4-31G or 6-31G representations. The latter representation has found its most important use as a starting point for the construction of the 6-31G* and 6-31G** polarization basis sets, representations which include functions of higher angular quantum number than required by the atom in its ground state. A full discussion is provided in the next section.

4.3.3. Polarization Basis Sets

One feature common to the basis sets discussed thus far is that they comprise functions constrained to be centered at the nuclear positions. Evidence suggests, however, that the description of highly polar molecules, and of systems incorporating small strained rings, requires that some allowance be made for the possibility of nonuniform displacement of charge away from the atomic centers. Without such allowance, comparisons of properties between, say, small-ring compounds and their acyclic isomers are likely to lead to unreliable results. This situation may be dealt with in one of two ways. Conceptually the simpler approach is to allow for the inclusion in the basis set of functions not associated with any particular center. Such a strategy is well illustrated by the work of Carlsen [19]. This author obtained equilibrium geometries for a number of polyatomic molecules in excellent agreement with those derived from near-limiting Hartree–Fock procedures, by using the 4-31G split-valence basis set [17] supplemented by gaussian s- and p-type primitives placed along the "bonds". Such an approach of allowing for functions to be placed away from the nuclear positions is not, however, without its disadvantages. In the first place, it is generally not *size consistent*. That is to say, the number of non-nuclear-centered functions, for example, those placed along bonds, does not necessarily increase in direct proportion to molecular size. Also, while the positions of non-nuclear-centered basis functions are easily and uniquely specified in molecules characterized by 2-center, 2-electron bonds, their placement is not obvious in molecules involving multi-center bonding. For example, where should one locate the off-center functions in ferrocene?

The only apparent recourse in such situations is to allow for bond functions between all pairs of atoms, a solution which results in an enormous increase in computational expense with increasing molecular size. Even then, the exact placement of the functions along the bonds would need to be specified in detail for each and every case. This is clearly undesirable.

Another way to allow for small displacements of the center of electronic charge away from the nuclear positions is simply to include functions of higher angular quantum number (d-type functions on heavy atoms and p-type functions on hydrogen) in the basis set. For example, the effect of mixing a p_x-type function into the valence s function of a hydrogen atom is to displace the center of the basis function along the x axis and away from the hydrogen nucleus.

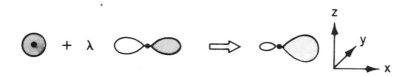

Mixtures of s- with p_y- or p_z-type functions result in analogous displacements along the y and z axes, respectively. Displacements of valence p orbitals may be realized in an analogous manner, by the admixture of d-type functions. For example, admixture of d_{xz} into p_z results in the displacement of charge along the x axis.

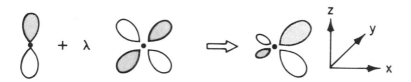

In a like manner, f-type orbitals effect off-center movement of valence d functions.

A basis set that incorporates functions of higher angular quantum number than are needed by the atom in its electronic ground state is called a *polarization basis set*; it provides for displacement of electronic charge away from the nuclear centers, that is, charge polarization.

TABLE 4.4.   Standard Polarization Function Exponents for 6-31G* and 6-31G**
Polarization Basis Sets

Atom	α_p	Atom	α_d	Atom	α_d
H	1.1	Li	0.2	Na	0.175
He	1.1	Be	0.4	Mg	0.175
		B	0.6	Al	0.325
		C	0.8	Si	0.45
		N	0.8	P	0.55
		O	0.8	S	0.65
		F	0.8	Cl	0.75
		Ne	0.8	Ar	0.85

a.   The 6-31G* and 6-31G** Polarization Basis Sets.   Among the simplest of
polarization basis sets are two representations originally proposed by Hariharan and
Pople [20] for first-row atoms and later extended to second-row elements [18c].
The simpler of the two, termed 6-31G*, is constructed by the addition of a set of
six second-order (d-type) gaussian primitives to the split-valence 6-31G basis set
description of each heavy (non-hydrogen) atom. d-orbital exponents, chosen on
the basis of energy optimization studies in molecules, are given in Table 4.4.

The 6-31G* basis contains no provision for polarization of the s orbitals on
hydrogen and helium atoms. As this feature is desirable for the description of the
bonding in many systems, particularly those in which hydrogen is a bridging atom,
a second, more complete basis set, termed 6-31G**, has been constructed. It is identical to 6-31G* except for the addition of a single set of gaussian p-type functions
to each hydrogen and helium atom. As before, an average radial exponent (1.1) has
been selected on the basis of optimization studies.

*b.   Larger Polarization Basis Sets. The 6-311G** Basis Set for Use with Correlated
Wavefunctions.*   A number of larger gaussian basis sets have been proposed in the
literature and applied to studies of chemically interesting systems. Although these
are generally more flexible than the simple 6-31G* and 6-31G** polarization basis
sets discussed in the previous section, their size has generally limited their application to only quite small molecular systems. One basis set which has been formulated
through the first row, and deserves special mention, is the 6-311G** representation
[21]. It comprises an inner shell of six s-type gaussians, and an outer (valence) region,
which has been split into three parts, represented by three, one, and one primitives,
respectively. The basis is supplemented by a single set of five d-type gaussian functions
for first-row atoms, and a single set of uncontracted p-type gaussians for hydrogen.
The 311 "triple" split has been chosen in place of the 31 separation used in the
previously discussed 6-31G* and 6-31G** polarization basis sets so as to increase
the overall flexibility of the representation, and to improve the description of the
outer valence region. The use of five d-type gaussians instead of the full set of six
second-order functions (x^2, y^2, z^2, xy, xz, and yz) results only in the loss of an

additional s-type gaussian function $(r^2 \exp(-\alpha r^2))$. This extra function would to some extent have duplicated the properties of the sp part of the basis.

One special feature of the 6-311G** basis set is that the gaussian exponents and expansion coefficients have been chosen so as to minimize the energy of the atomic ground state at the second-order Møller–Plesset (MP2) perturbation level, rather than at the corresponding Hartree–Fock level. This allows for the possibility of specific basis functions contributing to the electron correlation within the atomic valence region. It has been demonstrated that d-type functions are necessary for the proper description of electron correlation involving electrons in p-type atomic orbitals [22]. Details concerning the formulation and application of the 6-311G** basis set may be found in the original literature [21].

4.3.4. Efficient Basis Sets for Hypervalent Molecules

The simplest basis sets of practical use for compounds containing second-row elements generally do not incorporate functions of d-type symmetry. It should not be surprising, therefore, that such basis sets perform poorly in their description of the bonding in molecules containing a second-row element with an expanded valence shell. For example, while both the widely used minimal STO-3G and split-valence 3-21G basis sets are moderately successful in reproducing the experimental equilibrium geometries of a variety of normal-valent compounds, both fare very badly in their attempted descriptions of the structures of hypervalent species. The case of dimethylsulfoxide is typical. Here both STO-3G and 3-21G calculations suggest a description in terms of a zwitterionic structure, in which the sulfur atom is (formally) positively charged but remains normal valent. A more correct representation is, of course, a structure where the valence orbital manifold about sulfur accommodates 10 electrons.

$$ CH_3{\cdots}S^{+}{-}O^{-} \qquad CH_3{\cdots}S{=}O $$
$$ CH_3 \qquad\qquad\quad CH_3 $$

Numerous attempts have been made to construct simple, and hence widely applicable, atomic basis sets that allow for the description of hypervalent molecules. Among the most simple is the STO-3G* representation [23], formed from STO-3G by the addition of a set of five pure d-type gaussians. The limited number of applications that have been carried out at this level suggest that the STO-3G* basis set is moderately successful in accounting for the geometrical structures of hypervalent molecules. For example, the experimental geometry of dimethylsulfoxide is now well reproduced. Failings of the simple model are, however, apparent upon closer scrutiny. Thus, the calculated equilibrium structures of hypervalent molecules such as PF_5 and ClF_3 fail to reproduce the observed differences between axial and equatorial bond distances. Equally important, the STO-3G* basis set, like the minimal STO-3G representation from which it is constructed, performs poorly both in the description of the relative energies of isomers and in calculations of normal-mode vibrational frequencies. Favorable performance in both of these areas, as well as in the description of equilibrium geometries, are among the principal goals of any general structure theory.

We see in Chapter 6 that the 6-31G* polarization basis set does perform satisfactorily in the description of the equilibrium structures, relative energies, and normal vibrational frequencies of molecules incorporating second-row elements, both normal and hypervalent species alike. Unfortunately the 6-31G* basis set is too large, both in terms of the number of atomic basis functions and the number of primitive gaussians, to be widely applicable to even moderately sized molecules. In practice, its application is presently limited to molecules comprising at most four to five heavy (nonhydrogen) atoms. It is, therefore, highly desirable to have available a somewhat smaller and computationally more efficient basis set able to mimic the properties of the full 6-31G* polarization representation. In particular, such a basis set should be capable not only of accurately reproducing the known equilibrium geometries of molecules incorporating second-row elements with normal and expanded valence-shell electronic configurations, but it should also provide a satisfactory account of the relative energies and vibrational frequencies of both normal-valent and hypervalent species. It should be readily applicable to molecules comprising up to, say, ten heavy atoms.

a. The 3-21G$^{(*)}$ Basis Set for Second-Row Elements.   3-21G$^{(*)}$ basis sets for second-row elements [24] are constructed directly from the corresponding 3-21G representations by the addition of a complete set of six second-order gaussian primitives. Although the resulting representations contain the same number of atomic basis functions per second-row atom as the 6-31G* polarization basis sets previously described, these are made up of significantly fewer gaussians (three instead of six for each inner-shell atomic orbital, and two gaussians instead of three for the inner part of the valence description). In addition, descriptions for hydrogen, helium, and all first-row elements are in terms of unsupplemented 3-21G basis sets. Therefore, 3-21G$^{(*)}$ should not be viewed as a full polarized basis set; rather, it is best seen as a representation which, in as simple a manner as possible, is able to account for the participation of d-symmetry functions in the bonding about second-row atoms.

A simple example of the effect of the added d-type functions on the bonding in hypervalent compounds is provided in Figure 4.1, which depicts one member of the degenerate pair of highest-occupied molecular orbitals in trimethylphosphine oxide. It is this molecular orbital, together with its partner, which is largely responsible for the multiple-bond character of the PO linkage, the major feature of which basis sets unsupplemented by d-type functions are unable to protray accurately. The molecular orbital plotted on the right derives from a calculation using the unsupplemented 3-21G basis set, and that on the left from a 3-21G$^{(*)}$-level study. Note that the added basis functions effect a displacement of the wavefunction away from oxygen and into the region connecting the oxygen and phosphorus atoms.

The same average radial exponents, α, as used to define the set of d functions in the 6-31G* polarization basis set are also employed for the 3-21G$^{(*)}$ representations. Such a choice ensures that the radial description of the added d functions will closely approximate that appropriate for addition to a completely saturated s- and p-type basis set. Values of α may be found in Table 4.4.

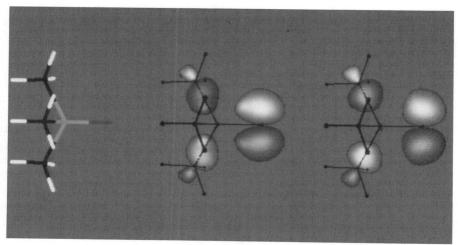

FIGURE 4.1. One member of the degenerate pair of highest-occupied molecular orbitals in trimethylphosphine oxide. 3-21G$^{(*)}$ basis set (left); 3-21G basis set (right). Optimum 3-21G$^{(*)}$ equilibrium structure.

Table 4.5 compares total energies, hydrogenation energies, equilibrium geometries, normal-mode vibrational frequencies, and electric dipole moments for sulfur dioxide calculated using both the 3-21G$^{(*)}$ and 6-31G* basis sets. The results are typical of many others that have been obtained [24]. With the exception of total energy, which in this instance differs by 2.66 hartrees (1672 kcal mol^{-1}!), the values of all properties calculated at the two levels are in close agreement. This is especially true

TABLE 4.5. Properties of SO_2 Derived with the 3-21G$^{(*)}$ and 6-31G* Basis Sets

Property	Basis Set	
	3-21G$^{(*)}$//3-21G$^{(*)}$	6-31G*//6-31G*
Total energy (hartrees)	−544.50373	−547.16901
Hydrogenation energy (kcal mol^{-1})^a	−75	−87
Equilibrium geometry r_{SO}	1.419	1.414
(distances in angstroms, <OSO	118.7	118.8
angles in degrees)		
Vibrational frequencies (cm^{-1}) a_1	1341	1357
a_1	602	592
b_1	1573	1568
Dipole moment (debyes)	2.29	2.19
Relative computational times^b	1	4

^{a}Energy of the reaction $SO_2 + 3H_2 \rightarrow H_2S + 2H_2O$. The experimental value (corrected for zero-point vibrational energy) is −58 kcal mol^{-1}.

^{b}Single-point energy calculation; the time difference for calculations involving energy derivatives would be significantly greater.

for calculated equilibrium geometries and vibrational frequencies, for which the two basis sets yield nearly identical results. The latter observation suggests that the $3\text{-}21\mathrm{G}^{(*)}$ basis set should find considerable use as an alternative to $6\text{-}31\mathrm{G}^*$ as a means of providing structures and normal-mode frequencies that closely approximate those of the larger basis representation.

The exact cost of application of the $3\text{-}21\mathrm{G}^{(*)}$ basis set, relative to $6\text{-}31\mathrm{G}^*$, is somewhat difficult to assess, and varies widely depending on the molecule, for example, number of first-row elements, and on the particular task at hand, for example, geometry optimizations or frequency calculations which are dominated by integral evaluation steps versus single-point runs which are often dominated by the SCF procedure; our experience indicates a decrease in computation time by a factor of between four and six. Cost differences arise both because of the different number of gaussian primitives that make up the individual functions, and because the $3\text{-}21\mathrm{G}^{(*)}$ basis set lacks d functions on first-row atoms. Size (memory) requirements will also differ between the two basis sets, and this too may influence the selection. While both $3\text{-}21\mathrm{G}^{(*)}$ and $6\text{-}31\mathrm{G}^*$ basis sets for hydrogen and for second-row atoms comprise the same number of atomic functions, they differ significantly in size for first-row elements (9 functions for $3\text{-}21\mathrm{G}^{(*)}$ versus 15 for $6\text{-}31\mathrm{G}^*$).

4.3.5.   Basis Sets Incorporating Diffuse Functions

The basis sets that have been discussed thus far, although developed with no intended bias, are more suitable for molecules in which the electrons are tightly held to the nuclear centers than they are for species with significant electron density far removed from those centers. Calculations involving anions pose special problems [25]. Since the electron affinities of the corresponding neutral molecules are typically quite low [26], the extra electron in the anion is only weakly bound. Even the largest basis sets considered thus far in this chapter do not incorporate functions with significant amplitude far distant from their center, and therefore do not provide a completely adequate description of molecules in which a large portion of the valence-electron density is allocated to diffuse lone-pair or to antibonding orbitals. For most stable anions, application of these basis sets yields positive energies for the highest-occupied molecular orbitals, indicating erroneously that the outermost valence electrons are unbound.

One way to overcome problems associated with anion calculations is to include in the basis representation one or more sets of highly diffuse functions [27]. These are then able to describe properly the long-range behavior of molecular orbitals with energies close to the ionization limit. It has been shown that the addition of diffuse functions has dramatic effects on calculated electron affinities [27e], proton affinities [27i,s] and inversion barriers [27a,c, 28a].

a.   The 3-21+G, 3-21+G$^{(*)}$ and 6-31+G* Basis Sets.   The 3-21+G and 6-31+G* basis sets for first-row elements, and 3-21+G, 3-21+G$^{(*)}$, and 6-31+G* basis sets for second-row elements [28] are constructed from the underlying 3-21G, 3-21G$^{(*)}$, and 6-31G* representations by the addition of a single set of diffuse gaussian *s-*

TABLE 4.6.   Diffuse Function Exponents for 3-21+G, 3-21+G$^{(*)}$, 6-31+G*, and Larger Basis Sets

Atom	$\alpha_{sp}{}^a$	Atom	α_{sp}	Atom	α_{sp}
H	0.0360	Li	0.0074	Na	0.0076
		Be	0.0207	Mg	0.0146
		B	0.0315	Al	0.0318
		C	0.0438	Si	0.0331
		N	0.0639	P	0.0348
		O	0.0845	S	0.0405
		F	0.1076	Cl	0.0483

[a]Hydrogen exponent for 3-21++G, 3-21++G$^{(*)}$, 6-31++G*, and larger basis sets. For most applications, adding diffuse functions to hydrogen does not lead to any significant improvement in results and quite little absolute energy lowering. However, such diffuse functions are absolutely essential to calculate H$^-$ and in situations, for example, transition structures, where hydrogen is expected to bear considerable negative charge.

and p-type functions. The gaussian exponents for both s and p functions, which are displayed in Table 4.6, have been determined as optimum for the deprotonated forms of the one-heavy-atom hydrides, for example, Li$^-$, BeH$^-$, BH$_2^-$, and so on, using the 3-21G basis set (and optimum 3-21G geometries) for both first-row and second-row elements.

Energy lowerings resulting from the addition of diffuse functions to a selection of one-heavy-atom hydrides and their deprotonated forms are displayed in Table 4.7. Two underlying basis sets have been investigated, 3-21G (3-21G$^{(*)}$ for second-row hydrides) and 6-31G*. Both sets of data clearly show the effect of the added diffuse functions to be far more significant for the negatively charged species than for the corresponding neutrals. Thus, added diffuse functions lower acidity. Note that the energy lowerings (both for neutral and negatively charged species) are significantly larger (usually by a factor of three or more) for first-row hydrides than they are for analogous second-row systems. This suggests that the effects of diffuse functions on calculated acidities (and probably also electron affinities) will be less important for compounds incorporating second-row elements than they will be for corresponding first-row systems. Finally, it should be noted that the supplementary functions affect the 3-21G total energies (3-21G$^{(*)}$ for second-row systems) considerably more than they do the energies of the larger 6-31G* representation. In fact, 6-31+G* energy lowerings for the neutral hydrides are uniformly small (less than 5 kcal mol^{-1} except for HF). This result is consistent with the fact that the valence region of the 6-31G* basis set is inherently more diffuse than that of 3-21G, and hence more able by itself to describe properly the long-range behavior of the weakly bonded electrons.

For first-row elements with lone pairs, the effects of diffuse and polarization functions are complementary to some extent. Hence, the energies of processes involving changes in the number of lone pairs, for example, protonation, hydrogen bonding, or other interactions, are improved at diffuse-orbital-augmented levels, even when large basis sets are used [28a,c,d].

TABLE 4.7.   Energy Lowerings (kcal mol^{-1}) due to Addition of Diffuse s- and p-type Functions[a]

XH	$3\text{-}21\text{G}^b \rightarrow 3\text{-}21\text{+G}^b$		$6\text{-}31\text{G}^* \rightarrow 6\text{-}31\text{+G}^*$	
	XH	X$^-$	XH	X$^-$
H_2	0^c	53^c		
LiH	0	6		
BeH_2	1	19		
BH_3	1	22		
CH_4	1	32	0	22
NH_3	13	57	3	27
H_2O	21	80	5	32
HF	29	101	8	44
NaH	1	5	0	5
MgH_2	2	8	0	5
AlH_3	3	8	1	4
SiH_4	2	10	1	6
PH_3	2	17	0	10
H_2S	3	16	0	9
HCl	3	15	1	9

[a]Geometries optimized using underlying 3-21G, 3-21G$^{(*)}$ and 6-31G* basis sets.
[b]3-21G$^{(*)}$, 3-21+G$^{(*)}$ for molecules incorporating second-row elements.
[c]Diffuse basis functions added to hydrogen (3-21++G and 6-31++G* basis sets). For the remaining hydrides, the addition of diffuse functions on hydrogen has little effect.

4.3.6.   Relative Computational Times of Hartree-Fock Models with Gaussian Basis Sets

Since a major factor in selecting a suitable basis set for a particular application is the availability of computational resources, estimates of the relative computational times involved in the application of each of these basis sets are useful. Note, however, that these estimates are dependent not only on the specific computer algorithms employed but also on the particular molecular system under investigation. Relative timings for methylamine and thiophene for several different basis sets are given in Table 4.8. For comparison, the same computations using semiempirical schemes such as CNDO, INDO, or MINDO would require approximately 5% of the time needed for an STO-3G calculation.

4.4.   ELECTRON CORRELATION METHODS

Because Hartree–Fock procedures do not adequately account for the correlation of electron motion, the next step is the replacement of RHF or UHF by a more elabo-

TABLE 4.8. Relative Timings for Hartree–Fock Calculations[a]

| Basis Set | Methylamine[b] | | Thiophene[c] | |
	Energy	Energy + Energy Gradient	Energy	Energy + Energy Gradient
STO-3G	1	1	1	1
3-21G[d]	2.6	1.9	6.5	3.8
6-31G*	9.1	13		

[a]Calculations performed using GAUSSIAN 85 on a Harris Corporation H800 digital computer. Actual times for energy and energy + energy gradient calculations on methylamine at the STO-3G level are 36 and 74 seconds, respectively; times for analogous calculations on thiophene are 152 and 441 seconds, respectively.
[b]C_s symmetry employed.
[c]C_{2v} symmetry employed.
[d]3-21G(*) for thiophene.

rate model using a multiple-determinant wavefunction. As previously noted, this is quite distinct from improvements in the basis set. However, it is important to recognize that the use of a small basis set may limit the fraction of the total correlation energy obtainable even with elaborate multiple-determinant techniques [22].

As with basis set selection, choice of the electron correlation method may ultimately be governed primarily by practical considerations. The technique should, however, satisfy the general requirements outlined in Chapter 1. In particular, the model should be well defined and size consistent, so that application to an assembly of isolated molecules will give additive results. The latter requirement is particularly important if comparisons are to be made among molecules of different size. The two techniques that have been widely used are limited configuration interaction (Section 2.9.2) and Møller–Plesset perturbation theory (Section 2.9.3). The first is variational but not size consistent, while the second is size consistent but not variational. In this section, we make some remarks on the practical aspects of these two methods that are of concern to the overall choice of theoretical model.

4.4.1. Limited Configuration Interaction

Two levels of limited configuration interaction will be considered in this book. The simpler of these, termed CID, utilizes a multiple-determinant wavefunction constructed from double substitutions only, Eq. (2.70), while the more complex, designated CISD, allows as well for inclusion of single substitutions, Eq. (2.71). The expansion coefficients are determined variationally according to methods described in Section 2.9. Both the CID and CISD energies are upper bounds to the full configuration interaction energy for a given basis set, that is, are variational. However, both methods suffer from not being size consistent. As mentioned previously (Section 2.9.2), corrections can be and have been introduced to make allowance for this lack of size consistency [29].

4.4.2. Møller–Plesset Perturbation Treatments

The most economical general correlation methods are based on the perturbation theory of Møller and Plesset [30]. As noted in Section 2.9.3, these methods treat the complete Hamiltonian as the sum of two parts, Eq. (2.79), the non-Hartree–Fock part being treated as a perturbation on the Hartree–Fock part using Rayleigh–Schrödinger theory. The energy expression, Eq. (2.82), may be terminated at any desired order, although practical programs are presently limited to fourth order. These energies have the property of size consistency, but are not variational, that is, do not represent rigorous upper bounds to the true energy. The complexity and cost of computing the energy terms in this expansion increase rapidly with the order.

The energy $E^{(0)} + E^{(1)}$ in Eq. (2.82), correct to first order in the Møller–Plesset (MP) expansion, is identical to the Hartree–Fock energy. This is the unrestricted (UHF) result for open-shell systems, although for closed-shell ground states this usually reduces to the restricted (RHF) function with doubly-occupied orbitals. The simplest approximation to the energy correction due to correlation is the second-order quantity $E^{(2)}$, Eq. (2.92). If the expansion is terminated here, the model may be described as MP2. Note, of course, that a complete model specification also requires a basis set. The third- and fourth-order methods may be denoted by MP3 and MP4, respectively. A partial fourth-order treatment, denoted by MP4SDQ, ignores the effects of triple substitutions.

Truncation of the Møller–Plesset series will be a satisfactory procedure only if convergence is fairly rapid. This appears to be so for some small molecules, as suggested by the data in Table 4.9, obtained for the ethylene molecule using the 6-31G** full polarization basis set and HF/6-31G* equilibrium geometries. While total energies are lowered by 22.5 kcal mol^{-1} from the MP2 to the full MP4 levels, the change in hydrogenation energy is only about 1 kcal mol^{-1}.

4.4.3. Relative Computational Times of Correlation Methods

The data in Table 4.10 provide some idea of the relative computational requirements for a single calculation on methylamine using the 6-31G* basis set at the Hartree–Fock and various correlation levels. As we see in Chapter 6, this basis set is perhaps the simplest representation suitable for use with the simple correlation energy treatments. Again, the relative timings should be taken only as rough estimates as they are highly dependent on the system at hand.

Solution of the CID or CISD equations (2.67) is usually carried out by an iterative procedure. The principal time-consuming steps are the initial full integral transformation and a subsequent large-scale matrix multiplication at each iteration. The latter requires $O(n^2 N^4)$ steps for n electrons and N basis functions.

MP2 theory requires only a partial transformation of the two-electron integrals, this being the most time-consuming part of a post-Hartree–Fock MP2 computation. The number of arithmetic steps required for this part is $O(n N^4)$, where n is the number of electrons and N the number of basis functions. Higher-order energy contributions in the Møller–Plesset expansion involve substantial further computation.

TABLE 4.9. Convergence of Møller–Plesset Perturbation Expansions for the Ethylene Molecule[a]

Level	Total Energy (hartrees)	Hydrogenation Energy[b] (kcal mol^{-1})
HF	-78.03884	-64.0
MP2	-78.31682	-60.9
MP3	-78.33998	-62.4
MP4SDQ	-78.34414	-61.3
MP4	-78.35306	-59.7

[a]Calculations performed with the 6-31G** basis set and optimum HF/6-31G* geometries. Frozen-core approximation used throughout.
[b]Energy of reaction, $CH_2 = CH_2 + 2H_2 \rightarrow 2CH_4$. The experimental value (corrected for zero-point vibrational energy) is -57.2 kcal mol^{-1}.

TABLE 4.10. Relative Timings for Single Calculations on Methylamine Using the 6-31G* Basis Set and Various Correlation Models[a]

Correlation Method[b]					
HF	CID	CISD	MP2	MP3	MP4SDQ
1	15.0	16.7	1.5	3.6	5.8

[a]Molecular symmetry has not been employed for the purpose of these comparisons. Actual time for an HF/6-31G* calculation on methylamine is 9 minutes, 50 seconds, performed using GAUSSIAN 82 on a Digital Equipment Corporation VAX 11/780 digital computer.
[b]Frozen-core approximation used throughout.

The rate-limiting step involved in obtaining the third-order energy $E^{(3)}$, Eq. (2.93), is the large-scale matrix multiplication, and requires $O(n^2 N^4)$ steps. This level of theory is equivalent to the first iteration of a CID calculation. The computational effort required to evaluate the fourth-order energy $E^{(4)}$ is comparable to $E^{(3)}$, if it is restricted to inclusion of single, double, and quadruple substitutions in the Hartree–Fock determinant. The triple-substitution contribution to $E^{(4)}$ requires $O(n^3 N^4)$ steps.

4.5. MOLECULAR GEOMETRY

Application of molecular orbital theory requires a specification of molecular geometry. A complete theoretical treatment of equilibrium structure would involve minimization of the energy with respect to each independent geometrical parameter, yielding a structure termed "optimized." A corresponding treatment of a transition structure would seek a saddle point on the potential surface. Complete optimization of geometrical structure generally requires considerable computational effort, and may not be necessary or even desired in all instances. It is sometimes satisfactory to use simpler models involving partial optimization of geometry or no optimization at all. The molecules and properties under investigation, as well as the available computational resources, may need to be considered in order to decide upon an appropriate geometrical model for a particular study.

4.5.1. Standard Molecular Geometries

The simplest and least-involved approach to obtaining suitable geometrical structures on which to perform calculations is to assume values for all bond lengths and bond angles. One recourse is to use experimental structural data, although such information may not be available for the particular system of interest (especially for short-lived intermediates or reaction transition structures). Even when available, experimentally determined structural parameters are subject to varying degrees of uncertainty, due to the use of different experimental techniques. More widely applicable are *standard molecular geometries* [9b, 31], which are constructed from standard bond lengths and bond angles that have been obtained by averaging large numbers of experimental and, in a few cases, the best available theoretical structural parameters. A listing of standard bond lengths and angles for acyclic molecules containing first- and second-row elements may be found elsewhere [9b, 31].

There are several reasons for selecting standard values of assumed structural parameters. This procedure eliminates the need for experimental structures. It also eliminates the subjectivity of any more arbitrary method of assigning bond lengths and angles, and enables comparisons among large sets of related molecules to be carried out at a well-defined level. With other models, variations in the properties being calculated may be obscured by the assumption of different geometric parameters. In other words, it may be desirable to use standard geometric parameters to isolate purely electronic effects from those caused by geometry changes.

There are shortcomings in the use of standard geometries. In the first place, the model has been defined only for a limited number of systems. At present, molecules incorporating elements beyond the second row may not be handled due to insufficient experimental structural data on which to base standard bond lengths and angles. Also, standard geometries are restricted to acyclic systems; model structures for molecules incorporating small rings remain to be defined. More importantly, many molecules within the scope of definition of the standard model are beset with large steric, conjugative, or other interactions, which would in practice lead to severe geometrical distortions. The use of standard model structures here may be completely unrealistic. Additionally, the bonding in many systems, including the majority of reactive intermediates, for example, pentacoordinate carbocations, as well as most transition structures, cannot be described in terms of classical valence forms. Here again, standard bond lengths and angles cannot be assigned satisfactorily.

A common variant on the standard model theme is to affix "standard model" substituents, that is, those defined using standard values for all geometric parameters, to some molecular skeleton, the geometry of which has been established by alternative means, for example, experiment or complete optimization. Such an approach is especially useful in situations where the skeletal structure itself may not easily be defined within the standard model framework. The utility of such an approach has been demonstrated in numerous published theoretical studies dealing with substituent effects [32].

In summary, assumed standard model geometries are best employed: (a) when the cost of an alternative, more refined, treatment is prohibitive; (b) in the study of properties that are not likely to be adversely affected by the choice of geometrical model; and (c) in comparative studies of large numbers of related systems, for which it might be desirable to have a uniform specification of geometry.

4.5.2. Partial Optimization of Molecular Geometry

In some situations it may be inadvisable to employ fixed molecular geometries, that is, standard model or experimental geometries, but at the same time economic or other considerations may rule out the use of completely optimized structures. For example, a poor choice of geometry for a calculation on *bis* (trifluoromethyl) ether might lead to an energy that is far too high, due to crowding of the two bulky CF_3 groups. Optimization of a single geometrical parameter, the central angle θ, would probably be sufficient to relieve much of the unfavorable steric interaction and would yield an energy that is more characteristic of the system than that using a standard model structure.

The cost of such a task should be far less than that of a complete optimization.

Alternatively, the system might comprise a rigid skeleton to which a flexible side chain is attached. Here optimization of only the side chain may be required. A number of published studies of substituted aromatic systems belong in this category [32]. For example, a reasonable approximate theoretical structure for aniline might be obtained by optimizing the four independent parameters (two bond lengths, r_{CN} and r_{NH}, and two bond angles, <HNH and <CNH) of the amino group, while constraining the ring skeleton to remain fixed. Again the limited optimization required here should be less costly computationally than complete structure variation.

A third example is the use of a partial optimization in studies of intermolecular interactions, for example, hydrogen bonding [12b, 33]. Here, the geometries of the interacting molecules in the complex are often taken to be those obtained by optimization of the isolated (noninteracting) species, and only intermolecular parameters are optimized.

Partial optimization of geometrical structure is often useful in the calculation of the rotational potentials of simple molecules [34]. In one type of *flexible-rotor approximation,* optimization is limited to those bond angles involving the heavy-atom skeleton. Other angles and all bond lengths are fixed either at standard model values, or at those appropriate for the equilibrium conformation of the molecule. For example, within the framework of this flexible-rotor model, the description of the torsion about the carbon–carbon single bond in 1-butene would require optimization of only the bond angles $<(C_1 C_2 C_3)$ and $<(C_2 C_3 C_4)$ for each choice of dihedral angle $\omega(C_1 C_2 C_3 C_4)$.

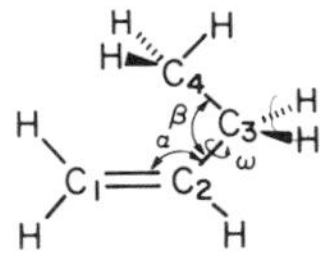

This type of treatment provides some geometrical freedom to cope with nonbonded repulsions present in the more crowded conformations, but at the same time it is inexpensive relative to the cost of complete structure optimization.

4.5.3. Complete Optimization of Molecular Geometry

Often the detailed structure itself is the point of interest, and there is no alternative to optimization of all geometrical parameters. Complete structure optimization should normally also be carried out for molecules poorly represented by conventional bonding and for which reasonable values for geometric parameters may not be easily assigned. Transition structures represent extreme examples. Here, molecular geometry may be completely uncertain, and optimization to give a structure at the saddle point on the potential surface is normally essential. Complete optimization of geometric parameters for minima and saddle points is becoming more common, and the emergence of computationally efficient analytical energy gradient and second-derivative methods will certainly accelerate this trend. Many examples of fully optimized structures are discussed in Chapters 6 and 7.

     Optimization is often carried out for a system constrained to a given symmetry. The resulting structure, while optimum within the confines of that symmetry, is not guaranteed to be a potential energy minimum outside of those confines. Such a procedure may sometimes be employed as an economical means of obtaining optimized geometries for transition structures. Simple examples include: (a) inversion of pyramidal ammonia through a planar (D_{3h}) structure,

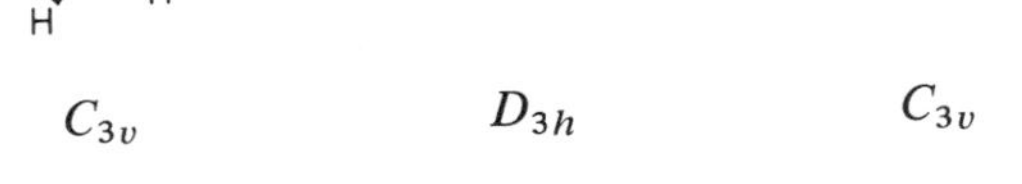

(b) rotation in ethane via a D_{3h} structure,

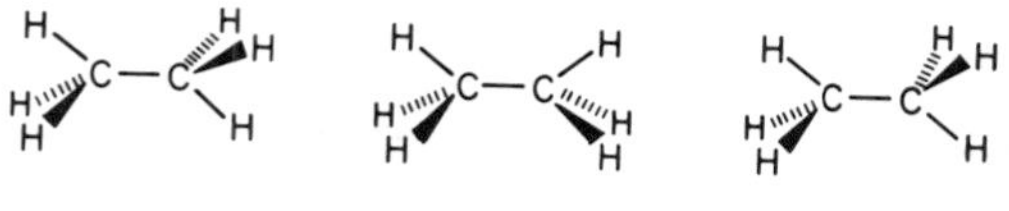

D_{3d} D_{3h} D_{3d}

and (c) degenerate 1,3-hydrogen migration in propene via a symmetry-allowed C_2 transition structure.

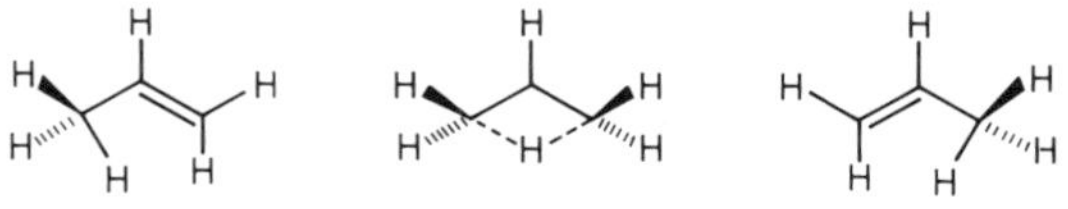

C_s C_2 C_s

The use of symmetry constraints can sometimes lead to incorrect classification of minima or saddle points on a surface. It is therefore important to characterize structures obtained in this manner by explicit evaluation and examination of the eigenvalues of the second-derivative matrix.

Often symmetry constraints are imposed in order to aid in the exploration of a complex potential surface. The possible structures for CH_5^+, among them

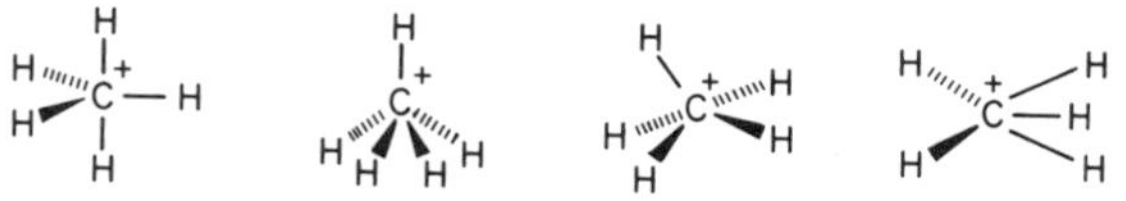

provide a good example. Note that not all (or any) of these structures necessarily correspond to minima on the CH_5^+ potential surface, and that examination of the second-derivative matrix is again necessary to establish clearly the nature of specific forms. Some discussion of the procedures involved is presented in Section 7.6.

4.5.4.   Use of Geometries from Lower Theoretical Levels

Complete or partial optimization of geometrical variables need not always be carried out at the same level of theory required for the calculation of other properties, for example, relative energies. A common practice has been to obtain a fully (or partially) optimized structure at a relatively low level of theory, for example, STO-3G or 3-21G, and then to use this geometry for a single calculation at a higher level, 6-31G* for example. The same general procedure has often been applied in studies incorporating some treatment of electron correlation, where the cost of geometry optimization is often prohibitive. Here, Hartree–Fock geometries are often employed for single-point correlation-energy calculations using the same basis set. This latter strategy is attractive since analytical gradient methods are far simpler for Hartree–

Fock schemes than they are for methods that take account of electron correlation [35].

A word of caution needs to be expressed. The position of a minimum or a saddle point is dependent upon the level of theory employed, and failure to reoptimize at the higher level may lead to inaccuracies. In some instances, for example, in deciding between bridged and open structures for the vinyl and ethyl cations, the effects of improved basis set and/or correlation levels may be extreme and may even change the qualitative conclusions. Nevertheless, in situations where full geometry optimization at the highest level is prohibitively expensive, the use of a lower-level structure may offer a valuable compromise. Chapter 6 addresses the question of differences in calculated equilibrium geometries, both for a variety of standard basis sets within the single-determinant Hartree–Fock framework, and for varying levels of treatment of electron correlation for a specific basis set. This material should be of aid to the practitioner in ascertaining under what conditions geometries obtained from lower theoretical levels might be appropriate for use in higher-level calculations.

4.6. NOMENCLATURE

Complete specification of a theoretical model requires designation of the Hartree–Fock or correlation procedure, as well as the basis set and geometrical structure employed. To indicate the level of calculation, we employ the notation: Hartree–Fock or correlation model/basis set. Thus, RHF/6-31G* denotes use of RHF wavefunctions and the 6-31G* basis set. Replacement of the RHF wavefunction by an unrestricted Hartree–Fock (UHF) procedure, for example, for a calculation on a radical or for a triplet state, results in a model described as UHF/6-31G*. Correlated wavefunctions are specified similarly. Thus, MP2/6-31G* indicates a calculation at the second-order Møller–Plesset level, using the 6-31G* basis set; CISD/6-31G** represents a calculation using a singles and doubles configuration interaction approach and the 6-31G** basis set. If specification of the nature of the underlying Hartree–Fock procedure (RHF or UHF) is also necessary, this may be combined with the designated correlation method. Thus, UMP3/6-31G* means a Møller–Plesset treatment complete through third order using an unrestricted Hartree–Fock wavefunction as a starting point, all orbitals being expanded in terms of the 6-31G* basis set. For the purpose of discussion in this book, failure to specify a Hartree–Fock or correlation procedure implies use of the appropriate HF model, that is, RHF for closed-shell systems and UHF for open-shell systems.

Further nomenclature is required to specify the geometry used for a calculation. We employ the // symbol for this purpose. Thus, RHF/6-31G*//RHF/STO-3G (which may in this case be abbreviated 6-31G*//STO-3G) designates a Hartree–Fock calculation with the 6-31G* basis set using a molecular geometry optimized at the Hartree–Fock level with the STO-3G basis set. Similarly, an MP3 correlation calculation carried out using the same geometry would be designated MP3/6-31G*//RHF/STO-3G or more simply MP3/6-31G*//STO-3G. Finally, the abbreviations "Std." and "Expt." following // indicate that the calculation has been performed using a standard geometry or an experimental geometry, respectively.

4.7. CONCLUSION

We have now set up a series of theoretical models that conform to the two-dimensional chart discussed in the introductory chapter (Figure 1.1). On the one hand we have a sequence of systematically improved basis sets: STO-3G, 3-21G, 3-21G$^{(*)}$, 6-31G*, 6-31G**, and 6-311G**. On the other, there is a similar sequence of improved correlation methods, starting with Hartree–Fock (HF) and continuing with the Møller–Plesset perturbation methods MP2, MP3, and MP4 (or alternatively the CID and CISD limited configuration interaction methods). The former procedures are size consistent but not variational, the latter the reverse.

Figure 4.2 illustrates the strategic use of a sequence of available theoretical models in a sample investigation, in this case, the study of the energy difference between two stable isomers, A and B. The same procedure is applied to both A and B. Initially each geometrical structure is optimized at the economical HF/STO-3G level. This may involve extensive searching of the potential surface if the locations of the minima are uncertain. These initial approximations of the geometry may then be used to refine the structures at the higher HF/3-21G and HF/6-31G* levels. These would not involve excessive computations if the HF/STO-3G geometry was reasonably close. The HF/6-31G* geometry is then used for single calculations with larger basis sets, for example, HF/6-31G**, and with methods including correlation to obtain improved estimates of the energy difference between A and B. It would, of course, be desirable to carry out correlation corrections with larger basis sets, but this may not always be feasible. If the route shown in Figure 4.2 is the best that can be executed, a "final" conclusion could be based on the assumption that the correlation energy correction (rather than the total energy) is unaltered by basis set improvement beyond the 6-31G* level. This is an approximation, and leads to an *estimate* of the MP4/6-31G** energy difference. As further computational resources

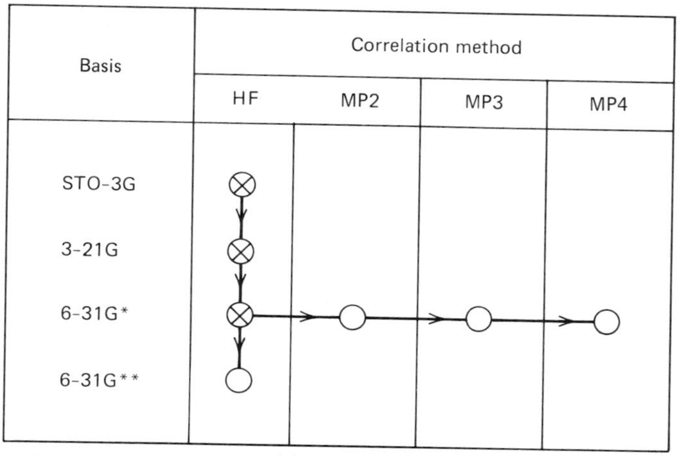

Basis	Correlation method			
	HF	MP2	MP3	MP4
STO-3G	⊗			
3-21G	⊗			
6-31G*	⊗	○	○	○
6-31G**	○			

FIGURE 4.2. Investigatory route for study of energy difference between two stable isomers A and B. ⊗ indicates full geometry optimization. ○ implies a single run using the preceding geometry.

become available, such predictions could be checked and confirmed, or possibly modified.

The final choice of detailed quantum mechanical model—Hartree–Fock method, basis set, treatment of electron correlation, and molecular geometry—to be used for the calculation of some specific property of a given molecular system, must ultimately depend not only on the established levels of performance, but also on practical considerations. Thus, while the performance of larger and more flexible atomic basis sets, or of more complete descriptions of electron correlation, is of great interest, the most sophisticated methods cannot presently be applied to larger systems. Fortunately, even Hartree–Fock calculations with small basis sets, for example, HF/STO-3G and HF/3-21G, often perform remarkably well in calculating certain properties. In these cases it is not necessary to employ more sophisticated treatments.

REFERENCES

1. (a) C. C. J. Roothaan, *Rev. Mod. Phys.,* **23,** 69 (1951); (b) G. G. Hall, *Proc. Roy. Soc. (London),* **A205,** 541 (1951).

2. J. A. Pople, *Int. J. Quantum Chem.,* **55,** 175 (1971).

3. R. B. Woodward and R. Hoffmann, *The Conservation of Orbital Symmetry,* Verlag Chemie/Academic Press, Weinheim (W. Germany), 1970.

4. (a) C. C. J. Roothaan, *Rev. Mod. Phys.,* **32,** 179 (1960); (b) J. S. Binkley, J. A. Pople, and P. A. Dobosh, *Mol. Phys.,* **28,** 1423 (1974).

5. J. A. Pople and R. K. Nesbet, *J. Chem. Phys.,* **22,** 571 (1954).

6. R. Seeger and J. A. Pople, *J. Chem. Phys.,* **66,** 3045 (1977); an algorithm to accomplish this has been incorporated into the GAUSSIAN 80 and GAUSSIAN 82 programs.

7. The computational efficiency of gaussians relative to Slater-type (exponential) functions was originally pointed out by Boys in 1950, although it was not widely appreciated until much later: S. F. Boys, *Proc. Roy. Soc. (London),* **A200,** 542 (1950).

8. See, for example: (a) T. H. Dunning and P. J. Hay, in *Modern Theoretical Chemistry,* H. F. Schaefer III, ed., Plenum Press, New York, 1977, vol. 3, p. 1; (b) *Gaussian Basis Sets for Molecular Calculations,* S. Huzinaga, ed., Elsevier, Amsterdam, 1984.

9. (a) first-row: W. J. Hehre, R. F. Stewart, and J. A. Pople, *J. Chem. Phys.,* **51,** 2657 (1969); (b) second-row: W. J. Hehre, R. Ditchfield, R. F. Stewart, and J. A. Pople, *ibid,* **52,** 2769 (1970); (c) third-row, main-group: W. J. Pietro, B. A. Levi, W. J. Hehre and R. F. Stewart, *Inorg. Chem.,* **19,** 2225 (1980); (d) fourth-row, main-group: W. J. Pietro, R. F. Hout, Jr., E. S. Blurock, W. J. Hehre, D. J. DeFrees, and R. F. Stewart, *ibid.,* **20,** 3650 (1981); (e) first- and second-row transition metals: W. J. Pietro and W. J. Hehre, *J. Comput. Chem.,* **4,** 241 (1983).

10. (a) E. Clementi and D. L. Raimondi, *J. Chem. Phys.,* **38,** 2686 (1963); (b) E. Clementi, D. L. Raimondi, and W. P. Reinhardt, *ibid.,* **47,** 1300 (1967); (c) E. Clementi and C. Roetti, Atomic and Nuclear Data Tables, **14,** (1974).

11. J. C. Slater, *Phys. Rev.,* **36,** 57 (1930).

12. (a) D. Poppinger, *Chem. Phys.,* **12,** 131 (1976); (b) J. Del Bene and J. A. Pople, *J. Chem. Phys.,* **52,** 4858 (1970).

13. (a) B. J. Ransil, *Rev. Mod. Phys.,* **32,** 239, 245 (1960); (b) E. Switkes, R. M. Stevens, and W. M. Lipscomb, *J. Chem. Phys.,* **51,** 5229 (1969).

14. (a) first-row: J. S. Binkley, J. A. Pople, and W. J. Hehre, *J. Am. Chem. Soc.,* **102,** 939 (1980); (b) second-row: M. S. Gordon, J. S. Binkley, J. A. Pople, W. J. Pietro, and W. J.

Hehre, *ibid.*, **104**, 2797 (1982); (c) third- and fourth-row main-group: K. D. Dobbs and W. J. Hehre, manuscript submitted to *J. Comput. Chem.*; (d) first- and second-row transition metals: K. D. Dobbs and W. J. Hehre, manuscript in preparation.

15. F. B. van Duijneveldt, *Gaussian Basis Sets for the Atoms H-Ne for Use in Molecular Calculations*, I.B.M. Publication RJ 945 (#16437).

16. Pulay and coworkers (P. Pulay, G. Fogarasi, F. Pang, and J. E. Boggs, *J. Am. Chem. Soc.*, **101**, 2550 (1979)) have proposed a 4-21G basis set for hydrogen and the first-row elements boron through fluorine. As an alternative to the optimization procedure outlined here, these authors have taken over without modification the inner-shell (1s) functions from 4-31G basis sets [17], and have obtained parameters for the inner and outer parts of the valence shell by limited optimization studies on simple hydrides. Equilibrium structures, force constants, dipole moments, and dipole moment derivatives for a series of small molecules appear to be given as well at this level as with the larger 4-31G basis set. The even smaller 3-21G basis sets discussed in this chapter yield results which are nearly identical to those arrived at from the 4-21G representations. For a discussion, see reference 14a.

17. (a) carbon to fluorine: R. Ditchfield, W. J. Hehre and J. A. Pople, *J. Chem. Phys.*, **54**, 724 (1971); (b) boron: W. J. Hehre and J. A. Pople, *ibid.*, **56**, 4233 (1972); (c) lithium, beryllium: J. D. Dill and J. A. Pople, *ibid.*, **62**, 2921 (1975); (d) phosphorus to chlorine: W. J. Hehre and W. A. Lathan, *ibid.*, **56**, 5255 (1972).

18. (a) carbon to fluorine: W. J. Hehre, R. Ditchfield, and J. A. Pople, *J. Chem. Phys.*, **56**, 2257 (1972); (b) lithium, beryllium, and boron: J. S. Binkley and J. A. Pople, *ibid.*, **66**, 879 (1977); (c) second-row: M. M. Francl, W. J. Pietro, W. J. Hehre, J. S. Binkley, M. S. Gordon, D. J. DeFrees, and J. A. Pople, *J. Chem. Phys.*, **77**, 3654 (1982).

19. N. R. Carlsen, *Chem. Phys. Lett.*, **51**, 192 (1977).

20. P. C. Hariharan and J. A. Pople, *Chem. Phys. Lett.*, **66**, 217 (1972).

21. R. Krishnan, M. J. Frisch, and J. A. Pople, *J. Chem. Phys.*, **72**, 4244 (1980).

22. See for example: (a) D. J. DeFrees, B. A. Levi, S. K. Pollack, W. J. Hehre, J. S. Binkley, and J. A. Pople, *J. Am. Chem. Soc.*, **101**, 4085 (1979); (b) C. E. Dykstra and H. F. Schaefer III in *The Chemistry of Ketenes and Allenes*, S. Patai, ed., Wiley, New York, 1980, p. 1.

23. J. B. Collins, P. v. R. Schleyer, J. S. Binkley, and J. A. Pople, *J. Chem. Phys.*, **64**, 5142 (1976).

24. W. J. Pietro, M. M. Francl, W. J. Hehre, D. J. DeFrees, J. A. Pople, and J. S. Binkley, *J. Am. Chem. Soc.*, **104**, 5039 (1982). 3-21G$^{(*)}$ basis sets for third- and fourth-row main-group elements have recently been formulated. See ref. 14c.

25. (a) L. Radom in *Modern Theoretical Chemistry*, H. F. Schaefer III, ed., Plenum Press, New York, 1977, vol. 4, p. 333; (b) A. C. Hopkinson in *Progress in Theoretical Organic Chemistry*, I. G. Csizmada, ed., Elsevier, New York, 1977, vol. 2, p. 194; (c) J. Simons, *Ann. Rev. Phys. Chem.*, **28**, 15 (1977).

26. (a) B. K. Janousek and J. I. Brauman in *Gas Phase Ion Chemistry*, M. T. Bowers, ed., Academic Press, New York, 1979, vol. 2, p. 53; (b) J. E. Bartmess and R. T. McIver, Jr., *ibid.*, vol. 2, p. 88.

27. (a) A. J. Duke, *Chem. Phys. Lett.*, **21**, 275 (1973); (b) R. Ahlrichs, *ibid.*, **15**, 609 (1972); **18**, 512, (1973); (c) F. Driessler, R. Ahlrichs, V. Staemmler, and W. Kutzelnigg, *Theoret. Chim. Acta*, **30**, 315 (1973); (d) B. Webster, *J. Phys. Chem.*, **79**, 2809 (1975); (e) T. H. Dunning and P. J. Hay in *Modern Theoretical Chemistry*, H. F. Schaefer, III, ed., Plenum Press, New York, 1977, vol. 3, p. 1; (f) C. E. Dykstra, M. Hereld, R. R. Lucchese, H. F. Schaefer III, and W. Meyer, *J. Chem. Phys.*, **67**, 4071 (1977); (g) R. B. Davidson and M. L. Hudak, *J. Am. Chem. Soc.*, **99**, 3918 (1977); (h) B. Jönsson, G. Karlstrom, and H. Wennerström, *ibid.*, **100**, 1658 (1978); (i) H. Kollmar, *ibid.*, **100**, 2665 (1978); (j) C. E. Dykstra, A. J. Arduengo, and T. Fukunaga, *ibid.*, **100**, 6007 (1978); (k) K. D. Jordan,

Acc. Chem. Res., **12**, 36 (1979); (l) J. K. Wilmshurst and C. E. Dykstra, *J. Am. Chem. Soc.,* **102**, 4668 (1980); (m) R. A. Eades, P. G. Gassman, and D. A. Dixon, *ibid.,* **103**, 1066 (1981); (n) R. Bonaccorsi, C. Petrongolo, E. Scrocco, and J. Tomasi, *Chem. Phys. Lett.,* **3**, 473 (1969); (o) G. T. Survatt and W. A. Goddard III, *Chem. Phys.,* **23**, 39 (1977); (p) D. S. Marynick and D. A. Dixon, *Proc. Natl. Acad. Sci., USA,* **74**, 410 (1977); (q) P. Cársky, R. Zahradnik, M. Urban, and V. Kello, *Chem. Phys. Lett.,* **61**, 85 (1979); (r) P. Cársky and M. Urban, *Lecture Notes in Chemistry, 16. Ab Initio Calculations. Methods and Applications in Chemistry,* Springer-Verlag, Berlin 1980, p. 50, and references cited therein; (s) J. Chandrasekhar, J. G. Andrade, and P. v. R. Schleyer, *J. Am. Chem. Soc.,* **103**, 5609 (1981); (t) G. W. Spitznagel, T. Clark, J. Chandrasekhar and P. v. R. Schleyer, *J. Comput. Chem.,* **3**, 353 (1982); (u) T. Clark, P. v. R. Schleyer, K. N. Houk, and N. G. Rondan, *J. Chem. Soc., Chem. Commun.,* 579 (1981); (v) P. v. R. Schleyer, J. Chandrasekhar, A. J. Kos, T. Clark, and G. W. Spitznagel, *ibid.,* 882 (1981); (w) J. Chandrasekhar, J. G. Andrade, and P. v. R. Schleyer, *J. Am. Chem. Soc.,* **103**, 5612 (1981); (x) E.-U. Würthwein, M.-B. Krogh-Jespersen and P. v. R. Schleyer, *Inorg. Chem.,* **20**, 3663 (1981); (y) N. G. Rondan, K. N. Houk, P. Beak, W. J. Zajdel, J. Chandrasekhar, and P. v. R. Schleyer, *J. Org. Chem.,* **46**, 4108 (1981); (z) J. Chandrasekhar, R. A. Kahn and P. v. R. Schleyer, *Chem. Phys. Lett.,* **85**, 493 (1982).

28. (a) T. Clark, J. Chandrasekhar, G. W. Spitznagel, and P. v. R. Schleyer, *J. Comput. Chem.,* **4**, 294 (1983); (b) W. J. Pietro, W. J. Hehre, and P. v. R. Schleyer, manuscript in preparation; (c) M. J. Frisch, J. A. Pople, and J. S. Binkley, *J. Chem. Phys.,* **80**, 3265 (1984); (d) Z. Latájka and S. Scheiner, *Chem. Phys. Lett.,* **105**, 435 (1984).

29. (a) S. R. Langhoff and E. R. Davidson, *Int. J. Quantum Chem.,* **8**, 61 (1974); (b) J. A. Pople, R. Seeger, and R. Krishnan, *ibid., Symp.,* **11**, 149 (1977).

30. (a) C. Møller and M. S. Plesset, *Phys. Rev.,* **46**, 618 (1934); for recent developments, see (b) J. S. Binkley and J. A. Pople, *Int. J. Quantum Chem.,* **9**, 229 (1975).

31. J. A. Pople and M. Gordon, *J. Am. Chem. Soc.,* **89**, 4253 (1967).

32. (a) W. J. Hehre, L. Radom, and J. A. Pople, *J. Am. Chem. Soc.,* **94**, 1496 (1972); (b) W. J. Hehre, L. Radom, and J. A. Pople, *Chem. Commun.,* 669 (1972); (c) W. J. Hehre, R. T. McIver, Jr., J. A. Pople, and P. v. R. Schleyer, *J. Am. Chem. Soc.,* **96**, 7162 (1974); (d) J. M. McKelvey, S. Alexandratos, A. Streitwieser, Jr., J. L. M. Abboud, and W. J. Hehre, *ibid.,* **98**, 244 (1976); (e) L. Radom, *Chem. Commun.,* 403 (1974); (f) A. Pross and L. Radom, *Prog. Phys. Org. Chem.,* **13**, 1 (1981).

33. For a review see: P. A. Kollman, in *Modern Theoretical Chemistry,* H. F. Schaefer III, ed., Plenum Press, New York, 1977, vol. 4, p. 109.

34. L. Radom and J. A. Pople, *J. Am. Chem. Soc.,* **92**, 4786 (1972).

35. J. A. Pople, R. Krishnan, H. B. Schlegel, and J. S. Binkley, *Int. J. Quantum Chem., Symp.,* **13**, 225 (1979).

5

PRACTICAL
CONSIDERATIONS:
INPUT AND OUTPUT

5.1. INTRODUCTION

The purpose of this chapter is to assist the reader in carrying out *ab initio* molecular orbital calculations using any member of the GAUSSIAN series of programs [1-5]. It is divided into three parts. The first gives some details of the required input, with particular emphasis on input of molecular geometry. The second illustrates normal output from the programs, including graphical output available from GAUSSIAN 85. The third part describes recently developed methods of permanently archiving the computational results and making them available for distribution [6]. Such a scheme will be incorporated into future releases of GAUSSIAN 82 and GAUSSIAN 85.

Our discussion in this chapter is primarily concerned with more general features of user–program interaction, and seeks to address some of the common practical points of concern to the user community. For full details of actual program input and output, the reader is referred to the appropriate user manuals [1-5].

5.2. INPUT: THE Z-MATRIX

The input for any member of the GAUSSIAN series of programs is separated into a number of sections. The first specifies the *task* to be performed, for example, single-

point energy calculation, optimization of geometry to an energy minimum, calculation of harmonic vibrational frequencies, and so forth, and includes details of the
method to be employed, that is, restricted or unrestricted Hartree–Fock, basis set,
and level of correlation treatment. This part of the input varies slightly from program to program. To carry out a single restricted Hartree–Fock calculation with the
STO-3G basis set, for example, the simple statement "# RHF/STO-3G" (on one
line of input) will suffice in GAUSSIAN 82. Here the "#" sign denotes initializa-
tion of input and "RHF/STO-3G" indicates the Hartree–Fock procedure and basis
set to be employed. The absence of mention of correlation treatment signifies that
the calculation will terminate at the single-determinant Hartree–Fock level. In
GAUSSIAN 85, the corresponding input statement is simply "STO-3G"; the RHF
procedure is specified by default.

The second section of input is a *title*. This provides some description of the molecule, its conformation and electronic state, and perhaps other features of the calculation. This part of the input is not processed by the programs; it is merely reproduced
in the program output.

The third section of input specifies the geometry of the nuclear framework, and
the number of electrons. In the case of structure optimization, the initial molecular
geometry together with the identities of geometrical parameters to be varied must be
given. The number of electrons of α and β spin is specified (implicitly) by first giving
the net *charge* (0 for neutral molecules, +1 for singly-charged cations, and so on)
followed by the *multiplicity* (1 for singlet, 2 for doublet, 3 for triplet, and so on).

The nuclear geometry input may comprise the *atomic numbers* and 3N *cartesian
coordinates* for the N atoms in a molecule. Alternatively, one of several possible
sets of 3N-6 *internal coordinates* may be input. One such set, termed *symmetry
coordinates* [7], is especially useful in computations of vibrational frequencies,
where the use of full molecular symmetry ensures maximum blocking of the matrix
of force constants. More commonly, the description of molecular geometry may be
given in terms of *bond lengths*, *bond angles*, and *dihedral angles*, information that
collectively is known as a *Z-matrix*. The Z-matrix procedure is a feature common to
all the GAUSSIAN programs, and it is the only aspect of the input to these programs
which we describe in detail.

The Z-matrix specifies the identity and position of a general atom, A_i, in a molecule by means of an identifying symbol, AN_i, explained in detail below, and three
pieces of geometric information: a bond length (in angstroms), a bond angle (in
degrees), and a dihedral angle (in degrees). (Alternative specification in terms of a
bond length and two bond angles can lead to difficulties and is not discussed further
here.) These data relate the atom A_i to previously defined atoms. The first three
atoms defined require less data. The first atom, A_1, is specified by its identifier
AN_1 alone, and is placed at the origin of a right-handed coordinate system, **1a**.

1a   1b   1c

The second atom, A_2, is specified by its identifier AN_2 and the bond length A_2A_1 it makes with A_1. It is placed on the positive z-axis, **1b**. The third atom, A_3, is specified by its identifier AN_3, the bond length to A_1 or to A_2, as the case may be, and the bond angle $A_3A_1A_2$ or $A_3A_2A_1$. It is positioned in the xz plane, **1c**, the x-coordinate being positive or zero for $A_3A_2A_1 \leqslant 180°$.

All subsequent atoms require full input information. For example, a general atom A_4 in a chain (as shown in **2**) is labeled by its identifier AN_4, and is specified as being joined to atom A_3 by a bond length A_4A_3, making an angle $A_4A_3A_2$ with atom A_2, and a dihedral angle $A_4A_3A_2A_1$, that is, angle between the planes $A_4A_3A_2$ and $A_3A_2A_1$, with A_1. The dihedral angle is readily visualized with the aid of a Newman projection, **3**, looking down the A_3A_2 bond. Here, the dihedral angle $A_4A_3A_2A_1$ is the angle swept out by the forward bond A_4A_3 to eclipse the rear bond A_2A_1, a positive sign corresponding to a clockwise rotation. Note that the signs of the dihedral angles $A_4A_3A_2A_1$ and $A_1A_2A_3A_4$ are identical.

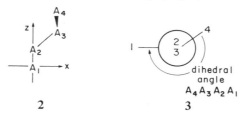

2   3

The schematic input is then:

AN_4 which is joined to

A_3 by a bond of length

A_4A_3 (angstroms), which makes an angle with A_2 of

$A_4A_3A_2$ (degrees), and a dihedral angle with A_1 of

$A_4A_3A_2A_1$ (degrees).

For branched systems, for example, **4**, A_4 can be defined in a similar manner, although in such cases the dihedral angle ($A_4A_2A_1A_3$, shown in Newman projection in **5**) is somewhat artificial.

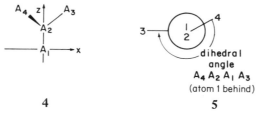

4   5

GAUSSIAN 82 and GAUSSIAN 85 specify A_i and AN_i by the elemental symbol of atom i, for example, H, He, Li, and so forth, followed, if necessary or desired, by a number. For example, the oxygen atom in water could simply be referred to by its elemental symbol "O," and the two hydrogen atoms by "H1" and "H2".

GAUSSIAN 85 input also allows for input of any characters, other than letters, following specification of the elemental symbol, in lieu of or in addition to numbers. Thus "H+" and "C*" are acceptable identifiers.

Input into the older GAUSSIAN 70 [1] and GAUSSIAN 76 [2] computer programs differs from the above in that the atom identifier, AN_i, is simply the atomic number of atom i; for example, "1" for hydrogen and "6" for carbon, and the identifier, A_i, is i, the number of the atom in the overall input list, for example, "2" for the second atom on the list, "10" for the tenth atom. This form of input, which is also acceptable for the more recent programs, is not, however, discussed further.

5.2.1. A Simple Triatomic: Water

We now turn to some examples of actual input to GAUSSIAN 82 and GAUSSIAN 85. We begin with a simple restricted Hartree–Fock calculation, with the STO-3G basis set, on water at its experimental geometry. The GAUSSIAN 82 input for such a task is as follows:

```
# RHF/STO-3G
(blank line)
WATER. EXPERIMENTAL GEOMETRY (GAUSSIAN 82 INPUT)
(blank line)
0  1
O
H1  O  0.958
H2  O  0.958   H1   104.5
(blank line)
```

Note that *blank lines* are included to terminate the job type and title sections. An additional blank line is needed at the end to terminate the list of nuclei. The corresponding GAUSSIAN 85 input is as follows:

```
STO-3G
WATER. EXPERIMENTAL GEOMETRY (GAUSSIAN 85 INPUT)
0  1
ENDPAR
O
H1  O  0.958
H2  O  0.958   H1   104.5
ENDZMA
```

Here, single lines of job type and title input are assumed. The delimiter "ENDPAR" indicates termination of the list of variable parameters (see below) and "ENDZMA" termination of Z-matrix input. No blank lines are required.

In both cases the input begins with the job type and title entries, followed by the charge and multiplicity (0 1) and then by the Z-matrix. Either input initially generates the structure **6**.

6

(As mentioned below, this may later be translated and/or rotated to take account of molecular symmetry.) The first atom to be specified is the oxygen. It is placed at the origin. The second atom, H1, is joined to O with a bond length of 0.958 Å. It is positioned on the positive z-axis. The third atom, H2, is also connected to O with a bond length of 0.958 Å. It makes a bond angle H2OH1 of 104.5° with H1. H2 is placed in the xz plane.

An alternative, equally acceptable, procedure is to specify the geometrical parameters in the Z-matrix by arbitrary names, the actual parameter values being given elsewhere (following the Z-matrix in the GAUSSIAN 82 program, preceding it in GAUSSIAN 85). The following GAUSSIAN 82 input is equivalent to that shown previously.

```
# RHF/STO-3G
(blank line)
WATER. EXPERIMENTAL GEOMETRY (GAUSSIAN 82 INPUT)
(blank line)
0  1
O
H1 O   ROH
H2 O   ROH  H1   HOH
(blank line)
ROH = 0.958
HOH = 104.5
(blank line)
```

An additional blank line is needed to terminate the list of parameter values.
Input to GAUSSIAN 85 is as follows:

```
STO-3G
WATER. EXPERIMENTAL GEOMETRY (GAUSSIAN 85 INPUT)
0  1
ROH = 0.958 HOH = 104.5
ENDPAR
O
H1   O   ROH
H2   O   ROH  H1   HOH
ENDZMA
```

This alternative form of input has the advantage that the body of the Z-matrix does not have to be altered if changes in bond lengths and bond angles are subsequently made. We use these two forms of input interchangeably in our discussion.

Both GAUSSIAN 82 and GAUSSIAN 85 take advantage of molecular symmetry (see Section 3.3.5). The C_{2v} symmetry of water is implicitly specified in the above input by requiring two identical OH bond lengths. To utilize symmetry efficiently, it is desirable to orient the molecule such that one or more molecular symmetry planes coincide with cartesian planes. Thus, the molecule is reoriented following initial specification of its geometry according to the Z-matrix input. In addition, the center of the cartesian coordinate system is translated so as to coincide with the center of positive charge of the molecule. This repositioning becomes apparent in our discussion of program output (Section 5.3).

5.2.2. A Tetraatomic Chain: Hydrogen Peroxide

GAUSSIAN 82 input for the STO-3G optimized structure for hydrogen peroxide, 7, is:

```
# RHF/STO-3G
(blank line)
HYDROGEN PEROXIDE. STO-3G OPTIMIZED GEOMETRY (GAUSSIAN 82 INPUT)
(blank line)
0  1
O1
O2   O1   1.396
H1   O2   1.001   O1   101.1
H2   O1   1.001   O2   101.1  H1   125.3
(blank line)
```

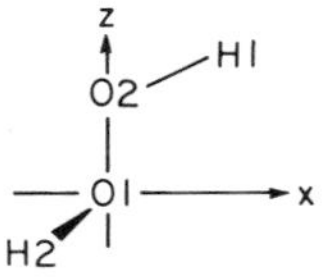

7

The first three atoms are defined like those of H_2O in the previous example. The fourth atom, H2, is joined to O1 by a bond length of 1.001 Å, and defines a bond angle with O2 (H2O1O2) of 101.1°. The final specification is the dihedral angle H2O1O2H1, that is, the angle between the planes H2O1O2 and O1O2H1, the numerical value of which is 125.3°. A Newman projection, 8, looking along the O1–O2 linkage shows the dihedral angle clearly:

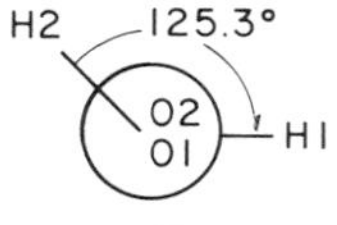

8

It is measured in a clockwise manner from H2O1 to O2H1. Note that for this molecule to possess a twofold symmetry axis, the two OH lengths and the two HOO bond angles must necessarily be equal.

5.2.3. Branched Structures: Fluoromethane

The same technique, that is, specification of an atom in terms of a bond angle and a dihedral angle, may also be used for branched structures. For example, in fluoromethane, **9**, the "branching" atoms H2 and H3 (where the atoms F, C, and H1 form the "chain") can be specified in terms of dihedral angles between the planes H2CF and CFH1 (H2CFH1 = 120.0°) and H3CF and CFH1 (H3CFH1 = −120.0°).

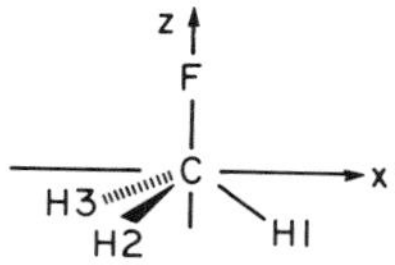

9

```
STO-3G
FLUOROMETHANE. EXPERIMENTAL GEOMETRY (GAUSSIAN 85 INPUT)
0   1
RCF = 1.383 RCH = 1.100 HCF = 108.3
ENDPAR
C
F    C   RCF
H1   C   RCH   F   HCF
H2   C   RCH   F   HCF   H1    120.0
H3   C   RCH   F   HCF   H1   -120.0
ENDZMA
```

The dihedral angles of ±120° are easily pictured as before with the aid of a Newman projection, **10**, looking along the C—F linkage.

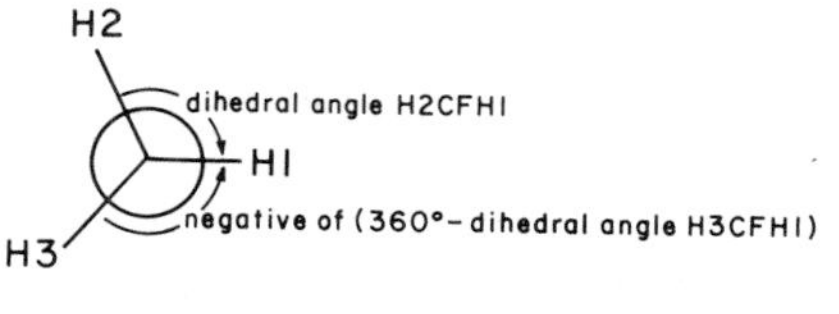

10

Note that, in this example, some geometrical parameters are defined in the body of the Z-matrix, while others are given outside.

5.2.4. Larger Systems

Larger systems present no special problems since the atoms are simply specified one at a time *with respect to previously defined atoms.* Consider below GAUSSIAN 82 input for a 6-31G* restricted Hartree–Fock calculation using the optimized STO-3G structure for the *anti* conformation of vinyl alcohol, **11**.

```
# RHF/6-31G*
(blank line)
ANTI VINYL ALCOHOL. STO-3G OPTIMIZED GEOMETRY (GAUSSIAN 82 INPUT)
(blank line)
0  1
C1
C2   C1  1.312
O    C2  1.390  C1  126.7
H1   C1  1.080  C2  122.0   O    0.0
H2   C1  1.077  C2  121.2   O  180.0
H3   C2  1.089  C1  122.2  H2    0.0
H4   O   0.990  C2  105.2  C1  180.0
(blank line)
```

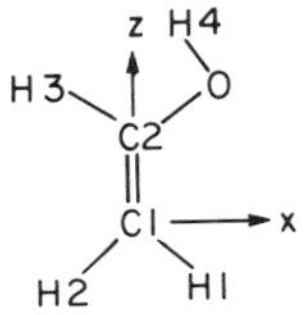

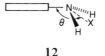

11

5.2.5. The Use of Dummy Atoms

Dummy atoms, indicated by the atom identifier "X" followed if need be or desired by an integer, for example, X1, X2, are introduced into a Z-matrix in a manner identical to that for normal elements. In the general case, each dummy atom requires specification of a "bond" length connecting the dummy to a previously defined (real or dummy) atom as well as one "bond" angle and a dihedral angle. *The cartesian coordinates generated for dummy atoms are not utilized in the quantum mechanical calculation.*

Dummy atoms have several uses in the construction of Z-matrices. Perhaps the most important of these is their role in defining specific sets of geometrical parameters suitable for use in structure optimizations. For example, a good choice of variables for optimization of the geometry of the amino group in aniline is the HNH bond angle and the angle (θ) that the HNH plane makes with the CN bond. These may be specified by placing a dummy atom, X, on the bisector of the HNH angle, **12**.

12

GAUSSIAN 82 input that illustrates this approach for the fluoroamine molecule, NH_2F, **13**, follows.

```
#  RHF/STO-3G
(blank line)
FLUOROAMINE.  STO-3G OPTIMIZED GEOMETRY (GAUSSIAN 82 INPUT)
(blank line)
0   1
N
F    N    1.387
X    N    1.000   F    113.5
H1   N    1.049   X     51.6   F     90.0
H2   N    1.049   X     51.6   F    -90.0
(blank line)
```

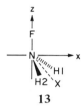

13

A slightly more complicated example of the use of dummy atoms in defining geometrical parameters is provided by the **GAUSSIAN** 85 input for methanol, **14**. Here the calculation is at the RHF/3-21G level and utilizes the experimental equilibrium geometry.

```
    3-21G
    METHANOL.  EXPERIMENTAL GEOMETRY (GAUSSIAN 85 INPUT)
    0   1
    RCO = 1.421, RCH1 = 1.094, RCH2 = 1.094, ROH = 0.963
    H1CO = 107.2, XCO = 129.9, H2CX = 54.25, HOC = 108.0
    ENDPAR
    C
    O    C    RCO
    H1   C    RCH1   O    H1CO
    X    C    1.0    O    XCO    H1    180.0
    H2   C    RCH2   X    H2CX   H1     90.0
    H3   C    RCH2   X    H2CX   H1    -90.0
    H4   O    ROH    C    HOC    H1    180.0
    ENDZMA
```

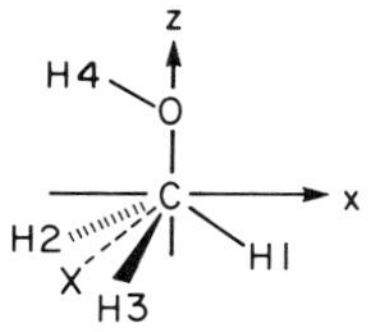

14

Note that in this example, "commas" rather than blanks have been used to separate the geometrical parameters. Either is acceptable.

A related important use of dummy atoms is to define the geometries of cyclic molecules and/or molecules with centers of symmetry. For example, dummy atoms placed at the center of each of the cyclopentadienyl rings in ferrocene, **15**, greatly facilitate its construction with the appropriate symmetry (D_{5h} point group) and aid in defining the extent to which the ring hydrogens bend away from or toward the central metal. GAUSSIAN 82 input is illustrated below.

```
# RHF/STO-3G
(blank line)
FERROCENE. STO-3G OPTIMIZED GEOMETRY (GAUSSIAN 82 INPUT)
(blank line)
0  1
FE
X1    FE    1.695
C1    X1    1.207   FE    90.0
C2    X1    1.207   C1    72.0    FE    90.0
C3    X1    1.207   C1    72.0    FE   -90.0
C4    X1    1.207   C2    72.0    FE    90.0
C5    X1    1.207   C3    72.0    FE   -90.0
X2    FE    1.000   X1    90.0    C1     0.0
X3    FE    1.695   X2    90.0    X1   180.0
C6    X3    1.207   FE    90.0    X2     0.0
C7    X3    1.207   C6    72.0    FE    90.0
C8    X3    1.207   C6    72.0    FE   -90.0
C9    X3    1.207   C7    72.0    FE    90.0
C10   X3    1.207   C8    72.0    FE   -90.0
H1    C1    1.076   X1   177.2    FE   180.0
H2    C2    1.076   X1   177.2    FE   180.0
H3    C3    1.076   X1   177.2    FE   180.0
H4    C4    1.076   X1   177.2    FE   180.0
H5    C5    1.076   X1   177.2    FE   180.0
H6    C6    1.076   X3   177.2    FE   180.0
H7    C7    1.076   X3   177.2    FE   180.0
H8    C8    1.076   X3   177.2    FE   180.0
H9    C9    1.076   X3   177.2    FE   180.0
H10   C10   1.076   X3   177.2    FE   180.0
(blank line)
```

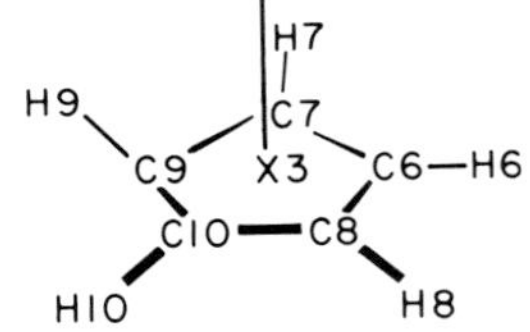

15

A third use of dummy atoms is illustrated by X2 in **15**, which is introduced to avoid using a bond angle of 180°, where the dihedral angle becomes ill defined.

5.2.6. Optimization of Molecular Geometry

One task commonly performed by the GAUSSIAN programs is the optimization of molecular geometry to a minimum on the potential surface (or section thereof). This task is specified by an additional keyword in the job-type instructions. Thus, "# OPT RHF/STO-3G" requests geometry optimization at the RHF/STO-3G level for the GAUSSIAN 82 program. The corresponding input statement for GAUSSIAN 85 is "GEOMOPT STO-3G" or optionally "GEOMOPT = ABC STO-3G" where "ABC" in the latter form denotes a file name on which intermediate information pertaining to the optimization is kept (in case it is necessary to restart the calculation). Optimization to a transition structure is handled similarly although different keywords are required. Again, full details are available in the appropriate program documentation [4, 5].

Geometry optimization requires that the parameters to be varied be distinguished from those that are to be held constant. Further, any constraints among variables imposed by symmetry, for example, the two equal OH bonds in water for the C_{2v} point group, need to be identified. In both the GAUSSIAN 82 and GAUSSIAN 85 programs, this is accomplished by replacing the numerical values for geometrical variables in the text of the Z-matrix by variable names. This input format is exactly that which has already been employed to specify geometries for single calculations on water, **6**, fluoromethane, **9**, and methanol, **14**.

As an actual example of optimization input, consider chlorine trifluoride, **16**, restricted to a planar (C_{2v}) structure.

16

If the experimental geometry is employed as a starting point for an RHF/3-21G$^{(*)}$ optimization, appropriate GAUSSIAN 82 input would be as follows:

```
# OPT RHF/3-21G*
(blank line)
CHLORINE TRIFLUORIDE. OPTIMIZATION (GAUSSIAN 82 INPUT)
(blank line)
0   1
F1
CL   F1   RCLF1
F2   CL   RCLF2   F1   F1CLF2
F3   CL   RCLF2   F1   F1CLF2   F2   180.0
(blank line)
RCLF1 = 1.598
RCLF2 = 1.698
F1CLF2 = 87.5
(blank line)
```

The corresponding GAUSSIAN 85 input differs only slightly.

```
GEOMOPT 3-21G*
CHLORINE TRIFLUORIDE. OPTIMIZATION (GAUSSIAN 85 INPUT)
0  1
RCLF1 = 1.598 RCLF2 = 1.698 F1CLF2 = 87.5
ENDPAR
F1
CL  F1  RCLF1
F2  CL  RCLF2  F1  F1CLF2
F3  CL  RCLF2  F1  F1CLF2  F2  TRANS
ENDZMA
```

Note the (optional) use of the word "TRANS" instead of the numerical value of 180.0 in the last line of Z-matrix input for the GAUSSIAN 85 program. Other optional allowable names are "CIS" for $0°$, "GAUCHE" for $60°$, "PERP" for $90°$, and "SKEW" for $120°$. *Unless they appear in the parameter list*, these names refer to *fixed* values and are not to be varied in the optimization procedure.

In the ClF_3 example, the three geometrical variables are the two unique ClF bond lengths (RCLF1 and RCLF2) and the equivalent pair of bond angles between them (F1CLF2). The dihedral angle for F3 is given as 180.0 (or TRANS) in the body of the Z-matrix, and is therefore held fixed. This ensures that the molecule is kept planar. If it were desired to test the planarity of ClF_3 at the RHF/3-21G$^{(*)}$ level, the 180.0 (TRANS) entry in the Z-matrix could be replaced by a new variable name, which would then appear in the parameter list. Assuming an initial value different from 180.0 (such as 150.0), the optimization procedure would then modify this angle and (eventually) return it to the $180°$ value (or something very close to this value) if the optimized RHF/3-21G$^{(*)}$ structure is indeed planar.

A further example of input (to GAUSSIAN 85) for geometry optimization is provided below for cyclopentadienyl nickel nitrosyl, **17**.

```
GEOMOPT STO-3G
CYCLOPENTADIENYL NICKEL NITROSYL (GAUSSIAN 85 INPUT)
0  1
PENTA = 72.0
ENDFIX
RNIX1 = 1.724, RXC1 = 1.206, RCH = 1.08, RNIN = 1.40,
RNIO = 2.72, ALPHA = 178.0
ENDPAR
NI
X1  NI  RNIX1
C1  X1  RCX1  NI  PERP
C2  X1  RCX1  C1  PENTA  NI  PERP
C3  X1  RCX1  C2  PENTA  NI  PERP
C4  X1  RCX1  C3  PENTA  NI  PERP
C5  X1  RCX1  C4  PENTA  NI  PERP
H1  C1  RCH   X1  ALPHA  NI  CIS
H2  C2  RCH   X1  ALPHA  NI  CIS
H3  C3  RCH   X1  ALPHA  NI  CIS
H4  C4  RCH   X1  ALPHA  NI  CIS
H5  C5  RCH   X1  ALPHA  NI  CIS
X2  NI  RX    X1  PERP   C1  CIS
```

```
N    NI   RNIN   X2   PERP   X1   TRANS
O    NI   RNIO   X2   PERP   X1   TRANS
ENDZMA
```

```
                          H2
                          |
              H3         C2
                \     /      \
                 C3   X I     C I — H I
                    \      /
                     C4 — C5
                   /          \
                 H4            H5
                          |
                        N I — X 2
                          |
                          N
                          ||
                          O
```

17

Note that an additional input section, terminated by the keyword "ENDFIX", has been included in this example. Variables explicitly mentioned here are assigned the specified values, for example, "PENTA" is assigned a value of 72.0°, and are held fixed throughout the optimization procedure. This form of input (in GAUSSIAN 85) is equivalent to specification of numerical values for variables inside the body of the Z-matrix. The names "CIS," "PERP," and "TRANS" are employed to indicate *fixed* angles of 0°, 90°, and 180°, respectively. The designator "RX" specifies a *fixed* bond length of 1.0 Å.

5.3. OUTPUT

Three sets of output obtained using the GAUSSIAN 85 program [5] are reproduced in full here. GAUSSIAN 82 output is similar and is not illustrated. The first example, resulting from an STO-3G calculation on formaldehyde, will be discussed in some detail. The other two outputs result from a 3-21G calculation on formaldehyde and a 6-31G* calculation on ammonia oxide, respectively. These are discussed only briefly, with an emphasis on differences compared with the first output. In addition to the three outputs, graphical presentations of the valence molecular orbitals and total electron densities of cyclopropane and of benzene, calculated at the 3-21G level, are provided. These have been obtained using the PHOTOMO program [8] attached to GAUSSIAN 85.

The overall objective of this section is to provide the reader with sufficient examples of actual output that he might become familiar not only with the presentation format (which will vary somewhat from program to program), but also with the type of information that might be obtained from the output. By so doing we hope to help bridge the considerable gap between the understanding of the formalism provided in the previous sections, and the practical experience of actually using molecular orbital calculations to aid in a chemical investigation.

5.3.1.   STO-3G Calculation on Formaldehyde

The output for an STO-3G calculation on formaldehyde is reproduced as Figure 5.1. The molecular geometry, which is optimum for this particular basis set, has been entered via a Z-matrix, which in turn is reproduced in the output. The cartesian coordinates resulting from this Z-matrix are also printed, as is the point group of the molecule (C_{2v} in this case). The latter provides a useful check on the input geometry. The result of the ensuing computation is a "Total energy" of -112.35435

```
FORMALDEHYDE:  STO-3G CALCULATION ON STO-3G GEOMETRY

        FORMALISM: RHF/STO-3G
        SINGLE POINT CALCULATION
  CHARGE = 0  MULTIPLICITY = 1
        POINT GROUP C2v

        Z-MATRIX COORDINATES                              CARTESIAN COORDINATES

A   B   AB    C   ABC    D   ABCD            ATOM     X          Y          Z

C                                            - 1-   0.000000   0.000000  -0.534048
O   C   CO                                   - 2-   0.000000   0.000000   0.682952
H1  C   CH    O   HCO                         - 3-   0.000000   0.925984  -1.129661
H2  C   CH    O   HCO    H1   TRANS          - 4-   0.000000  -0.925984  -1.129661

VARIABLE PARAMETER VALUES:

CO  =  1.2170000   CH  =  1.1010000   HCO  = 122.7500000


      Total energy =   -112.35434684 a.u.
    SCF convergence =     1.1816E-05


MOLECULAR ORBITALS ...

               1        2        3        4        5        6        7        8        9       10       11       12

IRRED REPS:   1A1      2A1      3A1      4A1      1B2      5A1      1B1      2B2      2B1      6A1      3B2      7A1

EIGENVALUES: -20.3127 -11.1250  -1.3373  -0.8079  -0.6329  -0.5455  -0.4431  -0.3545   0.2819   0.6291   0.7346   0.9126

 1  1 C  1S   -0.0005   0.9926   0.1225   0.1856   0.0000   0.0331   0.0000   0.0000   0.0000  -0.2081   0.0000  -0.0946
 2  1 C  2S    0.0072   0.0329  -0.2772  -0.5773   0.0000  -0.1070   0.0000   0.0000   0.0000   1.3042   0.0000   0.6298
 3  1 C  2Px   0.0000   0.0000   0.0000   0.0000   0.0000   0.0000   0.6094   0.0000   0.8210   0.0000   0.0000   0.0000
 4  1 C  2Pv   0.0000   0.0000   0.0000   0.0000   0.5332   0.0000   0.0000  -0.1820   0.0000   0.0000   1.1485   0.0000
 5  1 C  2Pz   0.0063   0.0005  -0.1577   0.2263   0.0000  -0.4475   0.0000   0.0000   0.0000  -0.4442   0.0000   1.1734

 6  2 O  1S   -0.9943   0.0001   0.2194  -0.0988   0.0000  -0.0938   0.0000   0.0000   0.0000   0.0282   0.0000   0.1157
 7  2 O  2S   -0.0259  -0.0057  -0.7691   0.4290   0.0000   0.4991   0.0000   0.0000   0.0000  -0.1622   0.0000  -0.8632
 8  2 O  2Px   0.0000   0.0000   0.0000   0.0000   0.0000   0.0000   0.6758   0.0000  -0.7673   0.0000   0.0000   0.0000
 9  2 O  2Pv   0.0000   0.0000   0.0000   0.0000   0.4421   0.0000   0.0000   0.8701   0.0000   0.0000  -0.3184   0.0000
10  2 O  2Pz   0.0056   0.0016   0.1701   0.1645   0.0000   0.6769   0.0000   0.0000   0.0000   0.2469   0.0000   0.9235

11  3 H  1S   -0.0002  -0.0065  -0.0318  -0.2646   0.3004   0.1589   0.0000  -0.3590   0.0000  -0.8894  -0.8402   0.1558

12  4 H  1S   -0.0002  -0.0065  -0.0318  -0.2646  -0.3004   0.1589   0.0000   0.3590   0.0000  -0.8894   0.8402   0.1558
```

FIGURE 5.1.   GAUSSIAN 85 output for an STO-3G calculation on formaldehyde.

.. OVER BASIS FUNCTIONS

				1	2	3	4	5	6	7	8	9	10	11	12
1	1	C	1S	2.0717	-0.0556	0.0000	0.0000	0.0000	0.0000	-0.0003	0.0000	0.0000	-0.0091	-0.0066	-0.0066
2	1	C	2S	-0.0556	0.8454	0.0000	0.0000	0.0000	-0.0001	-0.0637	0.0000	0.0000	0.1373	0.1398	0.1398
3	1	C	2Px	0.0000	0.0000	0.7428	0.0000	0.0000	0.0000	0.0000	0.1718	0.0000	0.0000	0.0000	0.0000
4	1	C	2Py	0.0000	0.0000	0.0000	0.6349	0.0000	0.0000	0.0000	0.0000	0.0323	0.0000	0.1758	0.1758
5	1	C	2Pz	0.0000	0.0000	0.0000	0.0000	0.5528	-0.0026	-0.0046	0.0000	0.0000	0.1834	0.0632	0.0632
6	2	O	1S	0.0000	-0.0001	0.0000	0.0000	-0.0026	2.1106	-0.1099	0.0000	0.0000	0.0000	0.0000	0.0000
7	2	O	2S	-0.0003	-0.0637	0.0000	0.0000	-0.0046	-0.1099	2.0506	0.0000	0.0000	0.0000	-0.0014	-0.0014
8	2	O	2Px	0.0000	0.0000	0.1718	0.0000	0.0000	0.0000	0.0000	0.9135	0.0000	0.0000	0.0000	0.0000
9	2	O	2Py	0.0000	0.0000	0.0000	0.0323	0.0000	0.0000	0.0000	0.0000	1.9049	0.0000	-0.0129	-0.0129
10	2	O	2Pz	-0.0091	0.1373	0.0000	0.0000	0.1834	0.0000	0.0000	0.0000	0.0000	1.0283	-0.0082	-0.0082
11	3	H	1S	-0.0066	0.1398	0.0000	0.1758	0.0632	0.0000	-0.0014	0.0000	-0.0129	-0.0082	0.6308	-0.0372
12	4	H	1S	-0.0066	0.1398	0.0000	0.1758	0.0632	0.0000	-0.0014	0.0000	-0.0129	-0.0082	-0.0372	0.6308

.. OVER ATOMS

		1	2	3	4
1	C	4.7364	0.4445	0.3722	0.3722
2	O	0.4445	7.7882	-0.0224	-0.0224
3	H	0.3722	-0.0224	0.6308	-0.0372
4	H	0.3722	-0.0224	-0.0372	0.6308

				BASIS FUNCTION POPULATIONS	ATOM POP.	NET CHARGES
1	1	C	1S	1.9936	5.9253	0.0747
2	1	C	2S	1.1429		
3	1	C	2Px	0.9147		
4	1	C	2Py	1.0188		
5	1	C	2Pz	0.8553		
6	2	O	1S	1.9981	8.1878	-0.1878
7	2	O	2S	1.8694		
8	2	O	2Px	1.0853		
9	2	O	2Py	1.9114		
10	2	O	2Pz	1.3235		
11	3	H	1S	0.9434	0.9434	0.0566
12	4	H	1S	0.9434	0.9434	0.0566

DIPOLE MOMENT

X   0.0000   TOTAL   1.5370
Y   0.0000
Z -1.5370

FIGURE 5.1.   (*Continued*)

hartrees (atomic units). The "SCF convergence" in this case is below a preset limit (5×10^{-5}), corresponding to the maximum deviation allowed for any density matrix element between the final two iterations of the SCF procedure. It is important to note that convergence is not always achieved, that is, the wavefunction does not always reach self-consistency. *Solutions which have not converged are meaningless!*

Printed next as column vectors are the "MOLECULAR ORBITALS." More precisely, what is actually tabulated are the linear coefficients defining the molecular orbitals, that is, the $c_{\mu i}$ in the expansions $\psi_i = \Sigma_\mu c_{\mu i}\phi_\mu$, Eq. (2.25). For example, the first column vector represents a molecular orbital which consists almost entirely of the $1s$ basis function on oxygen. Similarly, the second vector is well described as being almost totally a $1s$ function on carbon. The seventh molecular orbital comprises a sum of two basis functions, the $2p_x$ components of each of the carbon and oxygen atoms. It is the familiar π-bonding orbital of formaldehyde. The corresponding π^* (antibonding) orbital is given as number 9, and is the lowest unoccupied molecular orbital (the LUMO). All other molecular orbitals correspond to various bonding and antibonding combinations of the atomic functions among the σ system.

Printed above the molecular orbitals are "EIGENVALUES" or *one-electron energies*. According to Koopmans' theorem (Section 2.6.3), the negatives of the one-electron energies of filled molecular orbitals correspond to *vertical ionization potentials*. For example, the highest two vertical ionization potentials for formaldehyde are predicted on this basis to be 20.3127 and 11.1250 hartrees (552.7 and 302.7 eV). These correspond to electron removal from what are essentially $1s$ atomic orbitals on the oxygen and carbon atoms, respectively. A value of 0.3545 hartrees (9.6 eV) for the first ionization potential of the molecule derives from the energy of the highest filled molecular orbital. In the same manner, the negatives of the energies of the unfilled molecular orbitals correspond to *electron affinities*. It should be noted, however, that electron affinities calculated in this way are strongly dependent on the nature of the basis set, and extreme caution should be exercised in their use. In this example, the eigenvalues for all unfilled orbitals are positive, meaning that the corresponding anion is predicted to be unbound.

Also printed above each molecular orbital is a symmetry label, termed "IRRED REPS," obtained according to conventions originally set forth by Mulliken [9]. These characterize the symmetry properties of the particular molecular orbital. For formaldehyde, the molecular orbitals may be classified according to the C_{2v} point group as a_1 (orbitals 1–4, 6, 10, and 12), b_1 (orbitals 7 and 9), and b_2 (orbitals 5, 8, and 11). The single configuration presented here is $(1a_1)^2(2a_1)^2(3a_1)^2 \cdot (4a_1)^2(1b_2)^2(5a_1)^2(1b_1)^2(2b_2)^2$, yielding a state described as $^1A_1$. Solutions corresponding to other configurations can be generated by modifying the initial-guess configuration such that the desired orbitals are occupied. Details are provided in the appropriate program documentation [4, 5].

The molecular electronic charge distribution may be obtained following a method due to Mulliken [10] (Section 2.8). Although the shortcomings of such a procedure are well documented, and although the concept of atomic charge is somewhat arbitrary, the Mulliken approach is useful for assigning molecular charges and, in particular, for comparing charge distributions in series of related molecules. The diagonal elements of the matrix labeled "MULLIKEN POPULATION ANALYSIS. OVER BASIS FUNCTIONS" correspond to electrons assigned to specific basis functions (and are referred to as *net populations* in Section 2.8), while the off-diagonal elements correspond to electrons *shared* between basis functions. Note that this matrix provides a partitioning of all electrons (16 in the case of formaldehyde) among the basis functions. Other matrices could be constructed to take account of any subset of the total electron distribution. The Mulliken population analysis matrix is symmetrical, that is, elements (i, j) and (j, i) are equal. The sum of two such elements is called the *overlap population* between basis functions i and j (see Eq. 2.52). Of particular interest are π-overlap populations obtained by summing off-diagonal elements corresponding to π-symmetry basis functions. For example, the π-overlap population in the CO bond, obtained by adding the off-diagonal elements $(3, 8)$ and $(8, 3)$, is 0.34. This provides a measure of the double-bond character in the CO bond. Comparative (STO-3G) values for other systems include 0.22 for benzene and 0.39 for ethylene. The $2p_x$ "BASIS FUNCTION POPULATIONS" on C and O, **18a**, show a polarization in the π system consistent with that expected on intuitive grounds, **18b**.

18a   **18b**

The individual basis function populations are summed, and a "MULLIKEN POPULATION ANALYSIS" matrix is printed "OVER ATOMS." This matrix is again symmetrical, and the sum of the off-diagonal elements (k, l) and (l, k) gives the total overlap population between atoms k and l. Finally, a set of atom populations ("ATOM POP.") is displayed, see Eq. (2.56). When the atomic numbers of the individual nuclei are subtracted from the atom populations, "NET CHARGES" are obtained and these provide an estimate of the overall distribution of charge. In the case of formaldehyde, the oxygen atom is predicted to have an excess negative charge of 0.19 electrons, while carbon and the two hydrogen atoms are all predicted to be positively charged, **19**.

19

The final property printed is the electric dipole moment, calculated directly as the expectation value of the dipole moment operator. Note that the sign of a dipole moment component, which refers to the alignment of *positive* charge along a given axis, corresponds to the reverse of the commonly used $\longmapsto$ convention that corresponds to alignment of *negative* charge.

5.3.2. 3-21G Calculation on Formaldehyde

The output for a 3-21G calculation on formaldehyde is reproduced as Figure 5.2. The structural input is the same as that in the previous example (Section 5.3.1), which involved an STO-3G calculation.

Note that the number of molecular orbitals produced by the 3-21G calculation on formaldehyde is larger than that corresponding to the STO-3G treatment. This is because the *total number of molecular orbitals is always equal to the number of basis functions*. This is 12 for STO-3G and 22 for 3-21G. The number of occupied orbitals and the electronic configurations: $(1a_1)^2(2a_1)^2(3a_1)^2(4a_1)^2(1b_2)^2(5a_1)^2 \cdot (1b_1)^2(2b_2)^2$, $^1A_1$, are the same, however. The additional orbitals in the 3-21G calculation are all unoccupied or *virtual*.

The coefficients of the molecular orbitals indicate, among other things, the contributions of the inner and less-diffuse $(2s', 2p'_x, 2p'_y, 2p'_z)$ and outer and more-diffuse $(2s'', 2p''_x, 2p''_y, 2p''_z)$ parts of the split-valence basis functions to the particular orbital. The proportions are determined variationally in the SCF procedure, and lead to the anisotropy and size flexibility of the 3-21G basis set (see Section 4.3.2). For example, inner and outer components of the $2p_x$ basis function on carbon combine in a ratio of roughly 1 to 1 for use in the construction of the formaldehyde π orbital (column vector number 7), and approximately 0.5 to 1 in the building of the corresponding unfilled π^* orbital (vector number 9). Thus, the latter orbital is the more diffuse.

The "MULLIKEN POPULATION ANALYSIS" (the matrix "OVER BASIS FUNCTIONS" has been eliminated from this particular output) leads to an assignment of electron populations to each of the atomic basis functions of the 3-21G basis set. As was the case with the minimal basis set calculation, these are designated "BASIS FUNCTION POPULATIONS" in the output. For interpretive purposes, it is convenient to express the orbital populations for the split-valence basis set in terms of a set of "CONDENSED POPULATIONS". These have been formed simply by summing the inner and outer parts of the valence $2s$ and $2p$ functions. For example, the $2p'_x$ and $2p''_x$ populations are combined to produce a single $2p_x$ population.

There are quantitative differences between the results with the two basis sets. For example, the calculated total energy is lower (more negative), the charge separation greater, and the dipole moment larger with the 3-21G basis set. Detailed comments on differences in performance of the STO-3G and 3-21G basis sets are presented in Chapter 6.

5.3.3. 6-31G* Calculation on Ammonia Oxide

A third sample output, shown in Figure 5.3, is that for a 6-31G* calculation on ammonia oxide [11]. The structure employed has been optimized using this basis

FORMALDEHYDE: 3-21G CALCULATION ON STO-3G GEOMETRY

FORMALISM: RHF/3-21G
SINGLE POINT CALCULATION
CHARGE = 0 MULTIPLICITY = 1
POINT GROUP C2v

Z-MATRIX COORDINATES

A	B	AB	C	ABC	D	ABCD
C						
O	C	CO				
H1	C	CH	O	HCO		
H2	C	CH	O	HCO	H1	TRANS

CARTESIAN COORDINATES

ATOM	X	Y	Z
- 1-	0.000000	0.000000	-0.534048
- 2-	0.000000	0.000000	0.682952
- 3-	0.000000	0.925984	-1.129661
- 4-	0.000000	-0.925984	-1.129661

VARIABLE PARAMETER VALUES:

CO = 1.2170000   CH = 1.1010000   HCO = 122.7500000

Total energy =   -113.22120310 a.u.
SCF convergence =   5.7531E-06

MOLECULAR ORBITALS ...

				1	2	3	4	5	6	7	8	9	10	11	12
IRRED REPS:				1A1	2A1	3A1	4A1	1B2	5A1	1B1	2B2	2B1	6A1	3B2	7A1
EIGENVALUES:				-20.4856	-11.2866	-1.4117	-0.8661	-0.6924	-0.6345	-0.5234	-0.4330	0.1436	0.2718	0.3653	0.4512
1	1	C	1S	0.0001	-0.9864	0.1179	-0.1726	0.0000	-0.0259	0.0000	0.0000	0.0000	-0.1439	0.0000	0.0726
2	1	C	2S'	0.0005	-0.0932	-0.1313	0.2011	0.0000	0.0584	0.0000	0.0000	0.0000	0.0651	0.0000	0.0312
3	1	C	2Px'	0.0000	0.0000	0.0000	0.0000	0.0000	0.0000	0.2919	0.0000	0.3570	0.0000	0.0000	0.0000
4	1	C	2Py'	0.0000	0.0000	0.0000	0.0000	-0.3528	0.0000	0.0000	-0.1868	0.0000	0.0000	-0.4019	0.0000
5	1	C	2Pz'	0.0025	-0.0013	-0.1527	-0.1573	0.0000	0.3358	0.0000	0.0000	0.0000	-0.1827	0.0000	-0.0646
6	1	C	2S"	0.0174	0.0449	-0.1322	0.4993	0.0000	-0.0438	0.0000	0.0000	0.0000	1.8560	0.0000	-1.0875
7	1	C	2Px"	0.0000	0.0000	0.0000	0.0000	0.0000	0.0000	0.3160	0.0000	0.6922	0.0000	0.0000	0.0000
8	1	C	2Py"	0.0000	0.0000	0.0000	0.0000	-0.2844	0.0000	0.0000	-0.0306	0.0000	0.0000	-1.4304	0.0000
9	1	C	2Pz"	0.0138	-0.0079	0.0418	-0.1325	0.0000	0.0901	0.0000	0.0000	0.0000	-0.7204	0.0000	-1.6894
10	2	O	1S	0.9835	-0.0001	0.2147	0.0951	0.0000	0.0860	0.0000	0.0000	0.0000	0.0045	0.0000	-0.1217
11	2	O	2S'	0.0978	-0.0007	-0.2068	-0.0929	0.0000	-0.0786	0.0000	0.0000	0.0000	-0.0135	0.0000	0.0252
12	2	O	2Px'	0.0000	0.0000	0.0000	0.0000	0.0000	0.0000	0.3854	0.0000	-0.3342	0.0000	0.0000	0.0000
13	2	O	2Py'	0.0000	0.0000	0.0000	0.0000	-0.2819	0.0000	0.0000	0.4469	0.0000	0.0000	0.1416	0.0000
14	2	O	2Pz'	-0.0045	-0.0005	0.1269	-0.1033	0.0000	-0.4157	0.0000	0.0000	0.0000	0.0605	0.0000	-0.1839
15	2	O	2S"	-0.0471	0.0004	-0.6635	-0.3616	0.0000	-0.3612	0.0000	0.0000	0.0000	0.0332	0.0000	1.5720
16	2	O	2Px"	0.0000	0.0000	0.0000	0.0000	0.0000	0.0000	0.4359	0.0000	-0.5720	0.0000	0.0000	0.0000
17	2	O	2Py"	0.0000	0.0000	0.0000	0.0000	-0.2547	0.0000	0.0000	0.5321	0.0000	0.0000	0.3660	0.0000
18	2	O	2Pz"	0.0130	-0.0044	0.1316	-0.0831	0.0000	-0.4187	0.0000	0.0000	0.0000	0.0929	0.0000	-0.7438
19	3	H	1S'	-0.0007	0.0018	-0.0253	0.1699	-0.1663	-0.0817	0.0000	-0.1845	0.0000	-0.0503	-0.0099	-0.0008
20	3	H	1S"	0.0006	-0.0132	0.0170	0.0582	-0.1018	-0.0892	0.0000	-0.2780	0.0000	-1.3397	1.5175	-0.1822
21	4	H	1S'	-0.0007	0.0018	-0.0253	0.1699	0.1663	-0.0817	0.0000	0.1845	0.0000	-0.0503	0.0099	-0.0008
22	4	H	1S"	0.0006	-0.0132	0.0170	0.0582	0.1018	-0.0892	0.0000	0.2780	0.0000	-1.3397	-1.5175	-0.1822

FIGURE 5.2.   GAUSSIAN 85 output for a 3-21G calculation on formaldehyde.

					13	14	15	16	17	18	19	20	21	22
IRRED REPS:					3B1	4B2	8A1	9A1	5B2	10A1	4B1	11A1	6B2	12A1
EIGENVALUES:					0.9309	1.0085	1.0342	1.1557	1.2646	1.5678	1.8657	1.9076	1.9761	3.3110
1	1	C	1S		0.0000	0.0000	0.0791	0.0094	0.0000	0.0588	0.0000	0.0241	0.0000	-0.0050
2	1	C	2S'		0.0000	0.0000	0.0510	0.4065	0.0000	-1.6464	0.0000	0.5360	0.0000	0.0978
3	1	C	2Px'		-1.0855	0.0000	0.0000	0.0000	0.0000	0.0000	0.0048	0.0000	0.0000	0.0000
4	1	C	2Py'		0.0000	0.7653	0.0000	0.0000	-0.8013	0.0000	0.0000	0.0000	-0.0938	0.0000
5	1	C	2Pz'		0.0000	0.0000	-0.8396	0.8230	0.0000	0.2615	0.0000	0.1596	0.0000	0.2721
6	1	C	2S"		0.0000	0.0000	-0.7779	0.0186	0.0000	2.7189	0.0000	-0.0278	0.0000	1.1019
7	1	C	2Px"		1.0125	0.0000	0.0000	0.0000	0.0000	0.0000	-0.2942	0.0000	0.0000	0.0000
8	1	C	2Py"		0.0000	-1.8638	0.0000	0.0000	1.0069	0.0000	0.0000	0.0000	1.0159	0.0000
9	1	C	2Pz"		0.0000	0.0000	0.8728	-0.3511	0.0000	-0.5052	0.0000	0.1996	0.0000	0.7720
10	2	0	1S		0.0000	0.0000	-0.0500	0.0297	0.0000	0.0077	0.0000	0.0563	0.0000	-0.0642
11	2	0	2S'		0.0000	0.0000	0.0913	-0.0087	0.0000	-0.0211	0.0000	-0.2195	0.0000	1.7033
12	2	0	2Px'		-0.0077	0.0000	0.0000	0.0000	0.0000	0.0000	-1.0483	0.0000	0.0000	0.0000
13	2	0	2Py'		0.0000	0.2325	0.0000	0.0000	0.1153	0.0000	0.0000	0.0000	1.0100	0.0000
14	2	0	2Pz'		0.0000	0.0000	-0.2535	0.1767	0.0000	-0.2628	0.0000	-0.9750	0.0000	-0.2593
15	2	0	2S"		0.0000	0.0000	-0.0078	-0.0412	0.0000	-0.2791	0.0000	-0.1294	0.0000	-2.4968
16	2	0	2Px"		-0.1192	0.0000	0.0000	0.0000	0.0000	0.0000	1.0707	0.0000	0.0000	0.0000
17	2	0	2Py"		0.0000	0.2625	0.0000	0.0000	-0.1415	0.0000	0.0000	0.0000	-1.2744	0.0000
18	2	0	2Pz"		0.0000	0.0000	-0.3480	0.3098	0.0000	0.2700	0.0000	1.2729	0.0000	0.9398
19	3	H	1S'		0.0000	0.4800	0.5789	0.6952	0.7804	0.1711	0.0000	-0.0735	-0.2332	-0.0672
20	3	H	1S"		0.0000	0.4808	-0.0109	-0.5412	-1.1869	-0.9609	0.0000	0.1389	-0.3399	0.0239
21	4	H	1S'		0.0000	-0.4800	0.5789	0.6952	-0.7804	0.1711	0.0000	-0.0735	0.2332	-0.0672
22	4	H	1S"		0.0000	-0.4808	-0.0109	-0.5412	1.1869	-0.9609	0.0000	0.1389	0.3399	0.0239

FIGURE 5.2. (Continued)

set. Most of the output is very similar to that for the STO-3G and 3-21G calculations on formaldehyde in the previous examples, and requires no further comment. Note, however, that the version reproduced here has been abbreviated by printing only the valence molecular orbitals and the two lowest-energy unoccupied molecular orbitals. This is a standard procedure in GAUSSIAN 85. In addition, reproduction of the "MULLIKEN POPULATION ANALYSIS" matrix "OVER BASIS FUNCTIONS" has been restricted to that block of elements discussed below.

Both STO-3G and 3-21G basis sets predict very long NO bond lengths (1.607 and 1.531 Å, respectively) for the experimentally uncharacterized molecule ammonia oxide, NH_3O. These basis sets also substantially overestimate the NO length in trimethylamine oxide, Me_3NO, for which the structure is known, yielding bond lengths of 1.582 and 1.466 Å, respectively, compared with the experimental value, 1.404 Å. This suggests that the NO lengths in NH_3O are also overestimated by the STO-3G and 3-21G basis sets. Indeed, calculations with the 6-31G* basis set yield an optimized NO length of 1.377 Å. What role do the d functions in the 6-31G* basis set play in bringing about this bond length reduction?

A partial answer to this question is contained in the output listing. Consider the possibility that the d orbitals on nitrogen enable it to act as an acceptor for electrons donated by the oxygen lone pairs. For example, electron donation from the

```
MULLIKEN POPULATION ANALYSIS ...

... OVER ATOMS

                          1        2        3        4

     1   C          4.7244   0.4814   0.3316   0.3316
     2   O          0.4814   8.1020  -0.0506  -0.0506
     3   H          0.3316  -0.0506   0.6141  -0.0706
     4   H          0.3316  -0.0506  -0.0706   0.6141
```

			BASIS FUNCTION POPULATIONS		CONDENSED POPULATIONS	ATOM POPULATIONS	NET CHARGES
1	1	C	1S	1.9867	1S	1.9867	5.8689	0.1311
2	1	C	2S'	0.4241	2S	1.2573		
3	1	C	2Px'	0.3294	2Px	0.7765		
4	1	C	2Py'	0.5871	2Py	1.0913		
5	1	C	2Pz'	0.6065	2Pz	0.7571		
6	1	C	2S"	0.8331				
7	1	C	2Px"	0.4471				
8	1	C	2Py"	0.5043				
9	1	C	2Pz"	0.1506				
10	2	O	1S	1.9867	1S	1.9867	8.4822	-0.4822
11	2	O	2S'	0.4364	2S	1.9652		
12	2	O	2Px'	0.5012	2Px	1.2235		
13	2	O	2Py'	0.8766	2Py	1.8707		
14	2	O	2Pz'	0.6784	2Pz	1.4361		
15	2	O	2S"	1.5287				
16	2	O	2Px"	0.7223				
17	2	O	2Py"	0.9941				
18	2	O	2Pz"	0.7578				
19	3	H	1S'	0.4836	1S	0.8244	0.8244	0.1756
20	3	H	1S"	0.3408				
21	4	H	1S'	0.4836	1S	0.8244	0.8244	0.1756
22	4	H	1S"	0.3408				

```
            DIPOLE MOMENT

     X   0.0000     TOTAL   2.6857
     Y   0.0000
     Z  -2.6857
```

FIGURE 5.2. *(Continued)*

filled $2p_x$ orbital on oxygen into a d_{xz}-type function on nitrogen might be depicted as in **20**.

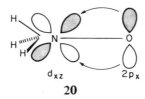

20

AMMONIA OXIDE: 6-31G* CALCULATION ON 6-31G* OPTIMIZED GEOMETRY
                    FORMALISM: RHF/6-31G*
                    SINGLE POINT CALCULATION
             CHARGE = 0 MULTIPLICITY = 1
                    POINT GROUP C3v

                    Z-MATRIX COORDINATES   CARTESIAN COORDINATES

A   B   AB   C   ABC   D   ABCD   ATOM   X   Y   Z
N   - 1-   0.000000   0.000000   -0.550405
O   N   ON   - 2-   0.000000   0.000000   0.827295
H1   N   NH   O   HNO   - 3-   0.812457   -0.469072   -0.921843
H2   N   NH   O   HNO   H1   SKEW   - 4-   -0.812457   -0.469072   -0.921843
H3   N   NH   O   HNO   H1   -SKEW   - 5-   0.000000   0.938144   -0.921843
VARIABLE PARAMETER VALUES:
ON   =   1.3777000   NH   =   1.0090000   HNO = 111.6000000

            Total energy =   -130.93388644 a.u.
            SCF convergence =   1.1275E-05

MOLECULAR ORBITALS

				3	4	5	6	7	8	9	10	11
IRRED REPS:				3A1	4A1	1E	1E	5A1	2E	2E	6A1	3E
EIGENVALUES:				-1.3595	-1.0775	-0.7536	-0.7536	-0.5642	-0.3688	-0.3688	0.1756	0.2629
1	1	N	1S	0.1694	-0.1019	0.0000	0.0000	-0.0447	0.0000	0.0000	-0.1244	0.0000
2	1	N	2S'	-0.3704	0.2229	0.0000	0.0000	0.1170	0.0000	0.0000	0.1229	0.0000
3	1	N	2Px'	0.0000	0.0000	0.0000	0.4979	0.0000	0.0000	-0.1096	0.0000	0.0000
4	1	N	2Py'	0.0000	0.0000	0.4979	0.0000	0.0000	-0.1096	0.0000	0.0000	0.3567
5	1	N	2Pz'	-0.0491	-0.2478	0.0000	0.0000	0.4456	0.0000	0.0000	-0.1625	0.0000
6	1	N	2S"	-0.2951	0.2791	0.0000	0.0000	0.1797	0.0000	0.0000	1.9779	0.0000
7	1	N	2Px"	0.0000	0.0000	0.0000	0.2891	0.0000	0.0000	-0.0363	0.0000	0.0000
8	1	N	2Py"	0.0000	0.0000	0.2891	0.0000	0.0000	-0.0363	0.0000	0.0000	1.1030
9	1	N	2Pz"	0.0148	-0.1191	0.0000	0.0000	0.3082	0.0000	0.0000	-0.3892	0.0000
10	1	N	Dxx	-0.0080	0.0171	-0.0220	0.0000	-0.0053	0.0088	0.0000	-0.0391	0.0062
11	1	N	Dyy	-0.0080	0.0171	0.0220	0.0000	-0.0053	-0.0088	0.0000	-0.0391	-0.0062
12	1	N	Dzz	-0.0188	-0.0198	0.0000	0.0000	0.0124	0.0000	0.0000	-0.0368	0.0000
13	1	N	Dxy	0.0000	0.0000	0.0000	-0.0254	0.0000	0.0000	0.0102	0.0000	0.0000
14	1	N	Dxz	0.0000	0.0000	0.0000	-0.0272	0.0000	0.0000	0.0445	0.0000	0.0000
15	1	N	Dyz	0.0000	0.0000	-0.0272	0.0000	0.0000	0.0445	0.0000	0.0000	-0.0077
16	2	O	1S	0.1230	0.1687	0.0000	0.0000	0.0974	0.0000	0.0000	-0.0068	0.0000
17	2	O	2S'	-0.2721	-0.3783	0.0000	0.0000	-0.1938	0.0000	0.0000	0.0174	0.0000
18	2	O	2Px'	0.0000	0.0000	0.0000	0.1288	0.0000	0.0000	0.6182	0.0000	0.0000
19	2	O	2Py'	0.0000	0.0000	0.1288	0.0000	0.0000	0.6182	0.0000	0.0000	-0.0612
20	2	O	2Pz'	0.0935	0.0075	0.0000	0.0000	-0.3729	0.0000	0.0000	0.0710	0.0000
21	2	O	2S"	-0.2359	-0.4063	0.0000	0.0000	-0.4964	0.0000	0.0000	0.0201	0.0000
22	2	O	2Px"	0.0000	0.0000	0.0000	0.0523	0.0000	0.0000	0.5142	0.0000	0.0000
23	2	O	2Py"	0.0000	0.0000	0.0523	0.0000	0.0000	0.5142	0.0000	0.0000	-0.1537
24	2	O	2Pz"	0.0460	0.0132	0.0000	0.0000	-0.1895	0.0000	0.0000	0.0983	0.0000
25	2	O	Dxx	0.0027	-0.0010	-0.0004	0.0000	-0.0009	0.0007	0.0000	-0.0009	0.0028
26	2	O	Dyy	0.0027	-0.0010	0.0004	0.0000	-0.0009	-0.0007	0.0000	-0.0009	-0.0028
27	2	O	Dzz	-0.0250	-0.0032	0.0000	0.0000	0.0516	0.0000	0.0000	-0.0004	0.0000
28	2	O	Dxy	0.0000	0.0000	0.0000	-0.0004	0.0000	0.0000	0.0008	0.0000	0.0000
29	2	O	Dxz	0.0000	0.0000	0.0000	-0.0124	0.0000	0.0000	-0.0237	0.0000	0.0000
30	2	O	Dyz	0.0000	0.0000	-0.0124	0.0000	0.0000	-0.0237	0.0000	0.0000	0.0014
31	3	H	1S'	-0.0908	0.1069	-0.1283	0.2221	-0.0548	0.0609	-0.1055	-0.0373	0.0330
32	3	H	1S"	0.0054	0.0070	-0.0484	0.0839	-0.0494	0.0654	-0.1132	-0.9573	0.8746
33	4	H	1S'	-0.0908	0.1069	-0.1283	-0.2221	-0.0548	0.0609	0.1055	-0.0373	0.0330
34	4	H	1S"	0.0054	0.0070	-0.0484	-0.0839	-0.0494	0.0654	0.1132	-0.9573	0.8746
35	5	H	1S'	-0.0908	0.1069	0.2565	0.0000	-0.0548	-0.1218	0.0000	-0.0373	-0.0661
36	5	H	1S"	0.0054	0.0070	0.0968	0.0000	-0.0494	-0.1307	0.0000	-0.9573	-1.7492

FIGURE 5.3. GAUSSIAN 85 output for a 6-31G* calculation on ammonia oxide.

122

... OVER BASIS FUNCTIONS

				13	14	15	16	17	18	19	20	21	22	23	24
1	1	N	1S	0.0000	0.0000	0.0000	0.0000	0.0000	0.0000	0.0000	-0.0003	0.0017	0.0000	0.0000	-0.0026
2	1	N	2S'	0.0000	0.0000	0.0000	0.0000	-0.0008	0.0000	0.0000	0.0151	-0.0276	0.0000	0.0000	0.0321
3	1	N	2Px'	0.0000	0.0000	0.0000	0.0000	0.0000	-0.0002	0.0000	0.0000	0.0000	-0.0093	0.0000	0.0000
4	1	N	2Py'	0.0000	0.0000	0.0000	0.0000	0.0000	0.0000	-0.0002	0.0000	0.0000	0.0000	-0.0093	0.0000
5	1	N	2Pz'	0.0000	0.0000	0.0000	-0.0001	0.0046	0.0000	0.0000	0.0544	-0.0451	0.0000	0.0000	0.0489
6	1	N	2S"	0.0000	0.0000	0.0000	0.0020	-0.0255	0.0000	0.0000	0.0273	-0.1176	0.0000	0.0000	0.0462
7	1	N	2Px"	0.0000	0.0000	0.0000	0.0000	0.0000	0.0036	0.0000	0.0000	0.0000	-0.0031	0.0000	0.0000
8	1	N	2Py"	0.0000	0.0000	0.0000	0.0000	0.0000	0.0000	0.0036	0.0000	0.0000	0.0000	-0.0031	0.0000
9	1	N	2Pz"	0.0000	0.0000	0.0000	0.0019	-0.0156	0.0000	0.0000	0.0398	-0.1284	0.0000	0.0000	0.0318
10	1	N	Dxx	0.0000	0.0000	0.0000	0.0000	-0.0002	0.0000	0.0000	-0.0001	-0.0009	0.0000	0.0000	-0.0006
11	1	N	Dyy	0.0000	0.0000	0.0000	0.0000	-0.0002	0.0000	0.0000	-0.0001	-0.0009	0.0000	0.0000	-0.0006
12	1	N	Dzz	0.0000	0.0000	0.0000	-0.0002	0.0046	0.0000	0.0000	0.0037	0.0043	0.0000	0.0000	0.0031
13	1	N	Dxy	0.0015	0.0000	0.0000	0.0000	0.0000	0.0000	0.0000	0.0000	0.0000	0.0000	0.0000	0.0000
14	1	N	Dxz	0.0000	0.0054	0.0000	0.0000	0.0000	0.0047	0.0000	0.0000	0.0000	0.0090	0.0000	0.0000
15	1	N	Dyz	0.0000	0.0000	0.0054	0.0000	0.0000	0.0000	0.0047	0.0000	0.0000	0.0000	0.0090	0.0000
16	2	O	1S	0.0000	0.0000	0.0000	2.0854	-0.0446	0.0000	0.0000	0.0000	-0.0478	0.0000	0.0000	0.0000
17	2	O	2S'	0.0000	0.0000	0.0000	-0.0446	0.5104	0.0000	0.0000	0.0000	0.4799	0.0000	0.0000	0.0000
18	2	O	2Px'	0.0000	0.0047	0.0000	0.0000	0.0000	0.7976	0.0000	0.0000	0.0000	0.3256	0.0000	0.0000
19	2	O	2Py'	0.0000	0.0000	0.0047	0.0000	0.0000	0.0000	0.7976	0.0000	0.0000	0.0000	0.3256	0.0000
20	2	O	2Pz'	0.0000	0.0000	0.0000	0.0000	0.0000	0.0000	0.0000	0.2957	0.0000	0.0000	0.0000	0.0753
21	2	O	2S"	0.0000	0.0000	0.0000	-0.0478	0.4799	0.0000	0.0000	0.0000	0.9344	0.0000	0.0000	0.0000
22	2	O	2Px"	0.0000	0.0090	0.0000	0.0000	0.0000	0.3256	0.0000	0.0000	0.0000	0.5342	0.0000	0.0000
23	2	O	2Py"	0.0000	0.0000	0.0090	0.0000	0.0000	0.0000	0.3256	0.0000	0.0000	0.0000	0.5342	0.0000
24	2	O	2Pz"	0.0000	0.0000	0.0000	0.0000	0.0000	0.0000	0.0000	0.0753	0.0000	0.0000	0.0000	0.0764
25	2	O	Dxx	0.0000	0.0000	0.0000	-0.0002	-0.0003	0.0000	0.0000	0.0000	0.0003	0.0000	0.0000	0.0000
26	2	O	Dyy	0.0000	0.0000	0.0000	-0.0002	-0.0003	0.0000	0.0000	0.0000	0.0003	0.0000	0.0000	0.0000
27	2	O	Dzz	0.0000	0.0000	0.0000	-0.0002	-0.0023	0.0000	0.0000	0.0000	-0.0258	0.0000	0.0000	0.0000
28	2	O	Dxy	0.0000	0.0000	0.0000	0.0000	0.0000	0.0000	0.0000	0.0000	0.0000	0.0000	0.0000	0.0000
29	2	O	Dxz	0.0000	0.0004	0.0000	0.0000	0.0000	0.0000	0.0000	0.0000	0.0000	0.0000	0.0000	0.0000
30	2	O	Dyz	0.0000	0.0000	0.0004	0.0000	0.0000	0.0000	0.0000	0.0000	0.0000	0.0000	0.0000	0.0000
31	3	H	1S'	0.0032	0.0041	0.0014	0.0000	0.0000	-0.0002	-0.0001	-0.0002	0.0005	-0.0051	-0.0017	-0.0020
32	3	H	1S"	0.0003	0.0006	0.0002	-0.0001	0.0009	-0.0032	-0.0011	-0.0022	0.0094	-0.0148	-0.0049	-0.0057
33	4	H	1S'	0.0032	0.0041	0.0014	0.0000	0.0000	-0.0002	-0.0001	-0.0002	0.0005	-0.0051	-0.0017	-0.0020
34	4	H	1S"	0.0003	0.0006	0.0002	-0.0001	0.0009	-0.0032	-0.0011	-0.0022	0.0094	-0.0148	-0.0049	-0.0057
35	5	H	1S'	0.0000	0.0000	0.0055	0.0000	0.0000	0.0000	-0.0003	-0.0002	0.0005	0.0000	-0.0068	-0.0020
36	5	H	1S"	0.0000	0.0000	0.0008	-0.0001	0.0009	0.0000	-0.0043	-0.0022	0.0094	0.0000	-0.0197	-0.0057

... OVER ATOMS

		1	2	3	4	5
1	N	6.5123	0.0177	0.3172	0.3172	0.3172
2	O	0.0177	8.7458	-0.0314	-0.0314	-0.0314
3	H	0.3172	-0.0314	0.3797	-0.0245	-0.0245
4	H	0.3172	-0.0314	-0.0245	0.3797	-0.0245
5	H	0.3172	-0.0314	-0.0245	-0.0245	0.3797

FIGURE 5.3. *(Continued)*

123

			BASIS FUNCTION POPULATIONS		CONDENSED POPULATIONS		ATOM POPULATIONS	NET CHARGES
1	1	N	1S	1.9954	1S	1.9954	7.4817	-0.4817
2	1	N	2S'	0.7837	2S	1.4870		
3	1	N	2Px'	0.8286	2Px	1.3240		
4	1	N	2Py'	0.8286	2Py	1.3240		
5	1	N	2Pz'	0.8213	2Pz	1.2128		
6	1	N	2S"	0.7034	Dxx	0.0249		
7	1	N	2Px"	0.4953	Dyy	0.0249		
8	1	N	2Py"	0.4953	Dzz	0.0224		
9	1	N	2Pz"	0.3915	Dxy	0.0086		
10	1	N	Dxx	0.0249	Dxz	0.0289		
11	1	N	Dyy	0.0249	Dyz	0.0289		
12	1	N	Dzz	0.0224				
13	1	N	Dxy	0.0086				
14	1	N	Dxz	0.0289				
15	1	N	Dyz	0.0289				
16	2	O	1S	1.9958	1S	1.9958	8.6692	-0.6692
17	2	O	2S'	0.9126	2S	1.9693		
18	2	O	2Px'	1.1244	2Px	1.9411		
19	2	O	2Py'	1.1244	2Py	1.9411		
20	2	O	2Pz'	0.5038	2Pz	0.7907		
21	2	O	2S"	1.0567	Dxx	-0.0011		
22	2	O	2Px"	0.8167	Dyy	-0.0011		
23	2	O	2Py"	0.8167	Dzz	0.0262		
24	2	O	2Pz"	0.2870	Dxy	0.0000		
25	2	O	Dxx	-0.0011	Dxz	0.0035		
26	2	O	Dyy	-0.0011	Dyz	0.0035		
27	2	O	Dzz	0.0262				
28	2	O	Dxy	0.0000				
29	2	O	Dxz	0.0035				
30	2	O	Dyz	0.0035				
31	3	H	1S'	0.5077	1S	0.6164	0.6164	0.3836
32	3	H	1S"	0.1087				
33	4	H	1S'	0.5077	1S	0.6164	0.6164	0.3836
34	4	H	1S"	0.1087				
35	5	H	1S'	0.5077	1S	0.6164	0.6164	0.3836
36	5	H	1S"	0.1087				

DIPOLE MOMENT

X   0.0000   TOTAL   5.6087
Y   0.0000
Z   -5.6087

FIGURE 5.3.   (*Continued*)

That this type of interaction actually takes place is clearly reflected in the calculated electron populations. The $2p_x$ and $2p_y$ orbitals on oxygen have each donated 0.059 electrons, that is, their electron populations are reduced from 2.000 in a localized structure to 1.941. The d_{xz} and d_{yz} orbitals on nitrogen accept electrons with calculated populations of 0.029. In addition, there is a significant overlap population between the d_{xz} orbital on nitrogen and the $2p'_x$ and $2p''_x$ orbitals on oxygen (and similarly between d_{yz} and $2p'_y$ and $2p''_y$) totaling 0.027 electrons. This is obtained by summing the (14, 18), (14, 22), (18, 14), and (22, 14) elements of the Mulliken population matrix.

The calculated charge distribution shows a net polarization (as reflected in the total atomic charges) in the N-O bond in the direction $H_3N \longrightarrow O$. Note that the nitrogen is actually predicted to be negatively charged: the total positive charge on the NH_3 group arises from the positive charges on the hydrogens. The resultant bonding can then be described pictorially as π donation and σ acceptance by oxygen, 21.

21

Finally, the lower calculated dipole moment at the 6-31G* level (5.6 D) compared with 6-31G (5.8 D) may also be a reflection of the back donation into the d orbitals on nitrogen.

5.3.4. Graphical Representation of Molecular Orbitals and Electron Densities

Pictorial representations of the complete set of valence molecular orbitals derived from a 3-21G calculation on cyclopropane are displayed in Figure 5.4. The corresponding total electron density is displayed in Figure 5.5. These figures have been obtained according to the procedures outlined in Section 3.3.6 using the PHOTOMO program [8] attached to GAUSSIAN 85 [5].

The molecular orbitals corresponding to different symmetries may now be clearly distinguished. Note especially that those designated a_1' and a_2'' exhibit the full threefold symmetry of the nuclear skeleton, while those designated e' and e'' do not. The latter correspond to degenerate representations. Note also the similarity of one of the members of the set of highest-occupied (Walsh) molecular orbitals of cyclopropane ($3e'$) to the π bond in a normal alkene, for example, ethylene. A similarity in the chemistry of the two species is to be expected. The use of valence-orbital descriptions such as these as a means of discussing molecular structure and reactivity has already been dealt with in some length [8b].

The valence molecular orbitals of benzene and corresponding total electron density are shown in Figures 5.6 and 5.7, respectively. Further examples of molecular orbital and density plots are provided in Chapters 6 and 7.

5.4. ARCHIVAL STORAGE

The widespread use of *ab initio* molecular quantum mechanics now requires manipulation of large amounts of numerical data. The results of general interest are *total energies, equilibrium and transition structures,* and *values of other physical properties.* At present, most access to the information is through the conventional chemical literature, where the results are published in a rather dispersed manner. Some

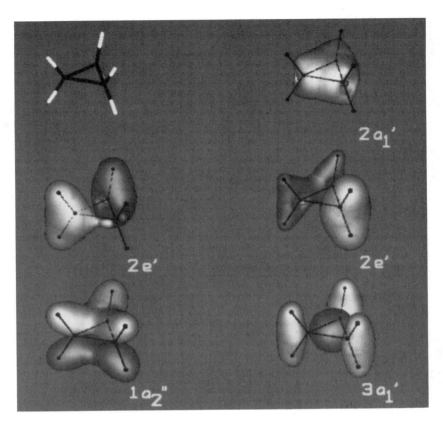

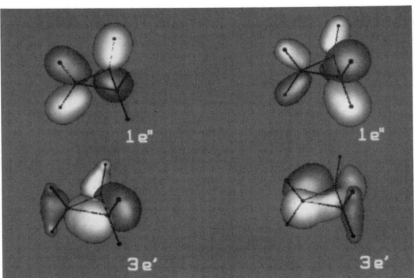

FIGURE 5.4. Valence molecular orbitals of cyclopropane from lowest energy to highest energy (left to right, top to bottom). 3-21G//3-21G.

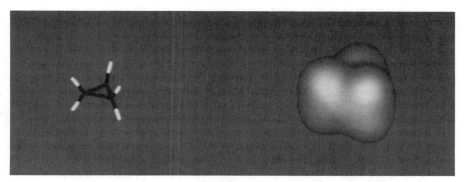

FIGURE 5.5. Total electron density for cyclopropane. 3-21G//3-21G.

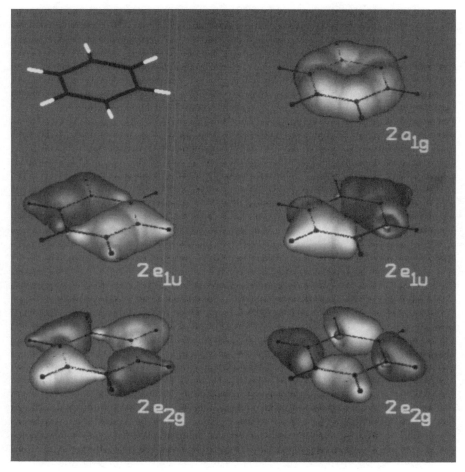

FIGURE 5.6. Valence molecular orbitals of benzene from lowest energy to highest energy (left to right, top to bottom). 3-21G//3-21G.

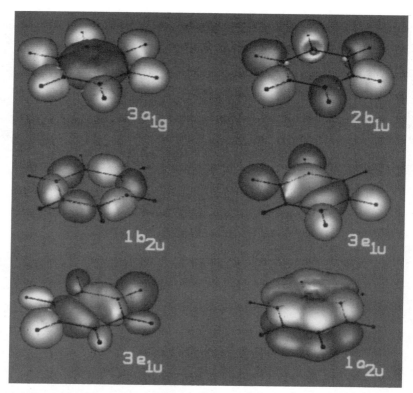

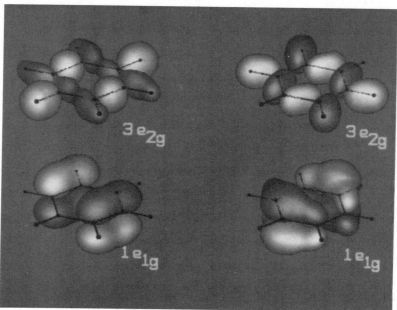

FIGURE 5.6. (*Continued*)

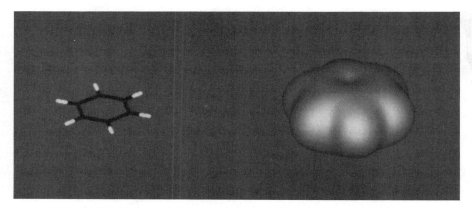

FIGURE 5.7.   Total electron density for benzene. 3-21G//3-21G.

attempts have been made to organize bibliographies of this material [12, 13], but the task is immense and the probability of error in the multiple transcription process is significant.

Efficiency in the distribution of chemical data would be significantly enhanced by making more of the communication steps computer based. The following three features are clearly desirable:

1.   The chemical information derived from a computation should be distributed in the form of *direct output* from the program, thereby avoiding transcription errors.

2.   The information should also be in a form that can be used as *direct input* into another computer for additional processing, avoiding further transcription errors.

3.   The information should be in a form that can be incorporated into large *archive files*, making access through computer searches relatively straightforward.

With these features in mind, an archive file associated with the GAUSSIAN 82 program has been developed. This is the *Carnegie–Mellon Quantum Chemistry Archive* (CMQCA) [6]. Each complete program run leaves behind an *archive entry*, giving key results of the calculations. This entry is in the form of standard GAUSSIAN 82 input data (quantum-mechanical technique, molecular and structural information) followed by selected numerical results (energies, dipole moments, harmonic frequencies, and so forth). If the calculation has involved structure optimization (finding a local minimum or saddle point on the potential surface), the final structural parameters are entered as the input part of the archive entry. Thus, most archive entries have the property of *self-reproduction:* if submitted as input to GAUSSIAN 82, they reproduce the same archived results together with additional information such as wavefunctions and population analysis. The archive file is therefore complementary to GAUSSIAN 82.

Compressed Archive Entry.

```
00100     4076\CMU\FOPT\UHF\STO-3G\H2N1 (2)\DEFREES\26-JUL-1979\5\\# STO-3(
00200     OPT ALTER\\2-A-1 NH2 AMINO RADICAL STO-3G GEOM OPT\\0,2\N\H,1,R\}
00300     ,1,R,2,A\\R=1.0151\A=131.29966\\\4 5\\HF=-54.7609407\S2=0.751\RM'
00400     D=0.136D-07\RMSF=0.361D-04\DIP=0.762\PG=CO2V \\
```

Expanded Archive Entry.

```
00100     4076
00200     CMU
00300     FOPT
00400     UHF
00500     STO-3G
00600     H2N1(2)
00700     DEFREES
00800     26-JUL-1979
00900     5
01000
01100     # STO-3G OPT ALTER
01200
01300     2-A-1 NH2 AMINO RADICAL STO-3G GEOM OPT
01400
01500     0,2
01600     N
01700     H,1,R
01800     H,1,R,2,A
01900
02000     R=1.0151
02100     A=131.29966
02200
02300
02400     4 5
02500
02600     HF=-54.7609407
02700     S2=0.751
02800     RMSD=0.136D-07
02900     RMSF=0.361D-04
03000     DIP=0.762
03100     PG=CO2V
03200
```

FIGURE 5.8.   Compressed and expanded archive entries for the $^2A_1$ excited state of NH_2.

Figure 5.8 shows a typical **CMQCA** entry, the STO-3G structure of the excited $^2A_1$ state of NH_2. It is given first in compressed form, as stored in the master file. The *back-slash* character "\" is used to delimit the various fields in the entry. The entry can be expanded to a more readable form by interpreting each back-slash as a "new-line" character. This leads to the expanded version, also shown in Figure 5.8. In this form, the entry can be submitted directly as input to **GAUSSIAN** 82.

Certain features of the example in Figure 5.8 differ from the input described in Section 5.2 and should be mentioned briefly. A more detailed description is provided in the published documentation [6]. Lines 100-1000 give information about the entry used in sorting and search routines. Line 100 is the serial number and line 200 the site at which the calculation was performed. Line 300, "FOPT," indicates that the entry corresponds to a *full optimization*, that is, all geometric parameters are optimized subject to the symmetry constraints imposed in the Z-matrix. (If only some parameters are optimized, this part of the entry would be "POPT," corresponding to *partial optimization*.) Lines 400–500 give the two parts of the theoretical model, here UHF/STO-3G. Line 600, H2N1(2), gives the stoichiometry; the (2) symbol indicates that a doublet state is involved. The next two lines give the name of the investigator and the date of the calculation. Line 900 is a code used in search algorithms. The actual GAUSSIAN 82 input begins on line 1100. It should be noted that the $^2A_1$ state of NH_2 is not the ground state and it is necessary to *alter* the electronic configuration used in the initial guess. This is requested by the keyword "ALTER" and the numbers "4 5" on line 2400, which indicate that ψ_4 and ψ_5 in the β initial guess must be interchanged. (The blank line 2300 means that no changes in the α guess orbitals are required.) Further, the Z-matrix on lines 1600–1800 is in a different format, acceptable to GAUSSIAN 82 and used in most of the present archive entries. In this format, N is atom number 1 (line 1600), a hydrogen atom H is atom 2 (line 1700) and this is attached to atom 1, that is, N, at a distance R. Finally, atom 3 (again hydrogen) is attached to atom 1 at the same distance. The angle it makes with atom 2, that is, HNH, is denoted by A. The optimized values of the parameters R and A are given on lines 2000–2100. The blank line terminates the part of the archive entry in GAUSSIAN 82 input format.

The remaining parts of the entry in Figure 5.8 give the Hartree–Fock energy (denoted by HF in line 2600), the expectation value of S^2 (for UHF calculations only), control information about the precision of the final density matrix and root-mean-square force on the nuclei (lines 2800–2900), the dipole moment (3000), and finally the point group of the nuclear framework (3100).

At the present time, there are two ways of accessing the information in CMQCA. An extensive tabulation of Hartree–Fock equilibrium and transition structures has been published [6]. This contains about 10,000 fully optimized structures at the HF/STO-3G, HF/3-21G, and HF/6-31G* levels. The second access mode is by telephone and computer terminal, which permits screen display of expanded archive entries. This mode connects the reader to the BROWSE program [14], which allows an interactive search of the entire archive file. It also enables the user to calculate harmonic frequencies and normal coordinates using the force constants stored in archive entries from calculations of analytical second derivatives of the energy. This is particularly valuable for studying vibrational isotope effects, where the user may supply the isotope information in an interactive session. Further details are given in the CMQCA documentation [6].

REFERENCES

1. GAUSSIAN 70, W. J. Hehre, W. A. Lathan, M. D. Newton, R. Ditchfield, and J. A. Pople, Program no. 236, Quantum Chemistry Program Exchange, Indiana University, Bloomington, Ind.

2. GAUSSIAN 76, J. S. Binkley, R. Whiteside, P. C. Hariharan, R. Seeger, W. J. Hehre, M. D.

Newton, and J. A. Pople, Program no. 368, Quantum Chemistry Program Exchange, Indiana University, Bloomington, Ind.

3. GAUSSIAN 80, J. S. Binkley, R. A. Whiteside, R. Krishnan, R. Seeger, D. J. DeFrees, H. B. Schlegel, S. Topiol, L. R. Kahn, and J. A. Pople, Program no. 406, Quantum Chemistry Program Exchange, Indiana University, Bloomington, Ind.

4. GAUSSIAN 82, J. S. Binkley, M. J. Frisch, D. J. DeFrees, K. Raghavachari, R. A. Whiteside, H. B. Schlegel, E. M. Fluder, and J. A. Pople. Department of Chemistry, Carnegie-Mellon University, Pittsburgh, PA.

5. GAUSSIAN 85, R. F. Hout, Jr., M. M. Francl, S. D. Kahn, K. E. Dobbs, E. S. Blurock, W. J. Pietro, D. J. DeFrees, S. K. Pollack, B. A. Levi, R. Steckler, and W. J. Hehre, Quantum Chemistry Program Exchange, Indiana University, Bloomington, Ind to be submitted.

6. R. A. Whiteside, M. J. Frisch, J. S. Binkley, D. J. DeFrees, H. B. Schlegel, K. Raghavachari, and J. A. Pople, *Carnegie-Mellon Quantum Chemistry Archive,* Department of Chemistry, Carnegie-Mellon University, Pittsburgh, Pa. 15213.

7. For a general discussion of symmetry coordinates, see: (a) E. B. Wilson, Jr., J. C. Decius, and P. C. Cross, *Molecular Vibrations,* McGraw-Hill, New York, 1955, Chapt. 6. See also: (b) R. F. Hout, Jr., B. A. Levi, and W. J. Hehre, *J. Comput. Chem.,* 4, 499 (1983).

8. For a description, see: (a) R. F. Hout, Jr., W. J. Pietro, and W. J. Hehre, *J. Comput. Chem.,* 4, 276 (1983); also (b) R. F. Hout, Jr., W. J. Pietro, and W. J. Hehre, *A Pictorial Approach to Molecular Structure and Reactivity,* Wiley, New York, 1984; (c) R. F. Hout, Jr., W. J. Pietro, and W. J. Hehre, Quantum Chemistry Program Exchange, Indianna University, Bloomington, Ind, to be submitted.

9. R. S. Mulliken, *J. Chem. Phys.,* 23, 1997 (1955).

10. R. S. Mulliken, *J. Chem. Phys.,* 23, 1833, 1841, 2338, 2343 (1955).

11. For further discussion, see: L. Radom, J. S. Binkley, and J. A. Pople, *Aust. J. Chem.,* 30, 699 (1977).

12. See, for example: (a) W. G. Richards, T. E. H. Walker, and R. K. Hinkley, *A Bibliography of Ab Initio Molecular Wavefunctions,* Clarendon Press, Oxford, 1971; (b) W. G. Richards, T. E. H. Walker, L. Farnell, and P. R. Scott, *Supplement for 1970-1973,* Clarendon Press, Oxford, 1974; (c) W. G. Richards, P. R. Scott, E. A. Colbourn, and A. F. Marchington, *Supplement for 1974-1977,* Clarendon Press, Oxford, 1978; (d) W. G. Richards, P. R. Scott, V. Sackwild, and S. A. Robins, *Supplement for 1978-1980,* Clarendon Press, Oxford, 1981.

13. (a) K. Ohno and K. Morokuma, *Quantum Chemistry Literature Data Base.* Bibliography of Ab Initio Calculations for 1978-1980, Elsevier, Amsterdam, 1982; (b) supplement for 1981, *J. Mol. Struct., Theochem.,* 8, 1 (1982); (c) supplement for 1982, *ibid.,* 15, 1 (1983); (d) supplement for 1983, *ibid.,* 20, 1 (1984).

14. R. A. Whiteside, unpublished work.

6

THE PERFORMANCE OF THE MODEL

6.1. INTRODUCTION

This chapter documents the performance of the various quantum mechanical models introduced previously. The successes and failures in calculating equilibrium geometries, relative energies, and vibrational frequencies will be emphasized, although some consideration will also be given to molecular conformations, rotation and inversion barriers, and to molecular charge distributions and electric dipole moments.

Each level of theory does not afford equal numbers of examples to compare with experiment. The availability and range of application of the various quantum mechanical models varies. Thus, only relatively small (minimal and split-valence) basis sets are generally used on molecules incorporating elements beyond the second row. Additionally, practical requirements often dictate the use of low levels of theory for large molecules. Still, a significant number of calculations at the various levels of theory previously discussed (see Chapter 4) are available for critical assessment. While the majority of quantum mechanical calculations carried out to date have been at the single-determinant (Hartree–Fock) level, the effects of electron correlation on calculated equilibrium structures, vibrational frequencies, and relative energies can be evaluated for small molecules.

We stress the obvious: Both experimental and theoretical methods are approximations of varying accuracy. Either (or both) can be in error. In assessing the performance of theory, the correctness of experimental results is assumed, but this reliance is not always certain. Theoretical results often reveal unexpected errors in experimental data!

Theoretical data for this chapter have been drawn both from the literature and from previously unpublished material from the authors' laboratories, as well as those of their collaborators. A small number of papers are frequently referred to in footnotes following the tables. These are cited below and have been assigned brief codes, which will be used in place of the full reference.

LHCP (1971). Tabulation of STO-3G equilibrium geometries for one-heavy-atom molecules containing first-row elements only. W. A. Lathan, W. J. Hehre, L. A. Curtiss, and J. A. Pople, *J. Am. Chem. Soc.,* **93,** 6377 (1971).

HP (1973). Formulation of 6-31G* and 6-31G** basis sets for first-row elements. Tabulation of energies of hydrogenation and related reactions for molecules containing first-row elements only. P. C. Hariharan and J. A. Pople, *Theor. Chim. Acta,* **28,** 213 (1973).

HP (1974). Tabulation of 6-31G* equilibrium geometries for one-heavy-atom molecules containing first-row elements only. P. C. Hariharan and J. A. Pople, *Mol. Phys.,* **27,** 209 (1974).

LCHLP (1974). Tabulation of STO-3G equilibrium geometries for two-heavy-atom molecules containing first-row elements only. W. A. Lathan, L. A. Curtiss, W. J. Hehre, J. B. Lisle, and J. A. Pople, *Prog. Phys. Org. Chem.,* **11,** 175 (1974).

DLPHBP (1979). Tabulation of 6-31G* and MP2/6-31G* equilibrium geometries for one- and two-heavy-atom molecules containing first-row elements only. D. J. DeFrees, B. A. Levi, S. K. Pollack, W. J. Hehre, J. S. Binkley, and J. A. Pople, *J. Am. Chem. Soc.,* **101,** 4085 (1979).

BPH (1980). Formulation of the 3-21G basis set for first-row elements. Tabulation of equilibrium geometries, relative energies and normal-mode vibrational frequencies for small molecules containing first-row elements only. J. S. Binkley, J. A. Pople, and W. J. Hehre, *J. Am. Chem. Soc.,* **102,** 939 (1980).

KBSP (1980). Formulation of the 6-311G** basis set for first-row elements. Tabulation of equilibrium geometries and energies for one-heavy-atom and selected two-heavy-atom molecules containing first-row elements only at the HF, MP2, MP3, and MP4SDQ levels. R. Krishnan, J. S. Binkley, R. Seeger, and J. A. Pople, *J. Chem. Phys.,* **72,** 650 (1980).

PLHS (1980). Formulation of the STO-3G basis set for third-row main-group elements. Tabulation of equilibrium geometries for small molecules containing third-row main-group elements. W. J. Pietro, B. A. Levi, W. J. Hehre, and R. F. Stewart, *Inorg. Chem.,* **19,** 2225 (1980).

PBHHDS (1981). Formulation of the STO-3G basis set for fourth-row main-group elements. Tabulation of equilibrium geometries for small molecules containing fourth-row main-group elements. W. J. Pietro, E. S. Blurock, R. F. Hout, Jr., W. J. Hehre, D. J. DeFrees, and R. F. Stewart, *Inorg. Chem.,* **20,** 3650 (1981).

PSKDBWHH (1981). Tabulation of 3-21G normal-mode frequencies for small polyatomic molecules containing first-row elements only. J. A. Pople, H. B. Schlegel, R. Krishnan, D. J. DeFrees, J. S. Binkley, R. A. Whiteside, R. F. Hout, Jr., and W. J. Hehre, *Int. J. Quantum Chem., Symp.,* **15,** 269 (1981).

WFBDSRP (1981). R. A. Whiteside, M. J. Frisch, J. S. Binkley, D. J. DeFrees, H. B. Schlegel, K. Raghavachari, and J. A. Pople. Published by Carnegie-Mellon University. Carnegie-Mellon Quantum Chemistry Archive. Tabulation of STO-3G, 3-21G and 6-31G* structures, energies, and dipole moments.

DKSP (1982). Tabulation of 6-31G*, MP2/6-31G*, MP3/6-31G*, and CID/6-31G* equilibrium geometries for one- and two-heavy-atom molecules containing first- and second-row elements only. D. J. DeFrees, R. Krishnan, H. B. Schlegel, and J. A. Pople, *J. Am. Chem. Soc.*, **104**, 5576 (1982).

FPHBGDP (1982). Formulation of 6-31G* and 6-31G** basis sets for second-row elements. Tabulation of equilibrium geometries, relative energies, normal-mode vibrational frequencies and electric dipole moments for small molecules containing first- and second-row elements only. M. M. Francl, W. J. Pietro, W. J. Hehre, J. S. Binkley, M. S. Gordon, D. J. DeFrees, and J. A. Pople, *J. Chem. Phys.*, **77**, 3654 (1982).

GBPPH (1982). Formulation of the 3-21G basis set for second-row elements. Tabulation of equilibrium geometries, relative energies and normal-mode vibrational frequencies for small molecules containing first- and second-row elements only. M. S. Gordon, J. S. Binkley, J. A. Pople, W. J. Pietro, and W. J. Hehre, *J. Am. Chem. Soc.*, **104**, 2797 (1982).

HLH (1982). Tabulation of 6-31G* and MP2/6-31G* normal-mode frequencies for small polyatomic molecules containing first-row elements only. R. F. Hout, Jr., B. A. Levi, and W. J. Hehre, *J. Comput. Chem.*, **3**, 234 (1982).

PFHDPB (1982). Formulation of the 3-21G$^{(*)}$ basis set for second-row elements. Tabulation of equilibrium geometries, relative energies, normal-mode vibrational frequencies and electric dipole moments for small molecules containing first- and second-row elements only. W. J. Pietro, M. M. Francl, W. J. Hehre, D. J. DeFrees, J. A. Pople, and J. S. Binkley, *J. Am. Chem. Soc.*, **104**, 5039 (1982).

CCSS (1983). Formulation of the 3-21+G basis set for first-row elements. Tabulation of equilibrium geometries of small anions and acidities of related protonated species. T. Clark, J. Chandrasekhar, G. W. Spitznagel, and P. v. R. Schleyer, *J. Comput. Chem.*, **4**, 294 (1983).

FPR (1983). Tabulation of structural predictions for open-shell systems and comparison and evaluation of RHF and UHF methods. L. Farnell, J. A. Pople, and L. Radom, *J. Phys. Chem.*, **87**, 79 (1983).

PH (1983). Formulation of the STO-3G basis set for first- and second-row transition metals. Tabulation of equilibrium geometries for small molecules containing first- and second-row transition metals. W. J. Pietro and W. J. Hehre, *J. Comput. Chem.*, **4**, 241 (1983).

Specific references to other sources are provided as footnotes to the tables. Much of the detailed information (complete structures, total energies, and dipole moments) for the molecules discussed in this chapter is available in computer-based archive files attached to the current versions of the GAUSSIAN 82 and GAUSSIAN 85 computer programs. For a discussion, see Section 5.4.

6.2. EQUILIBRIUM GEOMETRIES

This section is concerned with a comparison of theoretical and experimental geometries. Since theoretical equilibrium structures refer to isolated species, the most suitable experimental data for comparison involve measurements in dilute gases. Hence, whenever available, gas-phase experimental structural information has been quoted, generally from microwave spectroscopy, electronic spectroscopy, or electron diffraction. For charged species, such data are generally unavailable, and structural parameters have been taken from X-ray crystallographic studies.

We rely heavily on a number of recent tabulations of experimental structural data. Among these are the following:

General

1. J. H. Callomon, E. Hirota, K. Kuchitsu, W. J. Lafferty, A. G. Maki, and C. S. Pote, *Structure Data on Free Polyatomic Molecules*, Landolt–Börnstein, New Series, Group II, vol. 7, ed. K. H. Hellwege and A. M. Hellwege, Springer-Verlag, Berlin, 1976.

2. M. Hargittai and I. Hargittai, *The Molecular Geometries of Coordination Compounds in the Vapor Phase*, Elsevier, Amsterdam, 1977.

3. M. D. Harmony, V. W. Laurie, R. L. Kuczkowski, R. H. Schwendeman, D. A. Ramsay, F. J. Lovas, W. J. Lafferty, and A. G. Maki, *J. Phys. Chem. Ref. Data*, 8, 619 (1979).

4. K. P. Huber and G. Herzberg, *Molecular Spectra and Molecular Structure. IV. Constants of Diatomic Molecules*, Van Nostrand Reinhold, New York, 1979.

Microwave Spectroscopy

1. A. C. Legon and D. J. Millen in *Molecular Spectroscopy*, Specialist Periodical Report, The Chemical Society, London, vol. 2, p. 1 (1974); vol. 3, p. 1 (1975); N. J. MacDonald and J. Sheridan, *ibid.*, vol. 4, p. 1 (1976); vol. 5, p. 1 (1978); vol. 6, p. 1 (1979).

2. J. H. Carpenter in *Spectroscopic Properties of Inorganic and Organometallic Compounds*, Specialist Periodical Report, The Chemical Society, London, vol. 8, p. 194 (1975); vol. 9, p. 180 (1976); A. P. Cox, *ibid.*, vol. 10, p. 167 (1977); vol. 11, p. 168 (1979); vol. 12, p. 171 (1980); vol. 13, p. 183 (1980); S. Cradock, *ibid.*, vol. 14, p. 198 (1981); vol. 15, p. 159 (1982).

Electron Diffraction

1. B. Beagley, *Molecular Structure by Diffraction Methods*, Specialist Periodical Report, The Chemical Society, London, vol. 2, p. 5 (1974); D. W. H. Rankin, *ibid.*, vol. 3, p. 5 (1975); vol. 4, p. 5 (1976); vol. 5, p. 5 (1977); L. Schäfer, *ibid.*, vol. 6, p. 1 (1978).

Experimental structural and conformational data obtained from any of these sources have been used without further reference. The original literature is cited only where experimental results are more recent or where a particular investigation appears to be controversial.

Following a brief discussion of sources of error (Section 6.2.1), the calculated and experimental equilibrium geometries of one- and two-heavy-atom molecules are compared (Sections 6.2.2 and 6.2.3, respectively), followed by discussion of selected bond lengths and skeletal bond angles in a wider set of larger systems (Section 6.2.4). The structures of a number of special classes of molecules are dealt with next: hypervalent molecules (Section 6.2.5), compounds incorporating third- and fourth-row main-group elements (Section 6.2.6), transition-metal inorganic and organometallic molecules (Section 6.2.7) and short-lived species (Section 6.2.8). Free radicals, singlet carbenes, anions, and cations are considered. The equilibrium structures of weakly-bonded complexes, in particular donor-acceptor and hydrogen-bonded systems, are dealt with in Section 6.2.9.

For each of the classes of compounds, the performance of Hartree–Fock models is examined first and then, where data are available, the effect of electron correlation is discussed. It is important to recognize the situations in which electron correlation might significantly alter structures obtained within the framework of the single-determinant model.

The section concludes with a discussion of methods for improved structural predictions, and with a number of specific recommendations to guide future studies requiring the determination of equilibrium geometry.

6.2.1. Sources of Error in the Comparison of Theoretical and Experimental Equilibrium Geometries

What errors complicate comparison of theoretical and experimental structures? Computed structural parameters are subject to numerical (precision) errors as well as to systematic errors implicit in the underlying theory. Numerical errors in calculating geometries arise from rounding, insufficient accuracy in integral evaluation, incomplete convergence of the self-consistent-field procedures, and incomplete convergence of geometry optimization. While such errors may be eliminated by carrying out the computations at higher precision, there is little point in proceeding too far beyond the limits of accuracy imposed by the systematic errors of the methods used. The theoretical equilibrium geometries provided in this book are subject to numerical errors in the third decimal place for bond lengths (in angstroms) and in the first decimal place for angles (in degrees). Errors in calculated dihedral angles may be somewhat larger.

Experimental structural data are also subject to various sources of error. Whereas theoretical structures refer to the true equilibrium form, that is, the actual potential minimum in the absence of vibrational motion (bond lengths r_e and angles θ_e), experimental bond lengths and angles often refer to averages over zero-point vibrational motion [1]. These may differ from true equilibrium values by as much as 0.01 Å and 1°, respectively, and more often than not, sufficient experimental data to derive the appropriate r_e and θ_e values are unavailable. For many larger molecules, there are insufficient experimental data to allow complete structure determination; values for some parameters are often obtained by assuming values for others. In view of the combined theoretical and experimental uncertainties, *comparisons below the level of 0.01 Å for bond lengths and 1° for angles are seldom significant* except for very small molecules. Finally, when there is a discrepancy between a calculated and experimental value, there is always a possibility that it is the latter which is in error!

6.2.2. Structures of AH$_n$ Molecules

a. Hartree–Fock Structures. Table 6.1 compares calculated (STO-3G, 3-21G, 3-21G$^{(*)}$, 6-31G*, 6-31G**, and 6-311G**) and experimental bond lengths and bond angles for a set of AH$_n$ molecules, where A is hydrogen, or a first-, second-, third-, or fourth-row element. Only normal-valent hydrides, for example NH$_3$, are included in the comparison, and theoretical structures have been provided only where gas-phase experimental data exist. The equilibrium geometries of other one-

TABLE 6.1. Hartree–Fock and Experimental Equilibrium Structures for AH_n Molecules

Molecule	Point Group	Geometrical Parameter	STO-3G[a]	3-21G[b]	3-21G$^{(*)}$[c]	6-31G*[a]	6-31G**[e]	6-311G**[f]	Expt.
H_2	$D_{\infty h}$	r(HH)	0.712	0.735	0.735	0.730	0.732	0.735	0.742
LiH	$C_{\infty v}$	r(LiH)	1.510	1.640	1.640	1.636	1.623	1.607	1.596
CH_4	T_d	r(CH)	1.083	1.083	1.083	1.084	1.084	1.084	1.092
NH_3	C_{3v}	r(NH)	1.033	1.003	1.003	1.002	1.001	1.001	1.012
		<(HNH)	104.2	112.4	112.4	107.2	107.6	107.4	106.7
H_2O	C_{2v}	r(OH)	0.990	0.967	0.967	0.947	0.943	0.941	0.958
		<(HOH)	100.0	107.6	107.6	105.5	105.9	105.4	104.5
HF	$C_{\infty v}$	r(FH)	0.956	0.937	0.937	0.911	0.901	0.896	0.917
NaH	$C_{\infty v}$	r(NaH)	1.654	1.926	1.930	1.914	1.912		1.887
SiH_4	T_d	r(SiH)	1.422	1.487	1.475	1.475	1.476		1.481
PH_3	C_{3v}	r(PH)	1.378	1.423	1.402	1.403	1.405		1.420
		<(HPH)	95.0	96.1	95.2	95.4	95.6		93.3
H_2S	C_{2v}	r(SH)	1.329	1.350	1.327	1.326	1.327		1.336
		<(HSH)	92.5	95.8	94.4	94.4	94.4		92.1
HCl	$C_{\infty v}$	r(ClH)	1.313	1.293	1.267	1.267	1.266		1.275

KH	$C_{\infty v}$	$r(\text{KH})$	2.081	2.243
GeH$_4$	T_d	$r(\text{GeH})$	1.431	1.525
AsH$_3$	C_{3v}	$r(\text{AsH})$	1.457	1.511
		$<(\text{HAsH})$	94.0	92.1
SeH$_2$	C_{2v}	$r(\text{SeH})$	1.440	1.460
		$<(\text{HSeH})$	92.5	90.6
HBr	$C_{\infty v}$	$r(\text{BrH})$	1.412	1.414
RbH	$C_{\infty v}$	$r(\text{RbH})$	2.211	2.367
SnH$_4$	T_d	$r(\text{SnH})$	1.630	1.711
SbH$_3$	C_{3v}	$r(\text{SbH})$	1.644	1.704
		$<(\text{HSbH})$	94.4	91.6
TeH$_2$	C_{2v}	$r(\text{TeH})$	1.624	1.658
		$<(\text{HTeH})$	92.4	90.3
HI	$C_{\infty v}$	$r(\text{IH})$	1.599	1.609

[a]LHCP (1971); BPH (1980); PLHS (1980); PBHHDS (1981); GBPPH (1982).
[b]BPH (1980); GBPPH (1982).
[c]PFHDPB (1982).
[d]HP (1974); FPHBGDP (1982).
[e]FPHBGDP (1982).
[f]KBSP (1980).

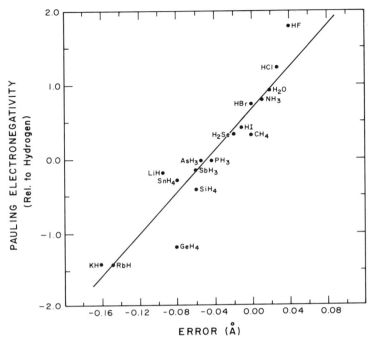

FIGURE 6.1. Correlation of errors (theory-experimental) in STO-3G bond lengths for one-heavy-atom hydrides with Pauling electronegativities. From: PBHHDS (1981).

heavy-atom systems, free radicals and carbenes, for example, are discussed in later sections.

Equilibrium AH bond lengths calculated at the STO-3G level are uniformly shorter than the corresponding experimental values when A is less electronegative, or only slightly more electronegative, than hydrogen, and longer than experimental values for elements of significantly greater electronegativity. In fact, as shown in Figure 6.1, a rough linear correlation exists between the signed deviation of the calculated bond length from experiment and the Pauling electronegativity [2] of the heavy atom. Substantial errors are noted for the hydrides of electropositive elements, for example, Na, K, Rb, Li, Ge, and Sn. Overall, the mean absolute deviation between STO-3G and experimental bond lengths is 0.060 Å (21 comparisons), but this is reduced to 0.037 Å if the highly polar diatomic alkali hydrides are eliminated. A graphical summary is presented in Figure 6.2. In this and in related plots which follow, the line drawn is of unit slope.

While STO-3G bond angles in ammonia and water are smaller than the experimental values (by 2.5° and 4.5°), the HAH bond angles in the corresponding second-, third-, and fourth-row hydrides are too large. The mean absolute deviation of STO-3G bond angles from experimental values is 2.2° for the eight compounds considered.

As may easily be seen from Figure 6.3, the 3-21G basis set offers an improved

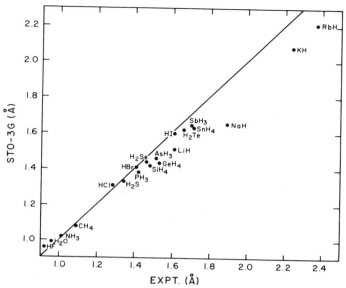

FIGURE 6.2. Comparison of STO-3G and experimental bond lengths in one-heavy-atom hydrides.

description of **AH** bond lengths. Nearly all bond lengths are now within 0.02 Å of the experimental values; LiH and NaH are exceptions, the calculated bond distances being, respectively, 0.045 and 0.039 Å longer than experimental values. Calculated **AH** bond lengths in most other systems are also too large. The mean absolute deviation of 3-21G **AH** bond lengths from experiment is 0.016 Å for the

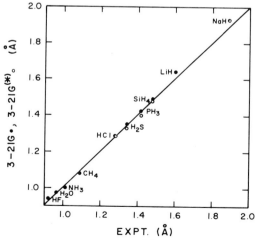

FIGURE 6.3. Comparison of 3-21G (•), 3-21G$^{(*)}$ (○) and experimental bond lengths in one-heavy-atom hydrides.

data in Table 6.1. 3-21G bond angles are uniformly larger than experimental values. This is a widely documented shortcoming of split-valence and larger basis sets containing only s- and p-type functions [3].

For the limited set of molecules incorporating second-row elements included in Table 6.1, the $3\text{-}21G^{(*)}$ basis set yields equilibrium structures which, on average, are no better than those obtained at the unsupplemented 3-21G level. Bond lengths are presented in Figure 6.3 alongside the 3-21G results.

Equilibrium structures obtained using the $6\text{-}31G^*$, $6\text{-}31G^{**}$, and $6\text{-}311G^{**}$ basis sets are generally somewhat better than those derived from the smaller representations. The calculated bond angles for ammonia and water are now both reasonably close to the experimental values. Equilibrium bond distances, with the exception of those in LiH and NaH, are shorter than experimental values. Differences among the $6\text{-}31G^*$, $6\text{-}31G^{**}$, and $6\text{-}311G^{**}$ results are generally quite small, the largest deviation being the decrease of 0.029 Å in the bond length of LiH. Overall, the mean absolute deviations from experiment of $6\text{-}31G^*$, $6\text{-}31G^{**}$, and $6\text{-}311G^{**}$ bond distances are 0.014, 0.014, and 0.013 Å, respectively. Figure 6.4 presents a comparison of $6\text{-}31G^*$ and experimental AH bond lengths.

b. Effect of Electron Correlation on the Structures of AH_n Molecules. Equilibrium geometries obtained for the one-heavy-atom hydrides of first- and second-row elements with various electron correlation models are compared with experimental structures in Table 6.2. The $6\text{-}31G^*$ polarization basis set has been employed throughout [4]. A graphical comparison of $MP2/6\text{-}31G^*$ and experimental bond lengths is presented in Figure 6.5.

For all systems, electron correlation leads to an increase in calculated bond lengths. Within a given row, the elongation increases monotonically with increasing atomic

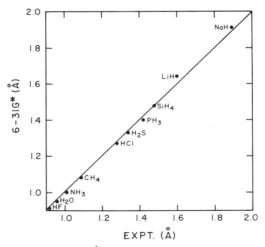

FIGURE 6.4. Comparison of $6\text{-}31G^*$ and experimental bond lengths in one-heavy-atom hydrides.

TABLE 6.2. Effect of Electron Correlation on 6-31G* Equilibrium Geometries of AH_n Molecules

Molecule	Point Group	Geometrical Parameter	HF[a]	MP2[b]	MP3[c]	MP4SDQ[d]	MP4[d]	CID[c]	Expt.
H_2	$D_{\infty h}$	$r(HH)$	0.730	0.738	0.742	0.744	0.744	0.746	0.742
LiH	$C_{\infty v}$	$r(LiH)$	1.636	1.640	1.643	1.648	1.648	1.649	1.596
CH_4	T_d	$r(CH)$	1.084	1.090	1.091	1.093	1.094	1.091	1.092
NH_3	C_{3v}	$r(NH)$	1.002	1.017	1.017	1.019	1.021	1.016	1.012
		$<(HNH)$	107.2	106.3	106.2	106.0	105.8	106.3	106.7
H_2O	C_{2v}	$r(OH)$	0.947	0.969	0.967	0.969	0.970	0.966	0.958
		$<(HOH)$	105.5	104.0	104.3	104.0	103.9	104.3	104.5
HF	$C_{\infty v}$	$r(FH)$	0.911	0.934	0.932	0.934	0.935	0.931	0.917
NaH	$C_{\infty v}$	$r(NaH)$	1.914						1.887
SiH_4	T_d	$r(SiH)$	1.475	1.484					1.481
PH_3	C_{3v}	$r(PH)$	1.403	1.415					1.420
		$<(HPH)$	95.4	94.6					93.3
H_2S	C_{2v}	$r(SH)$	1.326	1.340					1.336
		$<(HSH)$	94.4	93.3					92.1
HCl	$C_{\infty v}$	$r(ClH)$	1.266	1.280					1.275

[a] HP (1974); DLPHBP (1979); FPHBGDP (1982).
[b] DLPHBP (1979); W. J. Hehre and D. J. DeFrees, unpublished calculations.
[c] DKSP (1982).
[d] J. A. Pople, unpublished calculations.

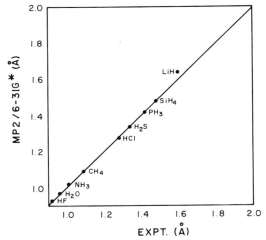

FIGURE 6.5.   Comparison of MP2/6-31G* and experimental bond lengths in one-heavy-atom hydrides.

number. In general, the bond lengths calculated with the correlated models are longer than experimental distances; as noted above, all HF/6-31G* lengths, except those for LiH and NaH, are smaller than the corresponding measured values. Electron correlation leads to a reduction in bond angles compared with Hartree–Fock values for the four molecules included in the comparison.

There are only small differences between the results obtained with the various correlated models. Interestingly enough, the changes in calculated bond lengths from third- to fourth-order theory are uniformly larger than those from the second- to the third-order treatments. Bond lengths calculated using the configuration interaction procedure with all double substitutions (CID) are generally close to those obtained from Møller–Plesset theory through third order.

The residual errors in the bond lengths displayed in Table 6.2 appear to be largely due to inadequacies in the basis set rather than to deficiencies in the electron correlation treatment. This is clearly shown by the results in Table 6.3. Here bond lengths calculated at the MP4 level using the more complete 6-311G** basis set are uniformly close to the experimental distances; the largest error is only 0.005 Å for the NH linkages in ammonia. Unfortunately, this level of treatment is costly and is not generally applicable at the present time to systems of much greater size.

Use of the 6-311G** basis set leads to bond angles (Table 6.3) in poorer agreement with experiment than those at the correlated 6-31G* levels (Table 6.2). The calculated angles are now too small. Better calculations show that the residual error is a basis set rather than a correlation effect [5].

A summary of mean absolute errors in calculated bond lengths in one-heavy-atom hydrides is presented in Table 6.4; the corresponding data for bond angles is provided in Table 6.5. Data for a common set of molecules from Tables 6.1 to 6.3 have been used. Sufficient comparisons between experimental and theoretical

TABLE 6.3. Comparison of Calculated and Experimental Equilibrium Geometries

Molecule	Point Group	Geometrical Parameter	MP4/6-311G**[a]	Expt.
H_2	$D_{\infty h}$	r(HH)	0.742	0.742
LiH	$C_{\infty v}$	r(LiH)	1.597	1.596
CH_4	T_d	r(CH)	1.094	1.092
NH_3	C_{3v}	r(NH)	1.017	1.012
		<(HNH)	105.6	106.7
H_2O	C_{2v}	r(OH)	0.959	0.958
		<(HOH)	102.4	104.5
HF	$C_{\infty v}$	r(FH)	0.913	0.917

[a]KBSP (1980); J. A. Pople, unpublished calculations.

TABLE 6.4. Mean Absolute Errors in Calculated Bond Lengths for AH_n Molecules

	HF	MP2	MP3	MP4
STO-3G	0.054[a]			
3-21G	0.016[a]			
3-21G[(*)]	0.017[a]			
6-31G*	0.014[a]	0.010[b]	0.013[b]	0.016[b]
6-31G**	0.014[a]			
6-311G**	0.013[b]			0.005[b]

[a]Data tabulated for 11 comparisons.
[b]Data tabulated for 6 comparisons.

TABLE 6.5. Mean Absolute Errors in Calculated Bond Angles for AH_n Molecules

	HF[a]	MP2[a]
STO-3G	2.3	
3-21G	3.8	
3-21G[(*)]	3.3	
6-31G*	1.5	0.9
6-31G**	1.8	
6-311G**	[b]	

[a]Data tabulated for 4 comparisons.
[b]Insufficient data.

bond lengths exist for a meaningful evaluation of both Hartree–Fock and correlated models. Comparisons of bond angles are more limited and significant statistics are presented only for Hartree–Fock and MP2 schemes.

6.2.3. Structures of Two-Heavy-Atom Molecules

a. Hartree-Fock Structures.   Table 6.6 provides a comparison of calculated and experimental equilibrium geometries for molecules containing two heavy (non-hydrogen) atoms. Uncharged closed-shell (singlet) systems containing single or multiple bonds between two first- or two second-row elements, or between a first- and a second-row element, for which gas-phase experimental structural data are available, are represented. Comparisons for molecules involving heavier elements are treated in Section 6.2.6.

Errors in lengths for bonds to hydrogen are consistent with corresponding results for AH_n systems (Section 6.2.2) and need not be discussed. For bonds connecting non-hydrogen atoms, the minimal basis set generally yields lengths to the electropositive elements, Li and Na, which are far too short. Examples are LiF (0.15 Å too short) and Na_2 (0.7 Å too short). An exception is the weakly bound Li_2 molecule, for which the calculated bond length is in good accord with experiment. Theory at this level also performs rather poorly for compounds incorporating bonds between two highly electronegative elements. The worst case is F_2, where the calculated bond length (1.315 Å) is 0.1 Å less than the experimental value (1.412 Å). In hydrogen peroxide and hypofluorous acid, the theoretical lengths are too short by 0.056 and 0.087 Å, respectively. Single bonds involving at least one carbon are generally well described by STO-3G calculations; the largest errors are only 0.02-0.03 Å. STO-3G lengths for multiple bonds are sometimes longer and sometimes shorter than experimental values; deviations of up to 0.09 Å are found. Figure 6.6 summarizes the performance of the STO-3G method for bond lengths between heavy atoms for the molecules listed in Table 6.6.

HXY and HXH bond angles calculated with the STO-3G basis set are in reasonable accord with experimental data, especially when X = C. Calculated bond angles are too small when hydrogen is attached to nitrogen or to oxygen (see Section 6.2.2a). Bond angles involving second-row elements are of comparable quality to those for the corresponding first-row compounds. The equilibrium dihedral angle in hydrogen peroxide at STO-3G is slightly larger than the experimental value, while for hydrogen disulfide there is good agreement with experiment.

For most two-heavy-atom hydrides incorporating first-row elements only, the 3-21G basis set provides a better account of AB bond lengths than does STO-3G. As is evident from Figure 6.7, this is particularly true for single-bond lengths between elements other than carbon. Note, however, that while the calculated distances for H_2O_2 and F_2 agree reasonably well with experiment, this is fortuitous. Near the single-determinant (Hartree–Fock) limit, both these bond lengths are seriously in error [6]. Calculated 3-21G bond distances in H_2S_2 and in Cl_2 are $\simeq 0.2$ Å too long, but the experimental bond lengths in disilane and in diphosphine are reproduced reasonably well.

TABLE 6.6. Hartree–Fock and Experimental Equilibrium Structures for $AH_m BH_n$ Molecules. First- and Second-Row Elements

Molecule	Point Group	Geometrical Parameter[a]	STO-3G[b]	3-21G[c]	3-21G(*)[d]	6-31G*[e]	6-31G**[f]	6-311G**[g]	Expt.
Li_2	$D_{\infty h}$	$r(LiLi)$	2.698	2.816	2.816	2.812	2.812	2.785	2.673
LiOH	$C_{\infty v}$	$r(LiO)$	1.432	1.537	1.537	1.592			1.582
		$r(OH)$	0.971	0.955	0.955	0.938			
LiF	$C_{\infty v}$	$r(LiF)$	1.407	1.520	1.520	1.555	1.555	1.557	1.564
LiCl	$C_{\infty v}$	$r(LiCl)$	1.933	2.112	2.091	2.072			2.021
B_2H_6	D_{2h}	$r(BB)$	1.805	1.786	1.786	1.778	1.778	1.783	1.763
		$r(BH_a)$	1.154	1.182	1.182	1.185	1.185	1.185	1.201
		$r(BH_b)$[h]	1.327	1.315	1.315	1.316	1.317	1.322	1.320
		$<(H_a BH_a)$	122.6	122.4	122.4	122.1	122.0	122.0	121.0
C_2H_2	$D_{\infty h}$	$r(CC)$	1.168	1.188	1.188	1.185	1.186	1.182	1.203
		$r(CH)$	1.065	1.051	1.051	1.057	1.057	1.056	1.061
C_2H_4	D_{2h}	$r(CC)$	1.306	1.315	1.315	1.317	1.316	1.316	1.339
		$r(CH)$	1.082	1.074	1.074	1.076	1.076	1.077	1.085
		$<(HCH)$	115.6	116.2	116.2	116.4	116.5	116.6	117.8
C_2H_6	D_{3d}	$r(CC)$	1.538	1.542	1.542	1.527			1.531
		$r(CH)$	1.086	1.084	1.084	1.086			1.096
		$<(HCH)$	108.2	108.1	108.1	107.7			107.8
HCN	$C_{\infty v}$	$r(CN)$	1.153	1.137	1.137	1.133	1.133	1.127	1.153
		$r(CH)$	1.070	1.050	1.050	1.059	1.059	1.058	1.065
HNC	$C_{\infty v}$	$r(NC)$	1.170	1.160	1.160	1.154	1.155	1.149	1.169
		$r(NH)$	1.011	0.983	0.983	0.985	0.984	0.984	0.994
CH_2NH	C_s	$r(CN)$	1.273	1.256	1.256	1.250	1.250	1.248	1.273
		$r(CH_{syn})$	1.091	1.081	1.081	1.084	1.085	1.086	1.103
		$r(CH_{anti})$	1.089	1.075	1.075	1.080	1.081	1.082	1.081
		$r(NH)$	1.048	1.015	1.015	1.006	1.006	1.006	1.023
		$<(H_{syn}CN)$	125.4	125.3	125.3	124.7	124.7	124.5	123.4

TABLE 6.6. (Continued)

Molecule	Point Group	Geometrical Parameter[a]	STO-3G[b]	3-21G[c]	3-21G(*)[d]	6-31G*[e]	6-31G**[f]	6-311G**[g]	Expt.
CH_3NH_2	C_s	$<(H_{anti}CN)$	119.1	119.2	119.2	119.2	119.2	119.3	119.7
		$<(HNC)$	109.1	114.9	114.9	111.6	111.5	111.3	110.5
		$r(CN)$	1.486	1.471	1.471	1.453	1.452	1.454	1.471
		$r(CH_{tr})$	1.093	1.090	1.090	1.091	1.091	1.092	1.099
		$r(CH_g)$	1.089	1.083	1.083	1.084	1.085	1.085	1.099
		$r(NH_a)$	1.033	1.004	1.004	1.001	1.000	1.000	1.010
		$<(NCH_{tr})$	113.7	114.8	114.8	114.8	114.8	114.6	113.9
		$<(NCH_{gg'})$	124.0	123.4	123.4	123.9	124.0	124.0	124.4
		$<(H_gCH_{g'})$	108.2	107.6	107.6	107.5	107.3	107.2	108.0
		$<(CNH_{a,a'})$	119.1	135.3	135.3	126.3	126.3	126.3	125.7
		$<(HNH)$	104.4	111.2	111.2	106.9	107.2	106.9	107.1
CO	$C_{\infty v}$	$r(CO)$	1.146	1.129	1.129	1.114	1.114	1.105	1.128
H_2CO	C_{2v}	$r(CO)$	1.217	1.207	1.207	1.184	1.184	1.179	1.208
		$r(CH)$	1.101	1.083	1.083	1.092	1.093	1.095	1.116
		$<(HCH)$	114.5	115.0	115.0	115.7	115.8	115.8	116.5
CH_3OH	C_s	$r(CO)$	1.433	1.441	1.441	1.400	1.399	1.399	1.421
		$r(CH_{tr})$	1.092	1.079	1.079	1.081	1.082	1.082	1.094
		$r(CH_g)$	1.095	1.085	1.085	1.087	1.088	1.088	1.094
		$r(OH)$	0.991	0.966	0.966	0.946	0.942	0.940	0.963
		$<(OCH_{tr})$	107.7	106.3	106.3	107.2	107.3	107.4	107.2
		$<(OCH_{g,g'})$	130.4	130.5	130.5	130.1	130.2	130.0	129.9
		$<(H_gCH_{g'})$	108.1	108.7	108.7	108.7	108.5	108.7	108.5
		$<(COH)$	103.8	110.3	110.3	109.4	109.7	109.4	108.0
CH_3F	C_{3v}	$r(CF)$	1.384	1.404	1.404	1.365	1.365	1.362	1.383
		$r(CH)$	1.097	1.080	1.080	1.082	1.083	1.081	1.100
		$<(HCH)$	108.3	109.5	109.5	109.8	109.8	110.1	110.6

Molecule	Symmetry	Parameter					
CH₃SiH₃	C_{3v}	$r(CSi)$	1.861	1.917	1.883	1.888	1.867
		$r(CH)$	1.082	1.085	1.087	1.086	1.093
		$r(SiH)$	1.423	1.490	1.477	1.478	1.485
		$<(HCH)$	107.6	108.3	107.8	108.3	107.7
		$<(HSiH)$	108.8	108.3	108.3	107.8	108.3
HCP	$C_{\infty v}$	$r(CP)$	1.472	1.548	1.513	1.515	1.540
		$r(CH)$	1.069	1.057	1.059	1.063	1.069
CH₂PH	C_s	$r(CP)$	1.615	1.683	1.645	1.652	1.67^i
		$r(CH_{syn})$	1.080	1.073	1.074	1.075	
		$r(CH_{anti})$	1.081	1.073	1.076	1.076	
		$r(PH)$	1.390	1.433	1.408	1.409	
		$<(H_{syn}CP)$	126.3	124.7	125.1	125.0	
		$<(H_{anti}CP)$	120.7	119.5	119.9	119.6	
		$<(HPC)$	97.0	98.8	99.0	98.9	
CH₃PH₂	C_s	$r(CP)$	1.841	1.908	1.855	1.861	1.862
		$r(CH_{tr})$	1.084	1.081	1.082	1.082	1.094
		$r(CH_g)$	1.083	1.082	1.084	1.084	1.094
		$r(PH_a)$	1.381	1.425	1.404	1.404	1.432
		$<(PCH_{tr})$	113.1	112.1	113.0	113.2	109.2
		$<(PCH_{g,g'})$	125.7	123.5	124.1	123.9	
		$<(H_gCH_{g'})$	107.4	108.9	107.7	107.6	
		$<(CPH_{a,a'})$	98.9	101.8	102.3	102.8	
		$<(HPH)$	93.7	95.6	94.6	95.1	
CS	$C_{\infty v}$	$r(CS)$	1.519	1.564	1.522	1.520	1.535
H₂CS	C_{2v}	$r(CS)$	1.574	1.638	1.594	1.597	1.611
		$r(CH)$	1.090	1.073	1.076	1.078	1.093
		$<(HCH)$	112.0	116.5	115.3	115.5	116.9
CH₃SH	C_s	$r(CS)$	1.798	1.895	1.823	1.817	1.819
		$r(CH_{tr})$	1.085	1.078	1.081	1.082	1.091
		$r(CH_g)$	1.087	1.077	1.080	1.081	1.091
		$r(SH)$	1.331	1.352	1.327	1.327	1.336

TABLE 6.6. (Continued)

Molecule	Point Group	Geometrical Parameter[a]	STO-3G[b]	3-21G[c]	3-21G(*)[d]	6-31G*[e]	6-31G**[f]	6-311G**[g]	Expt.
		$<(\mathrm{SCH_{tr}})$	108.5	105.6	106.9	106.7			
		$<(\mathrm{SCH_{g,g'}})$	130.1	126.6	128.9	129.3			
		$<(\mathrm{H_gCH_{g'}})$	108.1	111.4	110.1	110.0			109.8
		$<(\mathrm{CSH})$	95.4	97.9	97.5	97.9			96.5
$\mathrm{CH_3Cl}$	C_{3v}	$r(\mathrm{CCl})$	1.802	1.892	1.806	1.785			1.781
		$r(\mathrm{CH})$	1.088	1.073	1.076	1.078			1.096
		$<(\mathrm{HCH})$	110.1	112.6	110.8	110.5			110.0
$\mathrm{N_2}$	$D_{\infty h}$	$r(\mathrm{NN})$	1.134	1.083	1.083	1.078	1.078	1.070	1.098
$\mathrm{N_2H_2}$	C_{2h}	$r(\mathrm{NN})$	1.267	1.239	1.239	1.216	1.216	1.212	1.252
		$r(\mathrm{NH})$	1.061	1.021	1.021	1.014	1.015	1.013	1.028
		$<(\mathrm{NNH})$	105.3	109.0	109.0	107.5	107.6	107.8	106.9
$\mathrm{N_2H_4}$	C_2	$r(\mathrm{NN})$	1.459	1.451	1.451	1.413	1.411	1.412	1.449[i]
		$r(\mathrm{NH_{int}})$	1.037	1.003	1.003	0.999	0.998	0.998	1.021
		$r(\mathrm{NH_{ext}})$	1.040	1.007	1.007	1.003	1.001	1.001	1.021
		$<(\mathrm{NNH_{int}})$	105.4	109.0	109.0	107.9	108.3	108.3	106.0
		$<(\mathrm{NNH_{ext}})$	109.0	113.3	113.3	112.3	112.5	112.4	112.0
		$<(\mathrm{H_{int}NH_{ext}})$	104.6	111.8	111.8	108.2	108.5	108.4	
		$\omega(\mathrm{H_{int}NNH_{ext}})$	91.5	93.8	93.8	90.2	92.9	92.4	91.0
HNO	C_s	$r(\mathrm{NO})$	1.231	1.217	1.217	1.175	1.175	1.167	1.212
		$r(\mathrm{NH})$	1.082	1.036	1.036	1.032	1.033	1.032	1.063
		$<(\mathrm{ONH})$	107.6	109.4	109.4	108.8	108.9	109.2	108.6
$\mathrm{NH_2OH}$	C_s	$r(\mathrm{NO})$	1.420	1.472	1.472	1.403			1.453
		$r(\mathrm{NH})$	1.043	1.002	1.002	1.002			1.016
		$r(\mathrm{OH})$	1.001	0.959	0.959	0.946			0.962
		$<(\mathrm{ONH_{a,a'}})$	119.7	114.7	114.7	115.2			112.6
		$<(\mathrm{HNH})$	104.7	109.6	109.6	106.5			107.1
		$<(\mathrm{NOH})$	105.0	103.6	103.6	104.2			101.4

NP	$C_{\infty v}$	$r(\mathrm{NP})$	1.459	1.510	1.462	1.455			1.491
H_2O_2	C_2	$r(\mathrm{OO})$	1.396	1.473	1.473	1.393			1.452
		$r(\mathrm{OH})$	1.001	0.971	0.971	0.949			0.965
		$<(\mathrm{OOH})$	101.1	99.4	99.4	102.2			100.0
		$\omega(\mathrm{HOOH})$	125.3	180.0	180.0	115.2			119.1
HOF	C_s	$r(\mathrm{OF})$	1.355	1.439	1.439	1.376	1.376	1.362	1.442
		$r(\mathrm{OH})$	1.006	0.976	0.976	0.952	0.947	0.946	0.966
		$<(\mathrm{HOF})$	101.4	99.0	99.0	99.8	100.1	100.6	96.8
NaOH	$C_{\infty v}$	$r(\mathrm{NaO})$	1.763	1.870	1.865	1.922			1.95
		$r(\mathrm{OH})$	0.988	0.962	0.963	0.941			0.96
MgO	$C_{\infty v}$	$r(\mathrm{MgO})$		1.776	1.731	1.738			1.749
SiO	$C_{\infty v}$	$r(\mathrm{SiO})$	1.475	1.536	1.496	1.487			1.510
HPO	C_s	$r(\mathrm{PO})$	1.515	1.544	1.471	1.460			1.512
		$r(\mathrm{PH})$	1.410	1.447	1.429	1.431			
		$<(\mathrm{HPO})$	99.1	103.5	106.0	105.4			104.7
HOCl	C_s	$r(\mathrm{OCl})$	1.737	1.767	1.700	1.670			1.690
		$r(\mathrm{OH})$	1.004	0.975	0.973	0.951			0.975
		$<(\mathrm{HOCl})$	100.2	104.2	106.1	105.1			102.5
F_2	$D_{\infty h}$	$r(\mathrm{FF})$	1.315	1.402	1.402	1.345	1.345	1.331	1.412
NaF	$C_{\infty v}$	$r(\mathrm{NaF})$	1.753	1.863	1.831	1.885			1.926
SiH_3F	C_{3v}	$r(\mathrm{SiF})$	1.624	1.635	1.593	1.594			1.596
		$r(\mathrm{SiH})$	1.422	1.478	1.469	1.470			1.480
		$<(\mathrm{HSiH})$	109.6	109.8	109.9	110.2			110.6
ClF	$C_{\infty v}$	$r(\mathrm{ClF})$	1.677	1.689	1.636	1.613			1.628
Na_2	$D_{\infty h}$	$r(\mathrm{NaNa})$	2.359	3.228	2.651	3.130			3.078
NaCl	$C_{\infty v}$	$r(\mathrm{NaCl})$	2.221	2.421	2.392	2.397			2.361
Si_2H_6	D_{3d}	$r(\mathrm{SiSi})$	2.243	2.382	2.342	2.353			2.327
		$r(\mathrm{SiH})$	1.423	1.488	1.478	1.478			1.486
		$<(\mathrm{HSiH})$	108.0	108.8	108.6	108.5			107.8
SiH_3Cl	C_{3v}	$r(\mathrm{SiCl})$	2.089	2.191	2.056	2.067			2.048

TABLE 6.6. (Continued)

Molecule	Point Group	Geometrical Parameter[a]	STO-3G[b]	3-21G[c]	3-21G(*)[d]	6-31G*[e]	6-31G**[f]	6-311G**[g]	Expt.
		$r(SiH)$	1.423	1.475	1.467	1.468			1.481
		$<(HSiH)$	111.2	111.8	110.5	108.3			110.9
P_2	$D_{\infty h}$	$r(PP)$	1.808	1.930	1.853	1.859			1.893
P_2H_4	C_2	$r(PP)$	2.175	2.356	2.205	2.214			2.219
		$r(PH_{int})$	1.381	1.421	1.401	1.401			1.417
		$r(PH_{ext})$	1.379	1.419	1.400	1.402			1.414
		$<(PPH_{int})$	98.0	99.1	100.6	96.8			99.1
		$<(PPH_{ext})$	95.6	95.5	96.3	101.2			94.3
		$<(H_{int}PH_{ext})$	93.5	95.8	95.2	95.6			92.0
		$\omega(H_{int}PPH_{ext})$	79.0	79.1	73.0	77.3			74.0
H_2S_2	C_2	$r(SS)$	2.065	2.264	2.057	2.064			2.055
		$r(SH)$	1.334	1.352	1.327	1.327			1.327
		$<(SSH)$	96.9	96.7	99.0	99.1			91.3
		$\omega(HSSH)$	92.6	93.7	89.9	87.9			90.6
Cl_2	$D_{\infty h}$	$r(ClCl)$	2.063	2.193	1.996	1.990			1.988

[a] The notation $<(ABH_{d,e})$ is used to describe the angle between the AB bond and the H_dBH_e plane.
[b] LCHLP (1974); BPH (1980); GBPPH (1982).
[c] BPH (1980); GBPPH (1982).
[d] PFHDPB (1982).
[e] DLPHBP (1979); FPHBGDP (1982).
[f] D. J. DeFrees, W. J. Hehre, and J. A. Pople, unpublished calculations.
[g] KBSP (1980): J. A. Pople, unpublished calculations.
[h] H_b designates bridging hydrogens.
[i] Experimental C=P bond length estimated from investigation on CH_2=PCl, CF_2=PH and CH_2=PH. M. J. Hopkinson, H. W. Kroto, J. F. Nixon, and N. P. C. Simmons, *J. Chem. Soc., Chem. Commun.*, 513 (1976).
[j] Experimental structure from: K. Kohata, T. Fukuyama, and K. Kuchitsu, *J. Phys. Chem.*, **86**, 602 (1982).

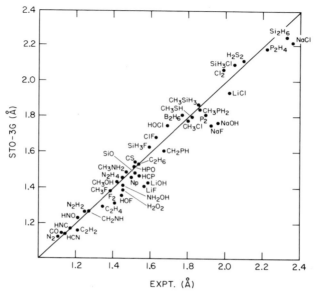

FIGURES 6.6. Comparison of STO-3G and experimental heavy-atom bond lengths in two-heavy-atom hydrides.

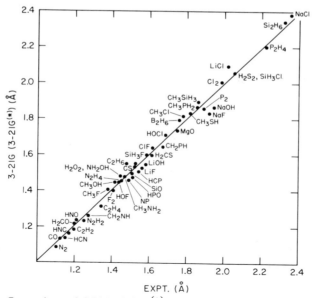

FIGURE 6.7. Comparison of 3-21G (3-21G$^{(*)}$ for molecules incorporating second-row elements) and experimental heavy-atom bond lengths in two-heavy-atom hydrides.

3-21G single-bond lengths for highly polar molecules, for example, LiF, LiOH, LiCl, NaF, and NaCl, are in moderate agreement with measured values and superior to STO-3G values. The opposite is true of single bonds involving carbon and oxygen, fluorine, silicon, phosphorus, sulfur, or chlorine, which are all generally too long. Overall, the mean deviation of 3-21G from experimental single-bond lengths (between heavy atoms) is 0.067 Å (30 comparisons), similar to that found for STO-3G calculations (0.082 Å, 30 comparisons).

With few exceptions, the lengths of multiple bonds at 3-21G are superior to those at STO-3G. The mean absolute deviation from experiment is only 0.017 Å for the 19 molecules considered; for only 6 molecules are errors greater than 0.02 Å.

HXY and HXH bond angles (with X a heteroatom) calculated using the 3-21G basis set are consistently larger than the experimental values (see Section 6.2.2a). The deviations are generally about as large as, but in the opposite direction to, those at STO-3G. Although the calculated dihedral angle in hydrogen disulfide is in reasonable accord with the measured value, the predicted conformation of hydrogen peroxide is *trans* rather than the *skew* result observed experimentally.

Errors in calculated AB single-bond lengths involving second-row elements are significantly reduced by the addition of a set of d functions to the 3-21G basis sets for these second-row elements alone, that is, by use of the 3-21G$^{(*)}$ basis set. Dramatic effects, such as bond length reductions of $\simeq 0.2$ Å, are noted for H_2S_2 and Cl_2; the 3-21G$^{(*)}$ structures for these compounds are now in good accord with experiment. The 3-21G$^{(*)}$ calculations also lead to a shortening of multiple bonds when compared with the 3-21G results, generally by approximately 0.04 Å (0.08 Å in P_2). However, these 3-21G$^{(*)}$ lengths are now uniformly shorter than those determined by experiment. A comparison of 3-21G$^{(*)}$ bond lengths (instead of 3-21G distances) with experimental values is also included in Figure 6.7.

Bond angles for molecules with second-row elements are similar at the 3-21G and 3-21G$^{(*)}$ levels. The 3-21G$^{(*)}$ dihedral angles, both in diphosphine and in hydrogen disulfide, are in better agreement with experiment than are the 3-21G values.

Calculations at 6-31G* usually lead to improved results over those obtained with the smaller basis sets. The major exceptions are molecules with bonds between two electronegative first-row atoms. The calculated NO, OO, OF, and FF bond lengths in hydroxylamine, hydrogen peroxide, hypofluorous acid, and molecular fluorine are all substantially smaller than the experimental values. These errors are due largely to the restriction to single-determinant wavefunctions; as the next section shows, even the simplest of correlation procedures leads to significant bond lengthening and much improved agreement with experiment.

Single bonds involving second-row elements are well described using the HF/6-31G* model. This is true even for diphosphine, hydrogen disulfide, and molecular chlorine, the central bond lengths of which are now all within 0.01 Å of experimental values. Evidently, *electron correlation effects are less significant in influencing the structures of these second-row systems than they are for the analogous first-row compounds.*

Lengths of multiple bonds are consistently underestimated by the HF/6-31G*

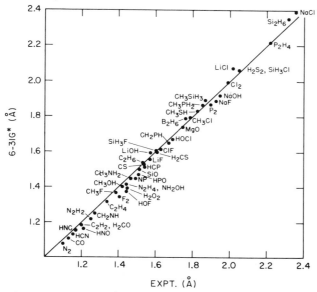

FIGURE 6.8.   Comparison of 6-31G* and experimental heavy-atom bond lengths in two-heavy-atom hydrides.

calculations, generally by 0.01–0.02 Å but sometimes by larger amounts, for example, in HNO, NP, HPO, and P_2. Electron correlation effects are partly responsible. A summary of the performance of the HF/6-31G* model in reproducing experimental AB bond distances is presented graphically in Figure 6.8.

Consistent with the results for one-heavy-atom hydrides (Section 6.2.2a), 6-31G* bond angles are generally in good accord with the measured values. Dihedral angles are also reproduced well. Hydrogen peroxide assumes its correct equilibrium structure and the known conformations of hydrazine, diphosphine, and hydrogen sulfide are properly represented.

Equilibrium structures for molecules containing first-row elements obtained using the more extensive 6-31G** and 6-311G** basis sets are generally very close to those derived from the 6-31G* representation. The largest bond-length discrepancy noted between 6-311G** and 6-31G* calculations is the 0.027 Å difference in the equilibrium distance in Li_2. Equilibrium bond angles calculated at all three levels are in close accord.

b.   Effect of Electron Correlation on Structures of Two-Heavy-Atom Molecules.
Theoretical equilibrium geometries calculated for a series of two-heavy-atom hydrides using Møller–Plesset perturbation theory through second and third order (MP2 and MP3), as well as configuration interaction including all double substitutions (CID), are compared with experimental structures in Table 6.7. The 6-31G* basis set has been used. Significant correlation effects are noted. At the MP2/6-31G* level, the mean absolute deviation between calculated and experimental AB (non-hydrogen)

TABLE 6.7. Effect of Electron Correlation on 6-31G* Equilibrium Geometries of Two-Heavy-Atom Molecules Containing First- and Second-Row Elements

Molecule	Point Group	Geometrical Parameter[a]	HF/6-31G*[b]	MP2/6-31G*[b]	MP3/6-31G*[c]	CID/6-31G*[c]	Expt.
Li_2	$D_{\infty h}$	$r(LiLi)$	2.812	2.782	2.760	2.724	2.673
LiOH	$C_{\infty v}$	$r(LiO)$	1.592	1.594			1.582
		$r(OH)$	0.938	0.960			
LiF	$C_{\infty v}$	$r(LiF)$	1.555	1.570	1.565	1.562	1.564
LiCl	$C_{\infty v}$	$r(LiCl)$	2.072	2.069			2.021
B_2H_6	D_{2h}	$r(BB)$	1.778	1.754	1.753	1.750	1.763
		$r(BH_a)$	1.185	1.190	1.192	1.190	1.201
		$r(BH_b)$[d]	1.316	1.311	1.312	1.310	1.320
		$<(H_aBH_a)$	122.1	121.7	121.6	121.5	121.0
C_2H_2	$D_{\infty h}$	$r(CC)$	1.185	1.218	1.206	1.202	1.203
		$r(CH)$	1.057	1.066	1.066	1.065	1.061
C_2H_4	D_{2h}	$r(CC)$	1.317	1.336	1.334	1.328	1.339
		$r(CH)$	1.076	1.085	1.086	1.084	1.085
		$<(HCH)$	116.4	116.6	116.4	116.3	117.8
C_2H_6	D_{3d}	$r(CC)$	1.527	1.527			1.531
		$r(CH)$	1.086	1.094			1.096
		$<(HCH)$	107.7	107.7			107.8
HCN	$C_{\infty v}$	$r(CN)$	1.133	1.177	1.158	1.154	1.153
		$r(CH)$	1.059	1.070	1.067	1.067	1.065
HNC	$C_{\infty v}$	$r(CN)$	1.154	1.187	1.174	1.171	1.169
		$r(NH)$	0.985	1.002	1.000	0.997	0.994
CH_2NH	C_s	$r(CN)$	1.250	1.282	1.275	1.268	1.273
		$r(CH_{syn})$	1.084	1.096	1.095	1.092	1.103
		$r(CH_{anti})$	1.080	1.090	1.090	1.087	1.081
		$r(NH)$	1.006	1.027	1.025	1.021	1.023

		A	B	C	D	E
CH$_3$NH$_2$	C_s					
	$<$(H$_{syn}$CN)	124.7	125.4	125.2	125.1	123.4
	$<$(H$_{anti}$CN)	119.2	116.1	116.1	116.1	119.7
	$<$(HNC)	111.6	109.7	110.1	110.4	110.5
	r(CN)	1.453	1.465	1.466	1.460	1.471
	r(CH$_{tr}$)	1.091	1.100	1.101	1.098	1.099
	r(CH$_g$)	1.084	1.092	1.093	1.091	1.099
	r(NH$_a$)	1.001	1.018	1.018	1.014	1.010
	$<$(NCH$_{tr}$)	114.8	115.4	115.2	115.2	113.9
	$<$(NCH$_{g,g'}$)	123.9	123.7	123.2	123.4	124.4
	$<$(H$_g$CH$_{g'}$)	107.5	107.5	107.5	107.4	108.0
	$<$(CNH$_{a,a'}$)	126.3	123.6	123.6	124.3	125.7
	$<$(HNH)	106.9	105.9	105.9	106.1	107.1
CO	$C_{\infty v}$					
	r(CO)	1.114	1.151	1.135	1.133	1.128
H$_2$CO	C_{2v}					
	r(CO)	1.184	1.221	1.210	1.205	1.208
	r(CH)	1.092	1.104	1.104	1.101	1.116
	$<$(HCH)	115.7	115.6	116.0	115.8	116.5
CH$_3$OH	C_s					
	r(CO)	1.400	1.424	1.421	1.415	1.421
	r(CH$_{tr}$)	1.081	1.090	1.091	1.088	1.094
	r(CH$_g$)	1.087	1.097	1.098	1.095	1.094
	r(OH)	0.946	0.970	0.967	0.963	0.963
	$<$(OCH$_{tr}$)	107.2	106.4	106.4	106.6	107.2
	$<$(OCH$_{g,g'}$)	130.1	130.7	130.5	130.4	129.9
	$<$(H$_g$CH$_{g'}$)	108.7	108.7	108.8	108.8	108.5
	$<$(HOC)	109.4	107.4	107.7	108.1	108.0
CH$_3$F	C_{3v}					
	r(CF)	1.365	1.392	1.388	1.382	1.383
	r(CH)	1.082	1.092	1.093	1.090	1.100
	$<$(HCH)	109.8	109.8	109.9	109.8	110.6
CH$_3$SiH$_3$	C_{3v}					
	r(CSi)	1.888				1.867
	r(CH)	1.086				1.093
	r(SiH)	1.478				1.485
	$<$(HCH)	108.3				107.7
	$<$(HSiH)	107.8				108.3

TABLE 6.7. (Continued)

Molecule	Point Group	Geometrical Parameter[a]	HF/6-31G*[b]	MP2/6-31G*[b]	MP3/6-31G*[c]	CID/6-31G*[c]	Expt.
HCP	$C_{\infty v}$	$r(CP)$	1.515	1.562			1.540
		$r(CH)$	1.063	1.076			1.069
CH_2PH	C_s	$r(CP)$	1.652				1.67[e]
		$r(CH_{syn})$	1.075				
		$r(CH_{anti})$	1.076				
		$r(PH)$	1.409				
		$<(H_{syn}CP)$	125.0				
		$<(H_{anti}CP)$	119.6				
		$<(HPC)$	98.9				
CH_3PH_2	C_s	$r(CP)$	1.861				1.862
		$r(CH_{tr})$	1.082				1.094
		$r(CH_g)$	1.084				1.094
		$r(PH_a)$	1.404				1.432
		$<(PCH_{tr})$	113.2				
		$<(PCH_gH_{g'})$	123.9				
		$<(H_gCH_{g'})$	107.6				
		$<(CPH_{a,a'})$	102.8				109.2
		$<(HPH)$	95.1				
CS	$C_{\infty v}$	$r(CS)$	1.520	1.546			1.535
H_2CS	C_{2v}	$r(CS)$	1.597	1.617			1.611
		$r(CH)$	1.078	1.090			1.093
		$<(HCH)$	115.5	116.0			116.9
CH_3SH	C_s	$r(CS)$	1.817	1.817			1.819
		$r(CH_{tr})$	1.082	1.091			1.091
		$r(CH_g)$	1.081	1.090			1.091

Molecule	Sym.	Parameter				
		$r(\mathrm{SH})$	1.327	1.341		1.336
		$<(\mathrm{SCH_{tr}})$	106.7	106.7		
		$<(\mathrm{SCH}_{g,g'})$	129.3	129.8		
		$<(\mathrm{H}_g\mathrm{CH}_{g'})$	110.0	109.9		109.8
		$<(\mathrm{CSH})$	97.9	96.8		96.5
CH$_3$Cl	C_{3v}	$r(\mathrm{CCl})$	1.785	1.778		1.781
		$r(\mathrm{CH})$	1.078	1.088		1.096
		$<(\mathrm{HCH})$	110.5	110.1		110.0
N$_2$	$D_{\infty h}$	$r(\mathrm{NN})$	1.078	1.116	1.103	1.098
N$_2$H$_2$	C_{2h}	$r(\mathrm{NN})$	1.216	1.250	1.242	1.252
		$r(\mathrm{NH})$	1.014	1.035	1.031	1.028
		$<(\mathrm{HNN})$	107.5	106.1	106.4	106.9
N$_2$H$_4$	C_2	$r(\mathrm{NN})$	1.413	1.440	1.430	1.449^f
		$r(\mathrm{NH_{int}})$	0.999	1.016	1.012	1.021
		$r(\mathrm{NH_{ext}})$	1.003	1.020	1.016	1.021
		$<(\mathrm{NNH_{int}})$	107.9	106.3	106.8	106.0
		$<(\mathrm{NNH_{ext}})$	112.3	111.3	111.6	112.0
		$<(\mathrm{H_{int}NH_{ext}})$	108.2	106.9	107.2	
		$\omega(\mathrm{H_{int}NNH_{ext}})$	90.2	90.9	90.8	91.0
HNO	C_s	$r(\mathrm{NO})$	1.175	1.213	1.206	1.212
		$r(\mathrm{NH})$	1.032	1.057	1.052	1.063
		$<(\mathrm{HNO})$	108.8	107.9	108.0	108.6
NH$_2$OH	C_s	$r(\mathrm{NO})$	1.403	1.444	1.433	1.453
		$r(\mathrm{NH})$	1.002	1.020	1.016	1.016
		$r(\mathrm{OH})$	0.946	0.968	0.964	0.962
		$<(\mathrm{ONH}_{a,a'})$	115.2	112.1	111.0	112.6
		$<(\mathrm{HNH})$	106.5	105.3	105.6	107.1
		$<(\mathrm{NOH})$	104.2	102.0	103.6	101.4
NP	$C_{\infty v}$	$r(\mathrm{NP})$	1.455			1.491
H$_2$O$_2$	C_2	$r(\mathrm{OO})$	1.393	1.467		1.452
		$r(\mathrm{OH})$	0.949	0.976		0.965

159

TABLE 6.7. (Continued)

Molecule	Point Group	Geometrical Parameter[a]	HF/6-31G*[b]	MP2/6-31G*[b]	MP3/6-31G*[c]	CID/6-31G*[c]	Expt.
HOF	C_s	<(HOO)	102.2	98.7			100.0
		ω(HOOH)	115.2	121.3			119.1
		r(OF)	1.376	1.444	1.433	1.420	1.442
		r(OH)	0.952	0.979	0.975	0.971	0.966
		<(HOF)	99.8	97.2	98.0	98.3	96.8
NaOH	$C_{\infty v}$	r(NaO)	1.922				1.95
		r(OH)	0.941				0.96
MgO	$C_{\infty v}$	r(MgO)	1.738				1.749
SiO	$C_{\infty v}$	r(SiO)	1.487	1.544			1.510
HPO	C_s	r(PO)	1.460	1.519			1.512
		r(PH)	1.431	1.453			
		<(HPO)	105.4	105.6			104.7
HOCl	C_s	r(OCl)	1.670				1.690
		r(OH)	0.951				0.975
		<(HOCl)	105.1				102.5
F$_2$	$D_{\infty h}$	r(FF)	1.345	1.421	1.415	1.399	1.412
NaF	$C_{\infty v}$	r(NaF)	1.885	1.920			1.926
SiH$_3$F	C_{3v}	r(SiF)	1.594	1.619			1.596
		r(SiH)	1.470	1.481			1.480
		<(HSiH)	110.2	110.0			110.6
ClF	$C_{\infty v}$	r(ClF)	1.613				1.628
Na$_2$	$D_{\infty h}$	r(NaNa)	3.130	3.170			3.078
NaCl	$C_{\infty v}$	r(NaCl)	2.397				2.361
Si$_2$H$_6$	D_{3d}	r(SiSi)	2.353	2.338			2.327
		r(SiH)	1.478	1.487			1.486
		<(HSiH)	108.5	108.6			107.8

Species	Symmetry	Parameter			
SiH$_3$Cl	C_{3v}	r(SiCl)	2.067		2.048
		r(SiH)	1.468		1.481
		<(HSiH)	108.3		110.9
P$_2$	$D_{\infty h}$	r(PP)	1.859	1.936	1.893
P$_2$H$_4$	C_2	r(PP)	2.214		2.219
		r(PH$_{int}$)	1.401		1.417
		r(PH$_{ext}$)	1.402		1.414
		<(PPH$_{int}$)	96.8		99.1
		<(PPH$_{ext}$)	101.2		94.3
		<(H$_{int}$PH$_{ext}$)	95.6		92.0
		ω(H$_{int}$PPH$_{ext}$)	77.3		74.0
H$_2$S$_2$	C_2	r(SS)	2.064	2.069	2.055
		r(SH)	1.327	1.344	1.327
		<(SSH)	99.1	99.0	91.3
		ω(HSSH)	87.9	90.3	90.6
Cl$_2$	$D_{\infty h}$	r(ClCl)	1.990	2.015	1.988

[a] See footnote a of Table 6.6.
[b] DLPHBP (1979); W. J. Hehre and D. J. DeFrees, unpublished calculations.
[c] DKSP (1982).
[d] See footnote h of Table 6.6.
[e] See footnote i of Table 6.6.
[f] See footnote j of Table 6.6.

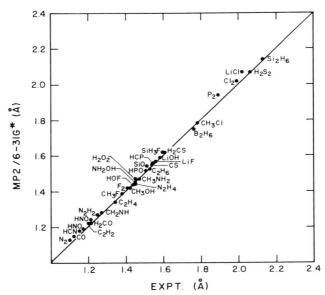

FIGURE 6.9. Comparison of MP2/6-31G* and experimental heavy-atom bond lengths in two-heavy-atom hydrides.

distances is only 0.019 Å (22 comparisons) for single bonds and 0.019 Å (16 comparisons) for multiple bonds. A graphical comparison of MP2/6-31G* and experimental bond lengths is presented in Figure 6.9.

In general, single bonds involving first-row elements only lengthen as a result of the second-order Møller–Plesset correction. The largest effects are noted when two highly electronegative elements are involved. The HF/6-31G* bond in F_2 is lengthened by 0.076 Å; the resulting MP2/6-31G* length is within 0.01 Å of the experimental value. Similar improvements are noted for the heavy-atom linkages in hydrazine, hydroxylamine, hydrogen peroxide, and hypofluorous acid. Smaller but significant bond lengthenings are noted for bonds involving only a single highly electronegative element, for example, Li—OH, Li—F, CH_3—NH_2, CH_3—OH, and CH_3—F. In all cases, the MP2/6-31G* bond lengths are closer to the experimental values than are those at the corresponding Hartree–Fock level. While the central bond in ethane remains unchanged as a result of the correlation corrections, those in Li_2 and in diborane decrease slightly (by 0.030 and 0.024 Å, respectively).

The changes in bond lengths arise from a mixing of excited-state wavefunctions into the Hartree–Fock ground-state electronic description. For molecules with a large complement of valence electrons, that is, those incorporating elements at the right of the Periodic Table, these excited-state wavefunctions generally have considerable antibonding character (relative to the ground state), and their admixture leads accordingly to bond lengthening. For compounds incorporating electropositive elements, the excited-state descriptions may have significant bonding character and linkages may actually tighten as a result of their admixture. For example, while the lowest vacant molecular orbital in diborane is BB antibonding (analogous to the familiar π^* orbital in ethylene), the next higher-energy unfilled function is strongly bonding between the two borons (see Figure 6.10). Promotion of an electron into the latter would be expected to lead to bond shortening.

All multiple bonds involving first-row elements lengthen as a result of the MP2 correction, generally by a few hundredths of an angstrom. As for single bonds, the largest effects are noted when two highly electronegative elements are involved, for example, the multiple bonds in HNO, H_2N_2, and N_2. Smaller lengthenings occur when only a single electronegative atom is involved. Even the bonds in ethylene and acetylene lengthen significantly. While the MP2/6-31G* bond lengths are greater than the corresponding experimental quantities, the discrepancies are generally smaller in magnitude than those for the corresponding HF/6-31G* values.

Improved treatment of electron correlation at the third-order Møller–Plesset level generally results in decreased bond lengths compared with the second-order theory. The effect is most dramatic for multiple linkages, where shortening on the order of 0.01–0.03 Å is common. Systems with double bonds between electronegative atoms change most. For example, the bond length in HNO decreases by 0.024 Å. Effects on single-bond lengths are far smaller and, except for the OF linkage in hypofluorous acid, are less than 0.01 Å. For most molecules, equilibrium MP3/6-31G* bond lengths are closer to the experimental values than are those at either the corresponding Hartree–Fock or MP2 levels. The CID bond lengths are similar to those obtained from the MP3 calculations. The mean absolute deviations of calculated from experimental AB distances are 0.016 Å (single bonds) and 0.005 Å (multiple bonds), nearly identical to the MP3/6-31G* errors (0.014 and 0.005 Å, respectively).

MP2/6-31G* bond angles often differ by 1–2° from the Hartree–Fock values. Significantly larger deviations are noted for the NOH and OOH angles in hydroxyl-

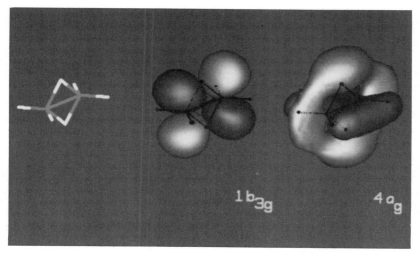

FIGURE 6.10.   Two lowest energy unoccupied molecular orbitals in diborane. STO-3G//STO-3G.

amine and hydrogen peroxide, and for the dihedral angle in H_2O_2. The third-order Møller–Plesset effects are generally small ($1°$ or less) and in the opposite direction to the second-order corrections. As with bond lengths, the bond angles calculated with the CID/6-31G* model are close to those obtained at MP3.

Less is known about the effects of electron correlation on the equilibrium structures of molecules incorporating second-row elements. As noted previously, HF/6-31G* structures for these molecules give better accord with the experimental data than for systems with first-row elements alone. It is to be expected, therefore, that correlation effects will be smaller than for the analogous first-row systems, and that equilibrium geometries, already well described within the framework of Hartree–Fock theory, will be altered only slightly. The available calculations (MP2/6-31G* only) support such expectations. The largest change in single-bond length is an increase of only 0.025 Å for Cl_2 (compared with 0.076 Å for F_2). Smaller increases are noted for the central-bond lengths of H_2S_2 and P_2H_4.

Multiple bonds involving second-row elements are lengthened considerably when correlation is included. However, the effects again tend to be somewhat smaller than those found for the analogous first-row compounds. Examples include the formal triple bond in CS, lengthened by 0.026 Å (compared with the increase of 0.037 Å for CO), and the double bond in H_2CS, lengthened by 0.020 Å (0.037 Å in H_2CO).

The mean absolute errors in AB bond lengths for two-heavy-atom hydrides are summarized in Tables 6.8 (single bonds) and 6.9 (multiple bonds). Data are presented for a common set of nine singly-bonded and ten multiply-bonded systems, as well as for the complete set of molecules examined in Tables 6.6 and 6.7.

6.2.4. Structures of Larger Molecules

Hartree–Fock schemes utilizing the computationally simple STO-3G, 3-21G, and 3-21G$^{(*)}$ basis sets may be readily applied to molecules of moderate size. Even the 6-31G* basis set is small enough to be suitable for calculations for molecules contain-

TABLE 6.8. Mean Absolute Errors in Lengths of Single Bonds Between Heavy Atoms in Two-Heavy-Atom Hydrides[a]

	HF	MP2	MP3	CID
STO-3G	0.049 (0.082)[b]			
3-21G	0.029 (0.067)[b]			
3-21G$^{(*)}$	0.029 (0.040)[b]			
6-31G*	0.044 (0.030)[b]	0.018 (0.019)[c]	0.013 (0.014)[d]	0.015 (0.016)[d]
6-31G**	0.046			
6-311G**	0.039			

[a]For a common set of 9 molecules. Values in parentheses correspond to average errors for larger sets of molecules as indicated.
[b]30 comparisons.
[c]22 comparisons.
[d]10 comparisons.

TABLE 6.9.   Mean Absolute Errors in Lengths of Multiple Bonds Between Heavy Atoms in Two-Heavy-Atom Hydrides[a]

	HF	MP2	MP3	CID
STO-3G	0.017 (0.027)[b]			
3-21G	0.012 (0.017)[c]			
3-21G(*)	0.012 (0.018)[c]			
6-31G*	0.023 (0.023)[c]	0.018 (0.019)[d]	0.005	0.005
6-31G**	0.021			
6-311G**	0.025			

[a]For a common set of 10 molecules. Values in parentheses correspond to average errors for larger sets of molecules, as indicated.
[b]18 comparisons.
[c]19 comparisons.
[d]16 comparisons.

ing 3–5 heavy atoms. The equilibrium structures of several hundred molecules and ions comprising more than two heavy atoms have already been determined using these models. As experimental structural data for many of these are available, it is possible to assess critically the performance of the theory. Furthermore, data for series of closely related molecules make it possible to test the ability of the simple molecular orbital models to account for subtle changes in structure.

a.   Equilibrium Bond Lengths.   Theoretical bond lengths connecting heavy atoms are compared with experimental distances in Table 6.10. These data have been abstracted from complete structure optimizations.

The errors in STO-3G multiple-bond lengths for a wide variety of molecules are relatively constant. Thus, the errors in C=C bond lengths deviate by more than 0.006 Å from a mean value of -0.024 Å $(r_{STO-3G} - r_{expt.})$ in only two cases, even though the experimental distances range from 1.257 Å for the central linkage in butatriene to 1.320 Å in but-1-yne-3-ene. The same is true for C≡C, C≡N, and C=O bonds. The calculations are not as successful in reproducing trends in single-bond lengths. For example, deviations $(r_{STO-3G} - r_{expt.})$ in carbon–carbon bonds range from $+0.036$ Å in acetaldehyde to -0.028 Å for the ring-fused linkage in bicyclo[1.1.0]butane. Bonds for which the calculated length is too short are often incorporated into small rings; calculated carbon–carbon bond lengths in acyclic compounds are generally greater than experimental values. Significant errors also occur for C—F and C—O bonds where extensive delocalization (conjugation) is possible. For example, the O—CH$_3$ bond length in methyl formate is well described, but the O—CO bond shortening is underestimated. The fluoromethanes provide another example. While the calculated CF length in fluoromethane is close to the experimental value, calculated lengths for the CF linkages in difluoromethane, trifluoromethane, and tetrafluoromethane are too long by successively increasing amounts. While single bonds involving carbon and a second-row element are generally well described by the STO-3G calculations, the theory is not completely successful

TABLE 6.10. Calculated and Experimental Equilibrium Bond Distances Connecting Heavy Atoms. Molecules Containing Two or More Heavy Atoms with First- and Second-Row Elements Only

Bond Type	Molecule	STO-3G[a]	3-21G[b]	3-21G$^{(*)b}$	6-31G*[b]	Expt.
Be—C	Dimethylberyllium	1.695	1.715	1.715		1.698
B—C	Trimethylborane	1.581	1.589	1.589		1.578
B$\equiv$N (aromatic)	Borazine	1.418	1.460	1.460		1.418
B—N	Difluoroaminoborane	1.403	1.397	1.397		1.402
B—O	Difluorohydroxyborane	1.358	1.354	1.354		1.344
B—F	Trifluoroborane	1.309	1.328	1.328	1.301	1.307
	Difluoroborane	1.302	1.337	1.337	1.312	1.311
	Difluorohydroxyborane	1.313	1.340	1.340		1.323
	Difluoroaminoborane	1.317	1.349	1.349		1.325
B—Cl	Trichloroborane	1.768	1.771	1.747		1.742
C$\equiv$C	Acetylene	1.168	1.188	1.188	1.185	1.203
	Cyanoacetylene	1.175	1.187	1.187		1.205
	Propyne	1.170	1.188	1.188	1.187	1.206
	But-1-yne-3-ene	1.171	1.190	1.190		1.208
	2-Butyne	1.171	1.189	1.189		1.214
C=C	Butatriene ($C_2 C_3$)	1.257	1.259	1.259		1.283
	Cyclopropene	1.277	1.282	1.282	1.276	1.300
	Allene	1.288	1.292	1.292	1.296	1.308
	Ketene	1.300	1.296	1.296		1.314
	1,1-Difluoroethylene	1.316	1.298	1.298		1.315

Bond	Molecule					
C=C	Butatriene (C$_1$C$_2$)	1.296	1.299	1.299		1.318
	Propene	1.308	1.316	1.316		1.318
	trans-1,2-Difluoroethylene	1.320	1.300	1.300		1.329
	Isobutene	1.311	1.318	1.318		1.330
	cis-1,2-Difluoroethylene	1.320	1.301	1.301		1.331
	Methylenecyclopropane	1.298	1.301	1.301		1.332
	Fluoroethylene	1.312	1.304	1.304		1.332
	Cyclobutene	1.314	1.326	1.326	1.322	1.332
	Acrylonitrile	1.315	1.319	1.319		1.339
	Ethylene	1.306	1.315	1.315		1.339
	But-1-yne-3-ene	1.320	1.320	1.320	1.317	1.341
	cis-Methyl vinyl ether	1.313	1.316	1.316		1.341
	Cyclopentadiene	1.319	1.329	1.329	1.329	1.345
	trans-1,3-Butadiene	1.313	1.320	1.320		1.345
	Vinylsilane	1.309	1.321	1.324		1.347
	Thiophene	1.334	1.335	1.348		1.369
C═══C (aromatic)	Benzene	1.387	1.385	1.385		1.397
C—C	Cyanoacetylene	1.409	1.370	1.370		1.378
	Acrylonitrile	1.461	1.427	1.427		1.426
	But-1-yne-3-ene	1.459	1.432	1.432		1.431
	Thiophene	1.454	1.448	1.438		1.433
	Methylenecyclopropane (C$_2$C$_3$)	1.474	1.472	1.472		1.457
	Acetonitrile	1.488	1.457	1.457		1.458
	Propyne	1.484	1.466	1.466	1.468	1.459
	Cyclopentadiene (C$_2$C$_3$)	1.490	1.485	1.485	1.476	1.468
	Malononitrile	1.493	1.461	1.461		1.468
	2-Butyne	1.483	1.467	1.467		1.468
	Oxirane	1.483	1.474	1.474		1.471
	Aziridine	1.491	1.497	1.497		1.481

TABLE 6.10. (Continued)

Bond Type	Molecule	STO-3G^a	3-21G^b	3-21G$^{(*)b}$	6-31G*^{b}	Expt.
C—C	trans-1,3-Butadiene	1.488	1.479	1.479		1.483
	Thiirane	1.507	1.463	1.490		1.484
	Bicyclo[1.1.0]butane (C$_1$C$_3$)	1.469	1.484	1.484	1.466	1.497
	Bicyclo[1.1.0]butane (C$_1$C$_2$)	1.501	1.513	1.513	1.489	1.498
	Acetaldehyde	1.537	1.507	1.507	1.504	1.501
	Propene	1.520	1.510	1.510	1.503	1.501
	Fluoroethane	1.547	1.521	1.521		1.505
	Cyclopentadiene (C$_4$C$_5$)	1.522	1.519	1.519	1.507	1.506
	Acetone	1.543	1.515	1.515		1.507
	Isobutene	1.526	1.516	1.516		1.507
	Cyclopropane	1.502	1.513	1.513	1.497	1.510
	Cyclopropene	1.493	1.523	1.523	1.495	1.515
	Cyclobutene (C$_1$C$_4$)	1.526	1.539	1.539	1.514	1.517
	Acetic acid	1.537	1.497	1.497		1.517
	Propane	1.541	1.541	1.541	1.528	1.526
	trans-Ethanethiol	1.541	1.531	1.541		1.529
	1,1,1-Trifluoroethane	1.562	1.490	1.490		1.530
	trans-Ethanol	1.542	1.525	1.525		1.530
	Ethane	1.538	1.542	1.542	1.527	1.531
	trans-Ethylsilane	1.543	1.550	1.555		1.540
	Neopentane	1.549	1.540	1.540		1.540
	Isobutane	1.545	1.541	1.541		1.541
	Methylenecyclopropane (C$_3$C$_4$)	1.522	1.545	1.545		1.542
	Cyclobutane	1.554	1.571	1.571	1.548	1.548
	Cyclobutene (C$_3$ C$_4$)	1.565	1.593	1.593	1.562	1.566

C≡N	Hydrogen cyanide	1.153	1.137	1.137	1.133	1.153
	Acetonitrile	1.154	1.139	1.139	1.133	1.157
	Cyanoacetylene	1.159	1.141	1.141		1.159
	Cyanogen fluoride	1.160	1.135	1.135		1.159
	Acrylonitrile	1.157	1.140	1.140		1.164
	Methyl isocyanide	1.170	1.160	1.160		1.166
	Malononitrile	1.155	1.138	1.138		1.167
	Formonitrile oxide	1.155				1.168
	Hydrogen isocyanide	1.170	1.160	1.160	1.154	1.169
	Nitrosyl cyanide	1.159				1.170
C=N	Isocyanic acid	1.246	1.160	1.160		1.209
	Isothiocyanic acid	1.226				1.216
	Methyleneimine	1.273	1.256	1.256	1.250	1.273
	Diazomethane	1.282	1.281	1.281		1.300
C—N	Formamide	1.436	1.353	1.353		1.376
	Nitrosyl cyanide	1.482			1.453	1.401
	Methyl isocyanide	1.447	1.432	1.432		1.424
	Trimethylamine	1.486	1.464	1.464		1.451
	Dimethylamine	1.484	1.466	1.466		1.462
	Methylamine	1.486	1.471	1.471		1.471
	Aziridine	1.482	1.491	1.491		1.475
	Diazirine	1.488	1.522	1.522		1.482
	Nitromethane	1.531	1.493	1.493		1.489
	Nitrosomethane	1.531	1.499	1.499		1.49
C=O	Carbonyl sulfide	1.176	1.140	1.147		1.160
	Ketene	1.183	1.162	1.162		1.161
	Carbon dioxide	1.188	1.156	1.156	1.143	1.162
	Isocyanic acid	1.183				1.166
	Formyl fluoride	1.210	1.180	1.180		1.181

TABLE 6.10. (Continued)

Bond Type	Molecule	STO-3G[a]	3-21G[b]	3-21G(*)[b]	6-31G*[b]	Expt.
C=O	Formamide	1.216	1.212	1.212		1.193
	Methyl formate	1.214	1.200	1.200		1.200
	cis-Formic acid	1.214	1.198	1.198		1.202
	Formaldehyde	1.217	1.207	1.207	1.184	1.208
	Acetic acid	1.216	1.202	1.202		1.212
	Acetaldehyde	1.217	1.209	1.209	1.188	1.216
	Acetone	1.219	1.211	1.211		1.222
C—O	Methyl formate (CH_3O—CHO)	1.388	1.344	1.344		1.334
	cis-Formic acid	1.385	1.350	1.350		1.343
	Acetic acid	1.391	1.360	1.360		1.360
	cis-Methyl vinyl ether (CH_3O—$CHCH_2$)	1.392	1.370	1.370		1.360
	Dimethyl ether	1.433	1.433	1.433	1.391	1.410
	Methanol	1.433	1.441	1.441	1.400	1.421
	trans-Ethanol	1.436	1.444	1.444		1.425
	cis-Methyl vinyl ether (CH_3—$OCHCH_2$)	1.434	1.437	1.437		1.428
	Oxirane	1.433	1.470	1.470		1.436
	Methyl formate (CH_3—OCHO)	1.441	1.456	1.456		1.437
C—F	Cyanogen fluoride	1.316	1.287	1.287		1.262
	Fluoroacetylene	1.318	1.297	1.297		1.279
	Tetrafluoromethane	1.366	1.325	1.325		1.317
	1,1-Difluoroethylene	1.350	1.334	1.334		1.323
	Trifluoromethane	1.371	1.345	1.345		1.332
	cis-1,2-Difluoroethylene	1.356	1.358	1.358		1.335
	Formyl fluoride	1.351	1.348	1.348		1.338
	trans-1,2-Difluoroethylene	1.356	1.360	1.360		1.344

Type	Compound					
	Fluoroethylene	1.354	1.363	1.363		1.348
	1,1,1-Trifluoroethane	1.375	1.352	1.352		1.348
	Difluoromethane	1.378	1.372	1.372		1.358
	Fluoromethane	1.384	1.404	1.404	1.365	1.383
	Fluoroethane	1.385	1.410	1.410		1.398
C—Al	Trimethylaluminum	1.899	1.997	1.981		1.957
C=Si	Methylenedimethylsilane	1.639	1.719	1.690		1.83[c]
C—Si	Vinylsilane	1.852	1.897	1.867		1.853
	Ethylsilane	1.869	1.920	1.886		1.866
	Methylsilane	1.861	1.917	1.883	1.888	1.867
	Dimethylsilane	1.862	1.917	1.885		1.867
	Trimethylsilane	1.862	1.918	1.887		1.868
	Tetramethylsilane	1.863	1.919	1.889		1.875
	Methylenedimethylsilane	1.856	1.908	1.876		1.91[c]
C—P	Trimethylphosphine	1.841	1.903	1.848		1.841
	Dimethylphosphine	1.841	1.905	1.851		1.848
	Methylphosphine	1.841	1.908	1.855	1.861	1.862
C=S	Carbon disulfide	1.532	1.626	1.542		1.553
	Carbonyl sulfide	1.548	1.579	1.565		1.560
	Isothiocyanic acid	1.542				1.561
	Thioformaldehyde	1.574	1.638	1.594	1.597	1.611
C—S	Thiophene	1.732	1.797	1.722		1.714
	Dimethyl sulfide	1.796	1.885	1.813		1.802
	Thiirane	1.774	1.934	1.817		1.815
	Methanethiol	1.798	1.895	1.823	1.817	1.819
	Ethanethiol	1.806	1.900	1.829		1.820

TABLE 6.10. (Continued)

Bond Type	Molecule	STO-3G[a]	3-21G[b]	3-21G(*)[b]	6-31G*[b]	Expt.
C—Cl	Trichloromethane	1.808	1.835	1.776		1.758
	Tetrachloromethane	1.818	1.832	1.778		1.767
	Dichloromethane	1.803	1.853	1.784		1.772
	Chloromethane	1.802	1.892	1.806	1.785	1.781
N≡N	Diazomethane	1.190	1.131	1.131		1.139
N=N	Diazirine	1.266	1.217	1.217		1.228
	trans-Difluorodiazene	1.373	1.211	1.211		1.214
	Diazene	1.267	1.239	1.239	1.216	1.252
N=O	Nitrosyl fluoride	1.222	1.152	1.152	1.128	1.136
	Nitrosyl chloride	1.203	1.131	1.149		1.139
	Formonitrile oxide	1.294				1.199
	Nitrosyl hydride	1.231	1.217	1.217	1.175	1.212
	Nitrosomethane	1.231	1.216	1.216		1.22
	Nitromethane	1.275	1.240	1.240		1.224
	Nitrosyl cyanide	1.235				1.228
N—F	Nitrogen trifluoride	1.386	1.402	1.402	1.328	1.365
	trans-Difluorodiazene	1.277	1.414	1.414		1.384
	Nitrosyl fluoride	1.381	1.460	1.460	1.384	1.512
N—Cl	Nitrogen trichloride	1.803	1.823	1.740		1.748
	Nitrosyl chloride	1.862	2.012	1.907		1.975
O—O	Ozone	1.285	1.308	1.308	1.204	1.278
O—O	Fluorine peroxide	1.392	1.308	1.308		1.217
	Hydrogen peroxide	1.396	1.473	1.473	1.393	1.452
O—F	Oxygen difluoride	1.358	1.427	1.427	1.384	1.405
	Hypofluorous acid	1.355	1.439	1.439	1.376	1.442
	Fluorine peroxide	1.358				1.575

Bond	Molecule					
O—Cl	Hypochlorous acid	1.737	1.767	1.700	1.670	1.690
	Oxygen dichloride	1.743	1.765	1.703		1.700
F—Mg	Magnesium difluoride	1.665	1.722	1.701		1.77
F—Al	Aluminum trifluoride	1.600	1.627	1.617		1.63
F—Si	Tetrafluorosilane	1.585	1.584	1.557		1.554
	Trifluorosilane	1.597	1.601	1.569		1.562
	Difluorosilane	1.610	1.618	1.581		1.577
	Fluorosilane	1.624	1.635	1.593	1.594	1.596
F—P	Trifluorophosphine	1.621	1.610	1.561		1.551
F—S	Difluorosulfide	1.641	1.652	1.592		1.592
Al—Cl	Trichloroaluminum	2.050	2.160	2.075		2.06
Si—Cl	Tetrachlorosilane	2.071	2.123	2.027		2.019
	Trichlorosilane	2.073	2.139	2.039		2.021
	Dichlorosilane	2.079	2.160	2.056		2.033
	Chlorosilane	2.089	2.191		2.067	2.048
P—Cl	Trichlorophosphine	2.104	2.220	2.039		2.043
S—Cl	Sulfur dichloride	2.077	2.222	2.019		2.014

[a] Among the major collections of STO-3G equilibrium geometries for three-heavy-atom and larger molecules are the following: M. D. Newton, W. A. Lathan, W. J. Hehre, and J. A. Pople, J. Chem. Phys., 52, 4064 (1970); L. Radom, W. A. Lathan, W. J. Hehre, and J. A. Pople, J. Am. Chem. Soc., 93, 5339 (1971); W. J. Hehre and J. A. Pople, ibid., 97, 6941 (1975). Most of the remaining structures are from unpublished calculations from the authors' laboratories. Results in this table come from WFBDSRP (1981) or from unpublished calculations from the authors' laboratories.

[b] Major collections of 3-21G and 6-31G* equilibrium geometries have not previously been published.

[c] P. G. Mahaffy, R. Gutowsky, and L. K. Montgomery, J. Am. Chem. Soc., 102, 2854 (1980). This result has recently been challenged (Y. Yoshioka, J. D. Goddard, and H. F. Schaefer III, J. Am. Chem. Soc., 103, 2452 (1981)) by the results of calculations which take partial account of electron correlation. The suggested $Si=C$ bond length is 1.705 ± 0.03 Å, in good agreement with the values given here. A crystal structure of a highly-substituted silaethylene has recently been reported and shows a $Si=C$ bond length of 1.764 Å: A. G. Brook, S. C. Nyburg, F. Abdesaken, B. Gutekunst, G. Gutekunst, R. K. M. R. Kallury, Y. C. Poon, Y. M. Chang, and W. W. Ng, J. Am. Chem. Soc., 104, 5667 (1982).

in reproducing subtle trends in bond lengths within a series of closely related compounds. The observed shortening of carbon–chlorine bond lengths in the chloromethanes is an example.

Average errors in bond distances are generally of comparable magnitude for the 3-21G and STO-3G basis sets. The larger and more flexible basis set is, however, usually more successful in reproducing the observed trends in bond lengths. For example, the 3-21G calculations do not exhibit the previously noted problems with C—F and C—O linkages.

3-21G lengths for carbon–carbon double and triple bonds are consistently shorter than the corresponding experimental values. Like the STO-3G calculations, errors at 3-21G for C=C double bonds are remarkably constant and range from a minimum of 0.006 Å (cyclobutene) to a maximum of 0.030 Å (1,2-difluoroethylene). With few exceptions, 3-21G calculations properly order the observed C=C bond distances. The lengths of carbon–carbon single bonds are reproduced more accurately with 3-21G than STO-3G; errors are typically less than 0.01 Å. The largest deviation is 0.040 Å for 1,1,1-trifluoroethane. The performance of the theory is commendable in view of the large range of experimental C—C distances (from 1.378 Å in cyanoacetylene to 1.566 Å for the C_3C_4 linkage in cyclobutene).

In general, 3-21G lengths of single and multiple bonds involving carbon and nitrogen, oxygen, or fluorine are better than those at STO-3G. This is especially true for CN single bonds; the extreme cases (formamide and nitromethane) are well described. Double bonds to oxygen are also well described with the 3-21G basis set, but tend to be too short. The mean absolute deviation (11 comparisons) is only 0.008 Å. The 3-21G basis set also describes CO and CF single bonds reasonably well. For example, compare the experimentally observed 0.068 Å reduction in CF bond distances from fluoromethane to the tetrafluoro compound with the shortening of 0.079 Å at 3-21G. Calculated NF, OF, and OO bonds are consistently longer than experimental values, generally by a few hundredths of an angström, although, as noted in Section 6.2.3, agreement with experiment is fortuitously better than obtained with larger basis Hartree-Fock calculations.

The 3-21G basis set performs poorly with bonds involving second-row elements (see also Section 6.2.3). Bonds connecting first- and second-row elements are consistently too long, often by as much as 0.1–0.2 Å. Only a few trends in bond lengths among related compounds are reproduced at this level, for example, the Si—F and Si—Cl lengths in the sets of halosilanes. In contrast, CS double-bond lengths in carbon disulfide and carbonyl sulfide are ordered incorrectly. Where second-row elements are involved, the 3-21G$^{(*)}$ basis set is more successful than 3-21G both in reproducing absolute bond lengths (see Section 6.2.3) and in describing bond length changes from one molecule to another. Absolute deviations between calculation and experiment seldom exceed 0.01–0.02 Å (see, however, MgF_2). Bond length trends in the methyl-, fluoro-, and chlorosilanes, the methylphosphines, the alkanethiols and sulfides, and the chloromethanes are well reproduced at this level.

6-31G* bond lengths are usually (but not always) shorter than those obtained at the 3-21G (3-21G$^{(*)}$) level. The differences are small ($\sim$0.01-0.02 Å) except where two highly electronegative elements are involved, e.g., a reduction of 0.074 Å in the NF bond lengths in NF_3.

TABLE 6.11. Mean Absolute Errors in Calculated Bond Lengths Between Heavy Atoms

Basis Set	Single Bonds	Multiple Bonds	All Bonds
STO-3G	0.030^a	0.027^b	0.029^c
3-21G	0.033^d	0.019^e	0.029^f
3-21G$^{(*)}$	0.015^g	0.018^h	0.016^i
6-31G*	0.020^j	0.021^k	0.020^l

[a] 128 comparisons.
[b] 67 comparisons.
[c] 195 comparisons.
[d] 126 comparisons.
[e] 60 comparisons.
[f] 186 comparisons.
[g] 124 comparisons.
[h] 60 comparisons.
[i] 184 comparisons.
[j] 33 comparisons.
[k] 18 comparisons.
[l] 51 comparisons.

A summary of mean absolute errors in calculated bond lengths between heavy atoms is presented in Table 6.11. These follow from the comparisons presented in Table 6.10.

b. Skeletal Bond Angles. More than any other feature, skeletal bond angles involving the heavy-atom framework characterize the gross structure of a molecule. The performance of the STO-3G, 3-21G, and 3-21G$^{(*)}$ Hartree–Fock models is presented in Table 6.12 and in Figures 6.11 and 6.12. Insufficient data are available at the 6-31G* level for meaningful comparisons. Trivial cases where bond angles are dictated by symmetry alone, for example, the 120° FBF angles in trifluoroborane, the tetrahedral ClCCl angles in tetrachloromethane or the 180° CCC bond angles in propyne, are omitted.

The STO-3G bond angles agree quite well with experimental values; the mean absolute error is only 1.3° (70 comparisons in Table 6.12). Only a few errors exceed 3°, for example, <(COC) in methyl vinyl ether (-4.6°), and <(FOO) in fluorine peroxide (-5.3°). The descriptions of systems including second-row elements are of comparable quality to those for first-row elements.

The 3-21G skeletal bond angles are also close to experimental values: the mean absolute deviation is 1.0°. In no instance does a calculated bond angle deviate by more than 3° from experiment. Trends in bond angles for related compounds are generally well reproduced at the 3-21G level, for example, known effects of methyl, fluorine, and chlorine substitution on the bond angles involving nitrogen, oxygen, phosphorus, and sulfur centers, and even the variations in FCC bond angles in fluoroethylene and the isomeric difluoroethylenes.

Bond angles involving one or more second-row elements usually improve further with the 3-21G$^{(*)}$ basis set, although the changes from 3-21G are slight (generally less

TABLE 6.12. Calculated and Experimental Equilibrium Skeletal Bond Angles. Molecules Containing Three or More Heavy Atoms with First- and Second-Row Elements Only

Angles	Molecule	STO-3G[a]	3-21G[b]	3-21G[(*)][b]	Expt.
F—B—F	Difluoroaminoborane		116.4	116.4	117.9
	Difluorohydroxyborane	119.6	117.6	117.6	118.0
	Difluoroborane	118.2	116.9	116.9	118.3
C=C—C	Acrolein	122.4	120.5	120.5	119.8
	Isobutene	122.4	122.6	122.6	122.4
	Acrylonitrile	122.9	122.9	122.9	122.6
	But-1-yne-3-ene	124.0	123.4	123.4	123.1
	trans-1,3-Butadiene	124.2	124.0	124.0	123.1
	Propene	125.1	124.7	124.7	124.3
	cis-1-Butene	127.0	126.6	126.6	126.7
C—C—C	Isobutane	110.9	110.4	110.4	110.8
	Malononitrile	111.0	111.9	111.9	110.9
	Propane	112.4	111.6	111.6	112.4
	cis-1-Butene	115.1	114.9	114.9	114.8
C=C—O	cis-Methyl vinyl ether	129.5	128.1	128.1	127.7
C—C=O	Glyoxal	122.4	121.4	121.4	121.2
	Acetone	122.4	122.5	122.5	121.4
	trans-Acrolein	124.0	124.1	124.1	123.3
	Acetaldehyde	124.3	124.8	124.8	124.0
	Acetic Acid	126.8	127.4	127.4	126.6
C—C—O	trans-Ethanol	107.7	106.2	106.2	107.3
C=C—F	trans-1,2-Difluoroethylene	122.3	120.4	120.4	119.3

176

Bond	Compound				
	Fluoroethylene	123.4	122.5	122.5	121.2
	cis-1,2-Difluoroethylene	123.9	123.3	123.3	122.1
	1,1-Difluoroethylene	124.7	125.3	125.3	125.5
C—C—F	Fluoroethane	111.2	108.9	108.9	109.7
C=C—Si	Vinylsilane	125.0	123.7	123.7	122.9
C—C—Si	Ethylsilane	113.8	113.0	113.2	113.2
C—C—S	Ethanethiol	110.7	108.4	109.1	108.6
N—C≡N	Nitrosyl cyanide	175.2			172.5
N—C=O	Formamide	124.3	125.3	125.3	124.7
O—C=O	Acetic acid	121.8	122.1	122.1	123.0
	Formic acid	123.7	124.6	124.6	124.6
	Methyl formate	124.8	124.6	124.6	125.9
O=C—F	Carbonyl fluoride	125.0	122.4	122.4	126.0
F—C—F	1,1,1-Trifluoroethane	107.7	107.2	107.2	106.7
	Difluoromethane	108.7	108.9	108.9	108.3
	Trifluoromethane	108.6	108.3	108.3	108.8
Cl—C—Cl	Trichloromethane	111.1	110.5	111.0	111.3
	Dichloromethane	112.7	110.8	112.0	111.8
C—N—C	Trimethylamine	110.3	113.1	113.1	110.9
	Dimethylamine	111.2	114.5	114.5	112.0
C—N=O	Nitrosomethane	111.5	112.7	112.7	112.6
	Nitrosyl cyanide	111.4			114.7
F—N=O	Nitrosyl fluoride	108.2	109.9	109.9	110.1

TABLE 6.12. (Continued)

Angles	Molecule	STO-3G[a]	3-21G[b]	3-21G(*)[b]	Expt.
Cl—N=O	Nitrosyl chloride	111.2	112.1	112.4	113.3
N=N—F	Difluorodiazene	111.5	105.4	105.4	114.5
O—N—O	Nitrous acid	108.3			110.7
	Nitromethane	125.2	126.2	126.2	125.3
	Nitric acid				
F—N—F	Nitrogen trifluoride	102.2	101.6	101.6	102.4
Cl—N—Cl	Nitrogen trichloride	107.2	109.0	107.6	107.4
C—O—C	Dimethyl ether	108.7	114.0	114.0	111.7
	Methyl formate	112.1	118.0	118.0	114.8
	cis-Methyl vinyl ether	113.7	119.2	119.2	118.3
O—O—O	Ozone	116.2	117.0	117.0	116.8
O—O—F	Fluorine peroxide	104.2			109.5
F—O—F	Oxygen difluoride	102.4	101.6	101.6	103.1

178

Angle	Molecule				
Cl—O—Cl	Oxygen dichloride	109.3	112.0	113.2	110.9
C—Si—C	Trimethylsilane	110.1	110.4	110.3	110.2
	Dimethylsilane	110.8	111.8	111.6	111.0
F—Si—F	Difluorosilane	106.3	105.3	106.4	107.9
	Trifluorosilane	107.5	106.8	107.6	108.3
Cl—Si—Cl	Trichlorosilane	109.2	109.2	109.6	109.4
	Dichlorosilane	109.3	109.9	110.4	109.4
C—P—C	Trimethylphosphine	98.3	98.9	99.5	98.9
	Dimethylphosphine	98.7	99.7	100.5	99.7
F—P—F	Trifluorophosphine	94.7	96.2	97.1	97.7
Cl—P—Cl	Trichlorophosphine	98.8	98.7	100.5	100.1
C—S—C	Dimethyl sulfide	98.3	99.3	99.4	98.9
F—S—F	Sulfur difluoride	94.2	97.3	98.3	98.2
Cl—S—Cl	Sulfur dichloride	100.3	99.9	102.5	102.7

[a]See footnote a of Table 6.10. See also W. J. Hehre, J. Am. Chem. Soc., 97, 5308 (1975).
[b]See footnote b of Table 6.10.

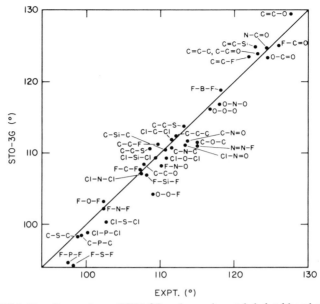

FIGURE 6.11. Comparison of STO-3G and experimental skeletal bond angles.

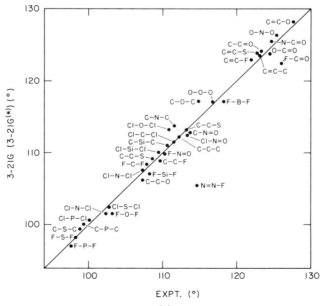

FIGURE 6.12. Comparison of 3-21G (3-21G[(*)] for molecules incorporating second-row elements) and experimental skeletal bond angles.

TABLE 6.13. Mean Absolute Errors in
Calculated Bond Angles Involving Heavy
Atoms

Basis Set	Error
STO-3G	1.3^a
3-21G	1.0^b
3-21G$^{(*)}$	0.9^b

^{a}70 comparisons.
^{b}67 comparisons.

than $1°$). This contrasts with the markedly superior performance of the 3-21G$^{(*)}$ basis set for the calculation of bond lengths.

A summary of mean absolute errors in calculated bond angles (involving the heavy-atom skeleton) is provided in Table 6.13.

6.2.5. Structures of Hypervalent Molecules

Compounds involving first-row atoms generally obey the classical octet rule. Although apparent exceptions exist, for example, Li_3O and F_3NO, most of these may be rationalized in terms of detailed electronic distributions that involve extensive separation of charge, such as

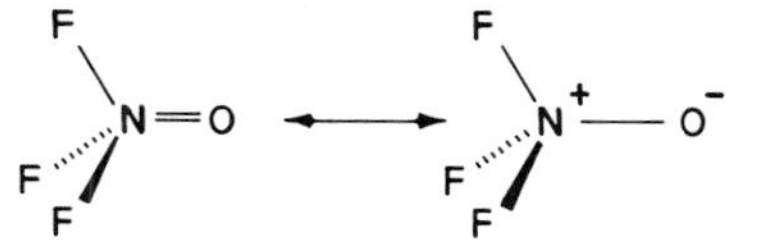

In contrast, the classical octet rule is frequently violated in compounds incorporating second-row elements. Lewis structures for "hypervalent" species like PF_5 and SF_6 imply valence manifolds of 10 and 12 electrons, respectively. The existence of such compounds, often with high thermochemical stability, is due to additional bonding involving low-lying d-type atomic orbitals on the second-row element [7].

As detailed in Section 4.3.4, the simplest basis sets for second-row atoms, STO-3G and 3-21G for example, do not incorporate functions of d-type symmetry. These models are unlikely to prove satisfactory in describing bonding in hypervalent compounds. The data in Table 6.14 and Figure 6.13 demonstrate the point. At the 3-21G level, bonds to the second-row element are consistently longer than the corresponding experimental quantities. The situation is extreme for multiple bonds, where errors in calculated lengths approaching one angström are noted, for example, ClO linkages in $FClO_2$ and $FClO_3$.

As is evident from the data in Table 6.14, both 3-21G$^{(*)}$ and 6-31G* basis sets are quite successful in accounting for the hypervalent structures [8]. The mean absolute deviations of calculated from experimental bond lengths between heavy atoms (Table 6.15) are only 0.015 and 0.014 Å, respectively. For comparison, the corre-

TABLE 6.14. Calculated and Experimental Equilibrium Geometries of Hypervalent Compounds

Molecule	Point Group	Geometrical Parameters	3-21G[a]	3-21G[(*)b]	6-31G*[c]	Expt.
PF_5	D_{3h}	$r(PF_{eq})$	1.580	1.538	1.535	1.534
		$r(PF_{ax})$	1.604	1.566	1.568	1.577
$(CH_3)_3PO^d$	C_{3v}	$r(PO)$	1.598	1.478	1.474	1.479
		$r(PC)$	1.849	1.805	1.819	1.813
		$r(CH_{tr})$	1.083	1.086	1.085	
		$r(CH_g)$	1.082	1.085	1.085	1.099
		$<(CPC)$	103.4	103.9	104.8	106.0
		$<(H_{tr}CP)$	110.7	112.1	112.1	
		$<(PCH_{g,g'})$	125.1	125.4	124.5	
		$<(H_gCH_{g'})$	108.4	107.6	107.9	
F_3PO	C_{3v}	$r(PO)$	1.506	1.427	1.425	1.436
		$r(PF)$	1.575	1.527	1.526	1.524
		$<(FPF)$	100.0	100.8	100.7	101.3
F_3PS	C_{3v}	$r(PS)$	2.042	1.855	1.874	1.866
		$r(PF)$	1.589	1.537	1.535	1.538
		$<(FPF)$	98.5	99.3	99.6	99.6
SO_2	C_{2v}	$r(SO)$	1.526	1.419	1.414	1.431
		$<(OSO)$	114.0	118.7	118.8	119.3
$(CH_3)_2SO^e$	C_s	$r(SO)$	1.678	1.490		1.485
		$r(SC)$	1.862	1.791		1.799
		$r(CH_a)$	1.079	1.082		1.054
		$r(CH_b)$	1.078	1.083		1.097
		$r(CH_c)$	1.078	1.082		1.093
		$<(CSO)$	105.0	107.8		106.7
		$<(CSC)$	98.1	96.6		96.6
		$<(SCH_a)$	108.0	108.0		108.3

Molecule	Sym.	Parameter				
		$<(SCH_b)$	109.2	110.2		108.2
		$<(SCH_c)$	106.7	110.3		109.6
		$<(H_aCH_b)$	112.0	108.6		113.6
		$<(H_aCH_c)$	109.8	109.2		110.6
		$\omega(H_aCSO)$	48.0	66.4		
SF_4	C_{2v}	$r(SF_{eq})$	1.616	1.550	1.544	1.545
		$r(SF_{ax})$	1.677	1.617	1.632	1.646
		$<(F_{eq}SF_{eq})$	107.4	101.7	102.7	101.6
		$<(F_{ax}SF_{ax})$	162.8	169.8	169.9	173.1
F_2SO	C_s	$r(SO)$	1.529	1.414	1.409	1.413
		$r(SF)$	1.633	1.569	1.571	1.585
		$<(FSO)$	105.8	106.8	106.9	106.8
		$<(FSF)$	91.8	92.0	92.4	92.2
NSF	C_s	$r(SN)$	1.568	1.440	1.431	1.448
		$r(SF)$	1.671	1.609	1.615	1.643
		$<(NSF)$	107.8	113.8	114.1	116.9
SO_3	D_{3h}	$r(SO)$	1.532	1.411	1.405	1.420
$(CH_3)_2SO_2{}^f$	C_{2v}	$r(SO)$	1.592	1.438	1.437	1.431
		$r(SC)$	1.830	1.756	1.774	1.777
		$r(CH_{tr})$	1.079	1.083	1.082	
		$r(CH_g)$	1.078	1.080	1.081	
		$<(OSO)$	119.1	119.5	120.1	121.0
		$<(CSC)$	102.7	102.8	104.3	103.3
		$<(H_{tr}CS)$	106.8	107.6	106.5	
		$<(SCH_{g,g'})$	124.4	127.9	126.8	
		$<(H_gCH_{g'})$	112.2	110.4	111.2	
SF_6	O_h	$r(SF)$	1.612	1.550		1.564
F_4SO	C_{2v}	$r(SO)$	1.549	1.413	1.404	1.403
		$r(SF_{eq})$	1.640	1.533	1.537	1.552
		$r(SF_{ax})$	1.597	1.578	1.582	1.575
		$<(F_{eq}SF_{eq})$	110.7	109.6	112.5	110.2
		$<(F_{ax}SF_{ax})$	169.2	171.6	164.5	178.6

TABLE 6.14. (Continued)

Molecule	Point Group	Geometrical Parameters	3-21G[a]	3-21G(*)[b]	6-31G*[c]	Expt.
ClF_3	C_{2v}	$r(ClF_{ax})$	1.757	1.676	1.672	1.698
		$r(ClF_{eq})$	1.673	1.601	1.579	1.598
		$r(F_{ax}ClF_{ax})$	167.2	171.0	172.6	175.0
ClF_5	C_{4v}	$r(ClF_{ax})$	1.696	1.599	1.590	1.65
		$r(ClF_{eq})$	1.715	1.624	1.630	1.65
		$<(F_{ax}ClF_{eq})$	83.4	84.4	84.9	86.5
$FClO_2$	C_s	$r(ClO)$	2.244	1.430	1.419	1.418
		$r(ClF)$	1.678	1.618	1.617	1.696
		$<(OClO)$	109.8	114.9	114.3	115.2
		$<(OClF)$	104.9	102.1	102.3	101.7
$FClO_3$	C_{3v}	$r(ClO)$	2.302	1.408	1.402	1.404
		$r(ClF)$	1.676	1.574	1.579	1.619
		$<(OClO)$	113.6	115.3	115.3	116.6

[a]GBPPH (1982); PFHDPB (1982).
[b]PFHDPB (1982).
[c]FPHBGDP (1982).
[d]Subscripts tr, g, and g' refer to relative orientations of the CH and PO bonds of 180° and approximately ±60°, respectively. $PCH_{g,g'}$ refers to the angle between the PC bond and the $H_gCH_{g'}$ plane.
[e]Subscripts a, b, and c refer to the projection below.

[f]Subscripts tr, g, and g' refer to relative orientations of the CH bond and the sulfur lone pair of 180° and approximately ±60°, respectively. $SCH_{g,g'}$ refers to the angle between the SC bond and the $H_gCH_{g'}$ plane.

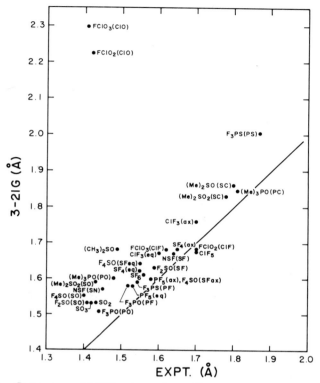

FIGURE 6.13. Comparison of 3-21G and experimental heavy-atom bond lengths in hypervalent molecules.

sponding mean absolute deviation for bond lengths calculated at the 3-21G level is 0.125 Å. A graphical comparison of 3-21G$^{(*)}$ and experimental bond lengths between heavy atoms is presented in Figure 6.14. A display of 6-31G* data would be similar.

The calculations with d functions accurately reproduce a number of subtle geometrical features, for example, the differences in axial and equatorial PF bond lengths

TABLE 6.15. Mean Absolute Errors in Bond Lengths Between Heavy Atoms for Hypervalent Molecules

Basis Set	Error
3-21G	0.125[a]
3-21G$^{(*)}$	0.015[a]
6-31G*	0.014[b]

[a]32 comparisons.
[b]30 comparisons.

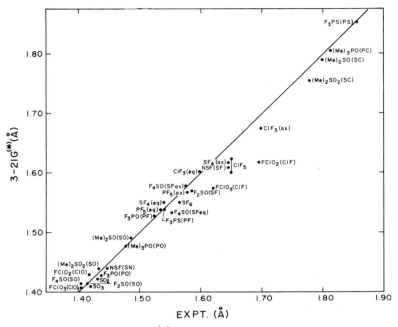

FIGURE 6.14. Comparison of 3-21G$^{(*)}$ and experimental heavy-atom bond lengths in hypervalent molecules.

in phosphorus pentafluoride, and in quasi-axial and quasi-equatorial SF bond distances in sulfur tetrafluoride and thionyl tetrafluoride. The theory also successfully accounts for the measured bond length differences in ClF_3. Predictions regarding the significant differences in "axial" and "equatorial" bond lengths in the pseudo-octahedral ClF_5 molecule remain untested due to the unavailability of refined experimental data. The 3-21G$^{(*)}$ calculations accurately mimic the observed progressive shortening in PO bond length as the methyl groups in trimethylphosphine oxide are replaced by fluorines. The highly electronegative fluorines evidently lower the energies of the d orbitals on phosphorus and make them more available for bonding. The SO bond length shows a similar shortening in going from dimethylsulfoxide to thionyl fluoride. Again the experimental and computational data are in agreement.

The 3-21G$^{(*)}$ equilibrium structures for this limited set of hypervalent molecules are very similar to those obtained with the 6-31G* basis set. The largest difference in bond length between the two levels of theory is 0.022 Å for the equatorial ClF distance in chlorine trifluoride. With the exception of thionyl tetrafluoride, bond angles calculated at the 3-21G$^{(*)}$ level are generally within $1°$ of 6-31G* values. Because the 3-21G$^{(*)}$ basis set is much less costly to apply than 6-31G*, it appears to be the method of choice for calculations of the equilibrium structures of hypervalent molecules.

6.2.6. Structures of Molecules Containing Third- and Fourth-Row Main-Group Elements

Calculated STO-3G equilibrium geometries for a selection of molecules incorporating third- and fourth-row main-group elements are compared with experimental structures in Table 6.16. A graphical comparison of calculated and experimental bond lengths between heavy elements is presented in Figure 6.15. With the exception of compounds incorporating bonds to the very electropositive group Ia elements (lithium, sodium, potassium, and rubidium), most of the theoretical structures presented are in reasonable accord with experimental geometries. Bond lengths are generally within 0.04 Å and are often better. A very large error is noted for the KC bond in KCN (too long by 0.25 Å); smaller but significant discrepancies are found for sodium iodide, lithium bromide, lithium iodide, rubidium hydroxide, and rubidium fluoride.

With the exception of potassium cyanide, bond lengths linking third- or fourth-row elements to carbon are underestimated, generally by 0.02–0.03 Å but sometimes by larger amounts, for example, 0.077 Å in GeH_3CN. The simple model does not properly mirror the observed ordering of CBr bond distances in the bromomethanes (see Section 6.2.4a for the similar failure of STO-3G to account properly for bond-length changes in the fluoro- and chloromethanes). In this case, the experimental data show little change in bond length as a result of increased substitution, while the STO-3G calculations depict a gradual increase in CBr distance. Some unexpected discrepancies arise. Thus, while the calculated InC distance in trimethylindium agrees well with experiment, a large error (0.201 Å) is found for the InC separation in cyclopentadienylindium.

Few comparisons for multiple bonds are available. The calculated SeC and TeC double-bond lengths in carbonyl selenide and thiocarbonyl telluride are 0.047 and 0.046 Å shorter, respectively, than experimental values.

The few comparisons of bond angles involving third- and fourth-row elements are satisfactory.

6.2.7. Structures of Transition-Metal Inorganics and Organometallics

STO-3G equilibrium structures for a number of inorganic compounds incorporating a first- or second-row transition metal are compared with experimental geometries in Table 6.17. A graphical comparison of calculated and experimental bond distances is provided in Figure 6.16. Highly polar molecules, for example, CuF and AgF, and molecules with formal multiple bonds, $VOCl_3$ for example, are poorly described. This parallels the behavior previously noted for the analogous main-group systems (see Section 6.2.6). Bonds of lower polarity, such as those to chlorine instead of fluorine, are better described, although errors in distances are seldom less than 0.03 Å and often approach 0.1 Å.

Table 6.18 compares STO-3G and experimental equilibrium structures for a number of simple organometallics with a single metal atom and with singlet electronic

TABLE 6.16. Calculated and Experimental Equilibrium Geometries for Molecules Containing Third- and Fourth-Row Main-Group Elements

Element	Molecule	Point Group	Geometrical Parameter	STO-3G[a]	Expt.
K	KCN	$C_{\infty v}$	$r(KC)$	2.541	2.294
			$r(CN)$	1.159	1.16
	KOH	$C_{\infty v}$	$r(KO)$	2.158	2.212
			$r(OH)$	0.988	0.912
K, I	KI	$C_{\infty v}$	$r(KI)$	3.014	3.048
Ge	GeH_3CH_3	C_{3v}	$r(GeC)$	1.910	1.945
			$r(GeH)$	1.434	1.529
			$r(CH)$	1.082	1.083
			$<(HGeH)$	109.3	109.3
			$<(HCH)$	107.4	108.4
	GeH_3CN	C_{3v}	$r(GeC)$	1.842	1.919
			$r(CN)$	1.154	1.155
			$r(GeH)$	1.435	1.529
			$<(HGeH)$	110.8	
	GeH_3F	C_{3v}	$r(GeF)$	1.705	1.732
			$r(GeH)$	1.436	1.532
			$<(HGeH)$	110.8	113.0
	GeH_3Cl	C_{3v}	$r(GeCl)$	2.158	2.150
			$r(GeH)$	1.435	1.537
			$<(HGeH)$	112.1	111.0
	Ge_2H_6	D_{3d}	$r(GeGe)$	2.360	2.403
			$r(GeH)$	1.429	1.541
			$<(HGeH)$	108.5	106.4
Ge, Br	GeH_3Br	C_{3v}	$r(GeBr)$	2.277	2.297
			$r(GeH)$	1.433	1.535
			$<(HGeH)$	110.9	106.9
As	AsF_3	C_{3v}	$r(AsF)$	1.712	1.710
			$<(FAsF)$	92.8	96.0
Se	CH_3SeH	C_s	$r(CSe)$	1.931	1.959
			$r(CH)$	1.085	1.088
			$r(SeH)$	1.441	1.473
			$<(HCH)$	109.0[b]	110.0[b]
			$<(CSeH)$	94.9	95.5
	Se=C=O	$C_{\infty v}$	$r(SeC)$	1.662	1.709
			$r(CO)$	1.168	1.157
Br	LiBr	$C_{\infty v}$	$r(LiBr)$	2.047	2.170
	CH_3Br	C_{3v}	$r(CBr)$	1.906	1.933
			$r(CH)$	1.087	1.086
			$<(HCH)$	108.9	111.2
	CH_2Br_2	C_{2v}	$r(CBr)$	1.917	1.927
			$r(CH)$	1.090	1.08

TABLE 6.16. (Continued)

Element	Molecule	Point Group	Geometrical Parameter	STO-3G[a]	Expt.
			<(BrCBr)	116.0	114.0
			<(HCH)	107.5	112.7
	$CHBr_3$	C_{3v}	r(CBr)	1.933	1.930
			r(CH)	1.094	1.068
			<(BrCBr)	112.6	110.8
	BrF	$C_{\infty v}$	r(BrF)	1.770	1.759
	SiH_3Br	C_{3v}	r(SiBr)	2.205	2.210
			r(SiH)	1.423	1.451
			<(HSiH)	109.9	111.0
	Br_2	$D_{\infty h}$	r(BrBr)	2.286	2.281
Br, Rb	RbBr	$C_{\infty v}$	r(RbBr)	2.920	2.945
Br, Sn	SnH_3Br	C_{3v}	r(SnBr)	2.461	2.469
			r(SnH)	1.630	1.767
			<(HSnH)	111.7	112.8
Br, Sb	$SbBr_3$	C_{3v}	r(SbBr)	2.490	2.490
			<(BrSbBr)	98.2	98.2
Br, Te	$TeBr_2$	C_{2v}	r(TeBr)	2.512	2.51
			<(BrTeBr)	98.0	98.0
Br, I	IBr	$C_{\infty v}$	r(IBr)	2.497	2.469
Rb	RbOH	$C_{\infty v}$	r(RbO)	2.308	2.301
			r(OH)	1.000	0.957
	RbF	$C_{\infty v}$	r(RbF)	2.365	2.270
	RbCl	$C_{\infty v}$	r(RbCl)	2.810	2.787
Rb, I	RbI	$C_{\infty v}$	r(RbI)	3.170	3.177
In	$In(CH_3)_3$	C_{3h}	r(InC)	2.092	2.093
			r(CH)	1.083	1.140
			<(HCH)	105.8	106.1
	$In(C_5H_5)$	C_{5v}	r(InC)	2.420	2.621
			r(CC)	1.414	1.427
			r(CH)	1.080	1.10
			<$(\alpha)^c$	179.4	175.5
Sn	$Sn(CH_3)_4$	T_d	r(SnC)	2.110	2.144
			r(CH)	1.082	1.118
			<(HCH)	107.0	106.8
	$SnCl_4$	T_d	r(SnCl)	2.293	2.281
Sb	$SbCl_3$	C_{3v}	r(SbCl)	2.352	2.333
			<(ClSbCl)	94.9	97.2
Te	Te=C=S	$C_{\infty v}$	r(TeC)	1.858	1.904
			r(CS)	1.517	1.557

TABLE 6.16. (Continued)

Element	Molecule	Point Group	Geometrical Parameter	STO-3G[a]	Expt.
I	LiI	$C_{\infty v}$	$r(LiI)$	2.281	2.392
	$CH_3 I$	C_{3v}	$r(CI)$	2.110	2.132
			$r(CH)$	1.084	1.084
			$<(HCH)$	108.2	111.2
	ICN	$C_{\infty v}$	$r(IC)$	1.991	1.995
			$r(CN)$	1.157	1.159
	IF	$C_{\infty v}$	$r(IF)$	1.962	1.910
	NaI	$C_{\infty v}$	$r(NaI)$	2.561	2.711
	$SiH_3 I$	C_{3v}	$r(SiI)$	2.438	2.437
			$r(SiH)$	1.432	1.457
			$<(HSiH)$	109.9	110.5
	ICl	$C_{\infty v}$	$r(ICl)$	2.367	2.321
	I_2	$D_{\infty h}$	$r(II)$	2.703	2.666

[a]PLHS (1980); PBHHDS (1981); W. J. Hehre, unpublished calculations.
[b]Average value.
[c]Angle between the CH bond and the plane of the ring. Values less than 180° indicate bending of CH toward the metal.

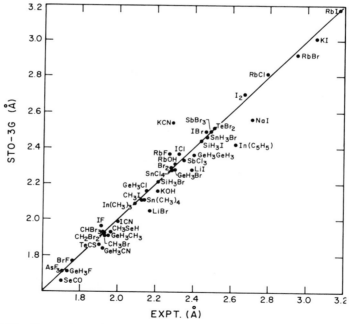

FIGURE 6.15. Comparison of STO-3G and experimental heavy-atom bond lengths in molecules incorporating third- and fourth-row main-group elements.

TABLE 6.17. Calculated and Experimental Equilibrium Geometries for Transition-Metal Inorganic Compounds

Molecule	Point Group	Geometrical Parameter	STO-3G[a]	Expt.
ScF_3	D_{3h}	$r(ScF)$	1.852	1.91
$TiCl_4$	T_d	$r(TiCl)$	2.167	2.170
VF_5	D_{3h}	$r(VF_{ax})$	1.648	1.734[b]
		$r(VF_{eq})$	1.616	1.708
$VOCl_3$	C_{3v}	$r(VO)$	1.465	1.570
		$r(VCl)$	2.108	2.142
		$<(ClVCl)$	110.1	111.3
CuF	$C_{\infty v}$	$r(CuF)$	1.592	1.745
$CuCl$	$C_{\infty v}$	$r(CuCl)$	2.028	2.051
$ZrCl_4$	T_d	$r(ZrCl)$	2.316	2.32
NbF_5	D_{3h}	$r(NbF_{ax})$	1.804	1.88[c]
		$r(NbF_{eq})$	1.788	1.88[c]
AgF	$C_{\infty v}$	$r(AgF)$	1.633	1.983
$AgCl$	$C_{\infty v}$	$r(AgCl)$	2.083	2.281

[a]PH (1983).
[b]K. Hagen, M. M. Gilbert, L. Hedberg and K. Hedberg, *Inorg. Chem.*, **21**, 2690 (1982).
[c]Axial and equatorial bonds assumed to be of equal length.

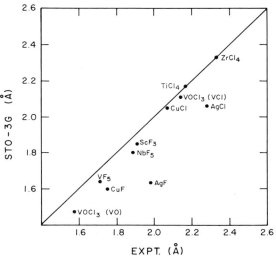

FIGURE 6.16. Comparison of STO-3G and experimental heavy-atom bond lengths for transition-metal inorganics.

TABLE 6.18. Calculated and Experimental Equilibrium Geometries for Transition-Metal Organometallics

Molecule	Point Group	Geometrical Parameter	STO-3G[a]	Expt.
$Ti(CH_3)_4$	T_d	$r(TiC)$	2.096	2.14[b]
		$r(CH)$	1.085	
		$<(HCH)$	107.2	
$Cr(C_6H_6)_2$	D_{6h}	$r(CrC)$	2.095	2.150
		$r(CC)$	1.426	1.423
		$r(CH)$	1.092	1.090
		$<(\alpha)^c$	178.6	175.3
$Fe(C_5H_5)_2$	D_{5h}	$r(FeC)$	2.144	2.064
		$r(CC)$	1.424	1.440
		$r(CH)$	1.076	1.104
		$<(\alpha)^c$	177.2	176.3
$(C_2H_4)Fe(CO)_4$	C_{2v}	$r(FeC_{ax})$	2.078	1.796
		$r(FeC_{eq})$	1.770	1.836
		$r(FeC)$	1.940	2.117
		$r(C_{ax}O_{ax})$	1.144	1.146
		$r(C_{eq}O_{eq})$	1.160	1.146
		$r(CC)$	1.497	1.46
		$r(CH)$	1.078	
		$<(C_{ax}FeC_{ax})$	171.8	180.0[d]
		$<(C_{eq}FeC_{eq})$	127.4	105.2
		$<(\beta)^e$	137.6	180.0[d]
		$<(HCH)$	109.1	
$(C_5H_5)NiNO$	C_{5v}	$r(NiC)$	2.081	2.11
		$r(NiN)$	1.424	1.626
		$r(CC)$	1.420	1.43
		$r(NO)$	1.268	1.165
		$r(CH)$	1.078	1.09[d]
		$<(\alpha)^c$	179.5	
$Zn(CH_3)_2$	D_{3d}	$r(ZnC)$	1.682	1.929
		$r(CH)$	1.083	1.090[f]
		$<(HCH)$	105.6	107.7
$Zr(CH_3)_4$	T_d	$r(ZrC)$	2.257	2.27[g]
		$r(CH)$	1.086	
		$<(HCH)$	107.3	
$Cd(CH_3)_2$	D_{3d}	$r(CdC)$	2.003	2.112
		$r(CH)$	1.084	1.090[f]
		$<(HCH)$	106.2	108.4

[a]PH (1983); W. J. Hehre and K. D. Dobbs, unpublished calculations.

[b]Average of four TiC bond distances in tetrabenzyltitanium. I. W. Bassi, G. Allegra, R. Scordamaglia, and G. Chioccola, *J. Am. Chem. Soc.*, **93**, 3787 (1971).

[c]Angle between the CH bond and the plane of the ring. Values less than 180° indicate bending of CH toward the metal.

[d]Assumed.

[e]Angle between HCH plane and ethylene CC bond.

[f]Assumed; error in HCH angle depends on validity of this assumption.

[g]Average of four ZrC bond distances in tetrabenzylzirconium. G. R. Davies, J. A. J. Jarvis, B. T. Kilbourn, and A. J. P. Pioli, *J. Chem. Soc., Chem. Commun.*, 677 (1971).

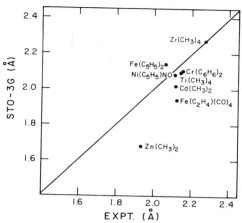

FIGURE 6.17. Comparison of STO-3G and experimental metal-carbon bond lengths for transition-metal organometallics.

ground states. Figure 6.17 compares calculated and experimental metal–carbon bond lengths. The theory describes certain structural aspects of these compounds reasonably well but there are significant discrepancies. The calculated metal-to-carbon distances in the three arene complexes (dibenzene chromium, ferrocene and cyclopentadienyl nickel nitrosyl) are within 0.08 Å of their experimental values. In addition, the geometries of the benzene and cyclopentadienyl moieties are accurately portrayed. The measured metal–carbon bond lengths in tetramethyltitanium, tetramethylzirconium, and dimethylcadmium are also reproduced reasonably well by the STO-3G calculations; on the other hand, the calculated Zn—C bond length in dimethylzinc is very poorly described. The metal–nitrogen distance in cyclopentadienyl nickel nitrosyl is underestimated by 0.2 Å.

These large discrepancies may be due to deficiencies in the basis set or to limitations inherent in the single-determinant approximation, and much further investigative work is needed. The systems are large, however, and even calculations using simple minimal basis sets are close to the present practical limit for complete structure optimization. Application of larger basis sets or procedures beyond the Hartree-Fock level has thus far been limited and is a direction for future research.

6.2.8. Structures of Short-Lived Reactive Species

Many molecules are short-lived, and complete characterization of their structures by experimental means is difficult. Theoretical calculations can yield information that is not readily available from measurement. In this section, we assess the ability of the theoretical models to reproduce the known geometrical structures of representative members of several classes of reactive intermediates, in particular, free radicals, carbenes and related divalent compounds, anions, and cations.

a. Open-Shell Systems: Free Radicals. Table 6.19 compares calculated (unrestricted Hartree–Fock) and experimental equilibrium geometries for a selection of one- and two-heavy-atom radicals, both doublets and triplets. The STO-3G, 3-21G, 6-31G*, 6-31G**, and 6-311G** basis sets have been examined.

In general, the level of agreement between theoretical and experimental structures is comparable to that for closed-shell species. Note, however, that some experimental parameters, for example, the bond angles in BH_2 and CH_2, are not known to sufficient accuracy to permit critical evaluation of the theoretical results.

Except for BeH and BH_2, STO-3G AH bond distances are longer than experimental values, typically by a few hundredths of an angström. The mean absolute deviation of calculated from experimental AH bond distances for the eight compounds considered is 0.027 Å. Except for the BO radical, STO-3G AB bond distances again are too long; the mean absolute deviation for the five molecules is 0.019 Å. STO-3G bond angles for BH_2, CH_2, and NH_2 are all smaller than experimental values.

The 3-21G basis set generally offers an improved description of AH bond distances in these species; the mean absolute deviation is reduced to 0.008 Å. Calculated bond lengths are consistently too long except for the three hydrocarbon radicals, for which the lengths are slightly (<0.01 Å) shorter than the experimental values. Calculated AB bond lengths are all too long, sometimes considerably so. For example, compare the 3-21G distance in NF, 1.390 Å, with the experimental value, 1.317 Å. Overall, the mean absolute deviation for the compounds is 0.040 Å, even poorer than the corresponding STO-3G result. Bond angles are both greater (NH_2) and smaller (BH_2 and CH_2) than experimental values.

Geometries calculated using the 6-31G* basis set are generally superior to the lower-level theoretical results. This is especially true for AB bond distances; the mean deviation from experiment is only 0.018 Å. Except for triplet BN, the calculated AB lengths are all shorter than the experimental distances. The theoretical AH lengths are also short, with the exception of those in BeH and BH_2. This behavior is consistent with previous observations that equilibrium geometries calculated using the 6-31G* basis set are close to Hartree–Fock limiting values, and that the effect of electron correlation is generally to increase bond length. 6-31G* equilibrium bond angles for BH_2 and CH_2 significantly underestimate the measured values while that for NH_2 is slightly (1.1°) too large.

Equilibrium structures obtained using the 6-31G** and 6-311G** basis sets are uniformly close to those resulting from 6-31G* calculations. The largest difference between 6-31G* and 6-31G** bond lengths is 0.004 Å; changes up to 0.011 Å in lengths and 1.3° in angles occur for calculations using the 6-311G** basis set.

Theoretical equilibrium structures for this same set of systems have also been obtained using Møller–Plesset perturbation theory complete through fourth-order, and configuration interaction including all double substitutions. The 6-31G* basis set has been used. The results are displayed in Table 6.20.

The effect of electron correlation on bond lengths is similar to that noted previously for closed-shell molecules. Thus, whereas UHF/6-31G* AH and AB bond lengths are generally shorter than experimental values, incorporation of electron correlation leads to bonds that are slightly too long. The differences among the

TABLE 6.19. Unrestricted Hartree–Fock (UHF) and Experimental Equilibrium Structures for One- and Two-Heavy-Atom Radicals

Molecule	Point Group	Electronic State	Geometrical Parameter	STO-3G[a]	3-21G[b]	6-31G*[c]	6-31G**[d]	6-311G**[e]	Expt.
BeH	$C_{\infty v}$	$^2\Sigma^+$	r(BeH)	1.301	1.356	1.348	1.347	1.343	1.343
BH₂	C_{2v}	$^2B_1$	r(BH)	1.161	1.185	1.185	1.185	1.185	1.181
			<(HBH)	123.5	127.7	126.5	127.0	127.6	131
CH	$C_{\infty v}$	$^2\Pi$	r(CH)	1.143	1.119	1.108	1.111	1.109	1.120
CH₂	C_{2v}	$^3B_1$	r(CH)	1.082	1.071	1.071	1.072	1.072	1.078
			<(HCH)	125.5	131.0	130.4	131.1	131.7	136
CH₃	D_{3h}	$^2A_2''$	r(CH)	1.080	1.072	1.073	1.073	1.074	1.079
NH	$C_{\infty v}$	$^3\Sigma^-$	r(NH)	1.082	1.046	1.024	1.024	1.023	1.036
NH₂	C_{2v}	$^2B_1$	r(NH)	1.058	1.026	1.013	1.012	1.012	1.024
			<(HNH)	100.2	106.0	104.4	104.3	104.1	103.3
OH	$C_{\infty v}$	$^2\Pi$	r(OH)	1.014	0.986	0.959	0.955	0.951	0.970
BO	$C_{\infty v}$	$^2\Sigma^+$	r(BO)	1.190	1.218	1.187	1.187	1.181	1.205
BN	$C_{\infty v}$	$^3\Pi$	r(BN)	1.305	1.319	1.293	1.293	1.288	1.281
CF	$C_{\infty v}$	$^2\Pi$	r(CF)	1.293	1.312	1.267	1.267	1.257	1.272
NF	$C_{\infty v}$	$^3\Sigma^-$	r(NF)	1.342	1.390	1.302	1.302	1.291	1.317
O₂	$D_{\infty h}$	$^3\Sigma_g^-$	r(OO)	1.217[f]	1.240[f]	1.168	1.168	1.157	1.208

[a]LHCP (1971); LCHLP (1974); FPR (1983).
[b]FPR (1983).
[c]HP (1974), DLPHBP (1979); FPR (1983).
[d]D. J. DeFrees, W. J. Hehre and J. A. Pople, unpublished calculations.
[e]KBSP (1980); J. A. Pople, unpublished calculations.
[f]Symmetry-constrained wavefunction.

195

TABLE 6.20. Effect of Electron Correlation on Equilibrium Structures of One- and Two-Heavy-Atom Free Radicals[a]

Molecule	Point Group	Electronic State	Geometrical Parameter	UHF	UMP2	UMP3	UMP4SDQ	UMP4	CID	Expt.
							/6-31G*			
BeH	$C_{\infty v}$	$^2\Sigma^+$	r(BeH)	1.348	1.348	1.351	1.353	1.353	1.355	1.343
BH$_2$	C_{2v}	$^2B_1$	r(BH)	1.185	1.188	1.191	1.193	1.193	1.193	1.181
			<(HBH)	126.5	127.6	127.9	128.0	128.0	127.9	131
CH	$C_{\infty v}$	$^2\Pi$	r(CH)	1.108	1.120	1.126	1.130	1.130	1.128	1.120
CH$_2$	C_{2v}	$^3B_1$	r(CH)	1.071	1.077	1.080	1.081	1.082	1.081	1.078
			<(HCH)	130.4	131.6	131.8	132.0	132.0	132.0	136
CH$_3$	D_{3h}	$^2A_2''$	r(CH)	1.073	1.079	1.081	1.082	1.083	1.081	1.079
NH	$C_{\infty v}$	$^3\Sigma^-$	r(NH)	1.024	1.040	1.044	1.046	1.047	1.045	1.036
NH$_2$	C_{2v}	$^2B_1$	r(NH)	1.013	1.029	1.031	1.033	1.034	1.030	1.024
			<(HNH)	104.4	103.3	103.2	103.0	102.9	103.1	103.3
OH	$C_{\infty v}$	$^2\Pi$	r(OH)	0.959	0.979	0.981	0.982	0.983	0.980	0.970
BO	$C_{\infty v}$	$^2\Sigma^+$	r(BO)	1.187	1.217	1.204			1.203	1.205
BN	$C_{\infty v}$	$^3\Pi$	r(BN)	1.293	1.317	1.312			1.287	1.281
CF	$C_{\infty v}$	$^2\Pi$	r(CF)	1.267	1.291	1.285			1.284	1.272
NF	$C_{\infty v}$	$^3\Sigma^-$	r(NF)	1.302	1.330	1.327			1.324	1.317
O$_2$	$D_{\infty h}$	$^3\Sigma_g^-$	r(OO)	1.168	1.242	1.211			1.206	1.208

[a]Theoretical data from: FPR (1983); J. A. Pople, unpublished calculations.

TABLE 6.21. Comparison of Calculated and Experimental Equilibrium Geometries for One-Heavy-Atom Radicals[a]

Radical	Point Group	Electronic State	Geometrical Parameter	UMP4/6-311G**	Expt.
BeH	$C_{\infty v}$	$^2\Sigma^+$	r(BeH)	1.343[b]	1.343
BH$_2$	C_{2v}	$^2B_1$	r(BH) <(HBH)	1.192 128.6	1.181 131
CH	$C_{\infty v}$	$^2\Pi$	r(CH)	1.128	1.120
CH$_2$	C_{2v}	$^3B_1$	r(CH) <(HCH)	1.082 132.4	1.078 136
CH$_3$	D_{3h}	$^2A_2''$	r(CH)	1.083	1.079
NH	$C_{\infty v}$	$^3\Sigma^-$	r(NH)	1.042	1.036
NH$_2$	C_{2v}	$^2B_1$	r(NH) <(HNH)	1.030 101.6	1.024 103.3
OH	$C_{\infty v}$	$^2\Pi$	r(OH)	0.970	0.970

[a]Theoretical data from: J. A. Pople, unpublished calculations.
[b]UMP4SDQ/6-311G** calculation.

various correlation levels (UMP2, UMP3, UMP4SDQ, UMP4, and CID) are usually quite small; O_2 is an exception with a large difference (0.036 Å) between UMP2 and UMP3.

Even at the highest correlation levels, calculated bond angles for BH$_2$ and CH$_2$ are significantly smaller than experimental values. For all correlation schemes, the theoretical angle in the amino radical is within a few tenths of a degree of the experimental value.

Table 6.21 compares experimental equilibrium geometries for one-heavy-atom radicals with those obtained using Møller–Plesset perturbation theory complete through fourth order with the 6-311G** basis set. All AH bond distances are close to, although generally somewhat larger than, experimental equilibrium values. The improved level of agreement over the MP4/6-31G* calculations parallels similar improvements for closed-shell molecules (Table 6.3).

Calculated bond angles at the MP4/6-311G** level for BH$_2$, CH$_2$, and NH$_2$ are still smaller than experimental angles, by 2.4°, 3.6°, and 1.7°, respectively. This is again consistent with the results in Table 6.3 for closed-shell systems.

The open-shell calculations discussed up to this point have involved the unrestricted Hartree–Fock (UHF) model. As discussed in Chapter 2, this is computationally more convenient than the restricted Hartree–Fock (RHF) alternative, both for calculating the gradient of the energy with respect to nuclear coordinates and also as a starting point for calculating the correlation energy using the Møller–Plesset perturbation expansion. However, the resultant UHF wavefunction is not a pure spin state, and the structural consequences of spin contamination must be explored.

Table 6.22 compares UHF and RHF structural predictions for all systems with

TABLE 6.22. Comparison of Unrestricted (UHF) and Restricted (RHF) Hartree–Fock and Møller–Plesset (UMP2, UMP3) Structures for Open-Shell Systems[a]

System	State	Parameter	UHF			RHF			UMP2/6-31G*	UMP3/6-31G*	Expt.
			STO-3G	3-21G	6-31G*	STO-3G	3-21G	6-31G*			
BeH	$^2\Sigma^+$	r(BeH)	1.301	1.356	1.348	1.300	1.354	1.346	1.348	1.351	1.343
BH$^+$	$^2\Sigma^+$	r(BH)	1.207	1.184	1.183	1.203	1.180	1.179	1.194	1.200	1.215[b]
CH	$^2\Pi$	r(CH)	1.143	1.119	1.108	1.143	1.119	1.107	1.120	1.126	1.120
NH	$^3\Sigma^-$	r(NH)	1.082	1.046	1.024	1.079	1.044	1.022	1.040	1.044	1.036
NH$^+$	$^2\Pi$	r(NH)	1.140	1.062	1.045	1.139	1.062	1.045	1.064	1.069	1.070
OH	$^2\Pi$	r(OH)	1.014	0.986	0.959	1.013	0.985	0.958	0.979	0.981	0.970
OH$^+$	$^3\Sigma^-$	r(OH)	1.084	1.049	1.013	1.080	1.047	1.011	1.035	1.038	1.028
FH$^+$	$^2\Pi$	r(FH)	1.034	1.043	1.006	1.033	1.043	1.005	1.030	1.029	1.001
BH$_2$	$^2A_1$	r(BH)	1.161	1.185	1.185	1.162	1.185	1.186	1.188	1.191	1.181
		<(HBH)	123.5	127.7	126.5	123.3	127.4	126.1	127.6	127.9	131
CH$_2$	$^3B_1$	r(CH)	1.082	1.071	1.071	1.083	1.071	1.072	1.077	1.080	1.078
		<(HCH)	125.5	131.0	130.4	123.6	128.7	128.4	131.6	131.8	136
NH$_2$	$^2B_1$	r(NH)	1.058	1.026	1.013	1.058	1.025	1.012	1.029	1.031	1.024
		<(HNH)	100.2	106.0	104.4	100.1	105.9	104.4	103.3	103.2	103.3
OH$_2^+$	$^2B_1$	r(OH)	1.033	1.011	0.988	1.032	1.010	0.988	1.011	1.011	0.999
		<(HOH)	109.8	117.8	111.7	109.8	117.7	111.7	109.9	109.7	110.5
CH$_3$	$^2A_2''$	r(CH)	1.080	1.072	1.073	1.080	1.070	1.072	1.079	1.081	1.079
		<(HCH)	118.3	120.0	120.0	116.9	120.0	119.4	120.0	120.0	120
HNF	$^2A''$	r(HN)	1.075	1.028	1.013	1.074	1.027	1.012	1.034	1.035	1.06[c]
		r(NF)	1.364	1.422	1.345	1.362	1.420	1.345	1.379	1.375	1.37
		<(HNF)	101.8	99.9	101.0	101.7	99.9	101.0	99.9	100.1	105

		1	2	3	4	5	6	7	8	9
HO$_2$	$^2A''$									
	r(HO)	1.004	0.973	0.953	1.004	0.973	0.953	0.982	0.978	0.977
	r(OO)	1.357	1.434	1.309	1.354	1.420	1.312	1.327	1.328	1.335
	<(HOO)	104.0	103.3	105.6	104.1	103.8	105.5	104.5	104.7	104.1
HCO	$^2A'$									
	r(HC)	1.101	1.095	1.106	1.112	1.094	1.103	1.124	1.122	1.125
	r(CO)	1.253	1.180	1.159	1.192	1.178	1.158	1.192	1.179	1.175
	<(HCO)	126.3	129.0	126.2	126.2	129.4	127.0	123.3	124.2	124.9
BeF	$^2\Sigma^+$ r(BeF)	1.297	1.376	1.364	1.298	1.376	1.365	1.378	1.375	1.361
BN	$^3\Pi$ r(BN)	1.305	1.319	1.293	1.296	1.314	1.291	1.317	1.312	1.281
BO	$^2\Sigma^+$ r(BO)	1.190	1.218	1.187	1.180	1.213	1.185	1.217	1.204	1.205[b]
C$_2^+$	$^2\Pi_u$ r(CC)	1.355[d]	1.303[d]	1.288[d]	1.358	1.347	1.317	1.269[d]	1.269[d]	1.301
CF	$^2\Pi$ r(CF)	1.293	1.312	1.267	1.293	1.313	1.267	1.291	1.285	1.272
CN	$^2\Sigma^+$ r(CN)	1.235	1.180	1.162	1.165	1.145	1.137	1.136	1.139	1.172
CO$^+$	$^2\Sigma^+$ r(CO)	1.206	1.123	1.098	1.151	1.102	1.088	1.103	1.101	1.115
N$_2^+$	$^2\Sigma_g^+$ r(NN)	1.513[d]	1.164[d]	1.117[d]	1.178	1.103	1.091	1.080[d]	1.085[d]	1.116
N$_2^+$	$^2\Pi_u$ r(NN)	1.251	1.153	1.138	1.251	1.153	1.139	1.204	1.182	1.175
NF	$^3\Sigma^-$ r(NF)	1.342	1.390	1.302	1.337	1.388	1.305	1.330	1.327	1.317
NO	$^2\Pi$ r(NO)	1.306	1.202	1.127	1.185	1.151	1.125	1.142	1.140	1.151
O$_2$	$^3\Sigma_g^-$ r(OO)	1.272[d]	1.291[d]	1.168	1.218	1.237	1.163	1.242	1.211	1.208
O$_2^+$	$^2\Pi_g$ r(OO)	1.211[d]	1.120	1.076	1.170	1.118	1.074	1.188	1.118	1.116
F$_2^+$	$^2\Pi_g$ r(FF)	1.407[d]	1.731[d]	e	1.266	1.327	1.231	1.184	1.185	1.322

[a]Theoretical data from: FPR (1983).
[b]Value for lowest observed vibrational state.
[c]Assumed value.
[d]Symmetry broken, see text.
[e]Unbound at this level.

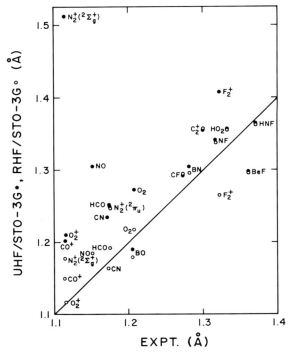

FIGURE 6.18. Comparison of UHF/STO-3G (●), RHF/STO-3G (○) and experimental heavy-atom bond lengths in radicals.

doublet or triplet ground states and no more than two first-row atoms for which good experimental structural data are available in the general references 3 and 4 listed at the beginning of Section 6.2. A graphical comparison of both UHF/STO-3G and RHF/STO-3G heavy-atom bond lengths in this series of compounds with experimental values is presented in Figure 6.18. Analogous comparisons using 3-21G and 6-31G* data are given in Figures 6.19 and 6.20, respectively. Finally, a statistical analysis of the errors for A—H lengths, A—B lengths and **HAB** or **HAH** bond angles is presented in Table 6.23.

Examination of the mean absolute errors shows that improvement of the basis set generally leads to improved results. The greatest improvement occurs in moving from STO-3G to 3-21G. With the UHF method, there is a substantial further improvement in A—B lengths in going to 6-31G*. Inclusion of electron correlation at the MP2 level produces a further general improvement, but only in the case of A—B lengths is there still further improvement at the MP3 level. F_2^+ is an exceptional case; it becomes more and more weakly bound at the UHF level as the basis set size is increased.

It is interesting that, whereas with the smaller basis sets UHF performs significantly worse than RHF, this is no longer the case with the 6-31G* representation, for which UHF/6-31G* and RHF/6-31G* generally give very similar results. Thus,

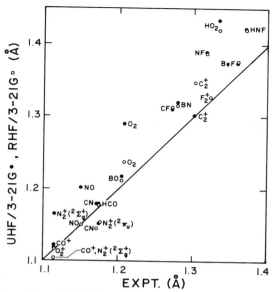

FIGURE 6.19. Comparison of UHF/3-21G (●), RHF/3-21G (○) and experimental heavy-atom bond lengths in radicals.

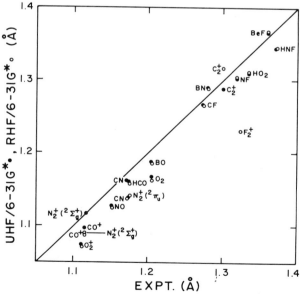

FIGURE 6.20. Comparison of UHF/6-31G* (●), RHF/6-31G* (○) and experimental heavy-atom bond lengths in radicals.

201

TABLE 6.23. Deviations of Theoretical from Experimental Geometrical Parameters[a]

		UHF/ STO-3G	UHF/ 3-21G	UHF/ 6-31G*	RHF/ STO-3G	RHF/ 3-21G	RHF/ 6-31G*	UMP2/ 6-31G*	UMP3/ 6-31G*
Mean absolute deviation	r(A—H)	0.031	0.014	0.013	0.030	0.014	0.014	0.007	0.008
	r(A—B)	0.078	0.035	0.019	0.033	0.029	0.023	0.024	0.014
	$<$(HAH), $<$(HAB)	3.5	3.5	2.4	4.0	3.8	2.9	1.9	1.8
Mean deviation	r(A—H)	0.019	0.002	-0.011	0.018	0.001	-0.012	0.004	0.006
	r(A—B)	0.068	0.032	-0.017	0.020	0.019	-0.019	0.008	-0.003
	$<$(HAH), $<$(HAB)	-3.2	0.0	-1.1	-3.6	-0.3	-1.4	-1.8	-1.7
Largest deviation	r(A—H)	0.070	0.042	0.032	0.069	0.042	0.036	0.029	0.028
	r(A—B)	0.397	0.099	0.040	0.076	0.085	0.045	0.072	0.033
	$<$(HAH), $<$(HAB)	10.5	7.3	5.6	12.4	7.3	7.6	5.1	4.9

[a] F_2^+, which is unbound at the UHF limit, is not included in this statistical analysis.

it would seem that, in contrast to conclusions based on minimal basis set results, structural predictions of the UHF method are generally quite satisfactory provided that an adequate basis set, for example, 6-31G*, is used. (This result is associated with the extent of spin contamination as discussed below.) The errors in UHF/6-31G* structures for the set of open-shell systems in Table 6.22 are comparable to the errors for RHF/6-31G* calculations on closed-shell systems. Likewise UMP2 and UMP3 errors are comparable to those for closed-shell systems.

Information regarding systematic errors in structural predictions is provided by comparing the mean errors and mean absolute errors displayed in Table 6.23. STO-3G calculations give bond lengths that are systematically too long, while 6-31G* calculations give lengths that are too short. While errors in RHF/3-21G calculations appear nonsystematic, the corresponding UHF/3-21G bond lengths are consistently longer than the experimental values. UHF A—B bond lengths are systematically longer than those obtained from RHF calculations, although differences decrease as the basis set is improved. The correlated calculations of bond lengths give small errors that are essentially random. Errors in calculated bond angles are similar for the UHF and RHF models; values are systematically too small with both STO-3G and 6-31G* basis sets.

Many of the poorer UHF results are associated with high spin contamination, the degree of which generally decreases quite markedly with increasing size of basis set (Table 6.24). This may account in part for the dramatically improved geometric

TABLE 6.24. Expectation Values of $\langle \hat{S}^2 \rangle$ from UHF Calculations[a]

System	State		STO-3G	3-21G	6-31G*
HO_2	$^2A''$		0.76	0.79	0.76
HCO	$^2A'$		1.19	0.78	0.76
BN	$^3\Pi$		2.13	2.08	2.04
BO	$^2\Sigma^+$		0.92	0.85	0.80
C_2^+	$^2\Pi_u$	sym.	0.75	0.75	0.75
		asym.	2.23	2.15	2.11
CN	$^2\Sigma^+$		1.56	1.25	1.09
CO^+	$^2\Sigma^+$		1.46	1.07	0.93
N_2^+	$^2\Sigma_g^+$	sym.	0.77	0.76	0.77
		asym.	2.53	1.51	1.24
NO	$^2\Pi$		1.45	1.07	0.78
O_2	$^3\Sigma_g^-$	sym.	2.00	2.02	2.03
		asym.	2.00	2.01	[b]
O_2^+	$^2\Pi_g$	sym.	0.75	0.76	0.76
		asym.	0.95	[b]	[b]
F_2^+	$^2\Pi_g$	sym.	0.75	0.76	0.76
		asym.	0.91	1.51	[c]

[a]For all systems in Table 6.22 not included in this table, values of $\langle \hat{S}^2 \rangle$ are in the range of 0.75–0.76 (doublets) or 2.00–2.02 (triplets). Data from: FPR (1983).
[b]Symmetric solution is stable at these levels.
[c]F_2^+ is unbound at this level.

predictions with the larger basis sets. When the degree of contamination is small, UHF and RHF results are very similar; when the degree of contamination is large, RHF results are to be preferred.

For some of the homonuclear diatomics considered in Table 6.22, the lowest energy UHF solution corresponds to a set of molecular orbitals that are asymmetric with respect to the inversion center. This means that the point group of the *molecular wavefunction* is reduced to $C_{\infty v}$ from $D_{\infty h}$. It is, however, possible to enforce symmetry. Geometries corresponding to such symmetry-constrained solutions, presented in Table 6.25, show a marked improvement over unconstrained results, and there is a concomitant reduction in UHF spin contamination to near zero (Table 6.24). In one sense these results are unfortunate. While enforcement of wavefunction symmetry can readily be achieved for symmetric systems, for example, homonuclear diatomics, it cannot be achieved for asymmetric systems, such as heteronuclear diatomics.

b.   Carbenes and Related Compounds.   Divalent compounds of carbon (carbenes) and of heavier group IVa elements form one class of reactive intermediates for which considerable structural information is available. Most of the data are for singlet carbenes, although equilibrium geometries for ground-state triplet methylene (discussed in the previous section) as well as for a number of divalent compounds of silicon and germanium substituted by halogens have now been determined. Table 6.26 compares theoretical and experimental structures for representative singlet systems.

Agreement between theoretical and experimental bond lengths for the singlet carbenes is quite reasonable at all levels of calculation. Both theoretical and experimental CH bonds are consistently longer than those found in "normal" tetravalent carbon compounds, for example, CH_4 (Table 6.1). On the other hand, both theory and experiment show CF linkages in fluorocarbene and difluorocarbene to be far shorter than those in the corresponding halomethanes (Table 6.10). Similarly, both calculated and experimental C—Cl linkages in chlorocarbene and in dichlorocarbene are shorter than those found in corresponding chloromethanes. Partial double bonding involving the out-of-plane carbon p orbital is responsible for the C—X bond shortening in these molecules. The theory also provides an adequate account of the bond angles for most of the divalent carbon compounds.

The equilibrium geometries of singlet silylenes are generally reproduced moderately well by the computational models. At the STO-3G level, however, calculated SiH bond lengths are consistently too short (the distance in silane also is 0.06 Å less than the experimental value), while the bonds to the halogens are generally longer than experiment. The calculated bond angle in SiH_2 is in good agreement with the experimental value; those in SiHCl, SiHBr, and SiHI are approximately $10°$ too small. 3-21G calculations lead to improvements in SiH bond distances, but give an SiCl bond distance in chlorosilylene nearly 0.2 Å greater than the experimental length. This result is improved by calculations at the $3\text{-}21G^{(*)}$ and $6\text{-}31G^*$ levels. The STO-3G equilibrium structure for GeF_2 is in only moderate agreement with experiment.

c.   Anions.   Experimental structural data on anions derives primarily from X-ray diffraction measurements in the solid state. The extent to which lattice interactions

TABLE 6.25. Comparison of Structural Predictions for Symmetry-Unconstrained and -Constrained UHF Wavefunctions[a]

System	State		UHF/STO-3G	UHF/3-21G	UHF/6-31G*	UMP2/6-31G*	UMP3/6-31G*	Expt.
C_2^+	$^2\Pi_u$	asym.	1.355	1.303	1.288	1.269	1.269	1.301
		sym.	1.358	1.346	1.316	1.319	1.307	1.301
N_2^+	$^2\Sigma_g^+$	asym.	1.513	1.164	1.117	1.080	1.085	1.116
		sym.	1.182	1.107	1.094	1.147	1.114	1.116
O_2	$^3\Sigma_g^-$	asym.	1.272	1.291	1.168	1.242	1.211	1.208
		sym.	1.217	1.240	1.168[b]	1.242[b]	1.211[b]	1.208
O_2^+	$^2\Pi_g$	asym.	1.211	1.120	1.076	1.188	1.118	1.116
		sym.	1.211	1.120[b]	1.076[b]	1.188[b]	1.118[b]	1.116
F_2^+	$^2\Pi_g$	asym.	1.407	1.731	∞	1.184	1.185	1.322
		sym.	1.266	1.327	1.232	1.411	1.320	1.322

[a]Theoretical data from: FPR (1983).
[b]Symmetric solution is not unstable at these levels; therefore constrained and unconstrained results are identical.

TABLE 6.26. Calculated and Experimental Equilibrium Geometries for Singlet Divalent Compounds of Carbon, Silicon, and Germanium[a]

Molecule	Point Group	Geometrical Parameter	STO-3G	3-21G	3-21G(*)	6-31G*	Expt.
CH_2	C_{2v}	$r(CH)$	1.123	1.102	1.102	1.097	1.111
		$<(HCH)$	100.5	104.7	104.7	103.1	102.4
CHF	C_s	$r(CH)$	1.142	1.107	1.107	1.104	
		$r(CF)$	1.312	1.339	1.339	1.295	1.314
		$<(HCF)$	102.3	103.1	103.1	102.8	101.8
CHCl	C_s	$r(CH)$	1.131	1.096	1.096	1.092	1.12
		$r(CCl)$	1.797	1.869	1.737	1.708	1.689
		$<(HCCl)$	100.2	100.2	103.9	103.5	103.4
CF_2	C_{2v}	$r(CF)$	1.323	1.321	1.321	1.283	1.304
		$<(FCF)$	102.7	104.0	104.0	104.5	104.8
CCl_2	C_{2v}	$r(CCl)$	1.803	1.848	1.737	1.711	1.304
		$<(ClCCl)$	106.8	107.1	100.0	110.3	100 ± 9

SiH$_2$	C_{2v}	r(SiH)	1.458	1.531	1.506	1.509	1.516
		<(HSiH)	91.5	93.8	93.4	93.2	92.1
SiHCl	C_s	r(SiCl)	2.119	2.240	2.080	2.093	2.064
		r(SiH)	1.460	1.523	1.502	1.505	1.561
		<(HSiCl)	92.6	94.3	95.8	95.5	102.8
SiHBr	C_s	r(SiBr)	2.230				2.231
		r(SiH)	1.461				
		<(HSiBr)	93.5				102.9
SiHI	C_s	r(SiI)	2.480				2.451
		r(SiH)	1.460				
		<(HSiI)	93.1				102.7
SiF$_2$	C_{2v}	r(SiF)	1.602	1.625	1.587	1.592	1.590
		<(FSiF)	93.2	96.2	99.2	99.6	100.8
GeF$_2$	C_{2v}	r(GeF)	1.650				1.732
		<(FGeF)	91.9				97.2

[a]Theoretical data from: W. J. Hehre, unpublished calculations.

207

influence the geometry is difficult to ascertain, but is indicated by the dependence of structural features on the nature of the counterion. There are many cases where the symmetry of an ion is perturbed because of its incorporation into an asymmetric crystal lattice; bond lengths that would be equal in the isolated species are different in the solid. As a result of lattice and counterion effects, it is normally not possible to assign unique values to the experimentally determined geometrical parameters of specific anions. Rather, ranges of experimental values corresponding to a number of crystal structures with different counterions are used for comparison.

Theoretical equilibrium geometries for a small selection of singly- and doubly-charged anions comprising first-row elements are compared with experimental structures in Table 6.27. At the STO-3G level, the largest errors occur for essentially pure single bonds. For example, the carbon–carbon length in the acetate anion is too long by 0.1 Å, while the boron–hydrogen linkages in BH_4^- are too short by almost the same amount. STO-3G bond angles are in reasonable accord with the experimental data, except in NH_2^- where the theoretical value is too small by 9°. Calculated bond lengths for the two dianions considered are of comparable quality to those for the singly-charged species. A graphical comparison of STO-3G and experimental bond lengths between heavy atoms is provided in Figure 6.21.

More accurate anion geometries are obtained with the 3-21G basis set. In particular, the BH_4^- and acetate anion geometries are significantly improved. Smaller improvements are noted for OH^-, N_3^-, and FHF^-. A graphical comparison of 3-21G bond lengths (connecting heavy atoms) with experiment is given in Figure 6.22.

3-21+G and 6-31G* geometries are generally superior to those obtained at lower theoretical levels, although the wide range of experimental values for some systems, NO_3^- for example, makes assessment difficult. The 3-21+G bond lengths provided in Figure 6.22 are usually closer to experimental values than the corresponding 3-21G values (also provided). The limited set of comparisons available between 6-31G* and experimental heavy-atom bond distances (Figure 6.23) is also favorable. Major effects of adding diffuse or d-type functions to the split-valence representation are noted for the distance in OH^- (decreased by 0.05–0.07 Å), that in NH_2^- (decreased by 0.037 Å) and the NO lengths in NO_2^- and NO_3^- (decreased by 0.06 Å at 6-31G* but not at 3-21+G). Surprisingly, the effects noted for the two dianions considered are not as large.

d. *Cations.* Very little experimental information is available on the equilibrium structures of isolated (gas-phase) cations [9]. The geometries of several *radical cations*, derived primarily from the electronic spectra of neutral closed-shell molecules near the ionization limit, were included in Table 6.22 and discussed in the context of open-shell RHF and UHF calculations.

Equilibrium structures for a wide variety of simple closed-shell carbocations are available from X-ray diffraction work in the solid state. As was the case for anion structures, these are subject to crystal packing forces and to counterion effects, and may differ significantly from true gas-phase geometries. Comparisons between theory and experiment for a number of representative closed-shell carbocations are provided in Table 6.28. Only bond lengths and bond angles involving the heavy-atom skeleton

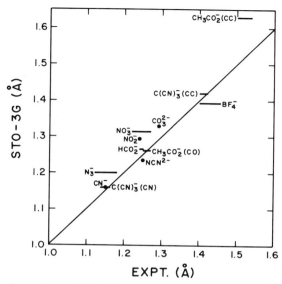

FIGURE 6.21.   Comparison of STO-3G and experimental heavy-atom bond lengths in anions.

are tabulated, although the theoretical equilibrium geometries have been obtained by complete optimization. Experimental bond lengths and angles formally equal by symmetry, for example, the three CN bond distances in $C(NH_2)_3^+$, represent averages of the actual measured values. Furthermore, experimental geometries quoted

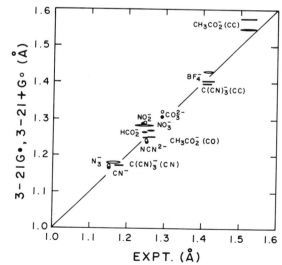

FIGURE 6.22.   Comparison of 3-21G ($\bullet$), 3-21+G ($\circ$) and experimental heavy-atom bond lengths in anions.

TABLE 6.27. Calculated and Experimental Equilibrium Geometries for Anions

Anion	Point Group	Geometrical Parameter	STO-3G[a,b]	3-21G[b]	3-21+G[b]	6-31G*[b]	Expt.[c]
OH^-	$C_{\infty v}$	$r(OH)$	1.068	1.029	0.978	0.962	0.97
NH_2^-	C_{2v}	$r(NH)$	1.080	1.066	1.029	1.029	1.03
		$<(HNH)$	95.2	98.0	105.9	99.2	104
FHF^-	$D_{\infty h}$	$r(FH)$	1.111	1.145	1.152	1.127	1.13–1.14
BH_4^-	T_d	$r(BH)$	1.176	1.241	1.244	1.243	1.25
BF_4^-	T_d	$r(BF)$	1.395	1.403	1.428	1.393	1.40–1.43
CN^-	$C_{\infty v}$	$r(CN)$	1.162	1.166	1.171	1.161	1.15
N_3^-	$D_{\infty h}$	$r(NN)$	1.202	1.176	1.177	1.156	1.12–1.18
NO_2^-	C_{2v}	$r(NO)$	1.294	1.286	1.285	1.228	1.24
		$<(ONO)$	114.3	116.4	116.5	116.7	115.4

NO_3^-	D_{3h}	$r(NO)$	1.315	1.283	1.286	1.226	1.22–1.27
HCO_2^-	C_{2v}	$r(CO)$	1.266	1.249	1.266	1.231	1.24–1.25
		$r(CH)$	1.152	1.125	1.103	1.127	
		$<(OCO)$	130.4	130.7	129.6	131.0	126.3
$CH_3CO_2^-$	C_s	$r(CC)$	1.631	1.576	1.546		1.50–1.54
		$r(CO)$	1.263	1.250	1.268		1.25–1.27
		$r(CH)$	1.087	1.085	1.085		
		$<(OCO)$	130.5	129.8	127.9		123–126
		$<(HCC)$	111.4	110.5	110.9		
$C(CN)_3^-$	D_{3h}	$r(CC)$	1.419	1.400			1.40–1.42
		$r(CN)$	1.162	1.150			1.14–1.16
NCN^{2-}	$D_{\infty h}$	$r(NC)$	1.234	1.236	1.242	1.231	1.25
CO_3^{2-}	D_{3h}	$r(CO)$	1.330	1.305	1.318	1.285	1.29

[a] L. Radom, *Aust. J. Chem.* **29**, 1635 (1976).

[b] CCSS (1983); W. J. Hehre, unpublished calculations.

[c] Experimental references are provided in the paper cited in footnote *a*.

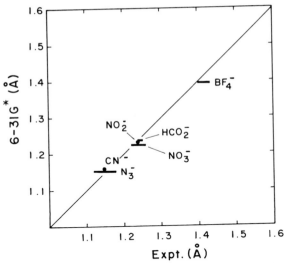

FIGURE 6.23.   Comparison of 6-31G* and experimental heavy-atom bond lengths in anions.

for the cyclopropenyl, tropylium, and benzenium cations derive from results for substituted systems. A graphical comparison of STO-3G and 3-21G carbon–carbon bond lengths with experimental values is provided in Figure 6.24.

Overall, the level of agreement between calculated and measured bond lengths and angles for both STO-3G and 3-21G levels of theory is reasonable. The STO-3G calculations consistently overestimate bond distances to the formal carbocation center, generally by 0.03–0.05 Å. Bond lengths derived from the 3-21G calculations reproduce the experimental data more satisfactorily. Errors larger than 0.02 Å are uncommon; in most cases agreement is far better. The only significant exception is the 3-21G CS bond distance in $(NH_2)_2CSH^+$ (0.07 Å too long). Similar deviations are found in uncharged molecules with single bonds to sulfur. The 3-21G$^{(*)}$ equilibrium structure for the same ion (given in footnote d of Table 6.28) is in far better accord with the experimental data. Bond angles obtained from both computational models are close (within 1–2°) to experiment.

6.2.9.   Structures of Intermolecular Complexes

The structures of weakly bonded intermolecular complexes, for example, donor–acceptor complexes, transition-metal carbonyl complexes, and hydrogen-bonded systems, are often difficult to establish accurately experimentally. In contrast, theoretical equilibrium geometries for such systems are easy to obtain. At present, sufficient experimental data are available to assess the performance of various theoretical models, although in some instances, hydrogen-bonded systems for example, these experimental results are tentative and may ultimately be subject to major reinterpretation.

a. *Donor–Acceptor Complexes.* Although donor–acceptor complexes are well known, the detailed geometries of only a small number have been determined experimentally. Some of these are presented in Table 6.29, and compared with calculated STO-3G, 3-21G, 3-21G$^{(*)}$, and 6-31G* structures.

The BC bond length in borane carbonyl, linking the formal two-electron donor (carbon monoxide) with the acceptor (borane), is nearly 0.1 Å too long with STO-3G, 3-21G, and 6-31G*. All three theoretical structures show both a lengthening of BH bonds (relative to those in borane), and a shortening of the CO linkage (relative to carbon monoxide) as a result of complex formation. Unfortunately, the equilibrium structure of free borane remains to be determined experimentally. However, the measured CO bond length in borane carbonyl is actually slightly longer rather than shorter than that in carbon monoxide.

Borane complexes involving ammonia and phosphine have also been investigated experimentally. All levels of calculation for both systems yield bond lengths linking donor and acceptor fragments that are longer than experimental values. Surprisingly, the STO-3G calculations give BN and BP bond lengths that are as close to the experimental distances as those obtained from higher-level treatments. All computational levels reproduce the observed shortening of PH bonds and the opening of the HPH angles relative to free phosphine in the borane-phosphine complex.

Comparisons between theoretical and experimental structures have also been made for two BF_3 complexes. Experimentally, the BC linkage in trifluoroborane carbonyl is 1.35 Å longer than the corresponding bond linking BH_3 and CO. All levels of theory also suggest a significant lengthening, 1.20 Å at STO-3G, 0.94 Å at 3-21G, and 1.34 Å at 6-31G*. On the other hand, the B–N distance in the trifluoroborane-ammonia complex, 1.60 Å, is virtually identical to that found in $BH_3 NH_3$ (1.56 and 1.60 Å from two X-ray studies). While STO-3G calculations show a significant lengthening (0.25 Å), 3-21G computations indicate a modest shortening (0.06 Å), and the BN distances in the two complexes are essentially identical at 6-31G*.

The effect of electron correlation on the theoretical structures of $BH_3 CO$ and $BH_3 NH_3$ has been examined (Table 6.30). At the MP2/6-31G* level, the theoretical structure for borane carbonyl is in excellent agreement with experimental values. The calculated geometry of ammonia–borane is also improved from the Hartree-Fock structure, although the BN bond is still somewhat too long. The limited comparisons suggest that electron correlation effects need to be considered in order to provide an accurate description of the equilibrium structures of donor–acceptor complexes.

b. *Metal-Carbonyl Complexes.* Calculated STO-3G equilibrium geometries for a selection of transition-metal carbonyls and carbonyl hydrides are compared with experimental structures in Table 6.31. The simple theory does not perform well; calculated bond lengths to carbon are usually too small (by more than 0.2 Å in $Ni(CO)_4$). The rather poor theoretical structure for iron pentacarbonyl is disturbing, and unfortunately not atypical. Experimentally, both axial and equatorial bonds to carbon are essentially identical; on the other hand, the calculations show a difference between axial and equatorial lengths of approximately 0.4 Å.

TABLE 6.28. Calculated and Experimental Equilibrium Geometries for Carbocations

Carbocation	Point Group	Geometrical Parameter	STO-3G[a]	3-21G[b]	Expt.[c]
	D_{3h}	$r(CN)$	1.355	1.324	1.323
	C_s	$r(CN_1)$	1.350	1.305	1.315
		$r(CN_2)$	1.343	1.314	1.312
		$r(CO)$	1.347	1.314	1.298
		$<(N_1CO)$	114.8	114.6	116.6
		$<(N_2CO)$	123.0	122.2	121.7
	C_s	$r(CN_1)$	1.346	1.306[d]	1.300
		$r(CN_2)$	1.348	1.307	1.312
		$r(CS)$	1.751	1.813	1.739
		$<(N_1CS)$	117.7	115.5	116.9
		$<(N_2CS)$	123.2	122.5	121.3
	C_s	$r(CC)$	1.530	1.481	1.480
		$r(CO_1)$	1.317	1.285	1.265
		$r(CO_2)$	1.308	1.273	1.272
		$<(O_1CC)$	125.5	124.5	123.5
		$<(O_2CC)$	118.7	119.0	118.2
	C_{3v}	$r(CC)$	1.516	1.449	1.452
		$r(CO)$	1.141	1.110	1.109
	D_{3h}	$r(CC)$	1.377	1.361	1.363–1.379[e]

214

TABLE 6.28. (Continued)

Carbocation	Point Group	Geometrical Parameter	STO-3G[a]	3-21G[b]	Expt.[c]
	D_{7h}	$r(CC)$	1.397	1.384	1.386–1.395[j]
	C_{2v}	$r(C_1 C_2)$	1.581	1.479	1.490[g]
		$r(C_2 C_3)$	1.356	1.352	1.365
		$r(C_3 C_4)$	1.422	1.407	1.407
		$<(C_2 C_1 C_6)$	114.6	115.5	115.4
		$<(C_1 C_2 C_3)$	122.1	121.6	121.1
		$<(C_2 C_3 C_4)$	119.0	119.0	119.6
		$<(C_3 C_4 C_5)$	123.2	123.1	122.8

[a]STO-3G structures for a number of the cations considered here have already been discussed: W. J. Hehre, in *Applications of Electronic Structure Theory*, H. F. Schaefer III, ed., Plenum Press, New York, 1977, vol. 4, p. 277ff. The remaining structures are from: W. J. Hehre and L. Radom, unpublished calculations.

[b]W. J. Hehre, unpublished calculations.

[c]Experimental data from: M. Sundaralingam and A. K. Chwang, in *Carbonium Ions*, G. A. Olah and P. v. R. Schleyer, ed., Wiley, New York, 1976, Volume 5, p. 2427.

[d]3-21G(*) structural parameters: $r(CN_1) = 1.313$ Å, $r(CN_2) = 1.313$ Å, $r(CS) = 1.747$ Å, $<(N_1 CS) = 116.3°$, and $<(N_2 CS) = 122.7°$.

[e]Experimental structural parameters from triphenylcyclopropenium ion and from tris(dimethylamino)cyclopropenium ion.

[f]Experimental structural parameters from 8-cyano-8-(diphenylcyclopropenyl) heptafulvenylium ion and from 8-cyano-8-cycloheptatrienylheptafulvenylium ion.

[g]Experimental structural parameters from heptamethylbenzenium ion.

c. Hydrogen-Bonded Complexes. The hydrogen-bonded dimers of water, hydrogen fluoride, and hydrogen chloride, as well as a number of mixed complexes including HF—H₂O, HF—HCl, and HF—HCN, have now been observed in the gas phase, and a number of structural features characterized [10]. The relative orientations of the two components have not generally been established experimentally, at least in an unambiguous manner. However, the equilibrium separations between heavy atoms have been determined, generally under the assumption that hydrogen bonding leaves the structures of the components unchanged.

STO-3G, 3-21G, and 6-31G* calculations give rise to a structure with a single hydrogen bond for the water dimer (Table 6.32), rather than a bifurcated structure with two such bonds. Evidently the favored near linearity of the hydrogen bond outweighs the benefits of interaction with more than a single hydrogen. The 6-31G* calculations yield an O—O distance in good agreement with experiment, but both

TABLE 6.29. Hartree–Fock and Experimental Equilibrium Geometries of Donor–Acceptor Complexes[a]

Complex	Point Group	Geometrical Parameter	STO-3G	3-21G	3-21G*	6-31G*	Expt.
	C_{3v}	r(BC)	1.632	1.615	1.615	1.630	1.540
		r(CO)	1.144	1.123	1.123	1.108	1.131
		r(BH)	1.162	1.203	1.203	1.203	1.194
		<(HBH)	114.9	114.3	114.3	115.1	113.9
	C_{3v}	r(BN)	1.659	1.740	1.740	1.690	1.56,1.6
		r(BH)	1.162	1.207	1.207	1.208	
		r(NH)	1.072	1.009	1.009	1.004	
		<(HBH)	114.1	113.9	113.9	114.1	
		<(HNH)	107.2	109.8	109.8	108.0	
	C_{3v}	r(BP)	2.011	2.213	2.008		1.937
		r(BH)	1.157	1.198	1.204		1.212
		r(PH)	1.374	1.407	1.390		1.399
		<(HBH)	114.6	116.8	114.4		114.6
		<(HPH)	100.1	100.5	100.7		101.3

216

Molecule	Symmetry	Parameter					
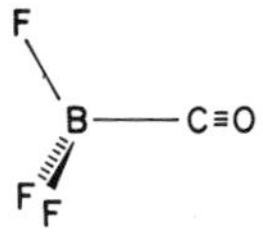 F₃B—C≡O	C_{3v}	r(BC)	2.833	2.551	2.551	2.966	2.886
		r(CO)	1.144	1.123	1.123	1.112	
		r(BF)	1.310	1.332	1.332	1.302	
		<(FBF)	120.0	119.8	119.8	120.0	120.0
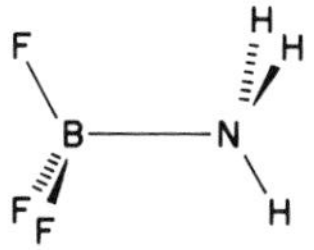 F₃B·NH₃	C_{3v}	r(BN)	1.905	1.683	1.683	1.695	1.60
		r(BF)	1.337	1.375	1.375	1.354	1.38
		r(NH)	1.031	1.012	1.012	1.005	
		<(FBF)	117.0	114.3	114.3	114.7	111.0
		<(HNH)	106.8	109.6	109.6	108.5	
BH₃	D_{3h}	r(BH)	1.160	1.188	1.188	1.188	
BF₃	D_{3h}	r(BF)	1.309	1.328	1.328	1.301	1.307
CO	$C_{\infty v}$	r(CO)	1.146	1.129	1.129	1.114	1.128
NH₃	C_{3v}	r(NH)	1.033	1.003	1.003	1.004	1.012
		<(HNH)	104.2	112.4	112.4	107.5	106.7
PH₃	C_{3v}	r(PH)	1.378	1.423	1.402	1.403	1.421
		<(HPH)	95.0	96.1	95.2	95.4	93.4

^{a} Theoretical data for complexes from: W. J. Hehre, unpublished calculations. For references to structural data on uncomplexed molecules, see footnotes to Tables 6.1, 6.6, and 6.10.

217

TABLE 6.30. Effect of Electron Correlation on the Equilibrium Geometries of Donor–Acceptor Complexes[a]

Complexes	Point Group	Geometrical Parameter	HF/6-31G*	MP2/6-31G*	Expt.
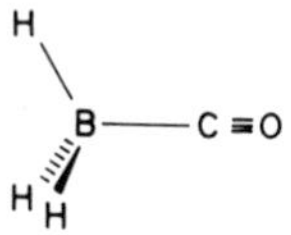 H₃B—C≡O	C_{3v}	r(BC)	1.630	1.550	1.540
		r(CO)	1.108	1.150	1.131
		r(BH)	1.203	1.207	1.194
		<(HBH)	115.1	114.3	113.9
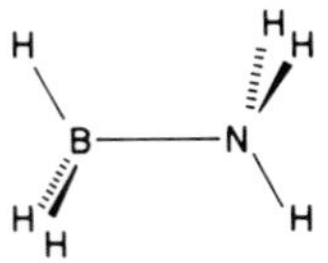 H₃B—NH₃	C_{3v}	r(BN)	1.689	1.666	1.56,1.6
		r(BH)	1.209	1.210	
		r(NH)	1.004	1.021	
		<(HBH)	114.1	113.9	
		<(HNH)	108.1	110.9	
BH₃	D_{3h}	r(BH)	1.188		
CO	$C_{\infty v}$	r(CO)	1.114	1.152	1.128
NH₃	C_{3v}	r(NH)	1.004	1.016	1.012
		<(HNH)	107.5	106.4	106.7

[a]Theoretical data from: W. J. Hehre, unpublished calculations. For references to structural data on uncomplexed molecules, see footnotes to Table 6.2 and 6.7.

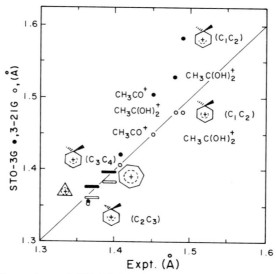

FIGURE 6.24. Comparison of STO-3G (●), 3-21G (○) and experimental heavy-atom bond lengths in cations.

TABLE 6.31. Calculated and Experimental Equilibrium Geometries for Transition-Metal Carbonyls and Carbonyl Hydrides

Molecule	Point Group	Geometrical Parameter	STO-3G[a]	Expt.
$Cr(CO)_6$	O_h	$r(CrC)$	1.789	1.92[b]
		$r(CO)$	1.167	1.16
$HMn(CO)_5$	C_{4v}	$r(MnC_{ax})$	1.732	1.823
		$r(MnC_{eq})$	1.723	1.823
		$r(MnH)$	1.624	1.50
		$r(C_{ax}O_{ax})$	1.162	1.139
		$r(C_{eq}O_{eq})$	1.163	1.139
		$<(C_{eq}MnH)$	72.7	83.6
		$<(MnC_{eq}O_{eq})$	170.9	
$Fe(CO)_5$	D_{3h}	$r(FeC_{ax})$	2.016	1.824
		$r(FeC_{eq})$	1.643	1.824
		$r(C_{ax}O_{ax})$	1.147	1.145
		$r(C_{eq}O_{eq})$	1.171	1.145
$Ni(CO)_4$	T_d	$r(NiC)$	1.583	1.82
		$r(CO)$	1.162	1.15

[a]PH (1983).
[b]L. E. Sutton, Spec. Pub., Chem. Soc. (London) 1958.

TABLE 6.32. Calculated and Experimental Equilibrium Geometries for Hydrogen-Bonded Complexes

Complex	Point Group	Geometrical Parameter	STO-3G[a,b]	3-21G[b,c]	3-21G(*)[b,d]	6-31G*[b,e]	Expt.[f]
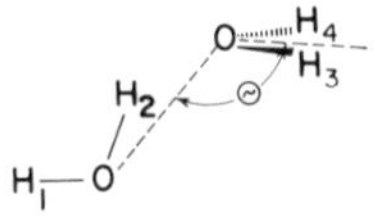	C_s	$r(OO)$	2.734	2.797	2.797	2.977	2.98
		$r(OH_1)$	0.990	0.966	0.966	0.946	
		$r(OH_2)$	0.988	0.973	0.973	0.952	
		$r(OH_3)$	0.987	0.967	0.967	0.948	
		$<(OOH_2)$	5.2	2.7	2.7	4.9	
		$<(\Theta)$	121.3	124.6	124.6	118.4	123
		$<(H_1OH_2)$	100.4	107.9	107.9	105.3	
		$<(H_3OH_4)$	100.9	108.8	108.8	105.7	
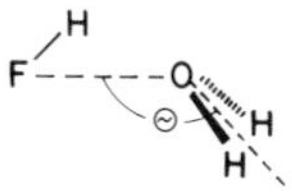	C_s	$r(OF)$	2.630	2.569	2.569	2.721	2.69
		$r(FH)$	0.956	0.953	0.953	0.921	
		$r(OH)$	0.987	0.964	0.964	0.946	
		$<(HFO)$	3.0	0.0	0.0	-0.5	
		$<(\Theta)$	128.2	179.9	179.9	145.3	134
		$<(HOH)$	101.3	110.7	110.7	106.9	
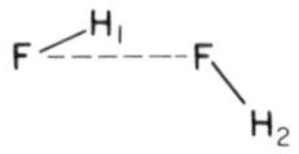	C_s	$r(FF)$	2.570	2.599	2.599	2.716	2.79
		$r(FH_1)$	0.954	0.942	0.942	0.915	
		$r(FH_2)$	0.953	0.941	0.941	0.915	
		$<(H_1FF)$	4.0	8.4	8.4	15.9	
		$<(H_2FF)$	109.1	110.9	110.9	98.4	108
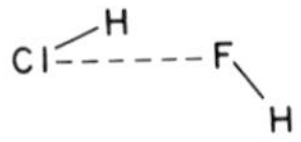	C_s	$r(ClF)$	2.785	3.128	3.125	3.358	3.37
		$r(ClH)$	1.352	1.301	1.273	1.268	
		$r(FH)$	0.953	0.940	0.940	0.913	
		$<(HClF)$	1.7	3.5	4.2	6.4	
		$<(HFCl)$	108.6	115.3	112.7	119.9	130

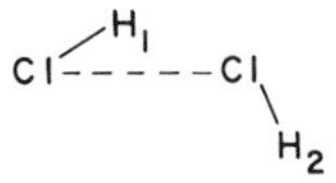 C_s	r(ClCl)	3.774	3.928	4.060	4.157	3.171[g]
	r(ClH$_1$)	1.317	1.295	1.269	1.267	
	r(ClH$_2$)	1.315	1.294	1.268	1.267	
	<(H$_1$ClCl)	5.0	5.0	6.6	9.6	
	<(H$_2$ClCl)	106.6	105.3	100.6	98.5	
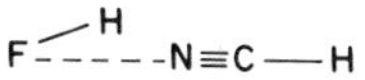 C_s	r(FN)	3.033	2.810	2.810	2.922	2.796
	r(FH)	0.953	0.942	0.942	0.917	
	r(NC)	1.152	1.135	1.135	1.130	
	r(CH)	1.071	1.052	1.052	1.060	
	<(HFN)	0.1	0.2	0.2	0.5	
	<(FNC)	179.7	179.7	179.7	179.8	
	<(NCH)	180.0	180.0	180.0	180.0	
H$_2$O C_{2v}	r(OH)	0.990	0.967	0.967	0.947	0.958
	<(HOH)	100.0	107.6	107.6	105.5	104.5
HF $C_{\infty v}$	r(FH)	0.956	0.937	0.937	0.911	0.917
HCl $C_{\infty v}$	r(ClH)	1.313	1.293	1.267	1.266	1.275
HCN $C_{\infty v}$	r(CH)	1.070	1.137	1.137	1.059	1.063
	r(CN)	1.153	1.050	1.050	1.132	1.154

Cl—H$_1$ - - - - Cl / H$_2$

H - - - - N≡C—H (F)

[a]LCHLP (1974); WFBDSRP (1981); W. J. Hehre, unpublished calculations.

[b]For reference to structural data on non-hydrogen-bonded systems, see footnotes to Tables 6.1 and 6.6.

[c]WFBDSRP (1981); W. J. Hehre, unpublished calculations.

[d]W. J. Hehre, unpublished calculations.

[e]WFBDSRP (1981); W. J. Hehre, unpublished calculations.

[f]Except where otherwise noted, experimental data is for vapor-phase species and is summarized in: Th. R. Dyke, *Topics Current Chem.*, **120**, 85 (1984).

[g]Cl–Cl separation in crystalline HCl. E. Sandor and R. F. C. Farrow, *Nature* **215**, 1265 (1967).

221

STO-3G and 3-21G basis sets give structures in which this separation is too small. This contrasts with the results for the borane–carbon monoxide and borane–ammonia complexes previously discussed. The 6-31G* basis results in an F—F distance in the HF dimer of 2.72 Å, in reasonable accord with the experimental value of 2.79 ± 0.05 Å. Similarly, the Cl—Cl distance in the dimer of HCl is underestimated by the STO-3G, 3-21G, and 3-21G$^{(*)}$ basis sets; the 6-31G* result is in reasonable accord with the experimental value.

In two of the three mixed hydrogen-bonded complexes involving hydrogen fluoride, this molecule acts as the hydrogen-bond donor.

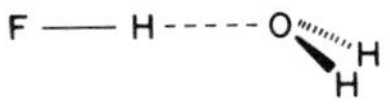

$$F \text{—} H \text{-----} O\substack{\diagdown H \\ H}$$

$$F \text{——} H \text{---} N \equiv C \text{—} H$$

In the third mixed system, HF acts as the proton acceptor.

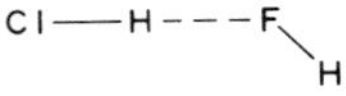

$$Cl \text{——} H \text{---} F\substack{\diagdown \\ H}$$

All the levels of calculation agree, yielding higher energies for the corresponding alternative forms.

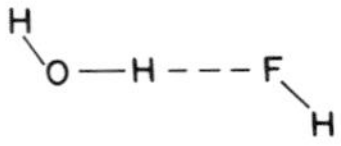

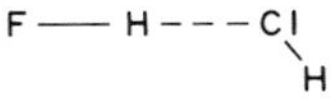

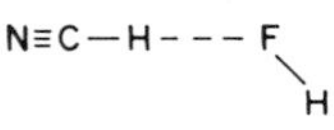

$$\substack{H \diagdown} O \text{—} H \text{---} F\substack{\diagdown \\ H}$$

$$F \text{——} H \text{---} Cl\substack{\diagdown \\ H}$$

$$N \equiv C \text{—} H \text{---} F\substack{\diagdown \\ H}$$

The STO-3G and 3-21G calculations underestimate the length of the hydrogen bond in the hydrogen fluoride-water complex; the 6-31G* value is slightly too long. The STO-3G Cl—F separation in the hydrogen chloride-hydrogen fluoride complex is 0.6 Å too short; 3-21G calculations fare better, although the calculated separation between monomers is still 0.1 Å shorter than the experimental value.

In summary, the 6-31G* basis set provides a somewhat better description of the geometries of hydrogen-bonded complexes than does STO-3G or 3-21G.

Optimization at the MP2/6-31G* level for the water dimer (Table 6.33) leads to shortening of the hydrogen bond. The O—O linkage is still 0.06 Å less than the experimental value; correction for zero-point vibrational motion might be significant.

TABLE 6.33. Effect of Electron Correlation on Equilibrium Geometries of Hydrogen-Bonded Complexes

Complex	Point Group	Geometrical Parameter	HF/6-31G*[a]	MP2/6-31G*[b]	Expt.[c]
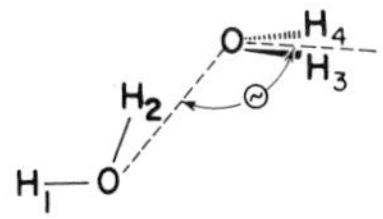	C_s	$r(OO)$	2.977	2.919	2.98
		$r(OH_1)$	0.946	0.968	
		$r(OH_2)$	0.952	0.975	
		$r(OH_3)$	0.948	0.971	
		$<(OOH_2)$	4.9	8.8	
		$<(\Theta)$	118.4	102.0	
		$<(H_1OH_2)$	105.3	104.2	
		$<(H_3OH_4)$	105.7	104.3	
H_2O	C_{2v}	$r(OH)$	0.948	0.969	0.957
		$<(HOH)$	105.5	104.0	104.5

[a]See footnote e to Table 6.32 for reference to theoretical data.

[b]W. J. Hehre, unpublished calculations.

[c]See footnote f to Table 6.32 for reference to experimental data.

223

6.2.10. Improved Structural Predictions

Even simple levels of *ab initio* molecular orbital theory can yield geometrical structures that are in reasonable accord with experimental values. While for many purposes the theoretical data may be sufficiently accurate, for some applications, for example, the assignment of an unknown microwave transition, the residual errors may simply be too large for the theoretical structures to be of value. To obtain the required accuracy directly, calculations using very large basis sets and incorporating electron correlation would be needed. This may present formidable computational problems.

Errors in theoretical geometrical parameters generally are *systematic* rather than *random*, as evidenced by the data in Tables 6.10 and 6.12, which compare calculated and measured bond lengths and angles for a series of related compounds. A further example is provided in Table 6.34, which shows that deviations from experiment in calculated CF bond distances, while often significant, are relatively constant for a given level of theory. Calculated equilibrium geometries may thus be *systematically corrected* to yield improved structures. Efforts in this direction show considerable promise. For example, Bouma and Radom [11] deduced rotational constants for the transient species vinyl alcohol to within 0.05 GHz ($<0.1\%$) of the spectroscopically determined values by correcting STO-3G and 4-31G equilibrium structures for systematic errors, the corrections being deduced from a number of related systems, methyl vinyl ether for example. In a similar manner, an accurate structure for vinylamine has been obtained by empirically correcting STO-3G, 3-21G, and 6-31G* structures utilizing data for $CH_2=CH_2$, $CH_3CH=CH_2$, $FCH=CH_2$, NH_3, NH_2-CN, NH_2-CH_3, and NH_2CHO. The rotational constants derived from the corrected structure ($A = 56.317$, $B = 10.013$, $C = 8.586$ GHz) agree well with experimental values ($A = 56.313$, $B = 10.035$, $C = 8.565$ GHz) [12].

Another example involved an attempt to obtain a very precise structure and a rotational transition frequency for HOC^+, postulated to exist in interstellar space [13]. Calculations were carried out at various levels of theory of increasing sophistication (HF/4-31G, HF/6-31G**, MP3/6-31G**, MP3/6-311G**, and MP3/6-311G***‡) on HOC^+ and on the related known systems HCN, HNC, and HCO^+. The corrections derived from the last three systems (Table 6.35) were used to obtain empirically corrected structures for HOC^+ (Table 6.36). Structures obtained with the various basis sets at the MP3 level show pleasing agreement; when averaged these yield a best estimate of the equilibrium structure of HOC^+ with $r_e(CO) = 1.155$ Å and $r_e(OH) = 0.988$ Å. The estimated uncertainty in each length is 0.003 Å. After correcting for zero-point vibrations, this structure leads to a predicted $J = 1 \rightarrow 0$ rotational transition frequency of 89.0 ± 0.8 GHz. Subsequent to the theoretical calculations, HOC^+ has been observed in the laboratory [14] and in interstellar space [15] with a frequency of 89.487 GHz.

---

‡This basis set is identical to the 6-311G** representation (Section 4.3.3b) except that two sets of d functions, rather than one, are provided for all heavy atoms.

TABLE 6.34. Errors in Calculated CF Bond Lengths[a]

Molecule	$r(CF)_{expt.}$	$r(CF)_{expt.} - r(CF)_{calc.}$		
		HF/STO-3G	HF/3-21G	HF/6-31G*
FCN	1.262	0.054	0.025	−0.009
FCCH	1.279	0.039	0.018	−0.010
CF_4	1.317	0.049	0.008	−0.015
CHF_3	1.332	0.039	0.012	−0.015
F_2CO	1.338	0.013	0.010	−0.048
$CHF{=}CH_2$	1.348	0.006	0.015	−0.019
CH_2F_2	1.358	0.020	0.014	−0.020
CH_3F	1.383	0.001	0.021	−0.018

[a]Data not already presented in Table 6.6 from: WFBDSRP (1981) and W. J. Hehre, unpublished calculations.

TABLE 6.35. Bond Lengths and Errors in Bond Lengths for HXY Systems[a]

Species	Level	$r_e(X{-}Y)$		$r_e(X{-}H)$	
		Value	Error[b]	Value	Error[b]
HCN	HF/4-31G	1.140	−1.1	1.051	−1.3
	HF/6-31G**	1.133	−1.8	1.059	−0.6
	MP3/6-31G**	1.158	+0.4	1.064	−0.1
	MP3/6-311G**	1.151	−0.2	1.067	+0.2
	MP3/6-311G***	1.146	−0.6	1.066	+0.1
	Expt.[c]	1.153			
HNC	HF/4-31G	1.163	−0.5	0.979	−1.5
	HF/6-31G**	1.155	−1.2	0.984	−1.0
	MP3/6-31G**	1.175	+0.5	0.994	0.0
	MP3/6-311G**	1.168	−0.1	0.996	+0.2
	MP3/6-311G***	1.163	−0.5	0.996	+0.2
	Expt.[d]	1.169		0.994	
HCO^+	HF/4-31G	1.098	−0.7	1.078	
	HF/6-31G**	1.087	−1.7	1.086	
	MP3/6-31G**	1.110	+0.4	1.090	
	MP3/6-311G**	1.099	−0.6	1.093	
	MP3/6-311G***	1.097	−0.8	1.093	
	Expt.[e]	1.105–1.106		1.086–1.098	
HOC^+	HF/4-31G	1.160		0.976	
	HF/6-31G**	1.142		0.975	
	MP3/6-31G**	1.161		0.989	
	MP3/6-311G**	1.152		0.988	
	MP3/6-311G***	1.148		0.990	

[a]Theoretical data from reference 13.
[b]Percent deviation from experimental value.
[c]G. Winnewisser, A. G. Maki, and D. R. Johnson, *J. Mol. Spectrosc.*, **39**, 149 (1971).
[d]R. A. Creswell and A. G. Robiette, *Mol. Phys.*, **36**, 869 (1978).
[e]L. E. Snyder, J. M. Hollis, F. J. Lovas, and B. L. Ulich, *Astrophys. J.*, **209**, 67 (1976).

TABLE 6.36.   Applied Correction Factors and Resultant Estimated
Structures of HOC^{+a}

	r_e(C—O)		r_e(O—H)	
Level	Correction Factor[b]	Corrected Value	Correction Factor[b]	Corrected Value
HF/4-31G	+0.8	1.169	+1.4	0.990
HF/6-31G**	+1.6	1.160	+0.8	0.983
MP3/6-31G**	−0.4	1.156	+0.1	0.990
MP3/6-311G**	+0.3	1.155	−0.2	0.986
MP3/6-311G***	+0.6	1.155	−0.2	0.988

[a]From reference 13.
[b]Percent correction to be applied to calculated values.

6.2.11.   Concluding Remarks and Recommendations

The equilibrium structures of small molecules comprising first- and second-row elements only may be obtained with good accuracy from molecular orbital theory. Use of moderately large basis sets, for example, 6-31G*, 6-31G**, and 6-311G**, and configuration interaction or higher-order Møller–Plesset electron correlation schemes generally guarantees that calculated structural parameters will be close to measured equilibrium values. However, the most sophisticated models are computationally expensive and their application may be limited. For many molecules, the single-determinant (Hartree–Fock) model yields bond lengths and angles to within 0.01–0.02 Å and 1–2° of the respective experimental values. The relatively small 3-21G basis set (3-21G$^{(*)}$ for molecules incorporating second-row elements) appears to be the method of choice because of its wide applicability to molecules of moderate size. Extensive comparisons between theoretical and experimental structures (both for stable neutral molecules and for a variety of short-lived neutral and charged species) document the value of calculations at this level as a generally applicable tool for the investigation of geometry. For very large molecules, or for molecules incorporating heavy elements, calculations at the 3-21G level may not be possible or may be prohibitive in cost. In these situations, even the STO-3G minimal basis set may provide a reasonable account of equilibrium structure.

6.3.   VIBRATIONAL FREQUENCIES AND THERMODYNAMIC PROPERTIES

The exploration of chemical reaction pathways provides a great challenge for theory. This involves a characterization of structures and relative energies not only of stable forms, but also of other stationary points on the potential surface, corresponding to reactive intermediates and transition structures. As experimental spectroscopic observations may be difficult or impossible for such species, theory has the opportunity of providing the major source of detailed information.

Calculated normal-mode vibrational frequencies play several important roles in the use of theory as a means of characterizing molecular potential surfaces. In the first place, they may be used to characterize *stationary points* on the surface, that is, to distinguish *local minima* which have *all real frequencies* from *saddle points* which have a *single imaginary frequency*. Secondly, for stable but highly reactive or otherwise short-lived molecules, they provide a means of identification. The appearance of only a single (hitherto unidentified) infrared line may be the only indication that a new molecule has actually been formed. Any theory capable of calculating vibrational frequencies with sufficiently high accuracy would obviously be of great value for assignment of spectra. Finally, calculated normal-mode vibrational frequencies provide thermodynamic properties of stable molecules by way of statistical mechanics. Thus, reaction entropies and equilibrium isotope effects, widely used by experimentalists to probe the structures of reactive intermediates, may be calculated from first principles. In addition, the corresponding quantities may also be obtained for reaction transition structures. Of particular interest are kinetic isotope effects and entropies of activation, which provide mechanistic information. Theoretical vibrational frequencies may also be employed to correct experimental thermochemical data to 0 K and to evaluate zero-point vibrational energies.

Our primary goal in this section is to provide an evaluation of simple quantum mechanical models for the calculation of normal-mode vibrational frequencies. The discrepancies between theoretical and experimental frequencies reveal systematic differences that may be corrected empirically. Following a brief introduction to the theory of normal-mode analysis, an outline of methods of frequency calculation (Section 6.3.1) and a discussion of errors (Section 6.3.2), comparisons are made of calculated and measured frequencies for diatomics (Section 6.3.3), one- and two-heavy-atom hydrides (Sections 6.3.4 and 6.3.5), and larger polyatomic molecules (Section 6.3.6). Section 6.3.7 deals with the calculation of frequencies for molecules poorly represented in terms of conventional valence structures, that is, hypervalent molecules, and Section 6.3.8 with comparisons of CH stretching frequencies for related molecules. Reaction entropies, equilibrium isotope effects and temperature and zero-point-energy corrections to thermochemical data derived from theoretical frequencies are compared with the corresponding experimental quantities, that is, those obtained from experimental frequencies, in Section 6.3.9. Only equilibrium thermodynamic properties have been considered here; application to non-minimum-energy forms requires the additional assumption of a transition-state model.

The performance of Hartree–Fock models is considered first. This is followed, where theoretical data are available, by assessment of electron correlation effects.

The published collections of vibrational frequencies listed below are employed without specific documentation. References to the primary literature are made only for more recent work or when a determination is in question.

Diatomic Molecules

1. K. P. Huber and G. Herzberg, *Molecular Spectra and Molecular Structure. Constants of Diatomic Molecules*, Van-Nostrand Reinhold, New York, 1979.

Polyatomic Molecules

1. T. Shimanouchi, *Tables of Molecular Vibrational Frequencies, consolidated volume I.,* NSRDS-NBS-39, National Bureau of Standards, Washington, D.C. (1972); T. Shimanouchi, *J. Phys. Chem. Ref. Data,* **6**, 993 (1977); T. Shimanouchi, H. Matsuura, Y. Ogawa, and I. Harada, *ibid.,* **7**, 1323 (1978); *ibid.,* **9**, 1149 (1980).
2. D. R. Stull and H. Prophet, *JANAF Thermochemical Tables,* 2nd. ed., NSRDS-NBS-32, National Bureau of Standards, Washington, D.C., 1971.

6.3.1. Calculation of Harmonic Vibrational Frequencies

The total energy of a molecule comprising N atoms near its equilibrium structure may be written as

$$E = T + V = \frac{1}{2} \sum_{i=1}^{3N} \dot{q}_i^2 + V_{eq} + \frac{1}{2} \sum_{i=1}^{3N} \sum_{j=1}^{3N} \left(\frac{\partial^2 V}{\partial q_i \partial q_j}\right)_{eq} q_i q_j. \qquad (6.1)$$

Here, the mass-weighted cartesian displacements, q_i, are defined in terms of the locations x_i of the nuclei relative to their equilibrium postiions $x_{i,\,eq}$ and their masses M_i,

$$q_i = M_i^{1/2}(x_i - x_{i,\,eq}). \qquad (6.2)$$

V_{eq} is the potential energy at the equilibrium nuclear configuration, and the expansion (6.1) of the vibrational energy in terms of a power series is truncated at second order [16]. For such a system, the classical-mechanical equations of motion take the form

$$\ddot{q}_j = - \sum_{i=1}^{3N} f_{ij} q_i \qquad j = 1, 2, \ldots, 3N. \qquad (6.3)$$

The f_{ij}, termed *quadratic force constants*, are the second derivatives of the potential energy with respect to mass-weighted cartesian displacements, evaluated at the equilibrium nuclear configuration, that is,

$$f_{ij} = \left(\frac{\partial^2 V}{\partial q_i \partial q_j}\right)_{eq}. \qquad (6.4)$$

The f_{ij} may be evaluated by numerical second differentiation,

$$\frac{\partial^2 V}{\partial q_i \partial q_j} \simeq \frac{\Delta(\Delta V)}{\Delta q_i \Delta q_j}, \qquad^{\ddagger} \qquad (6.5)$$

‡In practice, the right-hand-side of equation (6.5) is replaced by an expression of the form

$$\frac{V(q_i + \Delta q_i, q_j + \Delta q_j) - V(q_i + \Delta q_i, q_j) - V(q_i, q_j + \Delta q_j) + V(q_i, q_j)}{\Delta q_i \Delta q_j}$$

by numerical first differentiation of analytical first derivatives,

$$\frac{\partial^2 V}{\partial q_i \partial q_j} \simeq \frac{\Delta(\partial V/\partial q_j)}{\Delta q_i}, \tag{6.6}$$

or by direct analytical second differentiation, Eq. (6.4). The choice of procedure depends on the quantum mechanical model employed, that is, single-determinant or post-Hartree–Fock, and practical matters such as the size of the system.

Equation (6.3) may be solved by standard methods [16] to yield a set of $3N$ *normal-mode vibrational frequencies*. Six of these (five for linear molecules) will be zero as they correspond to translational and rotational (rather than vibrational) degrees of freedom.[‡]

6.3.2. Sources of Error in the Comparison of Theoretical and Experimental Vibrational Frequencies

Under ideal conditions, vibrational frequencies can be measured to within one wavenumber (cm^{-1}) or less. Assignment of a frequency to a particular vibrational motion is more tenuous; errors are possible and may be commonplace. Assignments are often based on prior experience, for example, $C=O$ stretching frequencies occur in the range $1600\text{--}1800$ cm^{-1}, and on measured frequency shifts due to specific isotopic substitution. Molecules comprising more than five or six atoms, and sometimes even smaller systems, present special difficulties. At room temperature, several rotational levels other than the lowest are significantly populated, and the experimentalist must contend with broad bands rather than sharp lines. For crowded portions of the spectrum, for example, the CH-stretching region for a hydrocarbon, the rotational envelopes corresponding to different vibrations overlap, and location of the vibrational transition is complicated. Gross errors are possible, and it is not uncommon to see differences in assigned frequencies of $30\text{--}50$ cm^{-1} or more.

The use of low-temperature vibrational spectroscopy, in particular matrix-isolation methods [17] and molecular beam techniques [18], will ultimately lead to precise frequency assignment even for complex molecules. However, at present, it should be recognized that frequency misassignments are possible. Some of the quoted experimental frequencies may thus be inaccurate.

Errors in calculated frequencies arise both from inherent inaccuracies of differentiation techniques required in the evaluation of the matrix of force constants, and from uncertainties in the selection of equilibrium geometry. Table 6.37 lists vibrational frequencies for ethylene calculated at the HF/6-31G* level using three different differentiation techniques. Units here and in all comparisons that follow are cm^{-1}. The first column of frequencies derives from force fields obtained by

---

[‡]In practice, it has been found that, while the three frequencies corresponding to translations are very nearly zero, the rotational degrees of freedom give non-zero values (up to 50 cm^{1}) unless the stationary point on the potential surface is determined very precisely. For large molecules with true low-frequency vibrations, this can lead to confusion. These difficulties can be avoided by a change of basis to coordinates which are linearly independent of translation and rotation, thereby reducing the order of the matrix of force constants from $3N$ to $3N - 6$ ($3N - 5$).

TABLE 6.37.   Dependence of Calculated Vibrational Frequencies for Ethylene on Method of Force Constant Evaluation

Symmetry of Mode	Description of Mode	Calculated Frequency[a]		
		Numerical Second Differentiation[b]	Numerical Differentiation of Analytical Gradient[b]	Analytical Second Differentiation[c]
a_g	CH_2 s-stretch	3344	3346	3344
a_g	CC stretch	1856	1849	1856
a_g	CH_2 scis.	1496	1494	1497
a_u	CH_2 twist	1149	1157	1155
b_{1g}	CH_2 a-stretch	3338	3390	3394
b_{1g}	CH_2 rock	1369	1349	1353
b_{1u}	CH_2 wag	1094	1095	1095
b_{2g}	CH_2 wag	1094	1100	1099
b_{2u}	CH_2 a-stretch	3418	3421	3420
b_{2u}	CH_2 rock	915	898	897
b_{3u}	CH_2 s-stretch	3321	3322	3321
b_{3u}	CH_2 scis.	1608	1610	1610

[a] HF/6-31G*level.
[b] HLH (1982).
[c] J. A. Pople, R. Krishnan, H. B. Schlegel, and J. S. Binkley, *Int. J. Quantum Chem., Symp.*, **13**, 225 (1979).

numerical second differentiation using symmetry coordinates. The next column was obtained by taking numerical differences of gradients obtained analytically. Normal-mode frequencies calculated by analytical second differentiation appear in the final column. The only point to be made is that the frequencies derived from all three methods are generally in close accord with one another. With but a single exception (the highest-frequency b_{1g} vibration), the largest deviation between frequencies obtained using strictly numerically derived force fields and either the hybrid numerical-analytical or straight analytical approaches is 18 cm^{-1}. Larger deviations have been noted in some comparisons involving larger polyatomics, although instances in which errors exceed 20 cm^{-1} appear to be relatively infrequent. Much smaller deviations (generally less than 5 cm^{-1}) are noted between the hybrid and direct analytical methods.

The data in Table 6.38 for ammonia illustrate the changes in vibrational frequencies that might be anticipated due to small errors in the exact choice of equilibrium geometry. A distortion in the equilibrium NH bond length of 0.005 Å leads to changes in frequencies associated with the symmetric and degenerate stretching modes on the order of 60 cm^{-1}. A smaller (25 cm^{-1}) change in the frequency associated with the nitrogen inversion motion is caused by the slight increase in equilibrium bond length. This particular frequency is also sensitive to slight changes in bond angle: an increase of 0.5° leads to a 24 cm^{-1} decrease in frequency. It should be emphasized that the frequency shifts depicted in Table 6.38 are *due to the anharmonicity of the potential function* in the vicinity of the minimum. They occur

TABLE 6.38. Dependence of Calculated Vibrational Frequencies for NH_3 on Choice of Equilibrium Geometry

Symmetry of Mode	Description of Mode	Calculated Frequency[a]		
		Equilibrium Structure[b]	+0.005 Å on r(NH)	+0.5° on <(HNH)
a_1	s-stretch	3695	3638	3694
a_1	s-deform	1203	1228	1179
e	d-stretch	3829	3769	3830
e	d-deform	1847	1856	1845

[a] HF/6-31G* level; from HLH (1982).
[b] r(NH) = 1.002 Å; <(HNH) = 107.2°.

irrespective of the choice of differentiation technique. Considerable care must be exercised in obtaining suitably converged equilibrium structures for use in frequency calculations.

The use of nonoptimized geometries, that is, structures that do not correspond to energy minima for the particular quantum mechanical model at hand, to obtain frequencies has been suggested [19]. While such an approach may appear reasonable, especially in those situations where the calculated equilibrium geometry is in poor agreement with the corresponding experimental structure, it is clearly unsatisfactory in several respects. Equation (6.1), expressing the energy in terms of a truncated Taylor series, *assumes that the first derivative terms, $(\partial V/\partial q_i)$, are zero*, that is, the structure corresponds to an energy minimum. Lack of a uniformly derived set of structures also thwarts attempts to unravel trends in frequencies among closely related molecules, for example, the effect of remote substituents on the CO-stretching frequencies in carbonyl compounds. Calculated frequencies are sufficiently sensitive to structure (see, for example, Table 6.38) that nonuniform choice of geometry is likely to completely disguise subtle effects. Even experimental geometries, where available, are not necessarily suitable replacements for theoretical structures. As discussed earlier (Section 6.2.1), uncertainties in measured equilibrium parameters are often on the order of 0.01 Å and 1° for bond lengths and angles, respectively; these may lead to variations in calculated frequencies of several tens of wavenumbers.

6.3.3. Frequencies of Diatomic Molecules

Table 6.39 and Figure 6.25 compare theoretical and experimental vibrational frequencies for H_2 and a selection of one- and two-heavy-atom diatomic molecules comprising first-row elements only. Three theoretical models have been investigated, the single-determinant HF/3-21G and HF/6-31G* levels and the correlated MP2/6-31G* model. For the compounds considered, sufficient spectroscopic data are available for the determination of purely harmonic vibrational frequencies. These are more appropriate for comparison with calculated frequencies than are directly

TABLE 6.39. Theoretical and Experimental Vibrational Frequencies for Diatomic Molecules

Molecule	Theoretical Frequency			Experimental Frequency	
	HF/3-21G[a]	HF/6-31G*[b]	MP2/6-31G*[b]	Harmonic[c]	Measured
H_2	4660	4647	4528	4401	4160
LiH	1424	1415	1393	1406	1360
BH	2451	2516	2453	2366	2268
FH	4062	4358	4038	4139	3962
Li_2	340	339	368	351	346
LiF	1082	1033	998	914	898
BeO	1729	1744	1385	1487	1464
BF	1476	1473	1401	1400	1377
CO	2312	2438	2113	2170	2143
N_2	2614	2763	2173	2360	2331
F_2	1299	1245	1008	923	891

[a]W. J. Hehre, unpublished calculations.
[b]HLH (1982).
[c]For a discussion of the derivation of harmonic frequencies from experiment, see: K. P. Huber and G. Herzberg, *Molecular Spectra and Molecular Structure. Constants of Diatomic Molecules,* Van Nostrand-Reinhold, New York, 1979.

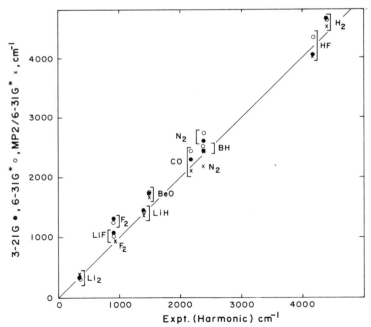

FIGURE 6.25. Comparison of 3-21G (•), 6-31G* (○), MP2/6-31G* (X) and experimental vibrational frequencies for diatomic molecules.

measured values. Note that the experimental harmonic frequencies are considerably larger than the actual measured fundamentals; deviations range from a maximum of 5.6% in H_2 to a minimum of 1.2% in CO and N_2. Overall, the mean percentage deviation of harmonic from actual vibrational frequencies is 2.8%. Unfortunately, experimental spectroscopic data sufficient to enable precise determination of harmonic frequencies are generally available only for diatomic and the simplest polyatomic species.

With the exception of Li_2 (both 3-21G and 6-31G* basis sets) and BH and HF (3-21G basis set), all calculated Hartree–Fock frequencies are larger than the corresponding harmonic experimental quantities. This suggests that single-determinant wavefunctions yield too steep a potential in the vicinity of the equilibrium structure. The mean absolute percentage deviations of the theoretical frequencies from experimental harmonic values are 9 and 11% for the HF/3-21G and HF/6-31G* models, respectively. Individually, the worst cases are F_2 and the highly polar molecules LiF and BeO. It was previously noted (Section 6.2.3) that the limiting Hartree-Fock bond length for F_2 is 0.1 Å shorter than the experimental value.

MP2/6-31G* frequencies generally are closer to the experimental harmonic values than are those at HF/6-31G*. This improvement is consistent with observations using other correlation energy schemes [20]. Overall, the mean absolute percentage deviation of MP2/6-31G* from experimental harmonic frequencies is 4.6%, roughly half that noted for either Hartree–Fock level. The most dramatic individual improvements are for F_2, BeO, CO, and N_2, the worst cases at the HF/6-31G* level. In general, MP2/6-31G* frequencies are larger than the corresponding experimental harmonic values, but this is not so for molecules with multiple bonds (BeO, CO, and N_2), or for the highly polar hydrides of lithium and fluorine. This parallels the observation that MP2/6-31G* bond lengths in these molecules (as well as those in other molecules incorporating double and triple bonds) are significantly larger than the corresponding experimental distances (see Section 6.2.3b).

6.3.4. Frequencies of One-Heavy-Atom Hydrides

A comparison of calculated and experimental vibrational frequencies for polyatomic one-heavy-atom hydrides comprising first- and second-row elements is provided in Table 6.40. Four levels of theory have been surveyed: Hartree–Fock methods utilizing 3-21G, 3-21G$^{(*)}$, and 6-31G* basis sets, and the correlated MP2/6-31G* method. Except for BH_3, sufficient experimental data are available to enable determination of harmonic vibrational frequencies. A graphical comparison of calculated and experimental (harmonic) frequencies corresponding to *symmetric* stretching vibrations is provided in Figure 6.26.

As commented previously for diatomics, the MP2/6-31G* model provides a better account of vibrational frequencies than any of the Hartree-Fock schemes considered. The mean absolute percentage deviation of calculated from experimental frequencies is 4.3%. Mean absolute errors of 3-21G and 6-31G* frequencies for first-row hydrides are comparable (7.1 and 8.0%, respectively) and, as before for diatomic systems, approximately twice those obtained from the correlated calcula-

TABLE 6.40. Theoretical and Experimental Vibrational Frequencies for One-Heavy-Atom Hydrides

Molecule	Symmetry of Vibration	Description of Mode	Theoretical Frequency				Experimental Frequency	
			HF/3-21G[a]	HF/3-21G(*)[b]	HF/6-31G*[c]	MP2/6-31G*[c]	Harmonic	Measured
BH_3	a_1'	s-stretch	2677	2677	2695	2647		2623[d]
	a_2'	out-of-plane deform	1211	1211	1224	1190		1125
	e'	d-stretch	2803	2803	2815	2789		2808
		d-deform	1293	1293	1304	1256		1604
CH_4	a_1	s-stretch	3187	3187	3197	3115	3137[e]	2917
	e	d-deform	1740	1740	1703	1649	1567	1534
	t_2	d-stretch	3280	3280	3302	3257	3158	3019
		d-deform	1520	1520	1488	1418	1357	1306
NH_3	a_1	s-stretch	3646	3646	3690	3504	3506[e]	3337
		s-deform	854	854	1207	1166	1022	950
	e	d-stretch	3802	3802	3823	3659	3577	3444
		d-deform	1858	1858	1849	1852	1691	1627
H_2O	a_1	s-stretch	3814	3814	4070	3772	3832[f]	3657
		bend	1800	1800	1826	1737	1648	1595
	b_1	a-stretch	3947	3947	4188	3916	3943	3756

SiH$_4$	a_1	s-stretch	2311	2398	2233	2337^g	2187
	e	d-deform	1046	1057	1052	975	975
	t_2	d-stretch	2285	2396	2385	2319	2191
		d-deform	974	1020	1016	945	914
PH$_3$	a_1	s-stretch	2404	2608	2666	2452^e	2323
		s-deform	1093	1145	1143	1041	992
	e	d-stretch	2398	2598	2602	2457	2328
		d-deform	1271	1288	1278	1154	1118
H$_2$S	a_1	s-stretch	2642	2903	2918	2722^h	2615
		bend	1323	1381	1368	1215	1183
	b_1	a-stretch	2656	2900	2930	2733	2626

[a]BPH (1980); GBPPH (1982); PFHDPB (1982); W. J. Hehre, unpublished calculations.
[b]PFHDPB (1982).
[c]HLH (1982); W. J. Hehre, unpublished calculations.
[d]A. Kaldor and R. F. Porter, J. Am. Chem. Soc., 93, 2140 (1971).
[e]J. L. Duncan and I. M. Mills, Spectrochim. Acta, 20, 523 (1964).
[f]G. Strey, J. Mol. Spectrosc., 24, 87 (1967).
[g]I. W. Levin and W. T. King, J. Chem. Phys., 37, 1375 (1962).
[h]H. C. Allen, Jr. and E. K. Plyler, J. Chem. Phys., 25, 1132 (1956).

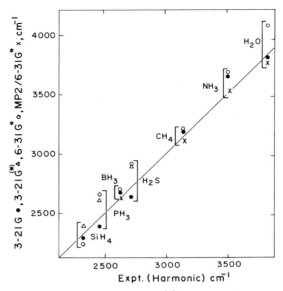

FIGURE 6.26.   Comparison of 3-21G (●), 3-21G$^{(*)}$ (△), 6-31G* (○), MP2/6-31G* (×) and experimental symmetric stretching frequencies for one-heavy-atom hydrides.

tions. The mean absolute error in 3-21G$^{(*)}$ frequencies for the second-row hydrides is similar (7.5%). A much smaller error (of 4.3%) results from calculations at the 3-21G level on second-row hydrides. This artificially low error is due to the fact that 3-21G frequencies in these compounds are both larger and smaller than experimental values, compared with 3-21G$^{(*)}$ and 6-31G* values (and 3-21G frequencies for the first-row hydrides) which are consistently larger. The difference in 3-21G and 3-21G$^{(*)}$ frequencies for second-row hydrides parallels differences already noted in bond lengths (see Section 6.2.2.a).

6.3.5.   Frequencies of Two-Heavy-Atom Hydrides

A comparison of theoretical and experimental frequencies of polyatomic two-heavy-atom hydrides comprising first- and second-row elements is presented in Table 6.41. A limited number of experimental harmonic values are available; these are included in the table in parentheses following the directly measured frequencies.

The same general conclusions as noted above apply. In particular, both HF/3-21G and HF/6-31G* frequencies are consistently larger than the experimental values, generally by 10–15%. Most, but not all, of the frequencies calculated using the MP2/6-31G* model are also larger than the corresponding experimental quantities, although the deviations here are generally about half those of the Hartree–Fock treatments. Overall, mean absolute percentage deviations of theoretical frequencies from experimental values are 12.8, 13.0, and 7.5% for the HF/3-21G, HF/6-31G*, and MP2/6-31G* methods, respectively, nearly the same as the errors observed for di-

TABLE 6.41. Theoretical and Experimental Vibrational Frequencies for Two-Heavy-Atom Hydrides

Molecule	Symmetry of Vibration	Description of Mode	Theoretical Frequency			Experimental Frequency[a]
			HF/3-21G[a]	HF/6-31G*[b]	MP2/6-31G*[c]	
LiOH	σ^+	OH stretch	4008	4258	3964	3666[e]
		OLi stretch	1161	1012	1000	630
	π	bend	464	376	295	362
B$_2$H$_6$	a_g	BH$_2$ s-stretch	2760	2754	2677	2524
		ring stretch	2256	2305	2312	2104
		BH$_2$ scis.	1287	1304	1201	1180
		ring deform.	812	829	840	794
	a_u	BH$_2$ twist	926	896	811	833
	b_{1g}	ring stretch	1915	1934	1769	1768
		BH$_2$ wag	886	900	956	850
	b_{1u}	BH$_2$ a-stretch	2864	2843	2767	2612
		BH$_2$ rock	1136	1126	899	950
		ring puckering	437	400	334	368
	b_{2g}	BH$_2$ a-stretch	2846	2829	2779	2591
		BH$_2$ rock	969	1001	895	915
	b_{2u}	ring stretch	1963	2069	2260	1915
		BH$_2$ wag	1073	1072	1186	973
	b_{3g}	BH$_2$ a-stretch	1233	1194	1330	1012
	b_{3u}	BH$_2$ a-stretch	2746	2735	2698	2525
		ring deform.	1819	1837	1746	1602
		BH$_2$ scis.	1278	1287	914	1177
HBNH	σ^+	NH stretch	4039	4120	3914	3700[f]
		BH stretch	3054	3030	2976	2800
		BH stretch	1937	1970	1834	1785
	π	BH bend	843	950	816	
		NH bend	709	667	526	460

237

TABLE 6.41. (Continued)

Molecule	Symmetry of Vibration	Description of Mode	Theoretical Frequency HF/3-21G[a]	HF/6-31G*[b]	MP2/6-31G*[c]	Experimental Frequency[d]	
HBO	σ^+	BH stretch	3094	3042	2982	2849[g]	
		BO stretch	1938	2022	1821	1817	
	π	bend	856	878	799	754	
C_2H_2	σ_g^+	CH stretch	3719	3719	3593	3374	(3497)[h]
		CC stretch	2234	2247	2006	1974	(2011)
		CH stretch	3596	3607	3516	3289	(3415)
	π_g	CH bend	918	794	444	612	(624)
	π_u	CH bend	902	883	783	730	(747)
C_2H_4	a_g	CH_2 *s*-stretch	3238	3344	3231	3026	(3153)[i]
		CC stretch	1842	1856	1724	1623	(1655)
		CH_2 scis.	1522	1499	1425	1342	(1370)
	a_u	CH_2 twist	1165	1155	1083	1023	(1044)
	b_{1g}	CH_2 *a*-stretch	3371	3394	3297	3103	(3232)
		CH_2 rock	1387	1353	1265	1236	(1245)
	b_{1u}	CH_2 wag	1115	1095	980	949	(969)
	b_{2g}	CH_2 wag	1157	1099	931	943	(959)
	b_{2u}	CH_2 *a*-stretch	3403	3420	3323	3106	(3234)
		CH_2 rock	944	897	873	826	(843)
	b_{3u}	CH_2 *s*-stretch	3305	3321	3222	2989	(3147)
		CH_2 scis.	1640	1610	1523	1444	(1473)
C_2H_6	a_{1g}	CH_3 *s*-stretch	3200	3201	3086	2954	(3043)[j]
		CH_3 *s*-deform	1571	1584	1493	1388	(1449)
		CC stretch	1004	1063	1040	995	(1016)
	a_{1u}	torsion	314	331	452	289	(303)

	Symmetry	Mode					
	a_{2u}	CH$_3$ s-stretch	3196	3194	3104	2986	(3061)
		CH$_3$ s-deform	1580	1546	1494	1379	(1438)
	e_g	CH$_3$ d-stretch	3241	3242	3228	2969	(3175)
		CH$_3$ d-deform	1677	1646	1520	1468	(1552)
		CH$_3$ rock	1351	1338	1264	1190	(1246)
	e_u	CH$_3$ d-stretch	3268	3271	3215	2985	(3140)
		CH$_3$ d-deform	1678	1652	1604	1469	(1526)
		CH$_3$ rock	921	894	783	822	(822)
HCN	σ^+	CH stretch	3691	3679	3517	3311[k]	(3442)[l]
		CN stretch	2395	2438	2038	2097	(2129)
	π	bend	990	889	702	712	(727)
HNC	σ^+	NH stretch	4018	4092	3844	3620[m]	(3842)[n]
		NC stretch	2254	2311	2038	2029	(2067)
	π	bend	721	519	389	477	(490)
CH$_2$NH	a'	NH stretch	3573	3719	3463	3297[o]	
		CH$_2$ s-stretch	3331	3347	3254	3036	
		CH$_2$ a-stretch	3231	3254	3116	2924	
		CN stretch	1845	1901	1724	1640	
		in-plane bend	1639	1628	1542	1453	
		in-plane bend	1484	1496	1412	1347	
		in-piane bend	1172	1164	1100	1059	
	a''	torsion	1308	1270	1159	1123	
		out-of-plane bend	1225	1223	1107	1063	
CH$_3$NH$_2$	a'	NH$_2$ s-stretch	3677	3730	3508	3361	
		CH$_3$ d-stretch	3230	3245	3155	2961	
		CH$_3$ s-stretch	3135	3156	3063	2820	
		NH$_2$ scis.	1851	1841	1745	1623	
		CH$_3$ d-deform	1680	1648	1539	1473	
		CH$_3$ s-deform	1614	1607	1469	1430	

239

TABLE 6.41. (Continued)

Molecule	Symmetry of Vibration	Description of Mode	Theoretical Frequency			Experimental Frequency[a]
			HF/3-21G[a]	HF/6-31G*[b]	MP2/6-31G*[c]	
CH$_3$NH$_2$		CH$_3$ rock	1256	1289	1237	1130
		CN stretch	1095	1149	1113	1044
		NH$_2$ wag	757	946	941	780
	a''	NH$_2$ a-stretch	3773	3813	3641	3427
		CH$_3$ d-stretch	3264	3281	3228	2985
		CH$_3$ d-deform	1702	1665	1596	1485
		NH$_2$ twist	1456	1479	1405	1419
		CH$_3$ rock	1047	1052	915	1195
		torsion	317	341	351	268
H$_2$CO	a_1	CH$_2$ s-stretch	3161	3159	3019	2783 (2944)[p]
		CO stretch	1916	2028	1786	1746 (1764)
		CH$_2$ scis.	1692	1680	1567	1500 (1563)
	b_1	CH$_2$ a-stretch	3233	3231	3064	2843 (3009)
		CH$_2$ rock	1378	1384	1249	1249 (1287)
	b_2	CH$_2$ wag	1337	1336	1194	1167 (1191)
CH$_3$OH	a'	OH stretch	3868	4117	3785	3681
		CH$_3$ d-stretch	3294	3305	3201	3000
		CH$_3$ s-stretch	3177	3185	3065	2844
		CH$_3$ d-deform	1698	1664	1552	1477
		CH$_3$ s-deform	1638	1638	1542	1455
		OH bend	1479	1508	1424	1345
		CH$_3$ rock	1152	1187	1120	1060
		CO stretch	1092	1164	1082	1033
	a''	CH$_3$ d-stretch	3217	3231	3140	2960
		CH$_3$ d-deform	1686	1652	1562	1477

Molecule	Symmetry	Assignment					
		CH₃ rock	1254	1289	1160	1165	
		torsion	360	348	250	295	
HCF	a'	CH stretch	2969	3056	2906	3000^k	
		bend	1520	1571	1498	1403	
		CF stretch	1268	1333	1257	1182	
CH₃F	a_1	CH₃ s-stretch	3228	3232	3110	2930	(3031)^q
		CH₃ s-deform	1663	1652	1549	1464	(1490)
		CF stretch	1143	1186	1102	1049	(1059)
	e	CH₃ d-stretch	3293	3312	3205	3006	(3132)
		CH₃ d-deform	1686	1653	1556	1467	(1498)
		CH₃ rock	1278	1312	1213	1182	(1206)
CH₃SiH₃	a_1	CH₃ s-stretch	3175			2898	
		SiH₃ s-stretch	2401			2169	
		CH₃ s-deform	1487			1260	
		SiH₃ s-deform	1044			940	
		CSi stretch	739			700	
	a_2	torsion	200			187	
	e	CH₃ d-stretch	3238			2982	
		SiH₃ d-stretch	2376			2166	
		CH₃ d-deform	1625			1403	
		SiH₃ d-deform	1055			980	
		CH₃ rock	992			869	
		SiH₃ rock	582			540	
CH₃SH	a'	CH₃ d-stretch	3301	3322		3000	
		CH₃ s-stretch	3224	3236		2931	
		SH stretch	2890	2908		2572	
		CH₃ s-deform	1662	1655		1475	
		CH₃ d-deform	1541	1521		1319	
		CH₃ rock	1233	1220		976	

TABLE 6.41. (Continued)

Molecule	Symmetry of Vibration	Description of Mode	Theoretical Frequency			Experimental Frequency[d]
			HF/3-21G[a]	HF/6-31G*[b]	MP2/6-31G*[c]	
CH₃SH		SH bend	891	873		803
		CS stretch	745	776		708
	a''	CH₃ d-stretch	3301	3325		3000
		CH₃ d-deform	1638	1623		1430
		CH₃ rock	1093	1080		1074
		torsion	240	265		
CH₃Cl	a₁	CH₃ s-stretch	3251			2937
		CH₃ s-deform	1546			1355
		CCl stretch	715			732
	e	CH₃ d-stretch	3354			3039
		CH₃ d-deform	1619			1452
		CH₃ rock	1141			1017
trans-N₂H₂	a_g	NH s-stretch	3419	3578	3386	3128[r]
		NH bend	1748	1911	1749	1583
		NN stretch	1658	1763	1523	1529
	a_u	torsion	1451	1472	1328	1359[s]
	b_u	NH a-stretch	3467	3613	3334	3131
		NH bend	1447	1472	1308	1286
N₂H₄	a	NH stretch	3751	3827	3621	3325[t]
		NH stretch	3634	3726	3500	3280
		NH bend	1870	1869	1755	1587
		NH bend	1399	1468	1382	1275
		NN stretch	1177	1226	1161	1098
		NN stretch	803	975	910	780
		torsion	496	447	426	377[k]

Molecule	Sym.	Mode					
	b	NH stretch	3757	3832	3630	3350	
		NH stretch	3618	3713	3477	3314	
		NH bend	1849	1858	1755	1628	
		NH bend	1389	1435	1363	1275	
		NH bend	946	1108	1069	966	
HNO	*a'*	NH stretch	3212	3551	2999	3039[u]	(3039)[u]
		bend	1678	1979	1586	1593	(1505)
		NO stretch	1629	1733	1479	1505	(1564)
trans-NH₂OH	*a'*	OH stretch	3860	4126	3787	3656[v, w]	
		NH stretch	3598	3718	3470	3297	
		NH₂ bend	1845	1855	1723	1605	
		OH bend	1495	1556	1439	1357	
		NH₂ wag	1196	1276	1200	1120	
		NO stretch	1065	1122	957	895	
	a''	NH stretch	3698	3810	3585	3350	
		NH₂ rock	1392	1482	1336		
		torsion	450	434	286	430	
H₂O₂	*a*	OH stretch	3827	4091	3710	3599	
		OH bend	1645	1630	1390	1402	
		OO stretch	1169	1161	926	877	
		torsion	106	397	329	371	
	b	OH stretch	3845	4095	3731	3608	
		OH bend	1278	1494	1294	1266	
HOF	*a'*	OH stretch	3752	4069	3710	3537	
		OF stretch	1490	1592	1407	1393	
		bend	1202	1176	986	886	
HOCl	*a'*	OH stretch	3796			3609	
		bend	1393			1242	
		OCl stretch	804			725	

TABLE 6.41. (Continued)

Molecule	Symmetry of Vibration	Description of Mode	Theoretical Frequency			Experimental Frequency[a]
			HF/3-21G[a]	HF/6-31G*[b]	MP2/6-31G*[c]	
SiH₃F	a_1	SiH₃ s-stretch	2460			2206
		SiH₃ s-deform	1156			990
		SiF stretch	998			872
	e	SiH₃ d-stretch	2434			2196
		SiH₃ d-deform	1037			956
		SiH₃ rock	825			728
SiH₃Cl	a_1	SiH₃ s-stretch	2451			2201
		SiH₃ s-deform	1062			949
		SiCl stretch	577			551
	e	SiH₃ d-stretch	2442			2195
		SiH₃ d-deform	1048			954
		SiH₃ rock	729			664
H₂S₂	a	SH stretch	2884	2904		2556
		SH bend	1022	1014		883
		SS stretch	549	568		509
		torsion	467	462		416
	b	SH stretch	2883	2905		2559
		SH bend	1029	1019		886

[a] 3-21G(*) for molecules containing second-row elements. Data from: BPH (1980); PSKDBWHH (1981); PFHDPB (1982); W. J. Hehre, unpublished calculations.

[b] FPHBGDP (1982); HLH (1982); W. J. Hehre, unpublished calculations.

[c] HLH (1982); W. J. Hehre, unpublished calculations.

[d] Harmonic frequencies where available are given in parentheses following directly measured values.

[e] M. W. Chase, J. L. Curnutt, A. T. Hu, H. Prophet, A. N. Syverud, and L. C. Walker, *J. Phys. Chem. Ref. Data,* **3,** 311 (1974).

[f] E. R. Lory and R. F. Porter, *J. Am. Chem. Soc.,* **95,** 1766 (1973).

[g] E. R. Lory and R. F. Porter, *J. Am. Chem. Soc.,* **93,** 6301 (1971).

[h] G. Strey and I. M. Mills, *J. Mol. Spectrosc.,* **59,** 103 (1976).

[i] J. L. Duncan, D. C. McKean, and P. D. Mallinson, *J. Mol. Spectrosc.,* **45,** 221 (1973).

[j] G. E. Hansen and D. M. Dennison, *J. Chem. Phys.,* **20,** 313 (1952).

[k] D. R. Stull and H. Prophet, JANAF Thermochemical Tables, National Bureau of Standards, Washington, D.C. (1971).

[l] G. Strey and I. M. Mills, *Mol. Phys.,* **26,** 129 (1973).

[m] D. E. Milligan and M. E. Jacox, *J. Chem. Phys.,* **47,** 278 (1967).

[n] R. A. Creswell and A. G. Robiette, *Mol. Phys.,* **36,** 869 (1978).

[o] M. E. Jacox and D. E. Milligan, *J. Mol. Spectrosc.,* **56,** 333 (1975).

[p] J. L. Duncan and P. D. Mallinson, *Chem. Phys. Lett.,* **23,** 597 (1973).

[q] J. L. Duncan, D. C. McKean, and G. K. Speirs, *Mol. Phys.,* **24,** 553 (1972).

[r] V. E. Bondybey and J. W. Nibler, *J. Chem. Phys.,* **58,** 2125 (1973).

[s] A. Trombetti, *J. Chem. Soc., A,* 1086 (1971).

[t] L. M. Sverdlov, M. A. Kovner, and E. P. Krainov, *Vibrational Spectra of Polyatomic Molecules,* Wiley, New York, 1974.

[u] S. Carter, I. M. Mills, and J. N. Murrell, *J. Chem. Soc., Faraday Trans. II,* **75,** 148 (1979).

[v] P. A. Giguere and I. D. Liu, *Can. J. Chem.* **30,** 948 (1952).

[w] The frequency of 765 cm^{-1} originally assigned to the a'' NH$_2$ rocking motion (see footnote v) has been challenged: K. Tamagake, Y. Hamada, J. Yamaguchi, A. Y. Hirakawa, and M. Tsuboi, *J. Mol. Spectrosc.,* **49,** 232 (1974).

atomics and one-heavy-atom hydrides. These errors would be significantly reduced were the data available to correct the measured frequencies for anharmonicity. Considering only molecules for which a complete harmonic analysis has been performed, the mean absolute percentage deviations in calculated frequencies from measured (uncorrected) values are 10.8, 10.6, and 5.3% at HF/3-21G, HF/6-31G*, and MP2/6-31G*, respectively. These errors are diminished to 8.0, 7.8, and 2.8% when harmonic frequencies are compared.

Effects of electron correlation on calculated frequencies are largest for molecules with multiple bonds. This parallels the effects already noted for the equilibrium structures of the same set of compounds (see Section 6.2.3b). For example, while correlation lowers the frequency primarily associated with the CC stretch in ethane by 2%, the corresponding effect is much larger in both ethylene (7%) and in acetylene (11%).

6.3.6. Frequencies of Larger Polyatomic Molecules

Frequencies calculated at the HF/3-21G level for larger polyatomic molecules follow the same general trends noted for smaller systems. That is, they are consistently larger than directly measured (uncorrected) values by 10–15%. A selection of results for hydrocarbons is provided in Table 6.42. These data have been abstracted from a larger collection of 3-21G frequencies for polyatomic molecules [PSKDBWHH (1981)].

6.3.7. Frequencies of Molecules Poorly Represented by Conventional Valence Structures

Data from the preceding tables suggest that, within the framework of the single-determinant Hartree–Fock model, the 3-21G basis set performs comparably with 6-31G* in reproducing experimental vibrational frequencies. It is to be expected, however, that problems may arise for molecules whose structures are not well described at the 3-21G level. In particular, frequencies calculated for molecules not adequately represented in terms of normal-valent structures are likely to be affected greatly by the addition of d functions to the basis; $3-21G^{(*)}$ and $6-31G^*$ calculations should provide a better description than those at the 3-21G level. The data in Table 6.43 confirm these expectations. Consistent with the performance of the theory for normal-valent species, $3-21G^{(*)}$ and $6-31G^*$ frequencies are uniformly higher than the measured values (and the available harmonic frequencies). The mean absolute percentage deviations of $3-21G^{(*)}$ and $6-31G^*$ frequencies from directly measured values are 12.4 and 13.9%, respectively. Comparison with experimental harmonic frequencies leads to a slight lowering of the errors (to 11.8 and 12.8% for the $3-21G^{(*)}$ and $6-31G^*$ levels), far smaller than the reductions previously noted for normal-valent compounds. It would appear that the harmonic frequencies for hypervalent compounds (at least those dealt with here) do not show upward shifts as large as previously noted for normal-valent systems. Frequencies calculated using the 3-21G basis set are (in the mean) closer to directly measured values (mean abso-

TABLE 6.42. Theoretical and Experimental Vibrational Frequencies for Three-Carbon and Larger Hydrocarbons

Molecule	Symmetry of Vibration	Frequency	
		HF/3-21G[a]	Expt.
CH_2CCH_2	a_1	3309	3015
		1664	1443
		1198	1073
	b_1	969	865
	b_2	3307	3007
		2224	1957
		1594	1398
	e	3381	3086
		1170	999
		1050	841
		412	355
CH_3CCH	a_1	3661	3334
		3203	2918
		2408	2142
		1586	1382
		954	931
	e	3263	3008
		1657	1452
		1205	1053
		912	633
		420	328
$CH_2CH_2CH_2$	a_1'	3332	3038
		1656	1479
		1216	1188
	a_1''	1269	1126
	a_2'	1272	1070
	a_2''	3421	3103
		920	854
	e'	3319	3025
		1622	1438
		1208	1029
		910	866
	e''	3397	3082
		1327	1188
		834	739
$HCCCCH$	σ_g	3657	3293
		2512	2184
		939	874
	σ_u	3654	3329
		2294	2020
	π_g	926	627
		598	482

247

TABLE 6.42. (Continued)

Molecule	Symmetry of Vibration	Frequency	
		HF/3-21G[a]	Expt.
	π_u	914	630
		243	231
trans-CH$_2$CHCHCH$_2$	a_g	3396	3087
		3324	3003
		3310	2992
		1873	1630
		1636	1438
		1461	1280
		1339	1196
		950	894
		578	512
	a_u	1180	1013
		1112	908
		588	522
		165	162
	b_g	1159	976
		1097	912
		869	770
	b_u	3397	3101
		3329	3055
		3316	2984
		1809	1596
		1576	1381
		1474	1294
		1135	990
		333	301
CH$_3$CCCH$_3$	a_1'	3198	2916
		2565	2240
		1590	1380
		737	725
	a_1''	28	0
	a_2''	3198	2938
		1583	1382
		1182	1152
	e'	3256	2973
		1659	1456
		1204	1054
		233	213
	e''	3256	2966
		1660	1448
		1208	1029
		596	371

[a] PSKDBWHH (1981).

TABLE 6.43. Theoretical and Experimental Vibrational Frequencies for Molecules Poorly Represented by Conventional Valence Structures

Molecule	Symmetry of Vibration	Description of Mode	Theoretical Frequency			Experimental Frequency[d]
			HF/3-21G[a]	HF/3-21G$^{(*)}$[b]	HF/6-31G*[c]	
PF_5	a_1'	PF_3 s-stretch	867	917		816
	a_2''	PF_2 s-stretch	786	784		648
		PF_2 a-stretch	1164	1170		947
	e'	PF_3 deform.	545	607		575
		PF_3 d-stretch	1180	1193		1024
	e''	PF_3 d-deform.	508	563		533
		PF bend	476	540		520
		PF bend	191	187		174
SO_2	a_1	s-stretch	1114	1341	1357	1151 (1167)[e]
	b_1	bend	1297	1573	1568	1362 (1381)
		a-stretch	497	602	592	518 (526)
SO_3	a_1'	s-stretch	926	1192	1227	1065 (1048)[f]
	a_2''	op.-deform.	356	543	561	498 (504)
	e'	d-stretch	1152	1537	1555	1391 (1409)
		d-deform.	440	588	584	530 (539)

[a]FPHBGDP (1982); W. J. Hehre, unpublished calculations.
[b]PFHDPB (1982); W. J. Hehre, unpublished calculations.
[c]FPHBGDP (1982).
[d]Harmonic frequencies where available are noted in parentheses following the directly measured values.
[e]R. D. Shelton, A. H. Nielson, and W. H. Fletcher, J. Chem. Phys., 21, 2178 (1953).
[f]A. J. Dorney, A. R. Hoy, and I. M. Mills, J. Mol. Spectrosc., 45, 253 (1973).

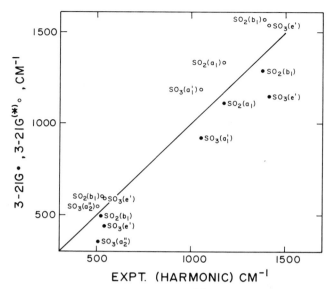

FIGURE 6.27. Comparison of 3-21G (•), 3-21G$^{(*)}$ (○) and experimental (harmonic) frequencies for SO_2 and SO_3.

lute error of 13.6%) than they are to corrected (harmonic) frequencies (error of 16.7%). At this level, calculated frequencies are not always larger than the corresponding measured quantities (as are 3-21G$^{(*)}$ and 6-31G* frequencies), and comparison with harmonic frequencies (which are larger than measured values) does not necessarily lead to improved agreement.

Figure 6.27 compares calculated frequencies for SO_2 and SO_3 with experimental harmonic values. Here it is easily seen that the scatter in the 3-21G data is much greater than that resulting from the 3-21G$^{(*)}$ calculations. This parallels the previously noted instability of the 3-21G basis set to account properly for the equilibrium structures of hypervalent compounds (see Section 6.2.5).

6.3.8. Isolated CH Stretching Frequencies

Although errors in calculated frequencies are not insignificant, they appear to be *systematic* rather than *random*. Therefore, errors should cancel, at least partially, when closely related systems are compared. CH stretching frequencies provide an example. For CH_2 and CH_3 groups, the vibrational level structure is complicated by bond–bond interactions and by the effects of Fermi resonance. These problems can be avoided experimentally by replacing all protons but one by deuterium. The properties of the "individual" CH bonds can be scrutinized, since the frequency associated with a single CH stretching motion is isolated. A plot of measured versus 3-21G CH stretching frequencies appears in Figure 6.28. Although the theoretical values are, as expected, consistently too large, variations in CH stretching frequencies from compound to compound are reasonably reproduced. Even subtle effects

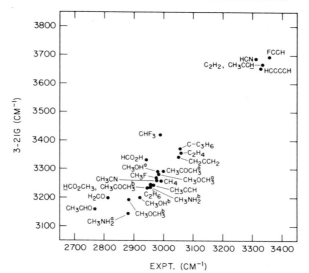

FIGURE 6.28. Comparison of 3-21G and experimental isolated CH stretching frequencies. From: PSKDBWHH (1981). Where more than one kind of hydrogen is present, data point corresponds to bond involving underlined atom. Superscripts *a* and *b* correspond to the unique and paired CH bonds in methyl rotors, respectively.

are well described. In methanol, for example, the CH bond *trans* to OH has a frequency that differs from that of the two CH linkages which are *gauche* to the OH bond. Both appear in the experimental spectrum, and the observed difference of 59 cm^{-1} is reasonably reproduced by the calculations, which give a separation of 88 cm^{-1}.

6.3.9. Calculation of Thermodynamic Properties

Calculated frequencies may be employed, along with the corresponding theoretical equilibrium geometries, to yield thermodynamic properties. Entropies of chemical reactions and equilibrium isotope effects are among the most important properties which may be calculated, and will be discussed in turn. The theoretical data may also be employed to correct experimental thermochemical information to zero Kelvin and to correct for the effects of zero-point vibrational energy.

a. ***Entropies.*** Absolute entropies, obtained according to the statistical mechanical relationships [21] (6.7-6.11):

$$S = S_{tr} + S_{rot} + S_{vib} + S_{el} - nR[\ln (nN_0) - 1] \tag{6.7}$$

$$S_{tr} = nR \left\{ \frac{3}{2} + \ln \left[\left(\frac{2 \pi MkT}{2} \right)^{3/2} \left(\frac{nRT}{P} \right) \right] \right\} \tag{6.8}$$

$$S_{rot} = nR \left\{ \frac{3}{2} + \ln \left[\frac{(\pi v_A v_B v_C)^{1/2}}{s} \right] \right\} \tag{6.9}$$

$$S_{\text{vib}} = nR \sum_i \{(u_i e^{u_i} - 1)^{-1} - \ln(1 - e^{-u_i})\} \tag{6.10}$$

$$S_{\text{el}} = nR \ln \omega_{\text{el}} \tag{6.11}$$

where
- n = moles of molecules
- R = gas constant
- N_0 = Avogadro's number
- M = mass of molecule
- k = Boltzmann's constant
- T = temperature
- h = Planck's constant
- P = pressure
- I_A, I_B, I_C = principal moments of inertia
- $v_A, v_B, v_C = h^2/8\pi I_A kT$, etc.
- s = symmetry number
- $u_i = h\nu_i/kT$
- ν_i = vibrational frequencies
- ω_{el} = electronic ground state degeneracy,

are displayed in Table 6.44. Standard entropy units (cal mol^{-1} deg^{-1}) are employed here and in all comparisons which follow. Experimental and theoretical (HF/3-21G, HF/6-31G*, and MP2/6-31G*) equilibrium structures and normal-mode vibrational frequencies have been employed consistently. Directly measured frequencies rather than corrected harmonic values have been utilized; entropies obtained using harmonic vibrational frequencies vary from the tabulated values by less than 1%. The spectroscopically derived entropies presented here differ slightly from the directly measured experimental values because of the neglect of residual (orientational) entropy present at 0 K in the crystal. Spectroscopically derived entropies are, however, more readily available, and may be compared directly with the theoretically calculated values.

Absolute entropies (Table 6.44) are predicted reasonably well by the three theoretical models. An exception is hydrogen peroxide, where the 3-21G basis set has difficulty in describing adequately the torsional motion (see Section 6.2.3a). Mean absolute deviations of the calculated entropies from the spectroscopic values are 0.36, 0.27, and 0.16 e.u. for HF/3-21G, HF/6-31G*, and MP2/6-31G* levels, respectively. Errors in calculated entropies for hydrogenation (Table 6.45), methane addition (Table 6.46) and disproportionation (Table 6.47) reactions are of comparable magnitude.

b. Equilibrium Isotope Effects. Equilibrium constants [22] for isotopic exchange processes (6.12), where A and A* differ only in isotopic composition,

$$A + B^* \overset{K_{\text{eq}}}{\rightleftharpoons} A^* + B \tag{6.12}$$

may be written in terms of the ratio of the *reduced isotopic partition function*

TABLE 6.44.   Theoretical and Experimental Absolute Entropies at 300 K^a

Molecule	Theoretical Entropy			Experimental Entropy
	HF/3-21G	HF/6-31G*	MP2/6-31G*	
H_2	31.31	31.28	31.32	31.34
LiH	40.94	40.93	40.94	40.84
BH	41.09	41.08	41.10	41.09
FH	41.63	41.52	41.61	41.54
Li_2	47.30	47.29	47.13	47.06
LiF	47.67	47.78	47.83	47.86
BeO	47.18	47.09	47.30	47.21
BF	48.04	47.91	47.98	47.92
CO	47.22	47.17	47.30	47.22
N_2	45.21	45.70	45.89	45.78
F_2	48.29	48.13	48.40	48.45
BH_3	44.96	44.96	45.00	44.80
CH_4	44.40	44.41	44.47	44.48
NH_3	46.00	45.87	45.97	46.02
H_2O	45.11	45.00	45.15	45.10
LiOH	49.10	49.96	50.77	50.42
B_2H_6	53.55	53.70	54.19	54.17
HBO	48.31	48.25	48.46	48.64
C_2H_2	47.09	47.28	48.73	48.02
C_2H_4	52.04	52.10	52.27	52.42
C_2H_6	54.24	54.13	53.89	54.57
HCN	47.75	47.85	48.29	48.22
HNC	48.12	48.78	49.67	49.03
CH_2NH	54.11	54.10	54.30	54.31
CH_3NH_2	57.41	57.03	57.20	57.73
H_2CO	52.16	52.12	52.28	52.29
CH_3OH	56.61	56.72	57.30	57.05
HCF	53.38	53.24	53.37	53.37
CH_3F	53.17	53.08	53.24	53.24
N_2H_2	52.04	51.96	52.21	52.31
N_2H_4	54.96	54.80	55.04	55.46
HNO	52.71	52.56	52.85	52.82
NH_2OH	56.01	55.76	56.67	56.29
H_2O_2	57.78	53.99	54.66	54.72
HOF	54.08	53.85	54.17	54.18

^{a}Calculated from Eqs. (6.7)–(6.11) using theoretical and experimental equilibrium structures (Tables 6.1, 6.2, 6.6, and 6.7) and normal-mode vibrational frequencies (Tables 6.39, 6.40, and 6.41). Some of these data have previously been tabulated: HLH (1982).

TABLE 6.45. Theoretical and Experimental Entropies of Hydrogenation Reactions at 300 K[a]

Hydrogenation Reaction	Theoretical Entropy			Experimental Entropy
	HF/3-21G	HF/6-31G*	MP2/6-31G*	
$Li_2 + H_2 \longrightarrow 2LiH$	3.3	3.3	3.4	3.3
$LiF + H_2 \longrightarrow LiH + HF$	3.6	3.4	3.4	3.2
$BF + H_2 \longrightarrow BH + HF$	3.4	3.4	3.4	3.4
$CH_3CH_3 + H_2 \longrightarrow 2CH_4$	3.4	3.4	3.7	3.1
$CH_3NH_2 + H_2 \longrightarrow CH_4 + NH_3$	1.7	2.0	1.9	1.4
$CH_3OH + H_2 \longrightarrow CH_4 + H_2O$	1.6	1.4	1.0	1.2
$CH_3F + H_2 \longrightarrow CH_4 + HF$	1.6	1.6	1.5	1.4
$NH_2NH_2 + H_2 \longrightarrow 2NH_3$	5.7	5.7	5.6	5.2
$NH_2OH + H_2 \longrightarrow NH_3 + H_2O$	3.8	3.8	3.1	3.5
$HOOH + H_2 \longrightarrow 2H_2O$	1.1	4.7	4.3	4.1
$HOF + H_2 \longrightarrow H_2O + HF$	1.4	1.4	1.3	1.1
$F_2 + H_2 \longrightarrow 2HF$	3.7	3.6	3.5	3.3

[a]Calculated from data in Table 6.44.

254

TABLE 6.46. Theoretical and Experimental Entropies of Methane Addition Reactions at 300 K[a]

Methane Addition Reaction	Theoretical Entropy			Experimental Entropy
	HF/3-21G	HF/6-31G*	MP2/6-31G*	
LiOH + CH$_4$ → CH$_3$OH + LiH	4.1	3.3	3.0	3.0
LiF + CH$_4$ → CH$_3$F + LiH	2.0	1.8	1.9	1.7
HB=O + CH$_4$ → H$_2$C=O + BH$_3$	4.4	4.4	4.4	4.0
BF + CH$_4$ → CH$_3$F + BH.	1.8	2.0	1.9	1.9
HC≡N + CH$_4$ → HC≡CH + NH$_3$	0.9	0.9	1.9	1.3
H$_2$C=NH + CH$_4$ → H$_2$C=CH$_2$ + NH$_3$	-0.5	-0.5	-0.5	-0.4
CH$_3$NH$_2$ + CH$_4$ → CH$_3$CH$_3$ + NH$_3$	-1.6	-1.4	-1.8	-1.6
C≡O + CH$_4$ → HC≡CH + H$_2$O	0.6	0.7	2.1	1.4
H$_2$C=O + CH$_4$ → H$_2$C=CH$_2$ + H$_2$O	0.6	0.6	0.7	0.8
CH$_3$OH + CH$_4$ → CH$_3$CH$_3$ + H$_2$O	-1.7	-2.0	-2.7	-1.9
CH$_3$F + CH$_4$ → CH$_3$CH$_3$ + HF	-1.7	-1.8	-2.2	-1.6
N≡N + CH$_4$ → HC≡N + NH$_3$	4.1	3.6	3.9	4.0
HN=NH + CH$_4$ → H$_2$C=NH + NH$_3$	3.7	3.6	3.6	3.5
NH$_2$NH$_2$ + CH$_4$ → CH$_3$NH$_2$ + NH$_3$	4.1	3.7	3.7	3.8
HN=O + CH$_4$ → H$_2$C=O + NH$_3$	1.1	1.0	0.9	1.0
HN=O + CH$_4$ → H$_2$C=NH + H$_2$O	2.1	2.1	2.1	2.1
NH$_2$OH + CH$_4$ → CH$_3$OH + NH$_3$	2.2	2.4	2.1	2.3
NH$_2$OH + CH$_4$ → CH$_3$NH$_2$ + H$_2$O	2.1	1.9	1.2	2.1
HOOH + CH$_4$ → CH$_3$OH + H$_2$O	-0.5	3.3	3.3	3.0
HOF + CH$_4$ → CH$_3$F + H$_2$O	-0.2	-0.2	-0.3	-0.3
HOF + CH$_4$ → CH$_3$OH + HF	-0.2	0.0	0.3	-0.1
F$_2$ + CH$_4$ → CH$_3$F + HF	2.1	2.1	2.0	1.9

[a]Calculated from data in Table 6.44.

TABLE 6.47. Theoretical and Experimental Entropies of Disproportionation Reactions at 300 K[a]

Disproportionation Reaction	Theoretical Entropy			Experimental Entropy
	HF/3-21G	HF/6-31G*	MP2/6-31G*	
$2H_2C{=}CH_2 \longrightarrow HC{\equiv}CH + CH_3CH_3$	-2.8	-2.8	-1.9	-2.3
$2H_2C{=}NH \longrightarrow H_2C{=}CH_2 + HN{=}NH$	-4.1	-4.1	-4.1	-3.9
$2H_2C{=}O \longrightarrow C{\equiv}O + CH_3OH$	-0.5	-0.3	0.0	-0.3
$2HN{=}NH \longrightarrow N{\equiv}N + NH_2NH_2$	-3.9	-3.4	-3.5	-3.4
$2LiOH \longrightarrow Li_2 + HOOH$		1.4	0.2	0.9
$2LiF \longrightarrow Li_2 + F_2$	0.3	-0.1	-0.1	-0.2
$2CH_3NH_2 \longrightarrow CH_3CH_3 + NH_2NH_2$	-5.6	-5.1	-5.5	-5.4
$2CH_3OH \longrightarrow CH_3CH_3 + HOOH$		-5.3	-6.1	-4.8
$2CH_3F \longrightarrow CH_3CH_3 + F_2$	-3.8	-3.9	-4.2	-3.5
$2NH_2OH \longrightarrow NH_2NH_2 + HOOH$		-2.7	-3.6	-2.4
$2HOF \longrightarrow HOOH + F_2$		-5.6	-5.3	-5.2

[a]Calculated from data in Table 6.44.

256

ratios, $(s_2/s_1)f$, for A and B:

$$K_{eq} = \frac{(s_2/s_1)f[A^*/A]}{(s_2/s_1)f[B^*/B]} \qquad (6.13)$$

which, in turn, are expressed in terms of the set of normal-mode frequencies, ν_i:

$$\left(\frac{s_2}{s_1}\right)f\left(\frac{A^*}{A}\right) = \prod_i \frac{u_i(A^*)}{u_i(A)} \frac{1 - e^{-u_i(A)}}{1 - e^{-u_i(A^*)}} e^{[u_i(A) - u_i(A^*)]/2}$$

$$u_i = \frac{h\nu_i}{kT}. \qquad (6.14)$$

By convention, A^* refers to the molecule substituted by the heavy isotope. The effect of symmetry numbers is of no particular interest, and is omitted from K_{eq}.

Table 6.48 compares theoretical and experimental reduced partition function ratios for some one- and two-heavy-atom hydrides. The theoretical quantities derive from the corresponding calculated normal-mode frequencies (based on harmonic force fields). The experimental data have been obtained in one of two ways: (a) by correcting observed frequencies for anharmonicity and fitting them to harmonic force fields (designated "harmonic" in the table), and (b) by fitting directly observed (uncorrected) frequencies to harmonic force fields (designated "anharmonic"). Since

TABLE 6.48. Comparison of Calculated and Spectroscopic Reduced Isotopic Partition Function Ratios $(s_2/s_1)f$ at 300 K

Molecule	Calculated[a]			Spectroscopic[b]	
	HF/3-21G	HF/6-31G*	MP2/6-31G*	Harmonic	Anharmonic
HD	3.86	3.85	3.71	3.56	3.32
LiD	1.76	1.76	1.74	1.74	1.71
CH_3D	14.04	13.81	12.44	11.63	10.09
NH_2D	15.89	17.23	14.65	13.65	11.60
HDO	15.76	16.39	13.22	13.23	11.50
DF	10.61	12.85	10.40	11.11	10.01
DCCH	12.20	11.63	9.35	9.02	8.30
CH_2CHD	14.49	14.19	12.53		10.19
CH_3CH_2D	15.71	15.25	13.64		11.01
DCN	13.25	11.81	9.31	8.89	8.10
H_2DCF	17.31	15.91	13.17	13.60	11.86

[a] R. F. Hout, Jr., M. Wolfsberg, and W. J. Hehre, *J. Am. Chem. Soc.*, **102**, 3296 (1980).
[b] Harmonic values derived from force constants obtained by fitting experimentally observed frequencies which have been corrected for anharmonicity. Anharmonic values derived from force constants obtained by fitting experimentally observed (uncorrected) frequencies. For details see: D. Z. Goodson, S. K. Sarpal, P. Bopp and M. Wolfsberg, *J. Phys. Chem.*, **86**, 659 (1982).

TABLE 6.49. Equilibrium Constants for Isotope Exchange Reactions $X-H + HD \rightarrow X-D + H_2$ at 300 K[a]

	Calculated			Spectroscopic[b]	
X—H	HF/3-21G	HF/6-31G*	MP2/6-31G*	Harmonic	Anharmonic
LiH	0.46	0.46	0.47	0.49	0.51
CH₄	3.6	3.6	3.4	3.3	3.0
NH₃	4.1	4.5	4.0	3.8	3.5
H₂O	4.1	4.3	3.6	3.7	3.5
HF	2.7	3.3	2.8	3.1	3.0

[a] Symmetry number effects are omitted. Theoretical data from: R. F. Hout, Jr., M. Wolfsberg, and W. J. Hehre, *J. Am. Chem. Soc.*, **102**, 3296 (1980).
[b] See footnote *b* of Table 6.48.

even slight errors in the observed frequencies can easily lead to large variations in the overall isotope effect, indirect procedures for obtaining "experimental" frequencies for isotopically related molecules are preferable to direct measurements on different isotopic species.

Table 6.49 compares theoretical and spectroscopic equilibrium constants for reactions

$$XH + HD \rightleftharpoons XD + H_2 \qquad (6.15)$$

where XH is a first-row hydride. Agreement between each of the three theoretical levels and the spectroscopically derived values is excellent over the entire range of equilibrium constants considered (experimentally from 0.51 for the LiH/H₂ exchange to 3.5 for NH₃/H₂). Overall, the MP2/6-31G* calculations fare best, although the performance of the two Hartree–Fock levels is not much worse. The performance of the HF/3-21G method is particularly encouraging.

Theoretical and spectroscopic equilibrium constants for exchange reactions

$$XH + CH_3D \rightleftharpoons XD + CH_4, \qquad (6.16)$$

involving CH bond cleavage, are compared in Table 6.50. The isotope effects here are smaller than those previously considered: spectroscopically derived values range from 0.76 for the HCN/CH₄ exchange to 1.17 for the reaction involving CH₃F and CH₄. Once again, all three levels of theory reproduce the spectroscopic equilibrium constants reasonably well. The MP2/6-31G* calculations are best overall, but the performance of the two Hartree–Fock models is only slightly inferior.

c. Temperature and Zero-Point-Energy Corrections to Experimental Thermochemical Data. Theoretical energies refer to isolated molecules at 0 K with stationary nuclei. Practical thermochemical measurements are carried out with vibrating molecules at finite temperatures (usually 298 K). Comparison of theoretical with experimental data therefore normally requires correction for zero-point vibration and for

TABLE 6.50. Equilibrium Constants for Isotope Exchange Reactions
$X—H + CH_3D \longrightarrow X—D + CH_4$ at 300 K[a]

X—H	Calculated			Spectroscopic[b]	
	HF/3-21G	HF/6-31G*	MP2/6-31G*	Harmonic	Anharmonic
C_2H_6	1.12	1.10	1.10		1.09
CH_3F	1.23	1.15	1.06	1.17	1.18
C_2H_4	1.03	1.03	1.01		1.01
C_2H_2	0.87	0.84	0.75	0.78	0.82
HCN	0.94	0.86	0.75	0.76	0.80

[a]Symmetry number effects are omitted. Theoretical data from: R. F. Hout, Jr., M. Wolfsberg, and W. J. Hehre, *J. Am. Chem. Soc.*, **102**, 3296 (1980).
[b]See footnote *b* of Table 6.48.

the finite temperature of the experimental determination. The corrections needed are straightforward, and require only the set of normal-mode vibrational frequencies, ν_i. From statistical mechanics [21], and assuming ideal gas behavior, the change in enthalpy from 0 K to some finite temperature T is given by Eqs. (6.17)-(6.20):

$$\Delta H(T) = H_{trans}(T) + H_{rot}(T) + \Delta H_{vib}(T) + RT \qquad (6.17)$$

$$H_{trans}(T) = \tfrac{3}{2} RT \qquad (6.18)$$

$$H_{rot}(T) = \tfrac{3}{2} RT \text{ (RT for a linear molecule)} \qquad (6.19)$$

$$\Delta H_{vib}(T) = H_{vib}(T) - H_{vib}(0)$$

$$= Nh \sum_{i}^{\substack{\text{normal} \\ \text{modes}}} \frac{\nu_i}{(e^{h\nu_i/kT} - 1)}, \qquad (6.20)$$

where the R, N, k, and h have the usual meanings. The zero-point energy is given by Eq. (6.21),

$$H_{vib}(0) = \epsilon_{zero\text{-}point} = \frac{1}{2} h \sum_{i}^{\substack{\text{normal} \\ \text{modes}}} \nu_i. \qquad (6.21)$$

While experimental vibrational frequencies are available for many simple molecules, systems comprising more than 10-15 atoms, as well as highly reactive or otherwise short-lived species, are generally not fully characterized. Of course, the vibrational spectra of transition structures are not amenable to measurement.

Given the uniformity of errors in calculated vibrational frequencies, even those obtained from fairly simple models, it is reasonable to use theoretical data in lieu of experimental frequencies as a basis for thermochemical corrections. This approach obviates the reliance on what might be scarce or even questionable experimental data, and allows comparisons to be made at a uniform level.

Table 6.51 compares calculated (HF/3-21G) and experimental enthalpy correc-

TABLE 6.51.   Calculated and Experimental Enthalpy Temperature Corrections and Zero-Point Vibrational Energies (kcal mol^{-1})

Molecule	$H_{300} - H_0$[a]		$E_{\text{zero-point}}$[b]	
	HF/3-21G	Expt.	HF/3-21G	Expt.
NNO	1.9	1.7	5.5	6.7
OF_2	2.0	2.0	12.2	12.9
NF_3	2.3	2.3	6.3	6.4
CO_2	1.7	1.7	6.6	7.2
CF_4	2.6	2.5	10.7	10.7
CHF_3	2.3	2.2	15.6	15.6
HCO_2H	2.1	2.0	20.1	20.4
HCCF	1.9	2.1	13.0	12.2
$trans$-$C_2H_2F_2$	2.5	2.5	22.7	22.7
CH_3CN	2.2	2.3	27.6	27.5
CH_3NC	2.4	2.5	27.2	27.5

[a] Calculated from Eqs. (6.17) to (6.20) using theoretical and experimental normal-mode vibrational frequencies from: PSKDBWHH (1981).
[b] Calculated from Eq. (6.21) using theoretical and experimental normal-mode vibrational frequencies from: PSKDBWHH (1981).

tions and zero-point energies for a selection of small molecules. The theoretical frequencies have been uniformly scaled by 0.9 since they are, on the average, approximately 10% higher than directly measured values (see Sections 6.3.5 and 6.3.6). Normal modes characterized by (scaled) frequencies less than 500 cm^{-1} and which correspond to torsions and inversions are treated as pure rotations. For each such mode, $\Delta H_{\text{vib}}(T)$ is replaced by $\frac{1}{2}RT$.

The tabulated results demonstrate the ability of the theory to reproduce both experimental enthalpy corrections and zero-point vibrational energies. Thus, even in the absence of experimental frequencies, thermochemical data may be corrected accurately and reliably.

6.3.10.   Concluding Remarks and Recommendations

Calculations of normal-mode vibrational frequencies have now become practical for molecules of moderate size. Within the framework of single-determinant Hartree–Fock theory, the 3-21G basis set is generally as successful as the larger 6-31G* representation in reproducing experimental frequencies of systems comprising first-row elements only, and appears to us to be the method of choice for widespread application. The 3-21G$^{(*)}$ basis set provides an improved description of vibrational frequencies in molecules incorporating second-row elements, and should be used in lieu of 3-21G for such systems.

At the 3-21G (3-21G$^{(*)}$) level, calculated frequencies are, in the mean, 11% larger than measured (anharmonic) values. Errors in frequencies corresponding to related motions, for example, CH stretching vibrations, are relatively constant from one molecule to another. The magnitude of the error is reduced by 3–5% when comparisons are made instead with experimental harmonic frequencies, that is, corrected for anharmonicity. Part of the remaining error is due to deficiencies in the single-determinant approximation; frequencies for one- and two-heavy-atom hydrides calculated at the MP2/6-31G* level deviate from experimental values by approximately half the amount of those evaluated from the corresponding Hartree–Fock model. Unfortunately, calculations at this level are not at present practical for molecules comprising more than 3–5 heavy atoms.

Entropies and equilibrium isotope effects, as well as temperature and zero-point-energy corrections to experimental thermochemical data, calculated using theoretical vibrational frequencies, are generally in good accord with values derived from experimental frequencies. The performance of the HF/3-21G (3-21G$^{(*)}$) model is reasonable, and it again appears to be the method of choice for general application.

6.4. MOLECULAR CONFORMATIONS AND BARRIERS TO ROTATION AND INVERSION

A complete specification of molecular geometry requires not only a description of internal bond lengths and angles, but also of conformation. This is normally defined in terms of one or more dihedral angles specifying the relative orientation of groups at opposite ends of a connecting bond. For example, different conformations of *n*-butane may be generated by rotation about the central carbon–carbon bond. The energy profile displayed in Figure 6.29 shows *trans* and *gauche* rotational isomers (*rotamers*) located at minima in the potential function. The energies of the *saddle points* connecting these stable forms correspond to *internal rotation barriers*. In the case of *n*-butane, two distinct barriers exist, one connecting equivalent *gauche* forms, the other connecting *gauche* and *trans* rotamers.

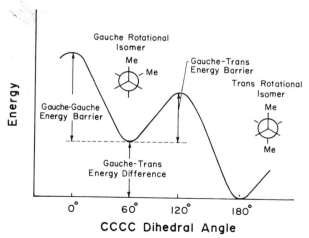

FIGURE 6.29. Energy profile describing internal rotation in *n*-butane.

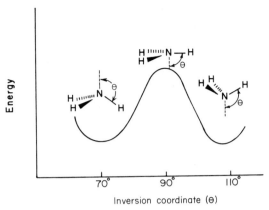

FIGURE 6.30.   Energy profile describing pyramidal inversion in ammonia.

Inversion about a pyramidal center may also lead to structural interconversion. A good example is ammonia, the energy profile for which is schematically displayed in Figure 6.30. The pyramidal forms [<(HNH) $\simeq$ 107°] are energy minima. Interconversion of equivalent structures takes place via the planar (<(HNH) = 120°) saddle point.

Relative thermochemical stabilities of stable conformers are derived principally from measurements of relative abundance under equilibrium conditions. The heights of energy barriers hindering interconversion may be obtained either from microwave or far-infrared spectroscopy for low-energy processes, for example, rotation about single bonds or inversion at nitrogen, or from nmr spectroscopy for higher-energy barriers, for example, inversion at phosphorus. Other techniques have found some application. The reader is referred elsewhere for a full account of experimental procedures [23].

Experimental rotation and inversion barriers and conformational energy differences (in kcal mol^{-1}) have been drawn from a number of sources. Specific references are provided in the tables.

A number of different geometrical models have been used to study conformational problems. For smaller systems, complete optimization of both energy minima and saddle points is feasible. For large molecules this may not always be practical, and the optimization may be restricted, that is, some parameters are fixed and only a few are varied, or dispensed with altogether. This section assesses the performance of only the two extreme geometric models: (a) complete structure optimization at all stationary points on the conformational profile, and (b) rigid rotation (inversion) in which no optimization is carried out [24].

6.4.1. Completely Optimized Rotors and Invertors

A complete investigation of the energy variation accompanying a conformational change requires relaxation of all other molecular coordinates. For example, the bond lengths and angles for *cis* and *trans* planar conformations of 1,3-butadiene

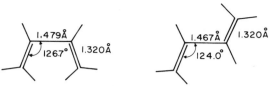

FIGURE 6.31. 3-21G heavy-atom bond lengths and skeletal bond angles for *cis*-planar (left) and *trans*-planar (right) conformers of 1, 3-butadiene.

differ significantly (Figure 6.31); in order to obtain a reliable description of the potential function for internal rotation about the central carbon–carbon bond, these should be allowed to vary. Furthermore, geometry optimization should be carried out not only for the stable conformers, but also for the transition structures separating them. Because of the additional cost of this approach (compared with the use of assumed geometries), only a limited number of studies of conformational potential energy surfaces have as yet been carried out in this manner. These generally involve relatively simple systems, a few of which are discussed below.

a. **Rotational Barriers.** Calculated and experimental barriers for rotation about single bonds are compared in Table 6.52. All four basis sets considered (STO-3G, 3-21G, 3-21G$^{(*)}$, and 6-31G*) reproduce observed trends in the methyl-group rotation barriers reasonably well. For example, the ordering of barriers in ethane, methylamine, and methanol,

$$CH_3CH_3 > CH_3NH_2 > CH_3OH,$$

as well as those of the second-row analogues, methylsilane, methylphosphine, and methanethiol,

$$CH_3SiH_3 < CH_3PH_2 > CH_3SH,$$

are given by all theoretical levels. However, none of the methods provides a completely satisfactory account of the magnitudes of the barrier heights. For example, while all basis sets satisfactorily reproduce the measured 2.9 kcal mol^{-1} rotation barrier for ethane, all yield barriers that are too high both for methanol and (except for 3-21G) for methylamine. Similarly, while all calculations underestimate the barrier in methylsilane (by as much as 0.6 kcal mol^{-1} at the 3-21G level), all reproduce the rotation barriers in both methylphosphine and methanethiol to a reasonable accuracy.

The calculated threefold rotation barriers in the borane–ammonia and borane–phosphine complexes are all smaller than the measured values. This is consistent with the fact that the calculated BN and BP bond lengths in these systems are longer than experimental values (see Section 6.2.9a).

For hydrogen peroxide (see also Section 6.2.3a), STO-3G calculations yield a *trans* barrier of only 0.1 kcal mol^{-1}, an order of magnitude smaller than the experimental value. The 3-21G basis set fares even worse, assigning a *trans* planar instead

TABLE 6.52. Calculated and Experimental Barriers to Internal Rotation[a]

Molecule	STO-3G//STO-3G	3-21G//3-21G	3-21G(*)//3-21G(*)	6-31G*//6-31G*	Expt.[b]
BH₃—NH₃	2.1	1.9	1.9	1.9	3.1
BH₃—PH₃	1.0	0.8	1.7		2.5
CH₃—CH₃	2.9	2.7	2.7	3.0	2.9
CH₃—NH₂	2.8	2.0	2.0	2.4	2.0
CH₃—OH	2.0	1.5	1.5	1.4	1.1
CH₃—SiH₃	1.3	1.1	1.4	1.4	1.7
CH₃—PH₂	1.9	1.7	2.0	2.0	2.0
CH₃—SH	1.5	1.1	1.4	1.4	1.3[c]
CH₃—CHO	1.1	1.1	1.1		1.2
HO—OH (*cis* barrier)	9.1	11.7[d]	11.7[d]	9.2	7.0
(*trans* barrier)	0.1			0.9	1.1
HS—SH (*cis* barrier)	6.1	5.7	8.8	8.5	6.8[e]
(*trans* barrier)	2.9	3.0	6.2	6.1	6.8[e]

^{a}Theoretical data from: W. J. Hehre, unpublished calculations.
^{b}Unless otherwise noted, experimental barriers summarized in: P. W. Payne and L. C. Allen, in *Modern Theoretical Chemistry*, H. F. Schaefer III, ed., Plenum Press, New York, 1977, vol. 4, p. 29.
^{c}T. Kojima, *J. Phys. Soc. (Japan)*, **15**, 1284 (1960).
^{d}The *trans* form of hydrogen peroxide is the only minimum on the 3-21G rotational potential surface. The "*cis* barrier" quoted corresponds to the calculated *cis-trans* energy difference.
^{e}Only a single barrier has been established; it is not known whether this corresponds to rotation through a *cis* or a *trans* structure.

of a near-orthogonal equilibrium conformation. Basis sets incorporating polarization functions are required for an adequate description of this rotational potential [6b, c]. At the 6-31G* level, both the calculated *trans* barrier (0.9 kcal mol^{-1}) and *cis* barrier (9.2 kcal mol^{-1}) are reasonably close to experimental values (1.1 and 7.0 kcal mol^{-1}, respectively). Only a single barrier has been established for hydrogen disulfide. This is slightly larger than the best theoretical estimate for the *trans* barrier and smaller than the estimate for the *cis* barrier.

b. Inversion Barriers. Calculated and experimental barriers hindering inversion about pyramidal nitrogen and phosphorus are compared in Table 6.53. The theoretical quantities derive from optimized equilibrium structures; a planar arrangement of bonds at nitrogen has been assumed for the transition structure for inversion of methylamine.

Neither the STO-3G nor 3-21G basis set reproduces known inversion barriers satisfactorily. The STO-3G barriers are consistently larger than the corresponding experimental quantities. This mirrors the tendency of the minimal basis set to overestimate the degree of pyramidalization at nitrogen and phosphorus (see Section 6.2.2a). The STO-3G calculations are, however, reasonably successful in reproducing the observed ordering of inversion barriers.

3-21G inversion barriers for nitrogen compounds are smaller than experimental values, again paralleling the tendency of calculations at this level to underestimate pyramidalization. The barrier calculated for phosphine is close to the experimental value.

It has previously been noted (Section 6.2.2a) that polarization-type functions are required in order to reproduce accurately the experimental bond angle in ammo-

TABLE 6.53. Calculated and Experimental Barriers to Inversion About Nitrogen and Phosphorus[a]

Molecule	STO-3G//STO-3G	3-21G//3-21G	6-31G*//6-31G*	Expt.[b]
NH$_2$—CN	3.3	0.0	1.1	1.9, 2.0
⬡—NH$_2$	2.7			$\simeq 2$
MeNH$_2$[c]	10.5	2.3	6.0	4.8
NH$_3$	11.3	1.6	6.5	5.8
▷NH	27.3	13.0		>11.6, > 12
PH$_3$	61.3	30.7 (38.8)[d]	37.8	31.5[e]

[a]Theoretical data from: W. J. Hehre, unpublished calculations.
[b]Experimental data for nitrogen compounds summarized in: J. M. Lehn, *Fort. Chem. Forsch.*, **15**, 311 (1970).
[c]Planar geometry about nitrogen assumed in the transition structure.
[d]3-21G(*)//3-21G(*) barrier.
[e]R. E. Weston, *J. Am. Chem. Soc.*, **76**, 2645 (1954).

TABLE 6.54. Calculated and Experimental Barriers to Ring Inversion in Cyclobutane

Method	Puckering Angle[a] (degrees)	Barrier to Ring Inversion
STO-3G[b]	173	0.01
6-31G*[c]	165	0.90
Experimental	160 $(+10, -20)^d$, 145[e] 143 $(\pm 6)^f$, 146 $(\pm 0.5)^g$, 145[h], 153[i]	1.14[f], 1.28[g], 1.44[h], 1.48[k]

[a] Angle between the planes made by carbons C_2, C_1, and C_4 and carbons C_2, C_3, and C_4.
[b] W. J. Hehre and J. A. Pople, *J. Am. Chem. Soc.*, **97**, 6941 (1975).
[c] D. Cremer, *J. Am. Chem. Soc.*, **99**, 1307 (1977); K. B. Wiberg, *ibid.*, **105**, 1227 (1983).
[d] J. D. Dunitz and V. Schomaker, *J. Chem. Phys.*, **20**, 1703 (1952).
[e] A. Almenningen, O. Bastiansen, and P. N. Skancke, *Acta Chem. Scand.* **15**, 711 (1961).
[f] D. A. Dows and N. Rich, *J. Chem. Phys.*, **47**, 333 (1967).
[g] T. Ueda and T. Shimanouchi, *J. Chem. Phys.*, **49**, 470 (1968).
[h] J. M. R. Stone and I. M. Mills, *Mol. Phys.*, **18**, 631 (1970).
[i] S. Meiboom and L. C. Snyder, *J. Chem. Phys.*, **52**, 3587 (1970).
[j] G. W. Rathjens, Jr., N. K. Freeman, W. D. Gwinn, and K. S. Pitzer, *J. Am. Chem. Soc.*, **75**, 5634 (1953).
[k] F. A. Miller and R. J. Capwell, *Spectrochim. Acta*, **27a**, 947 (1971).

nia. They are also needed for a satisfactory description of the inversion barrier; the 6-31G* result agrees reasonably well with the experimental barrier.

Table 6.54 compares calculated and experimental ring inversion barriers in cyclobutane. The STO-3G barrier is far too small. At this level of theory, the heavy-atom skeleton is calculated to be nearly planar. The 6-31G* ring-inversion barrier, while still too low, is in better agreement with experiment.

c. Effect of Electron Correlation on Rotation and Inversion Barriers.

The data in Table 6.55 suggest that effects of electron correlation on rotation barriers (as evaluated from the MP2/6-31G* model) are small, typically on the order of 0.1–0.2 kcal mol^{-1}. Except for the *trans* barrier in hydrogen peroxide, correlation effects lead to an increase in rotation barrier. This might have been anticipated in $BH_3 NH_3$, where introduction of correlation also leads to a decrease in the central bond length (see Table 6.30), but seems unexpected in the remaining systems, where correlation results instead in the lengthening of the central linkage, sometimes significantly so (see Table 6.7). Calculations at higher levels [25a] suggest that the residual errors in the methyl rotational barriers are due largely to deficiencies in the basis set rather than to the incomplete treatment of electron correlation.

The effects on inversion barriers are also small. Calculations near the Hartree-Fock limit for ammonia [25b] result in an inversion barrier that is too small (by

TABLE 6.55.   Effect of Electron Correlation on Barriers to Rotation and Inversion[a]

Molecule	6-31G*//6-31G*	MP2/6-31G*//MP2/6-31G*	Expt.[b]
Rotation			
BH_3—NH_3	1.9	2.1	3.1
CH_3—CH_3	3.0	3.1	2.9
CH_3—NH_2	2.4	2.6	2.0
CH_3—OH	1.4	1.5	1.1
HO—OH *cis* barrier	9.2	9.4	7.0
trans barrier	0.9	0.6	1.1
Inversion			
NH_3	6.5	6.6	5.8
CH_3NH_2[c]	6.0	6.5	4.8

[a] Theoretical data from: W. J. Hehre, unpublished calculations.
[b] See footnotes to Tables 6.52 and 6.53 for references to experimental data.
[c] Planar geometry about nitrogen assumed in the transition structure.

0.6 kcal mol^{-1}), implying a total correlation effect of this amount (acting to increase the barrier). While the HF/6-31G* model already overestimates the experimental ammonia inversion barrier (by 0.7 kcal mol^{-1}), the correlation effect is still in the direction of increasing this barrier.

d.   Conformational Isomers.   Calculated energies for the conformational isomers of a selection of small molecules, obtained from the STO-3G//STO-3G and 3-21G// 3-21G models, are compared with experimental values in Table 6.56. As expected, both models show the lowest-energy conformation for *n*-butane to be *trans*; the calculated *gauche–trans* energy differences are also in accord with experiment. Higher-level calculations have been applied to the investigation of the *gauche–trans* energy difference in *n*-butane (and to the connecting energy barriers) [25c]. At the HF/ 6-311G** level, the *trans* conformer is 1.0 kcal mol^{-1} more stable than the *gauche*, overestimating the experimental energy separation of 0.77 kcal mol^{-1}. On the other hand, the corresponding MP3/6-311G** calculations yield a difference of 0.6 kcal mol^{-1}, somewhat less than the experimental value.

Experimentally, the lowest-energy form of 1-butene is known to be *cis*; the energy of a second *skew* form is only slightly (0.2 kcal mol^{-1}) higher. The theoretical calculations are in qualitative agreement, although the calculated (*cis–skew*) energy differences are overestimated by 0.6 kcal mol^{-1} for both the STO-3G and 3-21G calculations.

Both experiment and theory assign the lowest-energy conformation of 1,3-buta-diene to a planar *trans* arrangement. A second minimum-energy form exists, although

TABLE 6.56. Calculated and Experimental Relative Conformational Energies

Molecule	Conformational Energy Difference	STO-3G//STO-3G[a]	3-21G//3-21G[b]	Expt.
n-Butane	trans/gauche	0.9	0.8	0.77[c]
1-Butene	skew/cis	0.8	0.8	0.2[d]
1,3-Butadiene	trans/gauche	1.8[e]	3.5	1.7[f], >2[g], 2.5[h]
Acrolein	trans/cis	0.5	0.0	2.0[i], 2.06[j]
Glyoxal	trans/cis	1.5	5.1	3.2[k]

[a] A. J. P. Devaquet, R. E. Townshend, and W. J. Hehre, *J. Am. Chem. Soc.*, **98**, 4068 (1976), and W. J. Hehre, unpublished calculations.

[b] W. J. Hehre, unpublished calculations.

[c] G. J. Szasz, N. Sheppard, and D. H. Rank, *J. Chem. Phys.*, **16**, 704 (1948).

[d] A. A. Bothner-By, C. Naar-Colin, and H. Gunther, *J. Am. Chem. Soc.*, **84**, 2748 (1962); A. A. Bothner-By, and H. Gunther, *Disc. Faraday Soc.*, **34**, 127 (1962); S. Kondo, E. Hirota, and Y. Morino, *J. Mol. Spectrosc.*, **28**, 471 (1968).

[e] Second minimum-energy form is *cis* planar.

[f] E. B. Reznikowa, V. I. Tulin, and W. M. Tatevskii, *Opt. Spektrosk.*, **13**, 200 (1962).

[g] S. Novick, J. M. Lehn, and W. Klemperer, *J. Am. Chem. Soc.*, **95**, 8189 (1973).

[h] J. G. Aston, G. Szasz, H. W. Woolley, and F. G. Brickwedde, *J. Chem. Phys.*, **14**, 67 (1946).

[i] A. C. P. Alves, J. Christoffersen, and J. M. Hollas, *Mol. Phys.*, **20**, 625 (1971).

[j] M. S. DeGroot and J. Lamb, *Proc. Roy. Soc. (London), Ser. A*, **242**, 36 (1957).

[k] G. N. Currie and D. A. Ramsay, *Can. J. Phys.*, **49**, 317 (1971).

it remains to be established experimentally whether its conformation is *cis* planar or whether the carbon skeleton is twisted slightly from the plane. The energy of this form (relative to the *trans* structure) has been established with reasonable certainty (2.5 ± 0.5 kcal mol^{-1} from calorimetric measurements, 1.7 ± 0.5 kcal mol^{-1} from the temperature dependence of the infrared spectrum, and >2 kcal mol^{-1} from molecular beam investigations). The STO-3G calculations suggest the second species to be the *cis* conformer with an energy 1.8 kcal mol^{-1} higher than that of the ground-state *trans* conformation. On the other hand, the 3-21G calculations suggest that the second energy minimum, which lies 3.5 kcal mol^{-1} above the lowest-energy conformer, adopts a geometry that is twisted slightly from a *cis* planar arrangement.

Both STO-3G and 3-21G calculations indicate the lowest-energy conformation of acrolein, like that of 1,3-butadiene, to be *trans* planar. Both levels of theory suggest that the second conformational minimum is *cis* planar rather than *gauche*, in contrast to results for 1,3-butadiene. Although the geometry of the second form is not known experimentally, its energy separation from the lowest-energy conformer is reasonably well established. Ultrasonic relaxation experiments indicate the *cis* (or *gauche*) structure to lie 2.0 kcal mol^{-1} above the *trans*, while a study of the temperature dependence of the ultraviolet spectrum indicates an energy difference of 2.06 kcal mol^{-1}. According to the 3-21G calculations, the two conformers are of equal stability; the STO-3G model suggests a small energy separation (0.5 kcal mol^{-1}), the *trans* form being preferred.

As was the case for acrolein, but not for 1,3-butadiene, both minimum energy conformers of glyoxal are calculated at both theoretical levels to have planar structures. Again, the *trans* form is more stable. The available experimental evidence concurs. The STO-3G (1.5 kcal mol^{-1}) and 3-21G (5.1 kcal mol^{-1}) energy differences respectively underestimate and overestimate the experimental separation (3.2 kcal mol^{-1}).

6.4.2. Rigid-Rotor Approximation

The computation required to map a conformational profile completely, taking care to obtain fully optimized structures along the way, may be significant for large molecules. There is ample incentive to utilize a simpler approach. In the rigid-rotor model, all bond lengths and bond angles are held fixed during rotation; they are assigned "standard" or experimental values, or alternatively, values derived from full optimization of the lowest-energy conformer. In the last instance, the energies of higher-energy conformers, as well as those of barriers interconnecting stable forms, necessarily represent upper bounds to values that would result from complete optimization of all structures.

a. Rotational Barriers. Rotational barriers for a small selection of substituted benzenes, obtained within the rigid-rotor approximation using STO-3G wavefunctions and standard model geometries, are compared with experimental values in Table 6.57. The theory is not very successful here (see, however, Section 6.4.2b). Not only are absolute barrier heights in error by as much as a factor of two, for example, nitrobenzene, but also the observed ordering of barriers is not reproduced.

TABLE 6.57.   Calculated and Experimental Rotational
Barriers in Substituted Benzenes: Rigid-Rotor Model[a]

Molecule	STO-3G//Std.	Expt.
C_6H_5—OH	5.2	3.3, 3.4
C_6H_5—CHO	6.6	4.9, 4.7
C_6H_5—NO	4.8	3.9
C_6H_5—NO$_2$	5.8	2.9

[a] Experimental and theoretical data summarized in: W. J. Hehre, L. Radom, and J. A. Pople, *J. Am. Chem. Soc.*, **94**, 1496 (1972).

It is not clear whether the errors here should be attributed to the limited basis set or to the rigid-rotor model.

b.   Remote Substituent Effects.   Although the absolute magnitudes of rotational and inversion barriers calculated using single-determinant, small basis set models are often significantly in error, even the simplest levels of theory appear capable of accurately reproducing *changes in barrier heights* due to remote substitution. For example, while the STO-3G basis set overestimates the twofold rotational barrier in phenol (5.2 kcal mol^{-1} compared with 3.3–3.4 kcal mol^{-1} experimentally, see Table 6.57), the calculated effects of *para* substituents on this barrier, that is, the energies of reactions

(6.22)

are in excellent accord with the corresponding experimental values (Table 6.58). An analysis of the substituent effects in these systems is presented in Section 7.2.2.

6.5.   THERMOCHEMICAL STABILITIES OF MOLECULES

This section documents the strengths and limitations of the theoretical models with respect to the description of a wide variety of energetic comparisons.

Following a brief discussion of sources of error (Section 6.5.1), we examine the performance of Hartree–Fock and correlated models in the calculation of *homolytic bond dissociation energies* by direct evaluation of the change in energy for bond-breaking processes (Section 6.5.2). These are reactions that result in the unpairing of electrons, and electron correlation effects would be expected to be large. We also

TABLE 6.58. Calculated and Experimental Rotational
Barriers in *para*-Substituted Phenols, Relative to Phenol[a]

Substituent R	Barrier Relative to Phenol	
	STO-3G//Std.	Expt.
OH	-0.95	-0.87
F	-0.53	-0.60
CH$_3$	-0.28	-0.32
H	0	0
CHO	+0.47	+0.87
CN	+0.66	+0.70
NO$_2$	+1.02	+0.98

[a] Experimental and theoretical data summarized in: L. Radom, W. J. Hehre, J. A. Pople, G. L. Carlson, and W. G. Fateley, *J. Chem. Soc., Chem. Commun.*, 308 (1972).

examine the energetics of complete *atomization* (dissociation of a molecule into its component atoms) in an attempt to assess the level of agreement between theory and experiment that might be expected given very large basis sets and extensive treatment of electron correlation.

While it is certainly desirable for a theory to give accurate bond dissociation energies, failure to do so does not preclude its use for the prediction of reaction thermochemistry. In this regard, it is particularly important to recognize that the absolute errors in theoretical total energies, that is, differences between the results of a particular molecular orbital model and the exact solutions of the Schrödinger equation, almost always exceed in magnitude the energy differences of chemical processes being investigated. Fortunately, much of the absolute error in the model calculation is relatively constant from one molecular environment to another, and quantitative comparisons can thus hope to be successful. Effective and yet practical theoretical investigations of thermochemistry require formulations that exploit the cancellations of such constant errors as much as possible. One way this may be accomplished is to deal with reactions in which changes in correlation energies would be expected to be small. These changes might then either be ignored altogether, that is, use of Hartree–Fock models, or be dealt with in an approximate manner. Conservation of the number of electron pairs is an important requirement. For hydrogenation reactions (Section 6.5.3), reactions relating multiple and single bonds (Section 6.5.4) and comparisons of stabilities of structural isomers (Section 6.5.5), it will be seen that the limiting behavior of Hartree–Fock models is generally satisfactory, that is, electron correlation effects are small. We note, however, that even here the performance of some of the simpler models, that is, HF/STO-3G and HF/3-21G, is still deficient, and that it is advantageous to seek out even more restrictive reactions. *Isodesmic* processes (Section 6.5.6), reactions in which not only the number of electron pairs is held constant but also formal chemical bond types are conserved, represent such a class. It will be shown that the energies of isodesmic reactions are generally (but not always) well described using even the simplest computational models, and the utility of such processes, both as interpretive tools and as a means of providing thermochemical data, is explored in detail.

The next two sections detail the performance of the theoretical models for two special but important general classes of reactions. Section 6.5.7 discusses the energies of hydrogen-transfer processes, that is, the evaluation of relative homolytic bond dissociation energies, and compares the performance of UHF and RHF models. Section 6.5.8 considers the energies of proton-transfer and related processes and, in particular, the calculation of gas-phase acidities and basicities.

A brief discussion of the additivity of basis set and electron correlation errors is presented in Section 6.5.9. Our purpose here is to illustrate the manner in which available theoretical data may be corrected for probable systematic errors.

One topic that is conspicuously missing in our treatment is the calculation of reaction barriers. The reason for this omission is the need for a (transition state) model to reinterpret the experimental rate data in terms of the relative stabilities of reactants and transition structures. It should be noted, however, that accurate descriptions of the energies of reaction transition structures are likely to be demanding, and that theoretical treatments beyond Hartree–Fock will often be required. Discussion is provided in Section 7.6.

Numerous compendia of experimental thermochemical data exist. These have been identified in footnotes to the tables. All comparisons are presented in units of kcal mol^{-1}

6.5.1. Sources of Error in Comparisons of Theoretical and Experimental Relative Thermochemical Stabilities

A potential source of error in the theoretical calculations is the choice of equilibrium geometry. While it may be desirable to utilize geometries optimized at the higher levels of theory, this may not always be practical. It is, therefore, fortunate that even simple theoretical models, that is, Hartree–Fock models with minimal and split-valence basis sets, are moderately successful in their description of equilibrium structure. Errors incurred from use of either experimental structures or geometries obtained from lower theoretical levels are likely to be small (on the order of 1–3 kcal mol^{-1}).

The quality of experimental thermochemical data is difficult to assess. While errors in measured heats of formation are generally placed at ±1 kcal mol^{-1} (or less), it is clear to some that existing experimental techniques often cannot achieve such an accuracy. Thus, McMillen and Golden [26] have recently observed:

The commonly invoked uncertainty of ±1 kcal/mol . . . cannot take into account unforeseen systematic errors and is therefore generally optimistic.

Clearly, comparisons of theoretical and experimental relative energies below the level of 1 kcal mol^{-1} are almost meaningless.

Finally, as already discussed in Section 6.3.9c, quantum chemical calculations refer to systems with fixed nuclei. They should therefore ideally be compared with experimental thermochemical data corrected to 0 K, and corrected for zero-point vibrational energy. Both of these corrections ultimately require knowledge of the complete set of normal-mode vibrational frequencies. While, in principle, these data are available from experiment, only for small molecules is the vibrational information complete.

In most of the following comparisons, experimental data at 298 K have been corrected to 0 K and also for zero-point vibrations. The vibrational corrections are made using theoretical (HF/3-21G or HF/3-21G$^{(*)}$) frequencies scaled by 0.9. As discussed earlier (Section 6.3.9c), such vibrational energy corrections are probably accurate to ±1 kcal mol^{-1}. Rotational and translational corrections to 0 K are based on classical statistical mechanics. Neglect of vibrational effects in this context would be expected to involve errors of less than ±1 kcal mol^{-1} (see Table 6.51).

6.5.2. Bond Dissociation Energies

As is widely recognized (see Section 2.9), Hartree–Fock theory with any basis set gives poor results for *direct* calculation of the energy of a homolytic dissociation process A—B → A + B. The correlation energy correction for the electrons forming the bond is a significant fraction of the total bond energy. If correlation is omitted, the error will be greater for the bonded system A—B than for separated A and B, and calculated dissociation energies will be too small. This is illustrated in the comparisons that follow.

a. AH Bond Energies.   Calculated AH bond dissociation energies, D_e, for a selection of one-heavy-atom hydrides comprising first-row elements only are compared with experimental values in Table 6.59. Experimental bond energies for BH_2 and BH_3 are poorly known, and have been included in the comparison only for completeness. Near the single-determinant Hartree–Fock limit, bond dissociation energies are typically in error by 25–40 kcal mol^{-1}. Energies calculated using the 6-31G** basis set are uniformly close to Hartree–Fock limiting values and thus are also poorly described; the largest deviation (from values at the Hartree–Fock limit) is 6 kcal mol^{-1} for the reaction HF → H$^{.}$ + F$^{.}$.

The theoretical results improve substantially upon treatment of correlation effects by the second-order Møller–Plesset model. The largest error is now 13 kcal mol^{-1} (for the bond dissociation energy of LiH); the corresponding HF/6-31G** error is 26 kcal mol^{-1}. Other dissociation processes are improved similarly, and most of the energies calculated at this level are between 5 and 10 kcal mol^{-1} lower than experimental findings. Only small further changes (up to 5 kcal mol^{-1}) in AH bond dissociation energies occur at higher Møller–Plesset orders. The maximum change from the MP3 to the MP4 model is only 2 kcal mol^{-1}. The perturbation series for the correlation energy appears to have nearly converged at the fourth-order level, and even the second-order calculations provide a reasonable description of the limit. The residual error in dissociation energy (the largest is 13 kcal mol^{-1} in HF), while due in part to insufficient treatment of electron correlation, is primarily a consequence of the incompleteness of the underlying Hartree–Fock basis set. This is explored in the next section.

TABLE 6.59. Calculated and Experimental A—H Bond Dissociation Energies, D_e

Bond Dissociation Reaction	HF/6-31G**//6-31G*	HF Limit[a]	MP2 /6-31G**//6-31G*	MP3 /6-31G**//6-31G*	MP4 /6-31G**//6-31G*	Expt.[b]
$H_2 \rightarrow H\cdot + H\cdot$	85	84	101	105	106	109
$LiH \rightarrow Li\cdot + H\cdot$	32		45	48	49	58
$BeH \rightarrow Be\cdot + H\cdot$	52		52	49	47	50, 56
$BH \rightarrow B\cdot + H\cdot$	62		77	79	80	82
$CH\cdot \rightarrow C + H\cdot$ ($^2\Pi$)	55	57	73	75	76	84
($^4\Sigma^-$)	62		68	66	66	67
$NH \rightarrow N\cdot + H\cdot$	50	51	71	72	73	79, $\leqslant$85
$OH \rightarrow O + H\cdot$	67	68	96	96	96	107
$FH \rightarrow F\cdot + H\cdot$	93	99	131	127	128	141
$BH_2\cdot \rightarrow BH + H\cdot$	84		88	87	85	99–130
$CH_2 \rightarrow CH\cdot + H\cdot$ ($^3B_1$)	101	99	109	108	107	105, 106
($^1A_1$)	70		89	90	91	98
$NH_2\cdot \rightarrow NH + H\cdot$	65	66	90	90	90	102
$OH_2 \rightarrow OH\cdot + H\cdot$	86	89	119	115	116	126
$BH_3 \rightarrow BH_2\cdot + H\cdot$	90		106	108	109	57–107
$CH_3\cdot \rightarrow CH_2 + H\cdot$	88	89	110	112	112	117
$NH_3 \rightarrow NH_2\cdot + H\cdot$	83	85	110	108	109	116
$CH_4 \rightarrow CH_3\cdot + H\cdot$	87	85	109	110	110	113

[a] Hartree–Fock limits summarized in: J. A. Pople and J. S. Binkley, *Mol. Phys.*, **29**, 599 (1975).
[b] Experimental bond dissociation energies have been corrected for zero-point vibrational energy, and are summarized in reference 28.

The calculations suggest that the bond energy for BH_2 is probably close to the low end of the range of experimental values (99 kcal mol^{-1}) while that for BH_3 appears to be near the high end (107 kcal mol^{-1}).

Finally, we note that experimental data are available for the $^4\Sigma^-$ excited state of CH in addition to the $^2\Pi$ ground state, and for the lowest singlet state of methylene ($^1A_1$) in addition to the ground-state triplet ($^3B_1$). These enable assessment of the performance of the theoretical models with regard to the calculation of energy differences between ground and excited states. In both cases, the HF/6-31G** calculations overestimate the stability of the state of higher spin multiplicity (relative to the state of lower spin). The theory at this level assigns the CH ground state as $^4\Sigma^-$, with the $^2\Pi$ state 7 kcal mol^{-1} higher in energy. Experimentally, the molecule possesses a doublet ground state with the lowest quartet 17 kcal mol^{-1} higher in enthalpy. The singlet–triplet separation in methylene is given as 31 kcal mol^{-1} (triplet ground state) at this level, compared with the experimental enthalpy difference of 8–9 kcal mol^{-1} [27]. The errors in these two pairs of systems at the Hartree–Fock level are approximately halved by consideration of electron correlation using the second-order Møller–Plesset expansion. MP3 and MP4 calculations lead to further improvements but significant differences with experiment remain, that is, the separation in CH is in error by 10 kcal mol^{-1} at MP4/6-31G** and that in CH_2 by 8–9 kcal mol^{-1}.

b.   Improved Calculations of CH_n Bond Dissociation Energies.

We have previously noted that errors in theoretical bond dissociation energies for one-heavy-atom hydrides are often sizable, even for calculations at the full fourth-order Møller–Plesset level using the 6-31G** basis set [28]. Here we examine the improvements both in treatment of correlation and in underlying basis set required to bring the theoretical data into better accord with experiment.

CH bond dissociation energies calculated directly from the differences of MP4/6-311G**//6-31G* total energies are provided in the first column of Table 6.60. These are similar to previously discussed results obtained using the (slightly) smaller 6-31G** basis set, and are in error from experimental bond energies (right-hand column of Table 6.60) by as much as 7.8 kcal mol^{-1}. The mean absolute error, 4.6 kcal mol^{-1}, is slightly smaller for this set of CH_n compounds than it would have been were the full selection of AH_n systems considered. The bond energies provided in the middle column of Table 6.60 result both from a change in strategy, and from systematic improvements both to the basis set and to the treatment of electron correlation, as described below. Here the largest single deviation is 2.7 kcal mol^{-1}, and the mean absolute error is 1.3 kcal mol^{-1}. Both of these are again slightly smaller than they would be were all AH_n systems to be considered. Note also that the energy separation between ground ($^2\Pi$) and first excited ($^4\Sigma^-$) states of CH is now reasonably well reproduced (19.1 kcal mol^{-1} versus 17 kcal mol^{-1} experimentally) as is the singlet–triplet splitting in methylene (7.7 kcal mol^{-1} versus 8–9 kcal mol^{-1} experimentally).

Three factors are responsible for the improvement of results.

(i) USE OF ISOGYRIC REACTIONS. Homolytic bond cleavage necessarily separates one pair of electrons. In the absence of any new electron pairs being

TABLE 6.60.   Theoretical and Experimental CH Bond Dissociation Energies, D_e, in CH_n Molecules

| | Bond Dissociation Energy $CH_n \rightarrow CH_{n-1} + H^{\textbf{.}}$ | | |
Molecule	MP4/6-311G**[a]	Corrected[a]	Expt.[b]
CH ($^2\Pi$)	76.4	83.6	84.2
($^4\Sigma^-$)	64.4	64.5	67.2
CH_2 ($^3B_1$)	104.9	104.5	105.6
($^1A_1$)	90.6	96.8	98.1
CH_3	111.6	117.8	117.4
CH_4	109.4	114.0	112.6

[a]Theoretical data from reference 28. See text for discussion.
[b]Corrected for differential zero-point vibrational energy. Experimental data summarized in reference 28.

formed, the correlation energy for the products (with one less electron pair) would be expected to be significantly less than for the reactants. More modest differences in correlation energy should result for *isogyric* reactions, processes in which the number of electron pairs is conserved. One would expect the theoretical models to fare better here, where differences in correlation effects are subtle, than for non-isogyric reactions, where they are extreme. The usual definitions of the bond energies of CH ($^4\Sigma^-$) and CH_2 ($^3B_1$) are themselves isogyric processes:

$$CH(^4\Sigma^-) \rightarrow C(^3P) + H(^2S) \tag{6.23}$$

$$CH_2(^3B_1) \rightarrow CH(^2\Pi) + H(^2S). \tag{6.24}$$

Isogyric reactions relating the bond energies in the remaining CH_n systems to that of H_2 are defined as follows:

$$CH(^2\pi) + H(^2S) \rightarrow C(^3P) + H_2(^1\Sigma_g^+) \tag{6.25}$$

$$CH_2(^1A_1) + H(^2S) \rightarrow CH(^2\Pi) + H_2(^1\Sigma_g^+) \tag{6.26}$$

$$CH_3(^2A_2'') + H(^2S) \rightarrow CH_2(^3B_1) + H_2(^1\Sigma_g^+) \tag{6.27}$$

$$CH_4(^1A_1) + H(^2S) \rightarrow CH_3(^2A_2'') + H_2(^1\Sigma_g^+). \tag{6.28}$$

Their use to establish absolute CH bond energies requires the bond energy, D_e, in H_2. This is well established experimentally as 109.49 kcal mol^{-1}.

   (ii) EXTENSION OF TREATMENT OF ELECTRON CORRELATION.   While cal-culations beyond full fourth order in Møller–Plesset perturbation theory are not yet practical even for the small systems dealt with here, it is reasonable to estimate the total (exact) correlation energy by way of the following expression [29]:

$$E^{(\text{corr})} = \frac{E^{(2)} + E^{(3)}}{1 - E^{(4)}/E^{(2)}}.$$

(6.29)

This assumes that the even and odd terms of the Møller–Plesset expansion both constitute geometric series, and furthermore that the ratio between terms is given by $E^{(4)}/E^{(2)}$. While untested at present, it does appear as reasonable a procedure as neglect of higher-order terms.

(iii)   IMPROVEMENT OF THE HARTREE–FOCK BASIS SET.   The 6-311G** basis set (described in Section 4.3.3b) may be supplemented by the addition of

1.   diffuse s- and p-type basis functions on heavy atoms,
2.   a second set of d-type polarization functions on heavy atoms, and
3.   a set of (seven) f-type functions on heavy atoms.

These improvements have been carried out independently and their relative contributions to the bond dissociation energies have been assumed to be additive.

The main conclusion that can be reached from comparison of theoretical and experimental data such as that given in Table 6.60 (and in reference 28) is that existing quantum chemical models are capable of reproducing experimental bond dissociation energies to high accuracy (± 1–2 kcal mol^{-1}). Thus, we do not agree with the skepticism on the part of many kineticists and thermochemists, as exemplified by the recent remark of Benson [30]:

Ab initio methods for solving the Schrödinger equation are not yet capable of providing results of ΔH_f to better than ± 4 kcal/mol even for relatively simple species and the prospect is not bright for any major breakthroughs in this area in the next decade or so.

While larger systems incorporating single and multiple bonds between heavy atoms have yet to be examined, and some of the conclusions made at this time may need to be modified, the results so far obtained are encouraging and open the possibility for accurate characterization of the thermochemistry of new molecules by theoretical procedures alone.

c.   AB Single-Bond Energies.   A selection of calculated AB single-bond dissociation energies for first-row two-heavy-atom hydrides are compared with experimental values in Table 6.61.

Hartree–Fock (6-31G** basis set) energies are in very poor agreement with the experimental values; errors in some cases exceed 50 kcal mol^{-1}! Both HOF and F_2 are calculated to be thermodynamically unstable with respect to OF and FF bond cleavage; a zero bond energy is calculated for the central linkage in hydrogen peroxide.

TABLE 6.61. Calculated and Experimental A—B Bond Dissociation Energies, D_e [a]

Bond Dissociation Reaction	6-31G**//6-31G*	MP2	MP3	MP4SDQ	MP4	Expt. [b]
		/6-31G**//6-31G*				
CH_3—$CH_3 \rightarrow CH_3^\cdot + CH_3^\cdot$	69	99	96	95	97	97
CH_3—$NH_2 \rightarrow CH_3^\cdot + NH_2^\cdot$	58	93	87	86	88	93
CH_3—$OH \rightarrow CH_3^\cdot + OH^\cdot$	58	98	91	91	92	98
CH_3—$F \rightarrow CH_3^\cdot + F^\cdot$	69	113	105	106	108	114
NH_2—$NH_2 \rightarrow NH_2^\cdot + NH_2^\cdot$	34	73	65	65	67	73
HO—$OH \rightarrow OH^\cdot + OH^\cdot$	0	53	43	44	47	55
HO—$F \rightarrow OH^\cdot + F^\cdot$	−11	48	38	40	43	54
F—$F \rightarrow F^\cdot + F^\cdot$	−33	35	24	27	30	38

[a]Theoretical data from: J. A. Pople, unpublished calculations. See also: J. S. Binkley and M. J. Frisch, *Int. J. Quantum Chem., Symp.*, **17**, 331 (1983).
[b]Experimental data from: D. D. Wagman, W. H. Evans, V. B. Parker, R. H. Schumm, I. Halow, S. M. Bailey, K. L. Churney and R. L. Nuttall, Selected Values for Inorganic and C_1 and C_2 Organic Substances in SI Units, *J. Phys. Chem. Ref. Data*, Suppl. no. 2 to vol. 11, 1982.

Even the simplest of the correlation schemes leads to significant improvement. At MP2/6-31G**, bond dissociation energies are typically in error by only a few kcal mol^{-1}; the largest deviation is 6 kcal mol^{-1} for the bond cleavage in HOF. The high level of agreement between calculated and experimental bond dissociation energies is (in part) fortuitous, and does not accurately represent the behavior of the Moller-Plesset model with the 6-31G** basis set. In particular, the convergence of the Møller-Plesset energies for this series of reactions is not as good as previously noted for the AH bond dissociation processes. Deviations of 5–10 kcal mol^{-1} between the MP2 and MP3 reaction energies are common; the largest single change is 11 kcal mol^{-1} for bond breaking in F$_2$. The MP3/6-31G**, MP4SDQ/6-31G**, and MP4/6-31G** results are much closer to one another; the largest change between the first two is only 3 kcal mol^{-1}, and between the MP4SDQ and full MP4 models, also 3 kcal mol^{-1}. Thus, it is again likely that the highest-order calculations (MP3 and MP4) approximate well the limiting behavior of the theory with the 6-31G** basis set, and that any residual errors (typically 5–10 kcal mol^{-1}) are due largely to the limitations of that basis set.

6.5.3. Energies of Hydrogenation Reactions

How effective is Hartree–Fock theory in describing the energies of reactions involving only closed-shell species? Are correlation energy changes smaller than those noted for homolytic bond-breaking processes (Section 6.5.2)? The formal hydrogenation reaction is typical: H$_2$ is added to the extent required for complete reduction of a given molecule to the set of simplest (one-heavy-atom) hydrides. The energies of such reactions relate the strength of bonds between heavy (non-hydrogen) atoms to bonds involving hydrogen. For example, the complete hydrogenation of methanol to give methane and water is achieved by a single hydrogen molecule,

$$CH_3—OH + H_2 \longrightarrow CH_4 + H_2O. \qquad (6.30)$$

Full reduction of the unsaturated system formaldehyde to yield the same products requires an additional molecule of H$_2$,

$$CH_2=O + 2H_2 \longrightarrow CH_4 + H_2O. \qquad (6.31)$$

Multiple hydrogenations can be divided into steps; for formaldehyde, the total process consists of two reactions,

$$CH_2=O + H_2 \longrightarrow CH_3—OH$$
$$CH_3—OH + H_2 \longrightarrow CH_4 + H_2O, \qquad (6.32)$$

the second of which is the hydrogenation of methanol. Similarly, three steps make up the total hydrogenation process for a molecule incorporating a triple bond.

In this and in the following sections, we require experimental thermochemical data for a subset of one- and two-heavy-atom hydrides incorporating first- and/or second-row atoms. These data, in the form of heats of formation at 298 K, are summarized in Table 6.62. In addition, the table provides *estimates* of zero-point vibrational energies for these molecules, obtained from calculated 3-21G vibrational frequencies (3-21G$^{(*)}$ for molecules incorporating second-row elements). These have been uniformly scaled by 0.9 in accordance with our previous discussion (Section 6.3.9c). Theoretical data have been used in lieu of experimental frequencies not only because of incompleteness of the latter set, but also to provide a more uniform treatment.

Corrections to 0 K are not tabulated. These could have been estimated directly from the relations given in Section 6.3.9c using 3-21G vibrational frequencies. Note, however, that the vibrational contribution (Eq. 6.20) depends almost exclusively on the very-low-frequency modes, which are not reproduced as accurately or as uniformly as the high-frequency vibrations which dominate the zero-point energy (Eq. 6.21). It was decided, therefore, to *approximate* the temperature corrections by their translational and rotational contributions alone: $\frac{7}{2} kT$ (2.1 kcal mol^{-1} at 298 K) for linear molecules, $4 kT$ (2.4 kcal mol^{-1}) for nonlinear molecules.

a. Two-Heavy-Atom Molecules.   Energies of complete hydrogenation obtained with Hartree–Fock models for a selection of two-heavy-atom molecules comprising first- and second-row elements only are presented in Table 6.63. The single-determi-

TABLE 6.62.   Experimental Heats of Formation (298 K) and Zero-Point Energy Estimates

Molecule	$\Delta H_f^\circ (298)^a$	Zero-Point Energy^b
H_2	0.0	6.0
LiH	33.6	1.8
CH_4	−17.8	27.1
NH_3	−11.0	20.4
H_2O	−57.8	12.3
HF	−64.8	5.2
NaH	29.6	1.5
SiH_4	8.2	19.0
PH_3	1.3	14.8
H_2S	−4.8	9.2
HCl	−22.0	4.1
Li_2	50.4	0.4
LiOH	−58.8	7.8
LiF	−79.6	1.4
LiCl	−47.1	0.8
CH_3CH_3	−20.1	45.2
CH_3NH_2	−5.5	38.4
CH_3OH	−48.2	30.8
CH_3F	−55.9	23.9
CH_3SiH_3	−6.9	37.0

TABLE 6.62. (Continued)

Molecule	$\Delta H_f^\circ(298)^a$	Zero-Point Energy[b]
CH_3SH	−5.5	28.0
CH_3Cl	−19.6	22.8
NH_2NH_2	22.8	
NH_2OH	−9.1	
NH_2F	−5.0	
HOOH	−32.5	15.3
HOF	−23.5	
NaOH	−50.2	6.9
HOCl	−17.8	7.7
F_2	0.0	1.7
NaF	−70.0	0.9
SiH_3F	−105.2	17.0
ClF	−12.1	1.1
Na_2	33.0	0.2
NaCl	−43.5	0.4
SiH_3SiH_3	19.1	29.9
SiH_3Cl	−48.0	16.1
PH_2PH_2	5.0	22.0
H_2S_2	3.8	11.4
Cl_2	0.0	0.7
$CH_2{=}CH_2$	12.5	31.0
$H_2C{=}NH$	26.5	24.2
$H_2C{=}O$	−26.0	16.4
$H_2C{=}S$	24.4	15.1
$HN{=}NH$	36.1	16.9
$HN{=}O$	23.8	8.4
O_2	0.0	2.0
$HC{\equiv}CH$	54.5	17.0
$HC{\equiv}N$	32.3	10.4
$C^-{\equiv}O^+$	−26.4	3.0
$HC{\equiv}P$	39.1	8.8
$C^-{\equiv}S^+$	55.0	1.8
$N{\equiv}N$	0.0	3.4
$P{\equiv}N$	23.9	2.0
$Si^-{\equiv}O^+$	−24.1	1.9
$P{\equiv}P$	40.6	1.2

[a] Data from one of the following sources, preference following the order of appearance. (a) J. B. Pedley and J. Rylance, CATCH Tables, University of Sussex, 1977; (b) S. W. Benson, F. R. Cruickshank, D. M. Golden, G. R. Haugen, H. E. O'Neal, A. S. Rogers, R. Shaw, and R. Walsh, *Chem. Rev.* **69**, 279 (1969); (c) S. W. Benson, *Thermochemical Kinetics,* 2nd ed., Wiley, New York, 1976.

[b] From 3-21G (3-21G(*)) vibrational frequencies. See text for discussion. Theoretical frequencies from: BPH (1980); PFHDPB (1982); W. J. Hehre, unpublished calculations.

TABLE 6.63. Hartree–Fock and Experimental Energies of Hydrogenation

Hydrogenation Reaction	STO-3G// STO-3G[a]	3-21G// 3-21G[a]	3-21G(*)// 3-21G(*)[b]	6-31G*// 6-31G*[c]	6-31G**// 6-31G*[c]	Expt.[a]
$Li_2 + H_2 \longrightarrow 2LiH$	19	19	19	20	22	20(17)
$LiOH + H_2 \longrightarrow LiH + H_2O$	36	36	36	24		35(35)
$LiF + H_2 \longrightarrow LiH + HF$	31	53	53	49	46	48(48)
$LiCl + H_2 \longrightarrow LiH + HCl$	69	72	59	60	58	60(59)
$CH_3CH_3 + H_2 \longrightarrow 2CH_4$	-19	-25	-25	-22	-21	-19(-16)
$CH_3NH_2 + H_2 \longrightarrow CH_4 + NH_3$	-20	-30	-30	-27	-28	-26(-23)
$CH_3OH + H_2 \longrightarrow CH_4 + H_2O$	-16	-28	-28	-27	-30	-30(-27)
$CH_3F + H_2 \longrightarrow CH_4 + HF$	-8	-22	-22	-23	-27	-29(-27)
$CH_3SiH_3 + H_2 \longrightarrow CH_4 + SiH_4$	-10	-11	-10	-13	-12	-6(-3)
$CH_3SH + H_2 \longrightarrow CH_4 + H_2S$	-17	-20	-23	-22	-23	-19(-17)
$CH_3Cl + H_2 \longrightarrow CH_4 + HCl$	-14	-21	-25	-22	-24	-22(-20)
$NH_2NH_2 + H_2 \longrightarrow 2NH_3$	-28	-47	-47	-46	-48	-48(-45)
$HOOH + H_2 \longrightarrow 2H_2O$	-31	-67	-67	-82	-87	-86(-83)
$NaOH + H_2 \longrightarrow H_2O + NaH$	15	16	25	10	9	21(22)
$HOCl + H_2 \longrightarrow H_2O + HCl$	-33	-54	-57	-64	-69	-65(-62)
$F_2 + H_2 \longrightarrow 2HF$	-29	-98	-98	-126	-134	-133(-130)
$NaF + H_2 \longrightarrow HF + NaH$	37	30	46	33	30	35(35)
$SiH_3F + H_2 \longrightarrow HF + SiH_4$	12	31	45	30	26	48(49)
$ClF + H_2 \longrightarrow HF + HCl$	-37	-63	-64	-73	-80	-77(-75)
$Na_2 + H_2 \longrightarrow 2NaH$	74	44	49	40	42	29(26)
$NaCl + H_2 \longrightarrow NaH + HCl$	122	73	62	60	58	52(51)
$SiH_3SiH_3 + H_2 \longrightarrow 2SiH_4$	-5	-8	-11	-12	-11	-5(-3)

Reaction					
$SiH_3Cl + H_2 \rightarrow SiH_4 + HCl$	23	15	17	16	13	33(34)
$PH_2PH_2 + H_2 \rightarrow 2PH_3$	-10	-8	-10	-11	-10	-4(-2)
$HSSH + H_2 \rightarrow 2H_2S$	-9	-19	-21	-21	-23	-14(-13)
$Cl_2 + H_2 \rightarrow 2HCl$	-25	-45	-51	-50	-55	-46(-44)
$CH_2{=}CH_2 + 2H_2 \rightarrow 2CH_4$	-91	-71	-71	-66	-64	-57(-48)
$CH_2{=}NH + 2H_2 \rightarrow CH_4 + NH_3$	-78	-68	-68	-61	-62	-64(-55)
$CH_2{=}O + 2H_2 \rightarrow CH_4 + H_2O$	-65	-64	-64	-54	-58	-59(-50)
$CH_2{=}S + 2H_2 \rightarrow CH_4 + H_2S$	-83	-62	-65	-64	-65	-54(-47)
$HN{=}NH + 2H_2 \rightarrow 2NH_3$	-75	-94	-94	-76	-80	-68(-58)
$HN{=}O + 2H_2 \rightarrow NH_3 + H_2O$	-78	-113	-113	-98	-105	-103(-93)
$O{=}O + 2H_2 \rightarrow 2H_2O$	-39	-98	-98	-94	-105	-125(-116)
$HP{=}O + 2H_2 \rightarrow H_2O + PH_3$	-72	-58	-41	-50	-55	-105(-90)
$HC{\equiv}CH + 3H_2 \rightarrow 2CH_4$	-154	-124	-124	-121	-118	-105(-90)
$HC{\equiv}N + 3H_2 \rightarrow CH_4 + NH_3$	-97	-85	-85	-78	-79	-76(-61)
$C^-{\equiv}O^+ + 3H_2 \rightarrow CH_4 + H_2O$	-72	-69	-69	-55	-59	-63(-49)
$HC{\equiv}P + 3H_2 \rightarrow CH_4 + PH_3$	-132	-89	-94	-99	-97	-67(-56)
$C^-{\equiv}S^+ + 3H_2 \rightarrow CH_4 + H_2S$	-134	-119	-114	-111	-112	-91(-78)
$N{\equiv}N + 3H_2 \rightarrow 2NH_3$	-36	-53	-53	-28	-33	-37(-22)
$P{\equiv}N + 3H_2 \rightarrow NH_3 + PH_3$	-91	-78	-74	-79	-82	-45(-34)
$Si^-{\equiv}O^+ + 3H_2 \rightarrow H_2O + SiH_4$	-82	-37	-37	-48	-51	-32(-25)
$P{\equiv}P + 3H_2 \rightarrow 2PH_3$	-98	-40	-56	-57	-56	-44(-38)

[a] BPH (1980); GBPPH (1982); W. J. Hehre, unpublished calculations.

[b] PFHDPB (1982); W. J. Hehre, unpublished calculations.

[c] FPHBGDP (1982).

[d] Data from Table 6.62. Experimental heats of reaction have been corrected for zero-point vibrational energy and to 0 K. Uncorrected heats are given in parentheses.

nant calculations utilize the STO-3G, 3-21G, 3-21G$^{(*)}$, 6-31G*, and 6-31G** basis sets. Except for the 6-31G** calculations (where 6-31G* structures have been used), optimized equilibrium geometries have been employed. Experimental thermochemical data and zero-point energies have been drawn from Table 6.62. The experimental reaction energies have been corrected for differential zero-point vibrational energies and to 0 K. The uncorrected heats, given in parentheses, are close to corrected values when the number of molecules is conserved, but are significantly different (by as much as 15 kcal mol^{-1}) for hydrogenation reactions involving unsaturated compounds where the number of molecules is no longer conserved.

STO-3G calculations provide a very nonuniform account of hydrogenation energies; errors range from 1 to over 100 kcal mol^{-1}. Some general trends are worthy of comment. While the STO-3G hydrogenation energy for ethane is in good accord with experiment, the results for methylamine, methanol, and methyl fluoride become progressively poorer. There is a simple explanation. While all four molecules possess 18 electrons (and all product hydrides, 10 electrons), the number of available basis functions with which to form the molecular orbitals decreases from 16 in ethane to 13 in methyl fluoride (and from 9 in CH_4 to 6 in HF). The deterioration in the overall quality of the molecular orbital description in going from the hydrocarbon to the halide results in a poorer account of the hydrogenation energies (see also discussion in Section 4.3.2).

The large errors in the STO-3G hydrogenation energies of ethylene and acetylene cannot be due to unequal apportionment of basis functions, for only carbon and hydrogen are involved. Rather, as detailed in Section 4.3.2, the large errors with the multiply-bonded systems are a consequence of the inability of a minimal basis set to account properly for *anisotropic* molecular environments. While the electron distributions in ethane and in its hydrogenation product, methane, are reasonably *isotropic*, those in the unsaturated systems, ethylene and acetylene, are highly anisotropic. As only the *p*-type basis functions in a minimal basis set are capable of describing anisotropic molecular charge distributions, and as the radial parts of all three components in the STO-3G representation are constrained to be the same, it might be expected that this model will have difficulty in handling highly anisotropic environments.

Both of the deficiencies mentioned above may be alleviated by increasing the number of valence basis functions available to all atoms, in particular, by providing two sets of valence *p* functions to all heavy atoms. This results in suitable numbers of basis functions in excess of the minimum. Furthermore, even though the radial parts of the individual components in each of the two sets of *p* functions are the same, that is, the basis is isotropic, the molecular orbital coefficients corresponding to the two components of each of the p_x, p_y, and p_z functions are determined *independently* in the variational calculation. Proper description of anisotropic molecular charge distributions is thus possible.

The 3-21G split-valence basis set does provide a somewhat improved description of hydrogenation energies, although its performance is not entirely satisfactory. Reactions that involve considerable changes in the degree of charge anisotropy are now better described. The 3-21G calculations fare best for molecules with at least one carbon, and poorest for systems containing two electronegative atoms. For saturated systems, the 3-21G hydrogenation energies are usually within 5–10 kcal mol^{-1} of experiment (hydrogen peroxide and F_2 are conspicuous exceptions). The calculations at this level are less successful in reproducing the known hydrogenation energies for unsaturated compounds: errors of 10–20 kcal mol^{-1} are common.

Hydrogenation energies calculated at 3-21G$^{(*)}$ for molecules incorporating second-row elements usually are in better agreement with experimental values than are those obtained using the unsupplemented 3-21G representation. Energy changes from the 3-21G to the 3-21G$^{(*)}$ basis set are usually small, that is, less than 5 kcal mol^{-1}, except for highly polar molecules, for example, LiCl, NaOH, NaF, SiH_3F, and NaCl, for which they are often greater than 10 kcal mol^{-1}.

For many of the systems tabulated, the 6-31G* and 6-31G** models yield hydrogenation energies that are within 3–5 kcal mol^{-1} of experimental values. Somewhat larger errors are noted for several of the molecules containing second-row elements and for some unsaturated systems such as acetylene. The hydrogenation energies of unsaturated molecules with second-row atoms, CS and SiO for example, are generally very poorly described, although it should be commented that in some cases the experimental thermochemical data are likely to be significantly in error.

b. Hypervalent Molecules. Calculated (Hartree–Fock) energies of complete hydrogenation of hypervalent compounds are compared with experimental values in Table 6.64. Data obtained from 3-21G calculations are uniformly too negative (often by 100 kcal mol^{-1} or more), suggesting that this level of theory significantly underestimates the thermal stabilities of hypervalent compounds, that is, relative to the normal-valent products of hydrogenation. This is consistent with previous observations that indicated the need for d-type functions in the basis set for the proper description of the structures (see Section 6.2.5) and vibrational frequencies (see Section 6.3.7) of hypervalent compounds. Hydrogenation energies obtained using the 3-21G$^{(*)}$ basis set are consistently more positive than 6-31G* or 6-31G** values; they are usually in better agreement with the experimental data. However, the quantitative description of the hydrogenation of hypervalent compounds is generally poor; calculated energies are in error by as much as 60 kcal mol^{-1}. While some of the discrepancy is no doubt due to uncertainties in the experimental data, the large errors associated with even the highest-level Hartree–Fock calculations suggests the probable importance of electron correlation effects. This is addressed in the next section.

c. Effect of Electron Correlation on Hydrogenation Energies. Two-heavy-atom systems are small enough to permit treatment of electron correlation at relatively high levels. The data in Table 6.65 for molecules comprising first-row elements only assess the effects of Møller–Plesset correlation expansions through full fourth order. The 6-31G** polarization basis set has been used throughout, with optimum HF/6-31G* equilibrium geometries.

In general, the changes in the hydrogenation energies from Hartree–Fock through

TABLE 6.64. Calculated and Experimental Energies of Complete Hydrogenation of Hypervalent Compounds

Hydrogenation Reaction	3-21G// 3-21G^a	3-21G(*)// 3-21G(*)^b	6-31G*// 6-31G*^{b}	6-31G**// 6-31G*^{b}	Expt.^{c}
PF$_5$ + 4H$_2$ → PH$_3$ + 5HF	47	135	60	40	46(54)
(CH$_3$)$_3$PO + 4H$_2$ → PH$_3$ + 3CH$_4$ + H$_2$O	-66	-24			(-9)
F$_3$PO + 4H$_2$ → PH$_3$ + H$_2$O + 3HF	5	98	43	26	36(46)
F$_3$PS + 4H$_2$ → PH$_3$ + H$_2$S + 3HF	2	74	10	14	31(39)
SO$_2$ + 3H$_2$ → H$_2$S + 2H$_2$O	-149	-75	-87	-100	-58(-49)
(CH$_3$)$_2$SO + 3H$_2$ → H$_2$S + 2CH$_4$ + H$_2$O	-123	-90			(-62)
SF$_4$ + 3H$_2$ → H$_2$S + 4HF	-126	-52	-97	-116	(-81)
F$_2$SO + 3H$_2$ → H$_2$S + H$_2$O + 2HF	-120	-47	-75	-91	-70(-62)
SO$_3$ + 4H$_2$ → H$_2$S + 3H$_2$O	-233	-107	-132	-150	-97(-84)
(CH$_3$)$_2$SO$_2$ + 4H$_2$ → H$_2$S + 2CH$_4$ + 2H$_2$O	-192	-94			(-67)
SF$_6$ + 4H$_2$ → H$_2$S + 6HF	-177	-26			(-102)
F$_4$SO + 4H$_2$ → H$_2$S + H$_2$O + 4HF	-210	-73	-135	-159	(-96)
ClF$_3$ + 2H$_2$ → HCl + 3HF	-288	-184	-207	-222	-164(-159)
ClF$_5$ + 3H$_2$ → HCl + 5HF	-405	-297	-348	-371	(-288)
FClO$_3$ + 4H$_2$ → HCl + 3H$_2$O + HF	-426	-307	-340	-363	-268(-255)

^{a}PFHDPB (1982).
^{b}FPHBGDP (1982).
^{c}Data for one-heavy-atom hydrides from Table 6.62. Heats of formation (298 K) for hypervalent molecules summarized in: FPHBGDP (1982). Heats of reaction have been corrected for zero-point vibrational energy and to 0 K. Uncorrected heats are given in parentheses.

TABLE 6.65. Effect of Electron Correlation on Energies of Hydrogenation Reactions[a]

Hydrogenation Reaction	6-31G**// 6-31G*	MP2	MP3	MP4SDQ	MP4	Expt.[b]
		/6-31G**//6-31G*				
$CH_3-CH_3 + H_2 \longrightarrow 2CH_4$	-21	-18	-19	-19	-18	-19(-16)
$CH_3-NH_2 + H_2 \longrightarrow CH_4 + NH_3$	-28	-25	-26	-26	-25	-26(-23)
$CH_3-OH + H_2 \longrightarrow CH_4 + H_2O$	-30	-28	-29	-30	-29	-30(-27)
$CH_3-F + H_2 \longrightarrow CH_4 + HF$	-27	-26	-27	-26	-25	-29(-27)
$NH_2-NH_2 + H_2 \longrightarrow 2NH_3$	-48	-46	-46	-46	-45	-48(-45)
$HO-OH + H_2 \longrightarrow 2H_2O$	-87	-83	-83	-82	-79	-86(-83)
$F-F + H_2 \longrightarrow 2HF$	-134	-126	-125	-123	-121	-133(-130)
$CH_2=CH_2 + 2H_2 \longrightarrow 2CH_4$	-64	-61	-62	-61	-60	-57(-48)
$CH_2=NH + 2H_2 \longrightarrow CH_4 + NH_3$	-62	-58	-60	-59	-57	-64(-55)
$CH_2=O + 2H_2 \longrightarrow CH_4 + H_2O$	-58	-55	-59	-55	-52	-59(-50)
$HN=NH + 2H_2 \longrightarrow 2NH_3$	-80	-76	-77	-75	-72	-68(-58)
$HN=O + 2H_2 \longrightarrow NH_3 + H_2O$	-105	-100	-103	-98	-95	-103(-93)
$O=O + 2H_2 \longrightarrow 2H_2O$	-105	-114	-116	-113	-109	-125(-116)
$HC\equiv CH + 3H_2 \longrightarrow 2CH_4$	-118	-111	-117	-114	-111	-105(-90)
$HC\equiv N + 3H_2 \longrightarrow CH_4 + NH_3$	-79	-71	-78	-74	-70	-76(-61)
$C^-\equiv O^+ + 3H_2 \longrightarrow CH_4 + H_2O$	-59	-58	-64	-58	-54	-63(-49)
$N\equiv N + 3H_2 \longrightarrow 2NH_3$	-33	-28	-36	-31	-26	-37(-22)

[a] Theoretical data from: HP (1973) and J. A. Pople, unpublished calculations.
[b] Data from Table 6.62. Heats of reaction have been corrected for zero-point vibrational energy and to 0 K. Uncorrected heats are given in parentheses.

full fourth-order Møller–Plesset perturbation theory range from 2 to 5 kcal mol^{-1}. Exceptions occur for H_2O_2 and F_2, and for the unsaturated systems N_2H_2, HNO, and HCN. Here, the exothermicity of reaction is increased by as much as 13 kcal mol^{-1} (in F_2). Generally, the MP2 calculations reproduce the full fourth-order result to within 1–2 kcal mol^{-1}; the largest differences are 4–5 kcal mol^{-1}. Corrections through third order do not lead to a consistent improvement over MP2.

Even the highest-level (MP4) calculations yield hydrogenation energies which often differ significantly from experimental values. Errors of over 10 kcal mol^{-1} are found for F_2 (12 kcal mol^{-1}), O_2 (12 kcal mol^{-1}), and N_2 (11 kcal mol^{-1}). Whether the large deviations are due to an inadequate basis set, contributions from higher-order perturbation terms, a poor choice of equilibrium geometry or a combination of these factors remains a subject for further research. Overall, the mean absolute error in MP4 hydrogenation energies is 6 kcal mol^{-1} (zero-point and 0 K corrections taken into account).

Far less consideration has been given to the effect of electron correlation on hydrogenation energies for molecules containing second-row elements. Calculations performed at the MP2/6-31G* level, using HF/6-31G* equilibrium geometries, are summarized in Table 6.66. In general, the magnitudes of the correlation corrections are significantly larger than those previously noted for the analogous first-row systems, in particular when multiple bonds are involved. For example, the exothermicity of complete hydrogenation of carbon monosulfide is lowered by 19 kcal mol^{-1} by the inclusion of correlation via the second-order Møller–Plesset expansion; by comparison, the corresponding lowering for carbon monoxide (Table 6.65) is only 1 kcal mol^{-1}. This result contrasts with the previous observation that effects of correlation on structure were smaller for molecules containing second-row elements than for those incorporating first-row atoms only (see Section 6.2.3b).

For systems incorporating multiple bonds, MP2/6-31G* hydrogenation energies are generally closer to the experimental quantities than are those derived from the corresponding Hartree–Fock treatments. The same conclusion applies to systems with only single bonds, although the correlation effects are much smaller. Note, however, that the experimental hydrogenation enthalpies for some of these molecules are subject to considerable uncertainty, and conclusions might ultimately need to be modified.

The correlation effect noted for SO_2, the only hypervalent compound considered, is huge (40 kcal mol^{-1}). The MP2/6-31G* result is still in error by 11 kcal mol^{-1} (after zero-point and 0 K corrections have been made), but is much closer to the experimental hydrogenation energy than the corresponding Hartree–Fock value.

Overall, the mean absolute difference between MP2/6-31G* and experimental hydrogenation energies is 7 kcal mol^{-1}, although it again needs to be noted that some of the experimental thermochemical data are probably not well established.

6.5.4. Energies of Reactions Relating Multiple and Single Bonds

As detailed in the previous section, hydrogenation energies calculated within the framework of single-determinant Hartree–Fock theory are subject to considerable

TABLE 6.66. Effect of Electron Correlation on Energies of Hydrogenation Reactions for Compounds Incorporating Second-Row Elements

Hydrogenation Reaction	6-31G*// 6-31G*[a]	MP2/631G*// 6-31G*[b]	Expt.[c]
$LiCl + H_2 \rightarrow LiH + HCl$	60	64	60(59)
$CH_3SiH_3 + H_2 \rightarrow CH_4 + SiH_4$	-13	-6	$-6(-3)$
$CH_3SH + H_2 \rightarrow CH_4 + H_2S$	-22	-15	$-19(-17)$
$CH_3Cl + H_2 \rightarrow CH_4 + HCl$	-22	-16	$-22(-20)$
$NaOH + H_2 \rightarrow NaH + H_2O$	10	19	21(22)
$HOCl + H_2 \rightarrow H_2O + HCl$	-64	-56	$-59(-62)$
$NaF + H_2 \rightarrow NaH + HF$	33	41	35(35)
$SiH_3F + H_2 \rightarrow SiH_4 + HF$	30	35	48(49)
$ClF + H_2 \rightarrow HCl + HF$	-73	-65	$-73(-75)$
$Na_2 + H_2 \rightarrow 2NaH$	40	42	29(26)
$NaCl + H_2 \rightarrow NaH + HCl$	60	62	52(51)
$SiH_3SiH_3 + H_2 \rightarrow 2SiH_4$	-12	-3	$-5(-3)$
$SiH_3Cl + H_2 \rightarrow SiH_4 + HCl$	16	21	33(34)
$PH_2PH_2 + H_2 \rightarrow 2PH_3$	-11	-3	$-4(-2)$
$HSSH + H_2 \rightarrow 2H_2S$	-21	-13	$-14(-13)$
$Cl_2 + H_2 \rightarrow 2HCl$	-50	-43	$-41(-44)$
$CH_2 = S + 2H_2 \rightarrow CH_4 + H_2S$	-64	-49	$-54(-47)$
$HC \equiv P + 3H_2 \rightarrow CH_4 + PH_3$	-99	-68	$-67(-56)$
$C^- \equiv S^+ + 3H_2 \rightarrow CH_4 + H_2S$	-111	-92	$-91(-78)$
$P \equiv N + 3H_2 \rightarrow PH_3 + NH_3$	-79	-40	$-45(-34)$
$Si^- \equiv O^+ + 3H_2 \rightarrow SiH_4 + H_2O$	-48	-18	$-32(-25)$
$P \equiv P + 3H_2 \rightarrow 2PH_3$	-57	-20	$-44(-38)$
$SO_2 + 3H_2 \rightarrow H_2S + 2H_2O$	-87	-47	$-58(-49)$

[a] FPHBGDP (1982).

[b] W. J. Hehre, unpublished calculations.

[c] Except for SO_2, data from Table 6.62. Data for SO_2 for FPHBGDP (1982). Heats of reaction have been corrected for zero-point vibrational energy and to 0 K. Uncorrected heats are given in parentheses.

errors. Significant differences between theoretical and experimental hydrogenation energies exist for some systems, for example N_2, even at the MP4/6-31G** level. This suggests that the large correlation energy changes associated with hydrogenation reactions are still not adequately described. While the total number of electron pairs is conserved in the hydrogenation processes, significant differences in bonding exist between reactant and product molecules. For example, in the reaction of acetylene with three hydrogen molecules to produce two molecules of methane, six new CH linkages have been formed at the expense of the original CC σ and π bonds and three HH bonds. If the changes in bonding were somewhat less dramatic, for example, by maintaining the total number of bonds involving a given pair of elements, differential correlation effects might be smaller, and reaction energies better described either at the Hartree–Fock level or by using one of the simpler electron correlation schemes.

Here we examine formal reactions relating the energies of multiple and single bonds, that is, converting unsaturated systems into molecules incorporating single bonds between the same pairs of atoms. For example, a carbonyl compound would be converted into two molecules each containing a carbon–oxygen single bond. Similarly, a molecule incorporating a cyano group would be converted into three molecules each of which possessed a carbon–nitrogen single bond. Reactions for formaldehyde and for hydrogen cyanide exemplify these conversions.

$$H_2C{=}O + CH_4 + H_2O \longrightarrow 2CH_3{-}OH \tag{6.33}$$

$$HC{\equiv}N + 2CH_4 + 2NH_3 \longrightarrow 3CH_3{-}NH_2. \tag{6.34}$$

The data in Table 6.67 support the expectation that differential correlation effects for reactions of this type should be less important than for hydrogenation processes. Nevertheless, the STO-3G basis again fares miserably: errors commonly range between 20 and 50 kcal mol^{-1}! This basis set is insufficiently flexible for the uniform description of reactants and products that differ significantly in charge anisotropy.

The 3-21G basis set (3-21G$^{(*)}$ for molecules incorporating second-row elements) is much more successful in describing the energies of reactions relating single and multiple bonds. Errors are typically 5–10 kcal mol^{-1}, although for some systems, for example O$_2$ and CS, the performance of the theory is much worse.

The 6-31G* basis set fares well for most reactions listed, where reactant and product molecules are ground-state singlets. Aside from the CS molecule, the largest error is 7 kcal mol^{-1} for P$_2$; reaction energies for most other systems are in much better accord with the experimental data. The high level of agreement suggests the relative unimportance of correlation effects on processes relating the energies of multiple and single bonds. However, the large error at 6-31G* for the reaction of triplet oxygen indicates the need for correlation corrections when different numbers of electron pairs are involved. This is supported by the data in Table 6.68, where the 6-31G** basis set has been employed in conjunction with Møller–Plesset models through full fourth order. Only molecules with first-row elements have been examined. Except for the reaction of oxygen molecule, the total correlation energies are small (less than 5 kcal mol^{-1}). The effect on the O$_2$ reaction is 21 kcal mol^{-1}. Overall, the mean absolute deviation of MP4/6-31G** from experimental reaction energies is 6 kcal mol^{-1} (after corrections for zero-point energy and temperature have been taken into account), very similar to the corresponding Hartree–Fock deviation.

6.5.5. Relative Energies of Structural Isomers

Among the most important energetic comparisons are those involving structural isomers. The spectrum of possibilities is enormous, ranging from consideration of the subtle differences between stereoisomers to major changes in skeletal architecture. One might anticipate that, while even the simplest models would probably prove of value for energetic comparisons among closely related systems, fairly sophisticated treatments will be necessary in order to account for gross differences in electronic structure. Table 6.69 surveys the performance of the STO-3G, 3-21G (3-21G$^{(*)}$ for

TABLE 6.67. Hartree–Fock and Experimental Energies of Reactions Relating Multiple and Single Bonds[a]

Reaction	STO-3G//STO-3G	3-21G//3-21G[b]	6-31G*//6-31G*	Expt.[c]
CH$_2$=CH$_2$ + 2CH$_4$ $\longrightarrow$ 2CH$_3$—CH$_3$	-53	-21	-22	-20(-17)
CH$_2$=NH + CH$_4$ + NH$_3$ $\longrightarrow$ 2CH$_3$—NH$_2$	-38	-12	-7	-12 (-9)
CH$_2$=O + CH$_4$ + H$_2$O $\longrightarrow$ 2CH$_3$—OH	-32	-7	1	1 (5)
H$_2$C=S + CH$_4$ + H$_2$S $\longrightarrow$ 2CH$_3$—SH	-52	-20(-22)	-20	-20(-13)
HN=NH + 2NH$_3$ $\longrightarrow$ 2NH$_2$—NH$_2$	-18	-1	16	28 (32)
O=O + 2H$_2$O $\longrightarrow$ 2HO—OH	29	31	69	47 (51)
HC≡CH + 4CH$_4$ $\longrightarrow$ 3CH$_3$—CH$_3$	-97	-49	-55	-49(-44)
HC≡N + 2CH$_4$ + 2NH$_3$ $\longrightarrow$ 3CH$_3$—NH$_2$	-37	5	3	4 (9)
C≡O$^+$ + 2CH$_4$ + 2H$_2$O $\longrightarrow$ 3CH$_3$—OH	-23	16	27	27 (33)
C≡S$^+$ + 2CH$_4$ + 2H$_2$S $\longrightarrow$ 3CH$_3$—SH	-88	-46(-58)	-45	-25(-26)
N≡N + 4NH$_3$ $\longrightarrow$ 3NH$_2$—NH$_2$	49	88	109	107(112)
P≡P + 4PH$_3$ $\longrightarrow$ 3PH$_2$—PH$_2$	-69	-26(-22)	-25	-32(-31)

[a]Theoretical data from: HP (1973); BPH (1980); GBPPH (1982); FPHBGDP (1982); PFHDPB (1982).
[b]3-21G(*)//3-21G(*) for molecules containing second-row elements. 3-21G//3-21G data given in parentheses.
[c]Data from Table 6.62. Heats of reaction have been corrected for zero-point vibrational energy and to 0 K. Uncorrected heats are given in parentheses.

TABLE 6.68. Effect of Electron Correlation on the Energies of Reactions Relating Multiple and Single Bonds[a]

Reaction	6-31G**//6-31G*	MP2	MP3	MP4SDQ	MP4	Expt.[b]
		/6-31G**//6-31G*				
$CH_2{=}CH_2 + 2CH_4 \longrightarrow 2CH_3{-}CH_3$	-22	-26	-25	-24	-24	-20(-17)
$CH_2{=}NH + CH_4 + NH_3 \longrightarrow 2CH_3{-}NH_2$	-7	-8	-8	-8	-7	-12 (-9)
$CH_2{=}O + CH_4 + H_2O \longrightarrow 2CH_3{-}OH$	1	2	0	2	6	1 (5)
$HN{=}NH + 2NH_3 \longrightarrow 2NH_2{-}NH_2$	16	16	15	17	18	28 (32)
$O{=}O + 2H_2O \longrightarrow 2HO{-}OH$	70	52	50	51	49	47 (51)
$HC{\equiv}CH + 4CH_4 \longrightarrow 3CH_3{-}CH_3$	-54	-58	-60	-58	-57	-49(-44)
$HC{\equiv}N + 2CH_4 + 2NH_3 \longrightarrow 3CH_3{-}NH_2$	4	4	0	3	5	4 (9)
$C^+{\equiv}O^+ + 2CH_4 + 2H_2O \to 3CH_3{-}OH$	30	27	24	27	33	27 (33)
$N{\equiv}N + 4NH_3 \longrightarrow 3NH_2{-}NH_2$	110	109	103	107	109	107(112)

[a]Theoretical data from: HP (1973) and J. A. Pople, unpublished calculations.
[b]Data from Table 6.62. Heats of reaction have been corrected for zero-point vibrational energy and to 0 K. Uncorrected heats given in parentheses.

molecules incorporating second-row elements) and 6-31G* Hartree–Fock models. Optimum geometries have been utilized for the minimal and split-valence basis set calculations; 3-21G (3-21G$^{(*)}$) structures have been employed for calculations at the 6-31G* level.

As in previous energetic comparisons, the experimental data have been systematically corrected for differential zero-point energies using 3-21G frequencies (3-21G$^{(*)}$ for molecules incorporating second-row elements). (Corrections to 0 K, as estimated for previous comparisons, are zero here.) Uncorrected heats are given in parentheses. While, in most cases, the zero-point energy corrections are only on the order of a few tenths of a kcal mol^{-1}, they are of greater significance for a few systems.

As expected, the STO-3G calculations fare poorly for the majority of systems considered. This is particularly evident for comparisons involving systems with differing numbers of single, double, and triple bonds, for example, one single and two double CC bonds in 1,3-butadiene vs. one triple and two single bonds in 2-butyne. Other failures involve comparisons between unsaturated acyclic molecules and their cyclic isomers, for example, propene vs. cyclopropane, and between molecules with heteroatoms, for example, formamide vs. nitrosomethane. The latter entails comparison of different types of bonds, for example, C=O in formamide vs. N=O in nitrosomethane.

Even though the overall performance of the minimal basis is poor, some energy differences, for example, 2-butyne vs. 1-butyne, and 1,3-butadiene vs. 1,2-butadiene, are reasonably well described. Most of these are for *isodesmic* processes, which are examined in greater detail in Section 6.5.6. Many of the STO-3G results are substantially in error (by more than 10 kcal mol^{-1}), and it is evident that the method is generally not well suited to the description of the relative thermochemical stabilities of structural isomers unless the molecules differ only in bond placement, for example 2-butyne vs. 1-butyne, rather than bond type.

The 3-21G basis (3-21G$^{(*)}$ for molecules incorporating second-row elements) provides a much improved description of relative isomer stabilities. The model performs reasonably well in comparisons among acyclic hydrocarbons or among cyclic hydrocarbons, although comparisons relating cyclic and acyclic systems often are significantly in error. While the relative energies of some systems involving heteroatoms are adequately described at the 3-21G level, for example, formaldehyde vs. hydroxymethylene, others are very poorly handled, glycine vs. nitroethane for example. Also disturbing is the fact that relative energies of the positional isomers of diazabenzene (pyridazine, pyrimidine, and pyrazine) are so badly reproduced. It is clear that HF/3-21G calculations may not be used blindly for energetic comparisons among structural isomers.

In most cases, the 6-31G* calculations provide better estimates of relative isomer stabilities than those at the 3-21G and 3-21G$^{(*)}$ levels. The most dramatic improvements involve comparisons between unsaturated acyclic molecules and their small-ring isomers, for example, propene vs. cyclopropane. Following the discussion in Section 4.3.3, it is to be expected that the energies of the strained-ring systems, in which the center of electron density may be significantly displaced from the nuclear positions, will be lowered preferentially by the inclusion of polarization

TABLE 6.69. Calculated and Experimental Relative Energies of Structural Isomers[a]

Formula	Molecule	STO-3G//STO-3G	3-21G//3-21G[b]	6-31G*//3-21G[b]	Expt.[c]
CHN	Hydrogen cyanide	0	0	0	0[d]
	Hydrogen isocyanide	19.3	9.0	12.4	14.5(15.1)
CH_2N_2	Diazomethane	0	0	0	0[e]
	Diazirine	-16.2	30.1		5.3(8)
CH_2O	Formaldehyde	0	0	0	0[f]
	Hydroxymethylene	47.7	47.4	52.6	54.9(54.2)
CH_3NO	Formamide	0	0	0	0[e]
	Nitrosomethane	23.9	54.8	65.3	62.4(60.5)
CH_3NO_2	Nitromethane		0	0[g]	0
	Methyl nitrite		-14.0	-2.7	(2.0)
$C_2H_2F_2$	cis-1,2-Difluoroethylene	0	0	0	0[h]
	trans-1,2-Difluoroethylene	-0.3	-1.4	-0.4	(1.1)
$C_2H_2Cl_2$	1,1-Dichloroethylene	0	0	0	0
	cis-1,2-Dichloroethylene	-1.0	-2.6	-2.5	0.2(0.5)
	trans-1,2-Dichloroethylene	-1.8	-2.8	-2.8	0.5(0.6)
C_2H_3N	Acetonitrile	0	0	0	0[i]
	Methyl isocyanide	24.1	20.5	20.8	20.9(20.5)
C_2H_4O	Acetaldehyde	0	0	0	0
	Oxacyclopropane	11.0	34.2	33.4	26.2(27.1)
$C_2H_4O_2$	Acetic acid	0	0	0	0
	Methyl formate	6.8	12.5	13.6	18.0(18.3)

Formula	Compound				
$C_2H_5NO_2$	Glycine	0	0		0
	Nitroethane	69.4	104.1		(69.7)
C_2H_6O	trans-Ethanol	0	0	0	0
	Dimethyl ether	-0.8	5.9	7.3	12.0(12.1)
$C_2H_6O_2$	trans-1,2-Ethanediol	0	0	0	0
	trans-Ethanehydroperoxide	13.9	41.1	55.2	(45.1)
	trans-Dimethylperoxide	11.4	45.6	60.3	(62.6)
C_2H_6Si	Vinylsilane	0	0	0	0^j
	1-Methylsilaethylene	25.8	12.0	17.0	(11)
C_2H_6S	trans-Ethanethiol	0	0	0	0
	Dimethylsulfide	-4.7	0.0	0.2	1.5(2.1)
C_2H_7N	Ethylamine	0	0	0	0
	Dimethylamine	2.3	5.8	5.4	6.9(6.9)
$C_2H_7PO_3$	Dimethylphosphonate		0		0^k
	Dimethylphosphorous acid		11.1		(6.5)
C_2H_8Si	Dimethylsilane	0	0	0	0^j
	Ethylsilane	11.7	14.8	10.5	(13)
C_3H_4	Propyne	0	0	0	0
	Allene	17.1	3.4	2.0	1.6(1.0)
	Cyclopropene	30.0	39.8	26.5	21.7(21.6)
C_3H_6O	Acetone	0	0	0	0^l
	Propanol	6.2	6.2	6.4	6.5(7.1)
	Prop-1-ene-2-ol	19.6	11.2	19.0	(7.7)
	Prop-2-ene-1-ol	31.8	22.4	29.9	21.6(22.2)
	Methyl vinyl ether	24.5	20.0	29.6	(25.6)
	Oxetane	1.6	30.6	35.6	30.8(32.7)

TABLE 6.69. (Continued)

Formula	Molecule	STO-3G//STO-3G	3-21G//3-21G[b]	6-31G*//3-21G[b]	Expt.[c]
$C_4H_4N_2$	Pyrazine	0	0	0	0
	Pyrimidine	-2.8	-5.5		(0.1)
	trans-1,2-Ethane-dinitrile	-6.5	-16.3		(3.3)
	Pyridazine	12.8	23.7		(19.7)
C_4H_6	trans-1,3-Butadiene	0	0	0	0
	2-Butyne	-12.8	3.6	6.5	6.9(8.5)
	Cyclobutene	-12.5	18.0	13.4	8.8(11.2)
	1,2-Butadiene	8.5	11.3	12.8	(12.5)
	1-Butyne	-5.3	9.2	13.0	(13.2)
	Methylenecyclopropane	5.8	25.6	20.5	(21.7)
	Bicyclo[1.1.0]butane	11.6	45.7	31.4	23.1(25.6)
	1-Methylcyclopropene	17.3	43.5	33.0	(32.0)
C_4H_8	Isobutene	0	0	0	0
	trans-2-Butene	-0.2	0.5	0.2	(1.1)
	cis-2-Butene	1.4	2.0	1.8	(2.2)
	cis-1-Butene	4.4	3.3	3.6	(3.9)
	Methylcyclopropane	-2.2	15.0	3.6	(9.7)[m]
	Cyclobutane[n]	-17.6	10.4	10.0	(10.8)
C_4H_{10}	Isobutane	0	0	0	0
	trans-n-Butane	0.1	1.3	0.4[o]	(2.1)
$C_7H_7^+$	Tropylium cation	0			0[p]
	Benzyl cation	9.7			(0-5)

[a] Theoretical data from: W. J. Hehre, unpublished calculations.

[b] 3-21G$^{(*)}$//3-21G$^{(*)}$, and 6-31G*//3-21G$^{(*)}$, respectively, for molecules containing second-row elements.

[c] Unless otherwise indicated, experimental data from: J. B. Pedley and J. Rylance, CATCH Tables, University of Sussex, 1977, corrected for zero-point vibrational energy. Uncorrected data are given in parentheses.

[d] C. F. Pau and W. J. Hehre, *J. Phys. Chem.*, **86**, 321 (1982).

[e] S. W. Benson, F. R. Cruickshank, D. M. Golden, G. R. Haugen, H. E. O'Neal, A. S. Rogers, R. Shaw, and R. Walsh, *Chem. Rev.*, **69**, 279 (1969).

[f] C. F. Pau and W. J. Hehre, *J. Phys. Chem.*, **86**, 1252 (1982).

[g] 6-31G*//6-31G*. From: M. M. Francl, Ph.D. Thesis, University of California, Irvine, 1983.

[h] N. C. Craig, L. G. Piper and V. L. Wheeler, *J. Phys. Chem.*, **75**, 1453 (1971), and references therein.

[i] Data for methylisocyanide from footnote *e*.

[j] Based on bond energies summarized in: R. Walsh, *Accounts Chem. Res.*, **14**, 246 (1981).

[k] Relative enthalpies of dimethylphosphonate and dimethylphosphorous acid from: W. J. Peitro and W. J. Hehre, *J. Am. Chem. Soc.*, **104**, 3594 (1982).

[l] Relative enthalpies of acetone and prop-1-ene-2-ol (enol of acetone) from: S. K. Pollack, Ph.D. Thesis, University of California, Irvine, 1980, and references therein. Data for methyl vinyl ether estimated from additivity rules given in footnote *e*.

[m] Gas-phase heat of formation for methyl cyclopropane estimated from liquid-phase data provided in CATCH tables, footnote *c*.

[n] Planar carbon skeleton assumed.

[o] 6-31G*//6-31G*, K. B. Wiberg, *J. Comput. Chem.*, **5**, 197 (1984).

[p] Relative enthalpies of tropylium and benzyl cations summarized in: D. H. Aue and M. T. Bowers, in *Gas Phase Ion Chemistry*, M. T. Bowers, ed., Academic Press, New York, vol. 2, 1979, p. 1.

functions in the basis. While the relative stabilities of all hydrocarbons considered are reasonably well described at this level of theory, results for some systems containing heteroatoms are still significantly in error, for example, the known 12 kcal mol^{-1} difference in energy between ethanol and dimethyl ether is underestimated by 5 kcal mol^{-1}. For only one comparison (that between *trans*-1,2-ethanediol and *trans*-ethanehydroperoxide) is the 6-31G* result in error by more than 10 kcal mol^{-1}. This level of theory is perhaps the simplest available with which to calculate relative isomer energies for a wide variety of systems to reasonable accuracy.

One pair of isomers that deserves special mention is *cis*- and *trans*-1,2-difluoroethylene. The *cis* isomer is known experimentally to be 1.1 kcal mol^{-1} more stable than the *trans* form. Both 3-21G and 6-31G* calculations indicate the opposite ordering of stabilities, with energy differences of 1.4 and 0.4 kcal mol^{-1}, respectively. A similar failure is found for the *cis* and *trans* isomers of 1,2-dichloroethylene, although here both calculations and experiment show the forms to be very close in energy. The failure of the simple quantum chemical models to describe adequately the relative energies of isomers as closely related as *cis* and *trans* 1,2-difluoroethylene is due to the inadequacy of the basis set. More extensive calculations (utilizing the 6-311G** and larger basis sets) lead to a preference for the *cis* isomer [25a, 31].

The effect of electron correlation on the relative energies of structural isomers is examined for a few systems in Table 6.70. The 6-31G* basis set has been employed (with 3-21G equilibrium geometries) and Møller–Plesset models through partial fourth order (single, double, and quadruple excitations only) have been examined. The noted correlation effects are all reasonably small ($<$5–10 kcal mol^{-1}) and, in most but not all cases, lead to improved agreement with experiment. In particular, the calculated energies of the small-ring compounds, oxacyclopropane and cyclopropene, relative to their acyclic isomers, acetaldehyde and propyne, are brought into good agreement with the known differences. The correlated results for some systems, for example, acetonitrile vs. methylisocyanide, are somewhat poorer than Hartree–Fock values. The reasons, insufficient treatment of correlation or incomplete basis set, remain a matter for further study.

6.5.6.   Energies of Isodesmic Reactions

Among the reactions discussed thus far are some that appear to be well described within the framework of single-determinant Hartree–Fock theory. Transformations relating the energies of multiple and single bonds (Section 6.5.4) provide one example; comparisons among positional isomers (Section 6.5.5.) provide another. What other reactions are likely to be treated successfully at the single-determinant level? Which reactions might be properly handled using split-valence or even minimal basis sets, the simplest and most widely applicable of the quantum mechanical models? The key to success is to find processes in which the initial reactants and final products are as similar as possible. *Isodesmic reactions, transformations in which the numbers of bonds of each formal type are conserved and only the relationships among the bonds are altered*, fit this description. For example, in the

TABLE 6.70. Effect of Electron Correlation on Relative Energies of Structural Isomers[a]

Formula	Molecule	6-31G*//3-21G	/6-31G*//3-21G			Expt.[b]
			MP2	MP3	MP4SDQ	
CHN	Hydrogen cyanide	0	0	0	0	0
	Hydrogen isocyanide	12.4	19.8	16.4	17.3[c]	14.5(15.1)
CH$_2$O	Formaldehyde	0	0	0	0	0
	Hydroxymethylene	52.6	61.6	57.4	58.7[c]	54.9(54.2)
CH$_3$NO	Formamide	0	0	0	0	0
	Nitrosomethane	65.3	60.7	59.4	57.9	62.4(60.5)
C$_2$H$_3$N	Acetonitrile	0	0	0	0	0
	Methyl isocyanide	20.8	27.4	24.9	24.6	20.9(20.5)
C$_2$H$_4$O	Acetaldehyde	0	0	0	0	0
	Oxacyclopropane	33.4	27.7	27.6	28.9	26.2(27.1)
C$_3$H$_4$	Propyne	0	0	0	0	0
	Allene	2.0	4.4	1.0	1.4	1.6(1.0)
	Cyclopropene	26.5	22.0	21.4	22.9	21.7(21.6)
C$_3$H$_6$	Propene	0	0	0	0	0
	Cyclopropane	8.2	4.2	5.4	6.3	6.9(6.7)

[a] Theoretical data from: J. A. Pople, unpublished calculations.
[b] See footnotes to Table 6.69 for references to experimental data. Corrected for zero-point vibrational energy. Uncorrected heats are given in parentheses.
[c] MP4SDTQ result.

formal reaction (6.35)

$$CF_3CHO + CH_4 \rightarrow CF_3H + CH_3CHO \qquad (6.35)$$

three C—F bonds, one C=O bond, one C—C bond, and five C—H bonds are present *in both reactants and products*; only the environment in which these bonds are located has been altered. It is reasonable to anticipate that errors inherent in the individual reactant and product molecules in isodesmic reactions will largely cancel, and that even simple levels of theory will provide an adequate description of the overall energetics.

a. Bond Separation Reactions.   An isodesmic process of considerable practical importance is the *bond separation reaction* [32]: all formal bonds between heavy (non-hydrogen) atoms are separated into the simplest *parent* (two-heavy-atom) *molecules* containing these same kinds of linkages. The set of parents involving H, C, N, O, and F consists of ethane, ethylene, acetylene, methylamine, methyleneimine, hydrogen cyanide, methanol, formaldehyde, fluoromethane, hydrazine, diazene, hydroxylamine, nitroxyl, fluoramine, hydrogen peroxide, and hypofluorous acid. Extension to second- (and higher-) row elements expands the set of parents. Stoichiometric balance is achieved by the addition of one-heavy-atom hydrides, methane, ammonia, and water for first-row elements, to the left-hand side of the reaction. For example, the bond separation reaction for methylketene,

$$CH_3-CH=C=O + 2CH_4 \rightarrow CH_3-CH_3 + CH_2=CH_2 + H_2C=O, \quad (6.36)$$

results in the formation of one molecule of ethane, one of ethylene and one of formaldehyde, that is, the simplest species containing the C—C, C=C, and C=O linkages, respectively. Two molecules of methane need to be added to the reactants in order to balance the overall process. The formal bond separation reaction for benzene,

$$\text{(benzene ring)} + 6CH_4 \rightarrow 3CH_3-CH_3 + 3CH_2=CH_2, \qquad (6.37)$$

requires the molecule to be represented in terms of one of its classical valence structures, that is, with alternating single and double bonds. Given such a description, the bond separation products (three molecules of ethane and three of ethylene) again represent the simplest molecules incorporating the same component linkages. As these two examples demonstrate, a unique bond separation reaction may be written for any molecule for which a classical valence structure may be drawn. This is advantageous since the bond separation energy necessarily *characterizes* a molecule written in terms of a given valence structure. It is also a liability, for it forces description in terms of classical bonding concepts. It is likely that the energies of bond separation reactions for molecules that are well represented in terms of a single classical valence structure, methylketene for example, will be better described using simple levels of theory than those for species that are not well described in this manner, benzene for example.

The energetics of bond separation reactions are important insofar as they characterize the interaction between neighboring bonds. For example, the positive bond separation energies for molecules such as 1,3-butadiene and benzene reflect stabilization due to conjugation and resonance of extended systems of formal single and multiple bonds when compared with the corresponding isolated linkages. Similarly, the ring strain present in a molecule such as cyclopropane displays itself in terms of a negative bond separation energy. Other effects may be identified and measured quantitatively by bond separation energies. Some discussion is provided below. Additional material of a more exploratory nature may be found in Section 7.2.2.

A selection of calculated (STO-3G, 3-21G, 3-21G$^{(*)}$, and 6-31G*) and experimental bond separation energies is provided in Table 6.71. The theoretical energies are based on optimized geometries, except for the 6-31G* calculations which utilize STO-3G structures. As in previous comparisons, most of the experimental heats of reaction have been corrected for zero-point vibrational energy (using 3-21G frequencies) and to 0 K. Uncorrected data are provided in parentheses. For bond separation reactions of acyclic systems, for which the number of product and reactant species are the same, the temperature corrections (based on translational and rotational contributions alone) are zero (except where the products are linear molecules), and differential zero-point energies are generally small, that is, on the order of 1 kcal mol^{-1} or less. Notable exceptions exist, for example, for the bond separation reaction of tetrafluoromethane where the differential effect of 3.5 kcal mol^{-1} improves agreement between calculated and experimental reaction energies. Differential zero-point energies and temperature corrections for bond separation reactions involving rings, where the number of product molecules is always less than the number of reactants, are much larger, that is, on the order of 2-4 kcal mol^{-1}. Again, the corrections improve agreement between theory and experiment.

The individual bond separation energies display a number of interesting features. The positive values for propane, isobutane and neopentane indicate that the theory reproduces (but significantly underestimates) the well-known stabilizing interaction between adjacent C—C bonds in paraffins (*branching effect*). Positive bond separation energies are also found both theoretically and experimentally for other saturated compounds. Very large interactions are noted in the fluoromethanes; the enhanced stabilization with increasing fluorine substitution is due to the *anomeric effect* (see Section 7.2.2a) and is also evidenced by carbon–fluorine bond shortening (see Section 6.2.4a).

The bond separation energies for molecules with methyl groups directly attached to unsaturated linkages, for example, propene, propyne, acetonitrile, and acetaldehyde, provide a measure of energy stabilization due to *hyperconjugation*. Stronger *conjugative* interactions are found for 1,3-butadiene and formamide.

Even the STO-3G basis set fares reasonably well in many cases, in contrast to its

TABLE 6.71. Calculated and Experimental Bond Separation Energies[a]

Molecule	Bond Separation Reaction	STO-3G//STO-3G	3-21G//3-21G[b]	6-31G*//STO-3G	Expt.[c]
Propane	$CH_3CH_2CH_3 + CH_4 \longrightarrow 2CH_3CH_3$	0.6	1.4	0.8	(2.6)
Isobutane	$CH(CH_3)_3 + 2CH_4 \longrightarrow 3CH_3CH_3$	1.3	4.0	1.9	(7.5)
Neopentane	$C(CH_3)_4 + 3CH_4 \longrightarrow 4CH_3CH_3$	1.6	7.3		(13.1)
Ethylsilane	$CH_3CH_2SiH_3 + CH_4 \longrightarrow CH_3CH_3 + CH_3SiH_3$	-1.8	-1.3(-1.0)	-1.8	
Dimethylsilane	$SiH_2(CH_3)_2 + SiH_4 \longrightarrow 2CH_3SiH_3$	0.4	0.0(0.3)	0.0	0.4(0.8)
Trimethylsilane	$SiH(CH_3)_3 + 2SiH_4 \longrightarrow 3CH_3SiH_3$	1.1	0.1(0.9)	0.0	(1.6)
Tetramethylsilane	$Si(CH_3)_4 + 3SiH_4 \longrightarrow 4CH_3SiH_3$	2.1	0.0(1.9)		(2.8)
Ethylamine	$CH_3CH_2NH_2 + CH_4 \longrightarrow CH_3CH_3 + CH_3NH_2$	2.3	3.4	2.5	2.9(3.6)
Dimethylamine	$CH_3NHCH_3\ NH_3 \longrightarrow 2CH_3NH_2$	1.2	2.3	2.1	3.8(4.4)
Trimethylamine	$(CH_3)_3N + 2NH_3 \longrightarrow 3CH_3NH_2$	2.6	6.1	5.2	9.5(11.1)
Ethanol (*trans*)	$CH_3CH_2OH + CH_4 \longrightarrow CH_3CH_3 + CH_3OH$	2.6	4.8	4.1	5.0(5.7)
Dimethyl ether	$CH_3OCH_3 + H_2O \longrightarrow 2CH_3OH$	0.9	2.0	2.8	4.4(5.4)
Ethane thiol (*trans*)	$CH_3CH_2SH + CH_4 \longrightarrow CH_3CH_3 + CH_3SH$	0.0	1.7(2.4)	1.1	2.7(3.3)
Dimethyl sulfide	$CH_3SCH_3 + H_2S \longrightarrow 2CH_3SH$	1.4	1.3(0.4)	1.6	2.2(2.9)
Difluoromethane	$CH_2F_2 + CH_4 \longrightarrow 2CH_3F$	10.2	13.9	12.5	13.3(13.9)
Trifluoromethane	$CHF_3 + 2CH_4 \longrightarrow 3CH_3F$	29.2	37.5	32.3	32.9(34.6)
Tetrafluoromethane	$CF_4 + 3CH_4 \longrightarrow 4CH_3F$	53.5	62.4	49.6	49.3(52.8)
Dichloromethane	$CH_2Cl_2 + CH_4 \longrightarrow 2CH_3Cl$	-3.8	-4.3(-7.2)	-4.3	0.9(1.5)
Trichloromethane	$CHCl_3 + 2CH_4 \longrightarrow 3CH_3Cl$	-10.8	-12.9(-19.1)	-13.8	-0.1(1.9)
Tetrachloromethane	$CCl_4 + 3CH_4 \longrightarrow 4CH_3Cl$	-20.9	-26.4(-34.8)		(-1.8)

Molecule	Reaction				
Propene	$CH_3CHCH_2 + CH_4 \rightarrow CH_3CH_3 + CH_2CH_2$	4.8	3.7	3.9	4.7(5.4)
Acetaldehyde	$CH_3CHO + CH_4 \rightarrow CH_3CH_3 + H_2CO$	7.7	10.3	9.9	10.7(11.4)
Propyne	$CH_3CCH + CH_4 \rightarrow CH_3CH_3 + HCCH$	8.4	7.8	8.1	7.2(7.6)
Acetonitrile	$CH_3CN + CH_4 \rightarrow CH_3CH_3 + HCN$	10.7	13.0	11.7	14.4(14.7)
Allene	$CH_2CCH_2 + CH_4 \rightarrow 2CH_2CH_2$	0.4	-3.3	-4.4	-3.9(-2.8)
Ketene	$CH_2CO + CH_4 \rightarrow CH_2CH_2 + H_2CO$	15.5	19.2	13.3	15.0(15.7)
Carbon dioxide	$CO_2 + CH_4 \rightarrow 2H_2CO$	54.3	59.3	57.7	60.9(59.9)
trans-1,3-Butadiene	$CH_2CHCHCH_2 + 2CH_4 \rightarrow CH_3CH_3 + 2CH_2CH_2$	12.5	10.8	11.2	11.3(14.2)
Formamide	$NH_2CHO + CH_4 \rightarrow CH_3NH_2 + H_2CO$	19.8	36.6	28.4	29.8(30.8)
Benzene	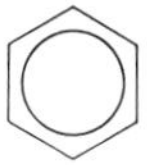 (benzene ring) $+ 6CH_4 \rightarrow 3CH_3CH_3 + 3CH_2CH_2$	70.3	60.2		(64.2)
Pyridine	 (pyridine ring) $+ 5CH_4 + NH_3 \rightarrow 2CH_3CH_3 + CH_3NH_2 + 2CH_2CH_2 + CH_2NH$	70.4	71.1		(71.9)
Pyridazine	 (pyridazine ring) $+ 4CH_4 + 2NH_3 \rightarrow 2CH_3CH_3 + CH_2CH_2 + 2CH_2NH + NH_2NH_2$	61.5	52.2		(74.6)[a]
Pyrimidine	 (pyrimidine ring) $+ 4CH_4 + 2NH_3 \rightarrow CH_3CH_3 + CH_2CH_2 + 2CH_3NH_2 + 2CH_2NH$	71.4	69.2		(80.4)[a]

TABLE 6.71. (Continued)

Molecule	Bond Separation Reaction	STO-3G// STO-3G	3-21G// 3-21G[b]	6-31G*// STO-3G	Expt.[c]
Pyrazine	 + $4CH_4$ + $2NH_3$ $\rightarrow$ CH_3CH_3 + CH_2CH_2 + $2CH_3NH_2$ + $2CH_2NH$	68.7	63.8		$(80.5)^d$
Cyclopropane	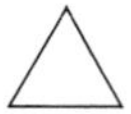 + $3CH_4$ $\rightarrow$ $3CH_3CH_3$	−45.1	−31.4	−26.2	−22.1(−19.6)
Azacyclopropane	 + $2CH_4$ + NH_3 $\rightarrow$ CH_3CH_3 + $2CH_3NH_2$	−39.7	−33.4	−22.3	−17.4(−14.7)
Oxacyclopropane	 + $2CH_4$ + H_2O $\rightarrow$ CH_3CH_3 + $2CH_3OH$	−35.3	−31.1	−19.2	−13.7(−10.5)
Thiacyclopropane	 + $2CH_4$ + H_2S $\rightarrow$ CH_3CH_3 + $2CH_3SH$	−35.3	−30.3(−19.4)	−16.2	−14.9(−10.1)
Cyclobutane	 + $4CH_4$ $\rightarrow$ $4CH_3CH_3$	−27.2	−23.7	−24.6	(15.9)

	Reaction			
Cyclopropene	 $\triangle$ + 3CH$_4$ → 2CH$_3$CH$_3$ + CH$_2$CH$_2$	−65.6	−60.4	−50.4	−43.9(−40.5)
Cyclobutene	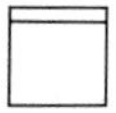 $\square$ + 4CH$_4$ → 3CH$_3$CH$_3$ + CH$_2$CH$_2$	−28.1	−28.0	−23.5	−17.3(−14.0)
Cyclopentadiene	 + 5CH$_4$ → 3CH$_3$CH$_3$ + 2CH$_2$CH$_2$	15.7	11.1	11.3	(22.4)
Thiophene	 S + 4CH$_4$ + H$_2$S → CH$_3$CH$_3$ + 2CH$_3$SH + 2CH$_2$CH$_2$	40.9	30.4(23.3)	29.2	(29.2)
Methylenecyclopropane	 + 4CH$_4$ → 3CH$_3$CH$_3$ + CH$_2$CH$_2$	−46.4	−35.5	−31.1	(−24.5)
Bicyclo[1.1.0]butane	 + 6CH$_4$ → 5CH$_3$CH$_3$	−105.3	−76.4	−63.8	−51.5(−45.5)

[a] Theoretical data from: WFBDSRP (1981) and W. J. Hehre, unpublished calculations.

[b] 3-21G(*)//3-21G(*) for molecules containing second-row elements; 3-21G//3-21G data given in parentheses.

[c] Data for one- and two-heavy-atom hydrides from Table 6.62. Data for larger molecules from: J. B. Pedley and J. Rylance, CATCH Tables, University of Sussex, 1977. Heats of reaction have been corrected for zero-point vibrational energy and to 0 K. Uncorrected heats are given in parentheses.

[d] Experimental heat of formation for methyleneimine from: D. J. DeFrees and W. J. Hehre, *J. Phys. Chem.*, **82**, 391 (1978).

performance for other types of reactions, for example, hydrogenation. There are, however, significant exceptions. The large error in the STO-3G bond separation energy for formamide is consistent with the notion that this species is inadequately described in terms of a single valence form, that is, **1**, and that account must be taken of alternative resonance contributors, for example, **2**.

$$
\begin{array}{cc}
\text{H—C} \diagup^{\displaystyle O} _{\diagdown \text{NH}_2} & \text{H—C} \diagup^{\displaystyle O^-} _{\diagdown \text{NH}_2^+} \\
\mathbf{1} & \mathbf{2}
\end{array}
$$

STO-3G bond separation energies for small-ring compounds are also significantly in error, typically by 10–20 kcal mol^{-1} but by over 50 kcal mol^{-1} in bicyclo [1.1.0] butane!

The 3-21G, 3-21G$^{(*)}$, and 6-31G* basis sets provide a better account of bond separation energies for the whole spectrum of compounds considered. In particular, many subtle effects are reproduced at these levels. For example, the 3-21G calculations reproduce the known interactions of cumulated double bonds, ranging from a small destabilizing effect in allene to a large stabilizing interaction in carbon dioxide. The calculated bond separation energies for three-.and four-membered rings are still more negative than the experimental values at the 3-21G and 6-31G* levels, although the errors are much reduced over those found in the STO-3G calculations. For example, the error in the bond separation energy for cyclopropane is 9 kcal mol^{-1} at the 3-21G level and 4 kcal mol^{-1} at 6-31G* (after differential zero-point energies have been taken into account) compared with an error of over 20 kcal mol^{-1} at STO-3G.

Note that according to all theoretical levels, the calculated bond separation energies for the chloromethanes are large and negative, indicating destablization of geminal C—Cl bonds. The experimental data do not concur, and suggest instead essentially zero bond separation energies for all three systems. The difference between experimental and theoretical bond separation energies appears to be a failure of the Hartree–Fock model. At the MP4/6-31G**//6-31G* level, the bond separation energy for dichloromethane is -0.1 kcal mol^{-1} [33], in reasonable accord with the experimental heat of reaction (0.9 kcal mol^{-1}). This is to be compared with the corresponding HF/6-31G**//6-31G* energy of 3.7 kcal mol^{-1}. It is likely that correlated calculations on CHCl$_3$ and CCl$_4$ would also result in improved bond separation energies. Further study is needed.

In summary, with suitable precautions, even the STO-3G basis set provides a reasonable description of the energetics of bond separation processes. The performance of the 3-21G (3-21G$^{(*)}$ for molecules incorporating second-row elements) and 6-31G* models is clearly superior; either of these is appropriate for the calculation of bond separation energies in most systems. Electron correlation effects, while probably of little importance in the majority of cases, do appear to play a significant role in others.

b. *Calculation of Heats of Formation.* The success of even simple levels of molecular orbital theory in reproducing the energies of isodesmic reactions can be exploited further. These calculated energy changes, combined with a limited number of experimental (or high-level theoretical) thermochemical data, yield direct estimates of heats of formation [34a]. For example, the calculated STO-3G//STO-3G energy change of 2.6 kcal mol^{-1} for the bond separation reaction (6.38),

$$CH_3 CH_2 OH + CH_4 \longrightarrow CH_3 CH_3 + CH_3 OH, \qquad (6.38)$$

may be used in conjunction with *experimental* (298 K) heats of formation for ethane (-20.1 kcal mol^{-1}), methanol (-48.2 kcal mol^{-1}), and methane (-17.8 kcal mol^{-1}) to yield a value of -53.1 kcal mol^{-1} for ΔH_f° (298 K) for ethanol.

$$\Delta H_f^{\circ} (CH_3 CH_2 OH) \simeq \Delta H_f^{\circ} (CH_3 CH_3) + \Delta H_f^{\circ} (CH_3 OH) - \Delta H_f^{\circ} (CH_4)$$

$$- \text{Bond Separation Energy of } CH_3 CH_2 OH \qquad (6.39)$$

The corresponding experimental value is -56.1 kcal mol^{-1}. (Such an approach neglects differences in zero-point vibrational energies between reactants and products and, therefore, might be expected to be less useful where comparisons of cyclic and acyclic systems are involved). An analogous treatment for dimethyl ether gives a heat of formation of -39.5 kcal mol^{-1} (experimental ΔH_f° = -44.0 kcal mol^{-1}). The heats obtained in this manner may then be used to determine an energy change of 13.6 kcal mol^{-1} for the non-isodesmic isomerization reaction

$$CH_3 CH_2 OH \longrightarrow CH_3 OCH_3, \qquad (6.40)$$

which compares favorably with the experimental difference of 12.1 kcal mol^{-1}. In contrast, the isomerization energy evaluated as the direct difference between STO-3G total energies (-0.8 kcal mol^{-1}) even fails to reproduce the correct ordering of stabilities of the two molecules.

In addition to bond separation reactions, other isodesmic processes may be employed to yield primary thermochemical data. These data may in turn be used to construct whatever energetic comparisons may be desired. For example, theoretical estimates of the heats of formation of carbocations may be obtained by combining calculated energies for the isodesmic hydride transfer reactions,

$$R^+ + CH_4 \longrightarrow RH + CH_3^+, \qquad (6.41)$$

with experimental heats of formation for methane, methyl cation and the hydrocarbon RH. Representative data are presented in Table 6.72. Even the STO-3G calculations are moderately successful in reproducing the known heats of formation. The 3-21G and 6-31G* basis sets fare even better.

In a similar manner, thermochemical data for carbanions, carbenes, and free radicals as well as other classes of compounds may be obtained.

TABLE 6.72. Calculated and Experimental Heats of Formation of Carbocations $(kcal \ mol^{-1})^a$

Formula	Cation	STO-3G// STO-3G	3-21G// 3-21G	6-31G*// 6-31G*	Expt.[b]
$C_3H_3^+$	Cyclopropenyl	241	260	252	256
$C_3H_5^+$	Allyl	223	230	230	226
$C_3H_7^+$	Isopropyl	198	203	203	192
$C_4H_9^+$	tert-Butyl	172		169	164
$C_7H_7^+$	Benzyl	213			212–217

[a] Theoretical heats of formation for carbocations R^+ from energies of reactions (6.41) and experimental heats for CH_4, CH_3^+ and hydrocarbons RH. Theoretical data from: WFBDSRP (1981) and W. J. Hehre, unpublished calculations.

[b] From: D. H. Aue and M. T. Bowers, in *Gas Phase Ion Chemistry*, M. T. Bowers, ed., Academic Press, New York, 1979, vol. 2, p. 1.

Other methods for the estimation of primary thermochemical data from theoretical calculations have been advanced. Particularly noteworthy is a procedure proposed by Wiberg [34b], who has fitted 6-31G*//6-31G* energies for a variety of hydrocarbons to experimental heats of formation in order to derive equivalents for CH_3, CH_2, CH, and C groups. An estimate of the heat of formation of any hydrocarbon may then be obtained as the difference of its calculated (6-31G*//6-31G*) energy and the sum of the appropriate groups. For the compounds considered, Wiberg has reported an uncertainty of ± 2 kcal mol^{-1} for heats of formation calculated in this manner.

6.5.7. Energies of Hydrogen-Transfer Reactions

Calculated and experimental energies of *hydrogen-transfer reactions* (6.42),

$$R^{\cdot} + CH_4 \rightarrow CH_3^{\cdot} + RH \qquad (6.42)$$

are compared in Table 6.73. RHF and UHF results are similar for saturated systems but are significantly different from one another in multiply-bonded systems. The differences decrease, however, with increasing size of basis set. Such behavior parallels that noted for RHF and UHF calculations of the geometry of open-shell systems (Section 6.2.8a) and appears to be associated with the extent of spin contamination. Unfortunately, the uncertainties in the experimental data are sufficiently large to preclude a definitive assessment of the relative performance of the RHF and UHF procedures in the calculation of reaction energies.

TABLE 6.73. Calculated and Experimental Energies of Hydrogen-Transfer Reactions[a]

Radical R•	STO-3G//STO-3G		3-21G//3-21G		6-31G*//6-31G*		Expt.[b]
	RHF	UHF	RHF	UHF	RHF	UHF	
$CH_3^•$	0	0	0	0	0	0	0
$CH_3CH_2^•$	4	4	3	3	3	3	7
$CH_3CH_2CH_2^•$	4	4	2	2	2	2	7
$CH_3CHCH_3^•$	6	7	3	3		4	10
$CH_2CH_2CH^•$	-1	-2	-6	-7	-4		-1
$CH_2{=}CH^•$	-2	7	-6	-2	-6	-3	-5
$CH_2{=}CH{-}CH_2^•$	3	28	7	21	7	21	19
$CH_2{=\!=}C{=\!=}CH^{•c}$	8	17		15		14	16
$CH_2{=\!=}C{=\!=}CH^{•d}$	25	34		18		16	17

[a] Energies for Eq. (6.42). Theoretical data from: J. S. Binkley, L. Farnell, J. A. Pople, L. Radom, and M. A. Vincent, unpublished calculations.
[b] From reference 26.
[c] Hydrogen transfer to give propyne.
[d] Hydrogen transfer to give allene.

6.5.8. Energies of Proton-Transfer and Related Reactions

Proton-transfer processes are among the most extensively studied of simple chemical reactions. Over many decades, the careful scrutiny of such processes has provided much insight into the nature of substituent–substrate interactions and into the role of the solvent in dictating molecular stability and reactivity. In addition, measurements of the acid and base strengths of isolated molecules, now made routine by the introduction of gas-phase techniques such as ion cyclotron resonance (ICR) spectroscopy and high-pressure mass spectrometry, make this a productive area with which to assess the performance of theoretical models.

The absolute *gas-phase basicity* of a molecule B is defined as the *negative* of the standard free energy for the process

$$B + H^+ \longrightarrow BH^+. \tag{6.43}$$

The absolute *proton affinity* of B is given as the negative of the enthalpy for the same reaction. In an analogous manner, the absolute *acidity* of a protonic species AH is defined as the free energy of the reaction

$$AH \longrightarrow A^- + H^+. \tag{6.44}$$

The corresponding enthalpy is termed the *heterolytic bond dissociation energy*, or *deprotonation energy*, of AH. Absolute acidities and basicities usually may not be measured directly; absolute free energy and enthalpy differences are very large (for example, the absolute proton affinity of ammonia is more than 200 kcal mol^{-1} and the heterolytic bond dissociation energy of hydrogen fluoride is nearly 380 kcal mol^{-1}!) and the equilibria, Eqs. (6.43) and (6.44), lie immeasurably close to the right and left, respectively. Nevertheless, knowledge of absolute acid and base strengths, useful for facilitating comparisons by placing all systems on a single scale, may be obtained by "pinning" relative acidity and basicity scales to those systems where direct measurements are possible.

a.   Absolute Acidities and Basicities.   Calculated and experimental absolute proton affinities for a number of simple neutral molecules are compared in Table 6.74. The experimental data have been corrected for differential zero-point energies and to 0 K. Uncorrected proton affinities (given in parentheses) often differ from the corrected data by 5 kcal mol^{-1} or more, primarily a consequence of differing numbers of reactant and product molecules in reactions (6.43).

Proton affinities calculated at STO-3G mirror experimental values poorly. With few exceptions, theoretical values are much larger than the measured quantities. Comparisons involving related systems are more successful; errors in calculated proton affinities for the three unsaturated hydrocarbons (acetylene, ethylene, and benzene) are similar, as are those associated with the first-row hydrides (HF, H$_2$O, and NH$_3$). We shall see shortly that even more subtle proton affinity comparisons fare reasonably well at the STO-3G level.

TABLE 6.74.   Hartree–Fock and Experimental Absolute Proton Affinities

	$\Delta E^{a,b}$			
Molecule	STO-3G//STO-3G	3-21G//3-21G^c	6-31G*//6-31G*	$\Delta H_{\text{Expt.}}^{a,d}$
O_2	161	126		(101)
N_2	141	126	118	122 (114)
HF	183	132	122	116 (112)
CH_4	120	115	121	130 (128)
CO	176	143	143	145 (139)
C_2H_6	150	143	142	(140)
HCl	141	128 (113)	130	144 (140)
C_2H_2	194	163	169	156 (152)
C_2H_4	210	170	175	168 (164)
H_2O	229	192	175	179 (173)
H_2S	223	170 (155)	172	183 (177)
C_6H_6	228	190		(186)
PH_3	244	194 (184)	197	197 (191)
NH_3	259	227	219	213 (205)

^{a}Energies (enthalpies) for reactions (6.43).

^{b}Theoretical data from: W. J. Hehre, unpublished calculations.

^{c}3-21G$^{(*)}$//3-21G$^{(*)}$ for molecules containing second-row elements. 3-21G//3-21G values given in parentheses.

^{d}Unless otherwise indicated, experimental proton affinities from: D. H. Aue and M. T. Bowers, in *Gas Phase Ion Chemistry*, M. T. Bowers, ed., Academic Press, New York, 1979, vol. 2, p. 1. Heats of reaction have been corrected for zero-point vibrational energy and to 0 K. Uncorrected heats are given in parentheses.

The 3-21G calculations are somewhat more successful in reproducing absolute gas-phase proton affinities. The calculations at this level generally overestimate the experimental affinities, exceptions being methane and the second-row hydrides of phosphorus, sulfur, and chlorine. The 3-21G$^{(*)}$ results for the latter compounds are better. While errors in 3-21G and 3-21G$^{(*)}$ proton affinities are typically much smaller than those resulting from STO-3G calculations, even these levels of theory do not provide a sufficiently uniform and reliable account of absolute proton affinities to be generally useful as investigative tools.

With the exception of acetylene and ethylene, and to a lesser extent the second-row hydrides, the 6-31G* calculations are quite successful in reproducing the experimental absolute proton affinities. Errors in the two unsaturated hydrocarbons are in part an artifact of the theoretical method employed for determination of equilibrium geometry. At the Hartree–Fock levels used here, both vinyl and ethyl cations, resulting from the protonation of acetylene and ethylene, respectively, are predicted to have "classical" open structures. However, when correlation is included, symmetrically-bridged structures are favored (see Section 7.3.1b for further discussion). If the ions are *constrained* to have bridged structures, the 6-31G* proton affinities are lowered by 7 and 1 kcal mol^{-1} for acetylene and ethylene, respectively.

Calculated absolute acidities (more precisely, deprotonation energies) for a selection of small molecules are compared with experimental enthalpies in Table 6.75. Minimal, split-valence, and polarized basis sets have been examined, in addition to two representations that include diffuse functions on the heavy atoms. Optimum equilibrium geometries have been employed, except for calculations at the 3-21+G, 3-21+G$^{(*)}$, and 6-31+G* levels; these utilize structures obtained from the 3-21G, 3-21G$^{(*)}$; and 6-31G* basis sets, respectively. Again, corrections for zero-point energy differences (between neutral and deprotonated forms) and to 0 K are significant, and have been added to the experimental acidities.

The STO-3G basis set performs very poorly. Absolute acidities are generally in error by 100-200 kcal mol^{-1}! The mean absolute deviation of theory from experiment is 137 kcal mol^{-1}! Even the ordering of acid strengths is often given incorrectly at this level. For example, the STO-3G ordering of acidities of first-row hydrides, $NH_3 > H_2O > HF$, is the reverse of the correct sequence. The STO-3G basis set does correctly order the acidities of the corresponding second-row hydrides, although the difference in acidities of H_2S and HCl is overestimated by 75 kcal mol^{-1}.

The 3-21G and 3-21G$^{(*)}$ calculations fare better, although even at this level acid strength is often in error (usually underestimated) by as much as 50 kcal mol^{-1} or more. Overall, the mean absolute deviation of 3-21G (3-21G$^{(*)}$ for compounds incorporating second-row elements) is 26 kcal mol^{-1}.

Calculations at the 6-31G* level also consistently underestimate acid strength. An error of approximately 30 kcal mol^{-1} is noted for the first-row hydrides methane, ammonia, water, and hydrogen fluoride; smaller errors are noted for acetylene and hydrogen cyanide (22 and 11 kcal mol^{-1}, respectively). The largest error for a second-row hydride is only 10 kcal mol^{-1} for SiH_4. Overall, the mean absolute deviation of calculated from observed acid strengths is 17 kcal mol^{-1}.

The 3-21+G (3-21+G$^{(*)}$ for molecules containing second-row elements) and 6-31+G* basis sets, which incorporate diffuse functions, provide the most satisfactory account of absolute acidities. The mean absolute deviation in calculated acidities for 3-21+G is only 9 kcal mol^{-1}, although sizable errors are noted in the acidities of HF (16 kcal mol^{-1}) and HCl (12 kcal mol^{-1}). The 6-31+G* basis set fares even better. The mean error is only 6 kcal mol^{-1}; the largest single deviation is 10 kcal mol^{-1} for hydrogen chloride.

Additional comparisons, provided in Table 6.76, confirm the generally satisfactory description of absolute acidities afforded by 3-21+G calculations. One conclusion is inescapable: *proper description of absolute acid strength requires basis sets which incorporate diffuse functions*. Anions generally have low ionization potentials; electrons are easily lost. These electrons, on average, are relatively far from the nuclei and require diffuse orbitals for bonding. Without diffuse functions, even moderately large basis sets, for example, 6-31G*, are not entirely successful either in their calculation of absolute acidities or in their ordering of the acidities. Supplemented by diffuse s- and p-type functions, even the relatively small 3-21G and 3-21G$^{(*)}$ basis sets perform quite well in both of these tasks.

TABLE 6.75. Hartree–Fock and Experimental Absolute Energies (Enthalpies) of Deprotonation

Molecule	$\Delta E^\circ_{\text{calc.}}$ [a,b]					$\Delta H^\circ_{\text{Expt.}}$ [a,e]
	STO-3G//STO-3G	3-21G//3-21G[c]	6-31G*//6-31G*	3-21+G//3-21G[d]	6-31+G*//3-21G	
CH_4	560	463	457	434	435	426 (417)
NH_3	547	463	444	420	421	409 (400)
H_2O	565	450	429	393	402	398 (391)
C_2H_2	496	405	403	377	382	381 (375)
SiH_4	510	390 (378)	388	383	384	378 (372)
HF	602	432	409	360	374	376 (372)
PH_3	525	387 (372)	383	372	374	376 (370)
H_2S	506	364 (349)	360	350	352	358 (353)
HCN	462	379	370	352	353	359 (353)
HCl	411	337 (324)	335	325	327	337 (333)

[a]Energies (enthalpies) for reactions (6.44).

[b]CCSS (1983); W. J. Hehre and P. v. R. Schleyer, unpublished calculations.

[c]3-21G(*)//3-21G(*) for molecules incorporating second-row atoms. 3-21G//3-21G values given in parentheses.

[d]3-21+G(*)//3-21G(*)//3-21G for molecules incorporating second-row atoms.

[e]Experimental data from: J. E. Bartmess and R. T. McIver, Jr., in *Gas Phase Ion Chemistry*, M. T. Bowers, ed., Academic Press, New York, 1979, vol. 2, p. 87. Corrected for differential zero-point vibrational energies and to 0 K. Uncorrected data are given in parentheses.

TABLE 6.76.   Calculated and Experimental Absolute Acidities

Molecule	3-21+G//3-21+G[a]	Expt.[b]
LiOH	447	448 ± 15[c]
CH_2CH_2	421	> 404
HBF_2	389	404 ± 31
CH_3NH_2	420	403
$HBCl_2$	372	398 ± 31
NaOH	462	386 ± 25
CH_3OH	391	379
CH_3CH_2OH	389	376
HCCH	382	375
CH_3CN	384	372
HOOH	366	368
CH_3CHO	372	366
CH_3SH	345	359
CH_3NO_2	342	359
CH_3COOH	350	349
HCOOH	344	345
HN_3	346	344 ± 2
HONO	334	338 ± 8
$HONO_2$	307	325

[a] CCSS (1983).

[b] Unless otherwise indicated, experimental data from: J. E. Bartmess and R. T. McIver, Jr., in *Gas Phase Ion Chemistry*, M. T. Bowers, ed., Academic Press, New York, 1979, vol. 2, p. 87. Acidities have not been corrected either for zero-point vibrational energy or to 0 K.

[c] JANAF Thermochemical Tables, 2nd Ed., D. R. Stull and H. Prophet, ed., NSRDS-NBS 37, National Bureau of Standards, Washington, D.C., 1971.

b.   Effect of Electron Correlation on Absolute Basicities.   The effect of electron correlation on absolute basicities is explored in Table 6.77. When the 6-31G** basis set is used, the effect is small even at the MP4 level; the largest change from the corresponding HF/6-31G** proton affinity is 6 kcal mol^{-1}. The correlation corrections do not always lead to improved agreement with experiment. It appears that the *sign* of the correlation correction to calculated proton affinities cannot be consistently predicted unless a basis set larger than 6-31G** is employed.

c.   Relative Acidities and Basicities.   A knowledge of absolute acidities and basicities is not always (nor even usually) needed; description of acid or base strength of one molecule *relative* to another may suffice. If the systems to be compared have similar structures, differential correlation effects are likely to be small, and simple levels of single-determinant theory may provide an adequate description. The data in Table 6.78 support such a hypothesis. Experimental proton affinities for a number of nitrogen bases, relative to ammonia as a standard, are compared with values calculated using the STO-3G//STO-3G, 3-21G//STO-3G, 3-21G//3-21G and 6-31G*//STO-3G models.

TABLE 6.77. Effect of Electron Correlation on Absolute Proton Affinities

Molecule	$\Delta E^{\circ}_{calc.}$[a]				$\Delta H^{\circ}_{Expt.}$[a,b]
	6-31G**//6-31G*	MP2/6-31G**//6-31G*	MP3/6-31G**//6-31G*	MP4/6-31G**//6-31G*	
N_2	121	124	125	125	122 (114)
HF	127	130	130	131	116 (112)
CO	144	152	150	151	145 (139)
C_2H_2	165	161	164	163	156 (152)
C_2H_4	177	172	174	172	168 (164)
H_2O	180	180	181	179	179 (173)
HCN	181	178	180	179	(179)
H_2CO	187	180	183	182	183 (177)
CH_3OH	196	194	196	195	(185)
NH_3	220	220	221	220	213 (205)
CH_2NH	225	221	222	221	212 (205)[c]
CH_3NH_2	230	230	231	230	222 (214)

[a]Energies (enthalpies) for reactions (6.43). Theoretical data from: J. E. Del Bene, M. J. Frisch, R. Krishnan, and J. A. Pople, *J. Phys. Chem.*, **86**, 1529 (1982).

[b]The experimental data have been corrected for differential zero-point energy and to 0 K. Uncorrected heats are given in parentheses. Unless otherwise indicated, the experimental data have been taken from: D. H. Aue and M. T. Bowers in *Gas Phase Ion Chemistry*, M. T. Bowers, ed., Academic Press, New York, 1979, vol. 2, p. 1.

[c]The proton affinities of $CH_2=NH$ and NH_3 have been established by ion cyclotron resonance spectroscopy to approximately the same. W. T. Huntress and J. L. Beauchamp, unpublished observations, cited in: J. F. Wolf, R. H. Staley, I. Koppel, M. Taagepera, R. T. McIver, Jr., J. L. Beauchamp and R. W. Taft, *J. Am. Chem. Soc.*, **99**, 5417 (1977).

TABLE 6.78.   Hartree-Fock and Experimental Relative Proton Affinities of Nitrogen Bases

Nitrogen Base, B	$\delta\Delta E^{a,b}$				$\delta\Delta H^{\circ}_{Expt.}{}^{a,c}$
	STO-3G// STO-3G	3-21G// STO-3G	3-21G// 3-21G	6-31G*// STO-3G	
N_2	-118	-107	-101	-104	$-94\ (-91)$
HCN	-57	-48	-44	-41	(-26)
CH_3CN	-38	-31	-28	-25	$-17\ (-14)$
NH_3	0	0	0	0	0 (0)
$CH_2=NH$	-1	2	4	4	-3 $(0)^d$
⬡—NH_2	8	7	1		(7)
CH_3NH_2	9	9	17	11	9 (9)
NH △	9		13	14	10 (11)
$CH_3CH_2NH_2$	13	12	17	13	12 (12)
$(CH_3)_2NH$	15	15	14	18	16 (16)
⬡N	18	14	21		(15)
$(CH_3)_3N$	20	19	10	22	19 (19)

[a] Energies (enthalpies) for reactions: $BH^+ + NH_3 \rightarrow B + NH_4^+$.
[b] Theoretical data from: W. J. Hehre and R. W. Taft, unpublished calculations.
[c] Unless otherwise indicated, experimental data from: D. H. Aue and M. T. Bowers, in *Gas Phase Ion Chemistry*, M. T. Bowers, ed., Academic Press, New York, 1979, Vol. 2, p. 1. Corrected for differential zero-point vibrational energy. Uncorrected data are given in parentheses.
[d] See footnote (c) of Table 6.77.

Even the minimal basis set calculations reproduce the majority of the experimental data accurately. Except for hydrogen cyanide, acetonitrile, and N_2, the electronic structures of which differ substantially from those of the remainder of the compounds, almost all of the calculated relative proton affinities are within 1–2 kcal mol^{-1} of experiment. The calculated proton affinities of HCN and CH_3CN, relative to N_2 as a standard, are also in good agreement with the experimental data.

Relative proton affinities calculated using the 3-21G//STO-3G model are also in close accord with the measured quantities. The calculated proton affinities for hydrogen cyanide, acetonitrile and molecular nitrogen, relative to ammonia, are still smaller than the corresponding experimental values, although the errors are reduced. The relative proton affinities of the two nitriles, referenced to nitrogen, again closely reproduce the experimental data.

The 3-21G//3-21G model does not fare quite as well. This parallels the known failure of calculations at this level to describe properly the geometry about tricoordinate nitrogen (see Section 6.2.2).

6-31G*//STO-3G calculations generally lead to modest improvements in relative proton affinities over 3-21G//STO-3G calculations. At this level, the affinities of N_2, HCN, and CH_3CN still deviate significantly from experiment, suggesting the necessity for further enhancement of the basis set and/or incorporation of electron correlation.

Calculated relative acidities for a selection of carbon acids are compared with experimental values in Table 6.79. At the 3-21G level, the experimental ordering of carbon acid strengths is reproduced (the placement of the "methyl" acidity of propyne is uncertain); however, calculated acidity differences are not in good accord with the data. Overall, the mean absolute deviation of 3-21G from experimental relative acid strengths is 22 kcal mol^{-1} (7 comparisons).

The 3-21+G basis set leads to improved agreement with experiment; the mean absolute deviation of theory from experiment is now 10 kcal mol^{-1} (5 comparisons). Still, the acid strengths of all systems (relative to methane) are overestimated, sometimes significantly so (by 31 kcal mol^{-1} in nitromethane). It is clear that the accurate description of relative acid strength requires the use of higher-level quantum chemical models than are needed for relative basicities.

d. Effect of Remote Substituents on Acid and Base Strength. Another important application of the theory is the description of effects of remote substituents on gas-phase acid and base strengths. Here, even the simplest Hartree–Fock models might be expected to perform favorably. For example, as shown by the data in Table 6.80, the agreement between experimental and STO-3G relative proton affinities of *meta*- and *para*-substituted pyridines, referenced to the parent com-

TABLE 6.79. Calculated and Experimental Acidities of Carbon Acids

Acid, CH_3X	$\delta\Delta E_{Theory}$[a,b]		$\delta\Delta H^\circ_{Expt.}$[a,c]
	3-21G//3-21G	3-21+G//3-21+G	
CH_4	0	0	0
$CH_3CH=CH_2$	47	28[d]	26
$CH_3C\equiv CH$	54		$\leqslant 37$[e]
$CH_3C_6H_5$	48		38
CH_3CF_3	59	47[d]	42 ± 3
CH_3CN	67	48	44
CH_3CHO	75	60	50
CH_3NO_2	100	89	58

[a] Energies (enthalpies) for reactions: $XCH_2^- + CH_4 \rightarrow XCH_3 + CH_3^-$.

[b] CCSS (1983); W. J. Hehre, unpublished calculations.

[c] Unless otherwise indicated, experimental data from: J. E. Bartmess and R. T. McIver, Jr., in *Gas Phase Ion Chemistry*, M. T. Bowers, ed., Academic Press, New York, 1979, vol. 2, p. 87. Reaction heats have not been corrected either for zero-point vibrational energy or to 0 K.

[d] 3-21+G//3-21G calculations.

[e] Experimental acidity corresponds to proton removal from the alkyne carbon.

TABLE 6.80.   Calculated and Experimental Relative Proton Affinities of *meta*

and *para* Substituted Pyridines, [a]

Substituent, X	meta		para	
	$\delta\Delta E$ STO-3G	$\delta\Delta H^{\circ}_{\text{Expt.}}$	$\delta\Delta E$ STO-3G	$\delta\Delta H^{\circ}_{\text{Expt.}}$
NO_2			−19.2	−11.9
CN	−12.4	−10.9	−11.4	−9.8
CF_3		−8.1	−6.8	−7.8
F	−6.3	−5.6	−0.2	−3.9
H	0.0	0.0	0.0	0.0
CH_3	2.5	2.4	5.0	3.3
OCH_3		3.0	8.5	6.7

[a]Energies (enthalpies) for reactions (6.45). Geometries based on partially optimized structures for pyridine and protonated pyridine, using standard model substituents. Theoretical data from: R. W. Taft, in *Proton Transfer Reactions*, E. F. Caldin and V. Gold, ed., Wiley-Halstead, New York, 1975, p. 31. Experimental proton affinities from: D. H. Aue and M. T. Bowers, in *Gas Phase Ion Chemistry*, M. T. Bowers, ed., Academic Press, New York, 1979, vol. 2, p. 1.

pound, that is, energies of reactions (6.45),

is very good; the only conspicuous exception is the *para* nitro compound. Further demonstration of the ability of even STO-3G calculations to account for the effects of remote substituents on base strength is provided by the data in Figure 6.32. Here, calculated effects of substituents β to the nitrogen center in methylamine, that is, energies for reactions (6.46),

$$XCH_2 NH_3^+ + CH_3 NH_2 \longrightarrow XCH_2 NH_2 + CH_3 NH_3^+ \qquad (6.46)$$

are compared with experimental values. The overall agreement is impressive.

The minimal basis set calculations are moderately successful in reproducing the effects of remote substituents on the acidity of phenol, that is, energy changes for reactions (6.47).

As the data in Table 6.81 show, the calculations mirror the observed effects of methyl and fluoro ring substituents.

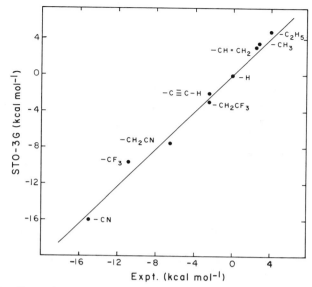

FIGURE 6.32. Comparison of STO-3G and experimental relative basicities in β-substituted methylamines, XCH_2NH_2. From M. Taagepera, W. J. Hehre, R. D. Topsom, and R. W. Taft, *J. Am. Chem. Soc.*, 98, 7438 (1976).

In summary, it appears that even the simplest theoretical levels are capable of accounting satisfactorily for many of the effects of remote substituents on the energies of protonation and deprotonation reactions within closely related series of molecules.

e. Lithium Cation Affinities. Related to gas-phase basicities are the heats of processes in which a general electrophile, E^+, is added to a molecule,

$$B + E^+ \rightarrow BE^+. \tag{6.48}$$

TABLE 6.81. Calculated and Experimental Relative Acidities of Substituted Phenols,[a] X—⟨ ⟩—OH

Substitutent X	ortho		meta		para	
	$\delta\Delta E_{STO-3G}$	$\delta\Delta H^\circ_{Expt.}$	$\delta\Delta E_{STO-3G}$	$\delta\Delta H^\circ_{Expt.}$	$\delta\Delta E_{STO-3G}$	$\delta\Delta H^\circ_{Expt.}$
Me	0.3	0.3	−0.6	−0.5	−1.0	−1.2
F	2.7	2.8	5.0	4.8	1.7	2.0

[a] Energies (enthalpies) for reactions (6.47). Geometries based on partially optimized structures for phenol and phenoxy anion, using standard model substitutents. Experimental data summarized in: R. W. Taft, in *Proton Transfer Reactions*, E. F. Caldin and V. Gold, ed., Wiley-Halstead, New York, 1975, p. 31. Theoretical data from: L. Radom, *J. Chem. Soc., Chem. Commun.*, 403 (1974).

Reactions analogous to those defining absolute acidity, that is, reactions (6.44), involve heterolytic cleavage to E^+,

$$AE \rightarrow A^- + E^+. \tag{6.49}$$

The same experimental techniques used to study proton-transfer processes in the gas phase, that is, ion cyclotron resonance spectroscopy and high-pressure mass spectrometry, have recently been employed to investigate the thermochemistry of reactions (6.48) where E^+ is Li^+, Na^+, Al^+, and Mn^+ [35]. Heats of reaction for other electrophilic species, for example CH_3^+, are available indirectly from proton affinity data [36], and direct measurements for other electrophiles will no doubt be forthcoming. A comparison of calculated and experimental lithium cation affinities, that is, energies (heats) of reactions (6.48) with $E^+ = Li^+$, are provided in Table 6.82. While the experimental data have been corrected for differential zero-point energies and to 0 K, these corrections are only of the order of 1 kcal mol^{-1}, far smaller than previously noted for absolute proton affinities (see Table 6.74).

All levels of calculation reproduce the fact that absolute lithium cation affinities are far smaller in magnitude than proton affinities. In addition, even the simplest

TABLE 6.82. Hartree–Fock and Experimental Lithium Affinities^a

Molecule	ΔE^b			$\Delta H^\circ_{Expt.}$ ^{e}
	STO-3G//STO-3G	3-21G//3-21G^c	6-31G*//3-21G^d	
CH_3Cl	45	23	22	24 (25)
CH_3F	69	49	35	31 (31)
$(CH_3)_2S$		32		(32)
H_2O	79	57	39	35 (34)
H_2CO	74	53	40	36 (36)
HCN	62	46	37	36 (36)
CH_3OH	78	58	41	37 (37)
C_6H_6	85	44		(38)
NH_3	75	56	45	40 (39)
$(CH_3)_2O$	77	58	41	40 (40)
CH_3CHO		59	45	(41)
CH_3NH_2	75	56	45	42 (41)
$(CH_3)_2NH$		55	44	(42)
$(CH_3)_3N$	72	54		(42)
CH_3CN	71	54	45	43 (43)

^{a} $-\Delta E$ ($-\Delta H$) for reactions (6.48) with $E^+ = Li^+$.
^{b} Theoretical data from: S. F. Smith, J. Chandrasekhar and W. L. Jorgensen, *J. Phys. Chem.*, **86**, 3308 (1982); W. J. Hehre, unpublished calculations.
^{c} 3-21G$^{(*)}$//3-21G$^{(*)}$ for molecules containing second-row elements.
^{d} 6-31G*//3-21G$^{(*)}$ for molecules containing second-row elements.
^{e} Experimental data from: R. H. Staley and J. L. Beauchamp, *J. Am. Chem. Soc.*, **97**, 5920 (1975); R. L. Woodin and J. L. Beauchamp, *ibid.*, **100**, 501 (1978). Heats of reaction have been corrected for zero-point vibrational energy and to 0 K. Uncorrected heats are given in parentheses.

quantum chemical models show little variation in the energetics of lithium cation addition to different substrates (relative to the large variation in proton-transfer energies), again in accord with experiment. The STO-3G calculations do not provide a successful account of variations in Li$^+$ affinities. For example, they fail to reproduce the known increase in affinity from water to methanol to dimethyl ether; they show instead the reverse trend. Similarly, the calculations at this level show a decrease in the lithium cation affinity of ammonia as a result of methyl substitution, the opposite of the experimentally observed trend.

The 3-21G (3-21G$^{(*)}$ for molecules incorporating second-row elements) calculations offer a somewhat improved description. The description of methyl substituent effects on the lithium cation affinities of water and ammonia given by the 3-21G model, however, is still not entirely satisfactory, and the calculations at this level fail to place properly methyl fluoride and benzene in the overall scale. The 6-31G* calculations fare much better. Not only are the absolute affinities in good agreement with experimental values (mean absolute deviation of 6-31G*//3-21G Li$^+$ affinities from experiment is only 3 kcal mol^{-1}), but relative effects are reasonably well reproduced.

6.5.9. The Additivity of Basis Set and Electron Correlation Effects in Estimating Relative Energies

We have seen that quantitatively reliable energy comparisons often require the use of large basis sets including polarization functions, in conjunction with methods which take into account the effects of electron correlation. However, correlated levels such as MP3 and MP4 (and to a lesser extent MP2) rapidly become prohibitively expensive as the number of functions in the basis set is increased, especially for larger systems of chemical interest. In many of these cases, polarization basis sets such as 6-31G* and 6-31G** may be practical only for HF and possibly MP2 calculations.

How can one overcome this problem and obtain accurate relative energies? One answer is to *assume* additivity of correlation and basis set enhancement effects. The adequacy of two levels of such an additivity approximation are examined here [37]. The first (designated (MP3/6-31G**)$_{HF, 6-31G}$) assumes that the effect on relative energies of addition of polarization functions to the basis set is the same at the HF and MP3 levels, while the second (designated (MP3/6-31G**)$_{MP2, 6-31G}$) assumes that this effect is the same at the MP2 and MP3 levels. Relative energies at these levels are defined by Eqs. (6.50) and (6.51), respectively.

$$\Delta E(\text{MP3/6-31G}^{**})_{\text{HF, 6-31G}} = \Delta E(\text{MP3/6-31G}) + \Delta E(\text{HF/6-31G}^{**})$$
$$- \Delta E(\text{HF/6-31G}) \qquad (6.50)$$

$$\Delta E(\text{MP3/6-31G}^{**})_{\text{MP2, 6-31G}} = \Delta E(\text{MP3/6-31G}) + \Delta E(\text{MP2/6-31G}^{**})$$
$$- \Delta E(\text{MP2/6-31G}). \qquad (6.51)$$

It is necessary to determine how relative energies obtained from Eqs. (6.50) and (6.51) compare with the directly calculated MP3/6-31G** values. In particular, do these additivity schemes lead to an improvement over the accessible raw values, namely ΔE(MP3/6-31G) and ΔE(HF/6-31G**) in the case of Eq. (6.50), and ΔE(MP3/6-31G) and ΔE(MP2/6-31G**) in the case of Eq. (6.51)?

Calculated relative energies for three isomers of CH_2O^+ are given in Table 6.83. Note that the relative energies obtained using the additivity schemes (6.50) and (6.51) are considerably closer to the actual MP3/6-31G** values than are those calculated at lower levels. For example, direct MP3/6-31G** calculations give an energy for $CHOH^{+\cdot}$ 4.1 kcal mol^{-1} higher than that of $CH_2O^{+\cdot}$. The (MP3/6-31G**)$_{HF, 6-31G}$ and (MP3/6-31G**)$_{MP2, 6-31G}$ estimates of 4.3 and 3.4 kcal mol^{-1}, respectively, are close to this value. In contrast, the relative energies that would be accessible in the absence of the additivity scheme show large deviations from the MP3/6-31G** value, for example, ΔE values of -1.3 and +7.1 kcal mol^{-1} at the MP2/6-31G** and MP3/6-31G levels, respectively. For the relative energies listed in Table 6.83, the maximum deviation from the MP3/6-31G** values is 10.0 kcal mol^{-1} at MP3/6-31G, 5.4 kcal mol^{-1} at MP2/6-31G** and 4.7 kcal mol^{-1} at HF/6-31G**, compared with only 3.0 kcal mol^{-1} for (MP3/6-31G**)$_{HF, 6-31G}$ and 0.7 kcal mol^{-1} at the (MP3/6-31G**)$_{MP2, 6-31G}$ level.

Correlation and polarization function effects are known to be particularly important in describing relative energies of singlet and triplet electronic states, and such situations provide an exacting test of the additivity hypothesis. Displayed in Table 6.84 are the calculated energy differences between CH_3X^+ ($^3A_1$) and CH_2XH^+ ($^1A'$) (X = O, S) at a number of theoretical levels. For both the O and S cases, the (MP3/6-31G**)$_{HF, 6-31G}$ energy differences represent a dramatic improvement over the HF/6-31G** and MP3/6-31G values. For X = O, the deviations from the calculated MP3/6-31G** results are respectively 0, 31, and 18 kcal mol^{-1} while for X = S, the deviations are 3, 18, and 21 kcal mol^{-1}. Similarly, the (MP3/6-31G**)$_{MP2, 6-31G}$ energies are closer to the MP3/6-31G** results than are the calculated MP2/6-31G** or MP3/6-31G values.

TABLE 6.83. Relative Energies of $CH_2O^{+\cdot}$ Isomers[a]

Level[b]	Energy (relative to $CH_2O^{+\cdot}$)	
	$CHOH^{+\cdot}$	$COH_2^{+\cdot}$
HF/6-31G	11.4	43.8
HF/6-31G**	8.8	56.7
MP2/6-31G	2.2	45.0
MP2/6-31G**	-1.3	54.3
MP3/6-31G	7.1	45.2
(MP3/6-31G**)$_{HF, 6-31G}$	4.3	58.2
(MP3/6-31G**)$_{MP2, 6-31G}$	3.4	54.5
MP3/6-31G**	4.1	55.2

[a]Data from reference 37a.
[b]4-31G structures.

TABLE 6.84. Energy Difference Between CH_3X^+ ($^3A_1$) and CH_2XH^+ ($^1A'$)[a]

Level[b]	ΔE	
	X = O	X = S
HF/6-31G	42.6	-16.6
HF/6-31G**	61.1	7.3
MP2/6-31G	82.8	5.6
MP2/6-31G**	102.8	28.8
MP3/6-31G	73.9	4.4
(MP3/6-31G**)$_{HF, 6-31G}$	92.4	28.3
(MP3/6-31G**)$_{MP2, 6-31G}$	93.9	27.6
MP3/6-31G**	92.3	25.7

[a] Data from reference 37a.
[b] 4-31G structures.

It is particularly striking that these additivity schemes enable the determination of good estimates of the MP3/6-31G** energy difference from uniformly poor values calculated at simpler levels of theory. For example, the CH_3S^+ ($^3A_1$) - CH_2SH^+ ($^1A'$) energy difference is calculated to be -16.6 kcal mol^{-1} (HF/6-31G), +7.3 kcal mol^{-1} (HF/6-31G**) and +4.4 kcal mol^{-1} (MP3/6-31G), all in very poor agreement with the MP3/6-31G** value of 25.7 kcal mol^{-1}. The (MP3/6-31G**)$_{HF, 6-31G}$ energy difference (28.3 kcal mol^{-1}), calculated with the aid of the additivity scheme *from the above three values*, is, on the other hand, quite close to the MP3/6-31G** value.

Although *strict* additivity of correlation and polarization function effects is neither expected nor observed, estimation schemes based on such assumptions are generally found to yield relative energies in much better accord with the results of full MP3 calculations with polarization basis sets than do calculations at lower levels of theory. Thus, in those cases where MP3 calculations with polarization basis sets are not feasible, an additivity scheme based on the effect of basis set improvement at either the HF or MP2 levels is likely to provide the most reliable relative energies.

6.5.10. Concluding Remarks and Recommendations

Except for *isodesmic* comparisons, the STO-3G minimal basis set appears to be of little value for establishing relative thermochemical stabilities. Thus, for the energy comparisons examined in this section, only bond separation energies (for compounds adequately represented in terms of classical valence forms), isodesmic relative isomer energies and relative basicities of closely related compounds are well described by STO-3G.

The performance of the 3-21G (3-21G$^{(*)}$ for molecules incorporating second-row elements) and 6-31G* basis sets is much better. While bond-dissociation energies are poorly described at these levels of theory (and in general at the Hartree–Fock limit), other reactions are treated satisfactorily. In particular, calculated hydrogenation energies and energies relating multiple to single bonds are reasonably close to experimental values. Significantly, the HF/3-21G and HF/6-31G* models generally provide realistic estimates of relative isomer stability. Both basis sets also provide a reasonable account of absolute and relative basicities, although they both fail to reproduce experimental acid strengths correctly. When anions are involved, basis sets supplemented by diffuse functions appear to be required.

Electron correlation effects are important for the accurate description of homolytic bond dissociation energies; they are less significant for heterolytic processes in which the number of electron pairs is held constant, although here too they almost always lead to improved quantitative agreement with experiment. For AH bonds, even the MP2/6-31G* model yields bond energies that are close to those resulting from higher-order perturbation treatments. Higher-order models, for example, MP3 and MP4, appear to be required for the proper description of the dissociation of bonds between heavy atoms.

6.6.   ELECTRIC DIPOLE MOMENTS AND MOLECULAR CHARGE DISTRIBUTIONS

The distribution of charge in molecular systems has great significance for molecular structure and reactivity [38]. Its description is a primary goal of any electronic structure theory. Experimentally, electron probability distributions may be obtained via X-ray diffraction, at least in principle. Careful experimental investigations have been carried out for a small number of systems, and have been compared with the results of theoretical calculations [39].

Some indication of the overall distribution of charge in molecules may be obtained from electric dipole moments. Furthermore, comparison of calculated and measured moments provides a means of assessing the performance of the theoretical models in describing charge distributions. Detailed comparisons are made in Section 6.6.1 for a wide variety of simple molecules, including systems of very low polarity, for example, hydrocarbons, where the moments arise from a delicate superposition of subtle effects, and nearly ionic compounds, for example, alkali-metal halides, where the moments can be properly mimicked by separated point charges.

Atomic charges cannot be obtained in a *unique* manner either from experiment or from theory. Assignment of electrons to specific atomic centers requires a model, for example, *Mulliken population analysis*, which, while it might appear reasonable, is necessarily arbitrary. Our comparisons of *Mulliken charges* (Section 6.6.2), calculated at different levels of theory but without reference to experiment, are not intended to provide an indication of the ability or inability of a particular model to describe charge distributions, but only to establish the relative behavior of the various theoretical levels.

6.6.1.   Electric Dipole Moments

The *electric dipole moment* is the principal experimental quantity related to charge for which comparisons with theory may be made. It is a *one-electron property*, defined according to Eq. (2.96) for closed-shell single-determinant wavefunctions (see Section 2.10).

The electric dipole moment is a *vector quantity*, and hence is characterized not only by a *scalar magnitude*, but also by a *sign* and a *direction*. Most experimental data on dipole moments relate only to magnitude, although microwave spectroscopy and related techniques provide some information about the direction and sign of moments in polyatomic molecules. We, likewise, focus primarily on the comparison of the magnitudes of calculated and experimental dipole moments. Experimental information relating to dipole moment directions and signs are scarce (and subject to considerable uncertainty); comparisons between theory and experiment are made only in a few cases where the experimental data appear to be well established.

In assessing the performance of various basis sets in reproducing experimental dipole moments, errors arising from the inability of the theory to describe correctly the charge distribution need to be distinguished from those resulting from an inappropriate choice of molecular geometry. For example, the dipole moment of ammonia is very sensitive to the degree of nonplanarity. Were the molecule planar, its dipole moment would be zero; ammonia actually is pyramidal and has a finite moment. Comparison of the 3-21G dipole moment calculated using the 3-21G optimized geometry (which underestimates the degree of nonplanarity) with the STO-3G dipole moment at the STO-3G optimized geometry (which makes ammonia too pyramidal) would reflect in part the effects of the different geometries. It is clear, therefore, that it is not always appropriate to employ optimized geometries for comparisons of calculated dipole moments. For the most part, we have employed either experimental geometries or a *single set* of theoretical geometries to compare calculated dipole moments. Standard model geometries may be appropriate and have been employed where a consistent set of experimental structural data is not available, and where the systems are too large for complete geometry optimization to be practical, for example, for monosubstituted benzenes.

Unless otherwise noted, experimental dipole moments have been selected from the following compendia:

1. R. D. Nelson, D. R. Lide, and A. A. Maryott, *Selected Values of Electric Dipole Moments for Molecules in the Gas Phase*, NSRDA-NBS 10, U.S. Government Printing Office, Washington, D.C., 1967.
2. A. L. McClellan, *Tables of Experimental Dipole Moments*, W. H. Freeman, San Francisco, 1963; vol. 2, Rahara Enterprises, El Ceritos, California, 1974.

Data from either of these sources will be employed without further reference. All theoretical and experimental moments tabulated are in units of debyes (D).

a. Diatomic and Small Polyatomic Molecules. Calculated electric dipole moments for a selection of diatomic and small polyatomic molecules are compared with experimental values in Table 6.85. Experimental equilibrium geometries have been used.

The STO-3G basis set generally, but not always, yields dipole moments that are smaller than the corresponding experimental quantities. The poorest agreement between theory and experiment occurs for compounds incorporating either very

TABLE 6.85. Calculated and Experimental Electric Dipole Moments for Diatomic and Small Polyatomic Molecules[a]

Molecule	STO-3G//expt.	3-21G//expt.	3-21G$^{(*)}$//expt.	6-31G*//expt.	Expt.
CO	0.17	0.39	0.39	0.33	0.11[b]
HCP	0.11	0.87	0.56	0.71	0.39
PH$_3$	0.53	1.16	0.79	0.77	0.58
ClF	0.50	1.18	1.24	1.25	0.88
H$_2$S	1.02	1.81	1.40	1.39	0.97
HCl	1.73	1.85	1.51	1.51	1.08
SiH$_3$F	1.04	1.83	1.46	1.51	1.27
NH$_3$	1.79	2.17	2.17	1.95	1.47
HF	1.29	2.16	2.16	1.98	1.82
H$_2$O	1.73	2.44	2.44	2.22	1.85
CH$_3$F	1.16	2.25	2.25	2.06	1.85
CH$_3$Cl	2.28	2.46	2.21	2.20	1.87
CS	0.93	0.99	1.38	1.26	1.98
H$_2$CO	1.51	2.63	2.63	2.75	2.34
HCN	2.44	3.07	3.07	3.23	2.99
LiH	4.86	5.89	5.89	5.77	5.83
LiF	3.60	5.91	5.91	6.05	6.28
NaH	6.59	6.92	6.85	6.85	6.96
LiCl	5.73	7.77	7.47	7.43	7.12
NaF	6.88	7.77	7.63	8.00	8.16
NaCl	9.84	9.94	9.39	9.38	9.00

[a]Theoretical data from: W. J. Hehre, unpublished calculations. Experimental equilibrium geometries given in Tables 6.1 and 6.6.
[b]The sign of the experimental moment for CO is actually the reverse of that given by the Hartree–Fock models.

326

electropositive or very electronegative first-row elements. Lithium fluoride represents an extreme example. Here the minimal basis description for lithium provides relatively too many basis functions (5 for only 3 electrons), while that for fluorine includes relatively too few (5 for 9 electrons). Therefore, lithium is able to accommodate more electrons than it really should, and conversely fluorine fewer electrons than its proper complement, leading to underestimation of the charge separation and too low a dipole moment. Calculated STO-3G moments for lithium hydride, lithium chloride, hydrogen fluoride, and sodium fluoride are likewise much smaller than experiment. Overall, the mean absolute deviation of STO-3G from experimental dipole moments is 0.65 D for the 21 compounds tabulated. For 11 of these compounds, individual deviations are in excess of 0.5 D. This places a clear limit on the reliability of STO-3G in assessing differing degrees of charge separation in unrelated molecules.

With few exceptions, dipole moments calculated using the 3-21G basis set are larger than the corresponding experimental values. The mean absolute deviation of theoretical from experimental moments is 0.49 D at this level, only marginally smaller than the average error noted for the STO-3G basis set. Calculated dipole moments for 10 of the compounds are in error by more than 0.5 D. The agreement between calculated and experimental moments for molecules incorporating second-row atoms is improved if the 3-21G$^{(*)}$ basis set is used. Differences of more than 0.5 D between calculated and experimental moments now occur only for ammonia, water, carbon monosulfide, and sodium fluoride. Overall, the mean absolute deviation of 3-21G$^{(*)}$ moments from experiment is 0.34 D.

Except for CS, all dipole moments calculated using the 6-31G* basis set are within 0.5 D of experimental values, and most are much closer. Again, with the exception of CS and some of the very polar molecules, 6-31G* moments are greater than the experimental values. The mean absolute error in calculated dipole moments is 0.30 D, making this the most successful of the theoretical models considered here. For compounds incorporating second-row elements, 6-31G* dipole moments are generally quite similar to those obtained with the smaller 3-21G$^{(*)}$ basis set.

6-31G* dipole moments are compared graphically with experimental values in Figure 6.33. A line of unit slope has been drawn. The plot clearly suggests that the theoretical model provides a reasonable description of the magnitudes of the electric dipole moments of widely differing compounds. Theoretical data derived from the smaller basis sets (not displayed) show somewhat greater scatter.

b. Hydrocarbons. As the data in Table 6.86 attest, the STO-3G, 3-21G, and 6-31G* basis sets are moderately successful in reproducing the observed, and usually small, electric dipole moments of hydrocarbons; mean absolute deviations are 0.10, 0.06, and 0.07 D for the three levels, respectively. The comparisons have all been made using STO-3G equilibrium geometries.

The signs and directions of the moments in a number of these systems have been determined experimentally. Here too, the calculations are generally in reasonable accord with the available experimental data. For example, the signs of the calculated

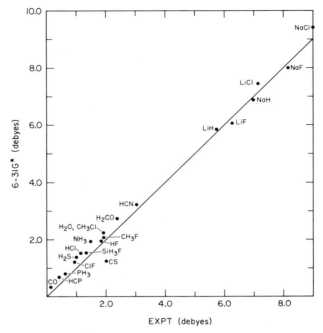

FIGURE 6.33. Comparison of 6-31G* and experimental electric dipole moments for diatomic and small polyatomic molecules.

moments in propane and in isobutane,

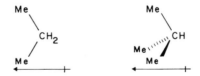

placing the methyl groups at the negative end of the dipole, are consistent with experimental assignments made on the basis of microwave investigations of deuterated species [40, 41]. The opposite situation prevails for propyne, where all levels of theory place the methyl group at the positive end of the dipole.

$$Me\!-\!C \equiv\!\equiv CH$$

Again, the experimental data concur [41].

TABLE 6.86. Calculated and Experimental Electric Dipole Moments for Hydrocarbons[a]

Formula	Molecule	STO-3G//STO-3G	3-21G//STO-3G	6-31G*//STO-3G	Expt.
C_3H_4	Propyne	0.50	0.69	0.62	0.75
	Cyclopropene	0.55	0.55	0.57	0.45
C_3H_6	Propene	0.25	0.30	0.29	0.36
C_3H_8	Propane	0.02	0.06	0.06	0.08
C_4H_4	But-1-yne-3-ene	0.37	0.45	0.46	0.4
C_4H_6	Cyclobutene	0.05	0.07	0.03	0.13
	1,2-Butadiene	0.33	0.39	0.35	0.40
	1-Butyne	0.54	0.67	0.65	0.80
	Methylenecyclopropane	0.22	0.35	0.40	0.40
	Bicyclo[1.1.0]butane	0.58	0.75	0.68	0.68
	1-Methylcyclopropene	0.83	0.90	0.89	0.84
C_4H_8	Isobutene	0.43	0.53	0.44	0.50
	cis-2-Butene	0.14	0.13	0.13	0.26
	cis-1-Butene	0.30	0.34	0.34	0.44
	Methylcyclopropane	0.11	0.11	0.10	0.14
C_4H_{10}	Isobutane	0.04	0.11	0.09	0.13

[a]Theoretical data from: L. Radom, W. A. Lathan, W. J. Hehre and J. A. Pople, *J. Am. Chem. Soc.*, **93**, 5339 (1971); ref. 42c; W. J. Hehre, unpublished calculations.

According to the calculations, the double bond in cyclopropene is at the positive end of the dipole.

This is the reverse of the direction originally assigned by Benson and Flygare on the basis of microwave measurements [42a], but is consistent both with experimental [42b] and theoretical [42c] work on the magnitude and direction of the moment in 1-methylcyclopropene. The original data have since been reinterpreted [42d].

c. Compounds Containing Heteroatoms.   Calculated dipole moments for several series of related compounds of nitrogen, oxygen, silicon, phosphorus, and sulfur are compared with experimental values in Table 6.87. STO-3G equilibrium geometries have been employed. The major point to be made is that *trends* in dipole moments within closely related sequences of molecules, for example, $NH_3 > MeNH_2 > Me_2NH > Me_3N$, $NH_3 > MeNH_2 > EtNH_2$, and $H_2O > MeOH > Me_2O$, are usually (but not always) well reproduced even at the simplest theoretical levels. However, variations in dipole moments accompanying larger structural changes, for example, $PhNH_2$ vs. $MeNH_2$, are not as well reproduced at the low levels; neither are the relative dipole moments of compounds of silicon, phosphorus and sulfur.

The performance of the 3-21G basis set is comparable to STO-3G in accounting for the relative moments of nitrogen- and oxygen-containing compounds. It is superior for systems containing silicon and sulfur, but fails to reproduce the experimentally observed ordering in the alkylphosphines. These compounds are, however, successfully treated by the calculations at the $3\text{-}21G^{(*)}$ and $6\text{-}31G^*$ levels (as are most of the other systems considered). As previously noted, dipole moments calculated at the $3\text{-}21G^{(*)}$ level for molecules incorporating second-row elements are very close to those obtained using the $6\text{-}31G^*$ basis set; less close are 3-21G and $6\text{-}31G^*$ moments for molecules with first-row elements only.

Overall, mean absolute deviations of STO-3G//STO-3G, 3-21G//STO-3G, $3\text{-}21G^{(*)}$//STO-3G ($3\text{-}21G$//STO-3G for molecules incorporating first-row elements only), and $6\text{-}31G^*$//STO-3G dipole moments are 0.43 (28 comparisons), 0.44 (28 comparisons), 0.34 (28 comparisons), and 0.30 D (26 comparisons), respectively. Comparison of relative errors among moments for related compounds points out more clearly differences in the theoretical models. For example, the mean absolute errors in the dipole moments for sulfur compounds relative to hydrogen sulfide are 0.73, 0.41, 0.25, and 0.21 D for the STO-3G//STO-3G, 3-21G//STO-3G, $3\text{-}21G^{(*)}$//STO-3G, and $6\text{-}31G^*$//STO-3G models, respectively.

d. Substituted Benzenes.   Dipole moments calculated with the STO-3G basis set for a series of monosubstituted benzenes are compared with experimental values in Table 6.88 and in Figure 6.34. Standard model geometries have been used. The theoretical values are generally smaller than experiment; the mean absolute deviation is 0.5 D. Although major trends are generally reproduced, the magnitudes of the calculated dipole moments for many of the compounds considered are substantially in error. Obviously, caution must be exercised in the use of STO-3G data for such comparisons of dipole moments.

e.   Hypervalent Molecules.   Calculated (3-21G, 3-21G$^{(*)}$, and 6-31G*) electric dipole moments for a selection of hypervalent molecules are compared with experimental values in Table 6.89. Molecular geometries have been optimized at their respective theoretical levels. The performance of the two larger basis sets is similar, but there are somewhat larger variations than in comparisons involving normal-valent molecules. Both levels generally overestimate the magnitudes of dipole moments, an observation that suggests that the description of the bonding provided by the 3-21G$^{(*)}$ and 6-31G* basis sets overemphasizes the contribution of the zwitterionic valence structures (see Section 6.2.5). The 3-21G$^{(*)}$ and 6-31G* basis sets, however, offer considerable improvement over the 3-21G basis set, which does not include d functions. Here, dipole moments are generally considerably larger than experimental values, consistent with the fact that 3-21G bond lengths in hypervalent compounds are often significantly longer than experimental values. This suggests that the theory at this level is unable to describe adequately the strong covalent interactions that hold hypervalent molecules together.

f.   Effect of Electron Correlation on Electric Dipole Moments.   It is well known that near the Hartree–Fock limit, the magnitudes of electric dipole moments are frequently in error by a few tenths of a debye [43]. Such errors are significant in view of the usual small range of moments (0–5 D). Problems are particularly conspicuous for some molecules with very small dipole moments, where an error of just one or two tenths of a debye can lead to an incorrect assignment of the sign of the moment. The classic failure of Hartree–Fock dipole moment predictions is carbon monoxide. The experimental dipole moment is 0.11 D in the direction $^-C{\equiv}O^+$, whereas near the Hartree–Fock limit the calculated moment is 0.27 D *in the reverse direction*, that is, $^+C{\equiv}O^-$.

Here, we examine the effect of electron correlation on the magnitudes of electric dipole moments, and point out the difference between calculation of the expectation value of the dipole moment operator μ (Eq. 2.96) and the direct evaluation of the full derivative expression $(dE/d\lambda)_{\lambda=0}$ for CI-type wavefunctions. Calculations (Table 6.90) have been performed on carbon monoxide and on formaldehyde using the CID/6-311G* and CISD/6-311G* models. Results for the corresponding Hartree–Fock model have also been obtained for comparison. Experimental equilibrium structures have been utilized.

It has been suggested that single substitutions are very important in the evaluation of one-electron properties. For dipole moment evaluation in the usual manner, that is, by calculation of the expectation value $\int \Psi^*_{CI} \hat{\mu} \, \Psi_{CI} \, d\tau$, this appears to be true. For example, for CO, both the Hartree–Fock moment and that obtained according to $\int \Psi^*_{CID} \hat{\mu} \, \Psi_{CID} \, d\tau$ have the incorrect sign; in contrast, the expectation value

TABLE 6.87. Calculated and Experimental Electric Dipole Moments for Heteroatom-Containing Compounds[a]

Heteroatom	Molecule	STO-3G//STO-3G	3-21G//STO-3G	3-21G$^{(*)}$//STO-3G	6-31G*//STO-3G	Expt.
Nitrogen						
	Me$_3$N	1.13	1.17	1.17	0.98	0.61
	Me$_2$NH	1.36	1.53	1.53	1.36	1.03
	EtNH$_2$	1.64	1.89	1.89	1.72	1.22
	MeNH$_2$	1.62	1.92	1.92	1.74	1.31
	NH$_3$	1.87	2.30	2.30	2.08	1.47
	PhNH$_2$	1.48	1.79	1.79	—	1.53
	△NH	1.82			2.09	1.90
	⬡N	1.95	2.50	2.50	—	2.19
Oxygen						
	Me$_2$O	1.33	1.93	1.93	1.71	1.30
	EtOH	1.44	2.04	2.04	1.89	1.69
	MeOH	1.51	2.21	2.21	2.04	1.70
	H$_2$O	1.71	2.50	2.50	2.33	1.85
	△O	1.46	2.67	2.67	2.43	1.89

	1	2	3	4	5
Silicon					
Me$_3$SiH	0.53	0.40	0.43	0.41	0.03
MeSiH$_3$	0.74	0.54	0.57	0.52	0.07
Me$_2$SiH$_2$	0.75	0.53	0.57	0.53	0.06
EtSiH$_3$	0.81	0.59	0.60	0.59	0.09
Phosphorus					
PH$_3$	0.58	0.88	0.87	1.25	0.60
MePH$_2$	1.10	1.23	1.17	1.39	0.53
EtPH$_2$	1.17	1.32	1.28	1.53	0.50
Me$_3$P	1.19	1.34	1.26	1.30	0.30
Me$_2$PH	1.23	1.34	1.26	1.37	0.46
Sulfur					
(thiophene)	0.55	0.94	0.81	1.10	0.10
H$_2$S	0.97	1.40	1.40	1.82	1.03
Me$_2$S	1.50	1.76	1.72	1.93	0.87
MeSH	1.52	1.73	1.70	1.98	0.96
EtSH	1.58	1.78	1.75	2.02	0.99
(thiirane)	1.85	2.08	1.93	2.15	0.75

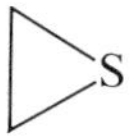

[a]Theoretical data from: BPH (1980); FPHBGDP (1982); GBPPH (1982); PFHDPB (1982); W. J. Hehre, unpublished calculations.

TABLE 6.88. Calculated and Experimental Electric Dipole Moments for Monosubstituted Benzenes[a]

Molecule	STO-3G//Std.	Expt.
$PhCH_3$	0.25	0.36
$PhCH_2CH_3$	0.28	0.59
$PhCH_2NH_2$	1.32	1.31
$PhCH_2OH$	1.37	1.71
$PhCH_2F$	1.02	1.77
$PhCF_3$	1.67	2.86
$PhNH_2$	1.44	1.53
$PhNHCH_3$	1.41	1.67
$PhNHNH_2$	2.54	1.67
$PhOH$	1.22	1.45
$PhOCH_3$	1.22	1.38
PhF	0.93	1.60
$PhCHO$	1.90	2.98
$PhCOCH_3$	1.95	3.02
$PhCONH_2$	2.81	3.77
$PhCOOH$	1.08	1.72
$PhNO$	2.38	3.17
$PhNO_2$	4.26	4.22
$PhCCH$	0.52	0.73
$PhCN$	3.65	4.18
$PhNC$	3.17	3.56

[a]Theoretical data from: W. J. Hehre, L. Radom, and J. A. Pople, *J. Am. Chem. Soc.,* **94**, 1496 (1972).

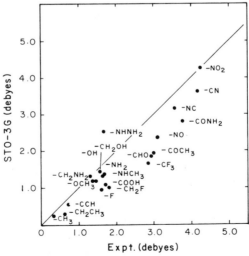

FIGURE 6.34. Comparison of STO-3G and experimental electric dipole moments for substituted benzenes.

TABLE 6.89. Calculated and Experimental Electric Dipole Moments
for Hypervalent Molecules

Molecule	$3\text{-}21G//$ $3\text{-}21G^a$	$3\text{-}21G^{(*)}//$ $3\text{-}21G^{(*)a}$	$6\text{-}31G^*//$ $6\text{-}31G^{*b}$	Expt.
$(CH_3)_3PO$	5.08	4.41	4.66	
F_3PO	1.48	1.75	1.96	1.76
F_3PS	1.03	1.23	1.42	0.64
SO_2	2.98	2.29	2.19	1.63
$(CH_3)_2SO$	4.90	4.30		3.96
SF_4	2.21	1.37	1.00	0.63
F_2SO	3.17	2.40	2.22	1.63
NSF	2.14	2.16	2.15	1.90^c
$(CH_3)_2SO_2$	6.15	4.98	5.11	4.49
F_4SO	1.54	1.75	1.48	
ClF_3	1.38	1.15	0.85	0.6
ClF_5	1.98	1.33	0.83	0.54^d
$FClO_2$	0.96	2.55	2.33	1.72^e
$FClO_3$	0.33	0.47		0.02

[a] PFHDPB (1982).
[b] FPHBGDP (1982).
[c] R. L. Cook and W. H. Kirchoff, *J. Chem. Phys.*, **47**, 4521 (1967).
[d] H. K. Bodench, W. Huttner, and P. Nowicki, *Z. Naturforsch.*, **31a**, 1638 (1976).
[e] C. I. Parent and M. C. L. Gerry, *J. Mol. Spectrosc.*, **49**, 343 (1974).

TABLE 6.90. Effect of Electron Correlation on Electric
Dipole Moments (6-311G* Basis Set)[a]

Model	CO^b	H_2CO
HF	-0.27	2.78
$\int \Psi_{CID}\, \hat{\mu}\, \Psi_{CID}\, d\tau$	-0.20	2.71
$\int \Psi_{CISD}\, \hat{\mu}\, \Psi_{CISD}\, d\tau$	0.12	2.43
$(dE_{CID}/d\lambda)_{\lambda=0}$	0.06	2.41
$(dE_{CISD}/d\lambda)_{\lambda=0}$	0.07	2.40
Expt.	0.11	2.34

[a] Theoretical data from reference 43.
[b] Positive dipole moment corresponds to $C^-\!\equiv\!O^+$.

obtained from the corresponding CISD wavefunction orients the dipole properly. While there is only a slight difference between dipole moments calculated from Hartree–Fock and CID wavefunctions (as obtained from the expectation value of $\hat{\mu}$), on the other hand, the moment in CO is lowered by 0.3 D upon going from $\int \Psi_{CID}^{*} \, \hat{\mu} \, \Psi_{CID} \, d\tau$ to $\int \Psi_{CISD}^{*} \, \hat{\mu} \, \Psi_{CISD} \, d\tau$. The latter value is in good accord with the experimental determination.

Both CID and CISD models yield comparable results for evaluation of the complete derivative $(dE_{CI}/d\lambda)_{\lambda = 0}$. For CO, both procedures yield a moment with the correct sign. The calculated dipole moments for formaldehyde are also similar. These observations (that the two methods yield comparable dipole moments) have been rationalized elsewhere [43].

6.6.2. Molecular Charge Distributions

An additional property of considerable interest is the charge distribution as derived from a *Mulliken population analysis* (see Section 2.8). As noted previously, the charge on an atom in a molecule can neither be defined uniquely nor is it subject to experimental measurement. Nevertheless, *Mulliken charges* are widely used as a basis for qualitative discussions of reactivity and bonding. It is useful, therefore, to provide a comparison of charge distributions calculated with various theoretical models.

Mulliken charges, obtained using the STO-3G, 3-21G, and 6-31G* basis sets, for a few diatomic and simple polyatomic molecules are given in Figure 6.35. Charges obtained for representative molecules incorporating second-row elements calculated using four different basis sets (STO-3G, 3-21G, 3-21G$^{(*)}$, and 6-31G*), are presented in Figure 6.36. Experimental equilibrium geometries have been used. These comparisons illustrate the extreme sensitivity of the calculated charges to the basis set. *It is obviously meaningless to compare calculated charges on different molecules if they have not been obtained with the same basis set.*

Charge separations obtained using the STO-3G minimal basis set usually are smaller than those derived from 3-21G, 3-21G$^{(*)}$, and 6-31G* calculations. There are conspicuous exceptions, for example, NaH. As we have seen (Section 6.6.1), STO-3G dipole moments are also generally smaller than values obtained at higher levels of theory. The 3-21G, 3-21G$^{(*)}$, and 6-31G* charge distributions are generally rather similar.

Which of the theoretical charge distributions is the most "realistic"? It is really not possible to say with certainty. If one could judge solely on the basis of calculated dipole moments (see Section 6.6.1), then it would be clear that the split-valence and polarization basis sets provide the better descriptions. Supporting this is the fact that basis set improvements beyond the split-valence level generally have little effect on the overall charge distribution, at least relative to the significant changes compared with STO-3G. One exception occurs for anionic species where substantial changes in charge distribution accompany the addition of diffuse functions to the basis set.

A final note of caution—the Mulliken charge partitioning scheme, while certainly

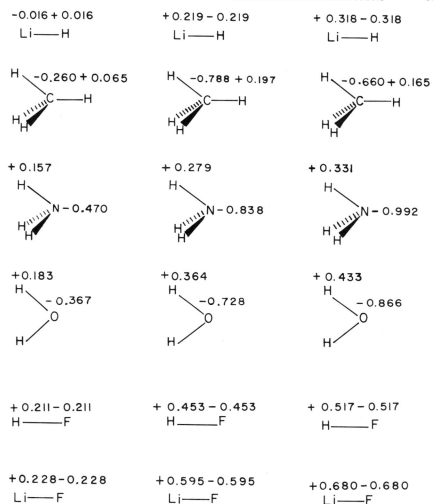

FIGURE 6.35. Comparison of STO-3G (left column), 3-21G (middle column) and 6-31G* (right column) Mulliken charges for molecules containing first-row elements. Experimental equilibrium geometries given in Tables 6.1 and 6.6. Data from: W. J. Hehre and L. Radom, unpublished calculations.

the most widely employed, is also the most criticized. Numerous attempts have been made to obtain more realistic descriptions of atomic charges. Among the most notable of recent efforts is work by Bader [44], who has proposed a unique definition of atomic boundaries in terms of surfaces of minimum electron density between atoms, by Streitwieser [45], who has employed direct (numerical) integration techniques in order to partition total electron density and by Wiberg and Wendoloski [46], who use as a basis of their analysis the electron population at the remote side

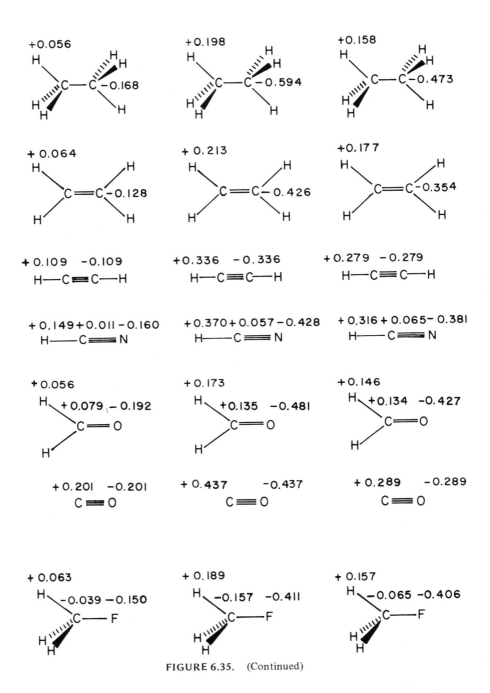

FIGURE 6.35. (Continued)

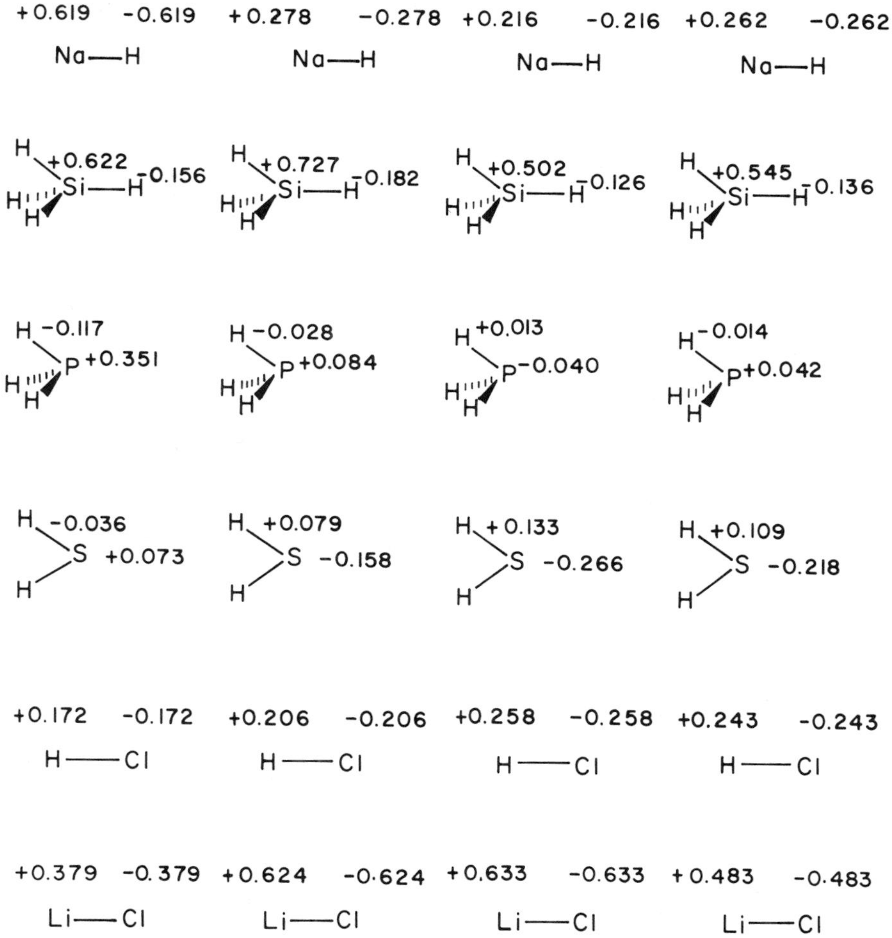

$+0.619$ -0.619 $+0.278$ -0.278 $+0.216$ -0.216 $+0.262$ -0.262

Na—H   Na—H   Na—H   Na—H

H $+0.622$ -0.156 H $+0.727$ -0.182 H $+0.502$ -0.126 H $+0.545$ -0.136
Si—H   Si—H   Si—H   Si—H

H -0.117 P $+0.351$ H -0.028 P $+0.084$ H $+0.013$ P -0.040 H -0.014 P $+0.042$

H -0.036 S $+0.073$ H $+0.079$ S -0.158 H $+0.133$ S -0.266 H $+0.109$ S -0.218

$+0.172$ -0.172 $+0.206$ -0.206 $+0.258$ -0.258 $+0.243$ -0.243

H——Cl   H——Cl   H——Cl   H——Cl

$+0.379$ -0.379 $+0.624$ -0.624 $+0.633$ -0.633 $+0.483$ -0.483

Li—Cl   Li—Cl   Li—Cl   Li—Cl

FIGURE 6.36.   Comparison of STO-3G (left column), 3-21G (second column), 3-21G$^{(*)}$ (third column) and 6-31G* (right column) Mulliken charges for molecules containing first- and second-row elements. Experimental equilibrium geometries given in Tables 6.1 and 6.6. Data from: W. J. Hehre, unpublished calculations.

of hydrogen in an X—H bond. All of these authors have criticized the Mulliken analysis. In particular, Wiberg challenges the commonly accepted C^{-}—H^{+} polarization and favors C^{+}—H^{-} for all except acetylenic C—H bonds. Both Bader and Streitwieser dispute the relatively high degree of covalent character the Mulliken analysis provides for compounds of lithium and of other highly electropositive elements. According to both the Bader and Streitwieser procedures, the lithium atom in a compound such as lithium hydride bears upward of 80% (or more) of a full positive charge (see also discussion in Section 7.3.2a).

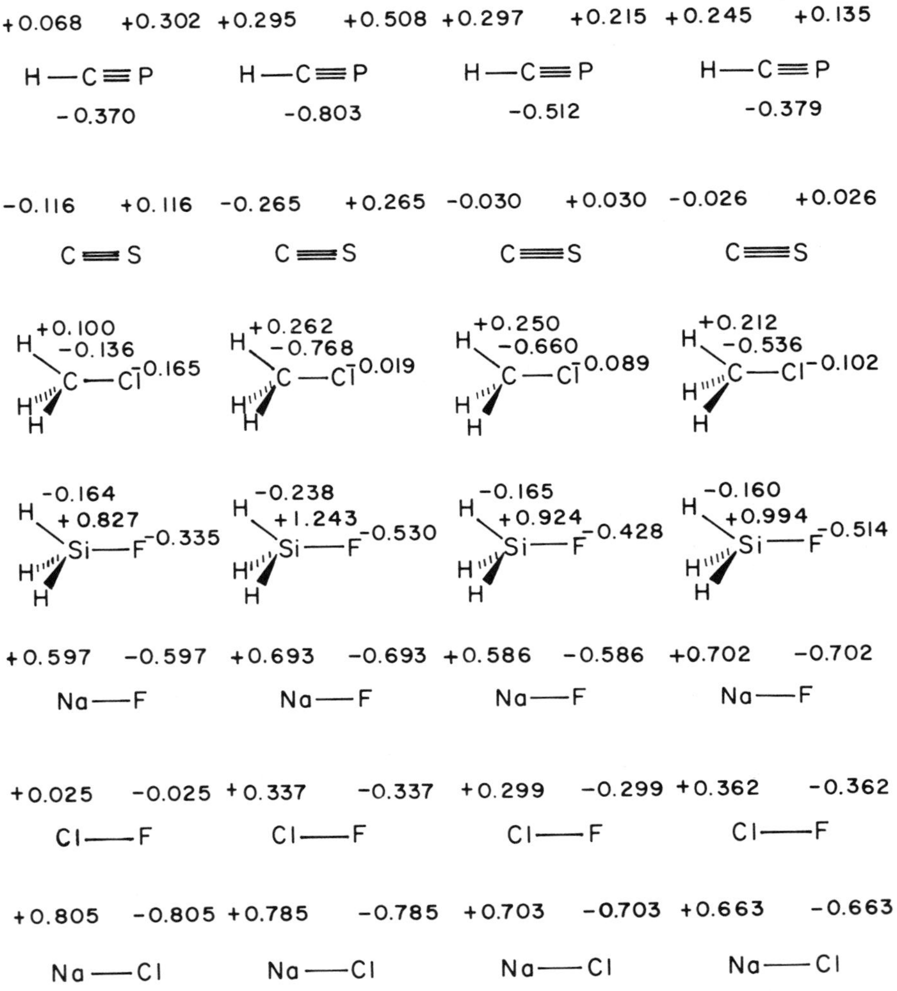

+0.068   +0.302 +0.295   +0.508 +0.297   +0.215 +0.245   +0.135

H—C≡P   H—C≡P   H—C≡P   H—C≡P
  -0.370   -0.803   -0.512   -0.379

-0.116   +0.116 -0.265   +0.265 -0.030   +0.030 -0.026   +0.026

C≡S   C≡S   C≡S   C≡S

+0.100   +0.262   +0.250   +0.212
H -0.136   H -0.768   H -0.660   H -0.536
 \C—Cl⁻0.165   \C—Cl⁻0.019   \C—Cl⁻0.089   \C—Cl⁻0.102
H'''   H'''   H'''   H'''
H   H   H   H

-0.164   -0.238   -0.165   -0.160
H +0.827   H +1.243   H +0.924   H +0.994
 \Si—F⁻0.335   \Si—F⁻0.530   \Si—F⁻0.428   \Si—F⁻0.514
H'''   H'''   H'''   H'''
H   H   H   H

+0.597   -0.597 +0.693   -0.693 +0.586   -0.586 +0.702   -0.702

Na—F   Na—F   Na—F   Na—F

+0.025   -0.025 +0.337   -0.337 +0.299   -0.299 +0.362   -0.362

Cl—F   Cl—F   Cl—F   Cl—F

+0.805   -0.805 +0.785   -0.785 +0.703   -0.703 +0.663   -0.663

Na—Cl   Na—Cl   Na—Cl   Na—Cl

FIGURE 3.36. (Continued)

340

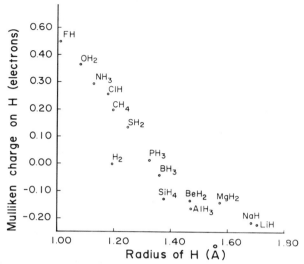

FIGURE 6.37. Relationship between "optimum" radii and Mulliken charge of H in one-heavy-atom hydrides, HX. 3-21G//3-21G (3-21G$^{(*)}$//3-21G$^{(*)}$ for second-row hydrides).

Further indication that the Mulliken method performs poorly for compounds incorporating electropositive elements may be inferred from the data in Figure 6.37. Here, the "optimum" radii for hydrogen in one-heavy-atom hydrides HX (obtained from a least-squares fitting of 3-21G electron density surfaces to a collection of spheres centered at the nuclei [47]) are plotted against the corresponding Mulliken charges. Except for H_2 and for the hydrides of electropositive elements, the two sets of theoretical data correlate as expected. Mulliken charges for the hydrides of lithium and sodium (and to a lesser extent magnesium) are much too close to zero to account for their large calculated radii. These apparent problems with the Mulliken analysis are not unanticipated. Here, the division of charge between atoms depends on the number of available basis functions. Hydrogen atoms appear to be short changed, while the alkali metals lithium and sodium appear to have an excess of functions.

REFERENCES

1. See references 1 and 3 on p. 136.

2. (a) L. Pauling, *The Nature of the Chemical Bond,* Cornell University Press, Ithaca, 1939, p. 58. The original Pauling scale has been redefined several times and has been extended to polyatomic systems. See: (b) J. K. Wilmshurst, *J. Chem. Phys.,* **27,** 1129 (1957); (c) A. L. Allred and E. G. Rochow, *J. Inorg. Nucl. Chem.,* **5,** 264 (1958); (d) P. R. Wells, *Prog. Phys. Org. Chem.,* **6,** 111 (1968); (e) R. J. Boyd and G. E. Marcus, *J. Chem. Phys.,* **75,** 5385 (1981); (f) N. Inamoto and S. Masuda, *Chem. Lett.,* 1003 (1982).

3. See, for example, A. Rauk, L. C. Allen, and E. Clementi, *J. Chem. Phys.,* **52,** 4133 (1970).

4. Previous work has demonstrated the need for the use of underlying polarization-type basis sets for methods which treat electron correlation. See, for example: (a) DLPHBP (1980); (b) C. E. Dykstra and H. F. Schaefer III, in *The Chemistry of Ketenes and Allenes,* S. Patai, ed., Wiley, New York, p. 1.

5. See, for example, (a) H_2O: R. J. Bartlett, I. Shavitt, and G. D. Purvis, *J. Chem. Phys.,* **71**, 281 (1979); (b) NH_3, W. R. Rodwell and L. Radom, *J. Am. Chem. Soc.,* **103**, 2865 (1981).

6. (a) A. C. Wahl, *J. Chem. Phys.,* **44**, 2600 (1964); (b) T. H. Dunning and N. W. Winter, *ibid.,* **63**, 1847 (1975); (c) W. R. Rodwell, N. R. Carlsen, and L. Radom, *Chem. Phys.,* **31**, 177 (1978).

7. The involvement of *d*-type orbitals in the bonding of hypervalent molecules has been repeatedly confirmed by quantitative molecular orbital calculations. For a discussion, see: E. E. Ball, M. A. Ratner, and J. R. Sabin, *Chem. Scr.,* **12**, 128 (1977), and references therein.

8. The STO-3G* basis set, formed from the minimal STO-3G basis set by the addition of a set of *d*-type functions to second-row elements only, is an even simpler representation which is also moderately successful in accounting for the equilibrium geometries of hypervalent compounds. For a discussion, see: J. B. Collins, P. v. R. Schleyer, J. S. Binkley, and J. A. Pople, *J. Chem. Phys.,* **64**, 5142 (1976).

9. The recent work of Woods, Saykally and coworkers shows considerable promise for the structural characterization of small ions in the gas phase. See: (a) R. C. Woods, in *Molecular Ions. Geometric and Electronic Structure,* J. Berkowitz and K. O. Groeneveld, eds., Plenum Press, New York, 1980, p. 11; (b) R. J. Saykally and R. C. Woods, *Ann. Rev. Phys. Chem.,* **32**, 403 (1981); (c) C. S. Gudeman and R. J. Saykally, *Ann. Rev. Phys. Chem.,* **35**, 387 (1984).

10. There has been extensive theoretical and experimental work on the structures of hydrogen-bonded systems. For reviews, see: (a) P. A. Kollman, in *Applications of Electronic Structure Theory,* H. F. Schaefer III, ed., Plenum Press, New York, vol. 4, 1977, Chapt. 3; (b) A. Beyer, A. Karpfen, and P. Schuster, *Topics Current Chem.,* **120**, 1 (1984); (c) C. Sandorfy, *ibid.,* **120**, 41 (1984); (d) Th. R. Dyke, *ibid.,* **120**, 85 (1984).

11. W. J. Bouma and L. Radom, *J. Mol. Struct.,* **43**, 267 (1978).

12. (a) S. Saebo and L. Radom, *J. Mol. Struct., Theochem,* **89**, 227 (1982); see also: (b) S. Saebo and L. Radom, *ibid.,* **105**, 119 (1983).

13. R. H. Nobes and L. Radom, *Chem. Phys.,* **60**, 1 (1981), and references therein. See also Section 7.6.4a.

14. C. S. Gudeman and R. C. Woods, *Phys. Rev. Lett.,* **48**, 1344 (1982).

15. R. C. Woods, C. S. Gudeman, R. L. Dickman, P. F. Goldsmith, G. R. Huguenin, W. M. Irvine, A. Hjalmarson, L. A. Nyman, and H. Olofsson, *Astrophys. J.,* **270**, 583 (1983).

16. For a discussion, see: E. B. Wilson, J. C. Decius, and P. C. Cross, *Molecular Vibrations,* McGraw-Hill, New York, 1955.

17. For a discussion, see: H. E. Hallam, *Vibrational Spectroscopy of Trapped Species,* Wiley, New York, 1973.

18. For recent examples, see: (a) T. E. Grough, R. E. Miller, and G. Scoles, *Faraday Disc. Chem. Soc.,* **71**, 77 (1981); (b) D. L. Snavely, S. D. Colson, and K. B. Wiberg, *J. Chem. Phys.,* **74**, 6975 (1981).

19. (a) R. H. Schwendeman, *J. Chem. Phys.,* **44**, 2115 (1966); For a discussion, see: (b) L. S. Bartell, S. Fitzwater, and W. J. Hehre, *ibid.,* **63**, 4750 (1975); (c) P. Pulay, J. G. Lee, and J. E. Boggs, *ibid.,* **79**, 3382 (1983).

20. See HLH (1982) for a discussion.

21. For a discussion, see: D. McQuarrie, *Statistical Mechanics,* Harper and Row, New York, 1976.

22. For an account of the theory of equilibrium isotope effects, see: J. Bigeleisen, in *Isotopes and Chemical Principles,* P. A. Rock, ed., American Chemical Society Symposium Series no. 11, American Chemical Society, Washington, D.C., 1975, p. 1.

23. See, for example, (a) *Internal Rotation in Molecules,* W. J. Orville-Thomas, ed., Wiley, New York, 1974; (b) D. G. Lister, J. N. Macdonald and N. L. Owen, *Internal Rotation and Inversion,* Academic Press, London, 1978. For an excellent review of theoretical studies of internal rotation and inversion see (c) P. W. Payne and L. C. Allen, in *Modern Theoretical Chemistry,* H. F. Schaefer, ed., Plenum Press, New York, 1977, vol. 4, p. 29.

24. Examples of the use of the *flexible rotor model* (some geometrical variables optimized while others are held fixed) may be found in L. Radom and J. A. Pople, *J. Am. Chem. Soc.,* 92, 4786 (1970).

25. (a) L. Radom, J. Baker, P. M. W. Gill, R. H. Nobes and N. V. Riggs, *J. Mol. Struct., Theochem,* 126, 271 (1985); (b) W. R. Rodwell and L. Radom, *J. Chem. Phys.,* 72, 2205 (1980); (c) R. Krishnan, *ibid.,* 81, 1383 (1984).

26. D. F. McMillen and D. M. Golden, *Ann. Rev. Phys. Chem.,* 33, 493 (1982).

27. Very recent determinations of the singlet–triplet splitting in methylene are described in: (a) A. R. W. McKellar, P. R. Bunker, T. J. Sears, K. M. Evenson, R. J. Saykally, and S. R. Langhoff, *J. Chem. Phys.,* 79, 5251 (1983); (b) T. J. Sears and P. R. Bunker, *ibid.,* 79, 5265 (1983), and references therein.

28. J. A. Pople, M. J. Frisch, B. T. Luke, and J. S. Binkley, *Int. J. Quantum Chem., Symp.,* 17, 307 (1983).

29. Expression (6.29) is related to a series of extrapolation procedures put forward by Bartlett and Shavitt: R. J. Bartlett and I. Shavitt, *Chem. Phys. Lett.,* 50, 190 (1977).

30. S. W. Benson, *J. Phys. Chem.,* 85, 3375 (1981).

31. (a) J. S. Binkley and J. A. Pople, *Chem. Phys. Lett.,* 45, 197 (1977); (b) D. Cremer, *ibid.,* 81, 481 (1981); (c) S. R. Gandhi, M. A. Benzel, C. E. Dykstra, and T. Fukunaga, *J. Phys. Chem.,* 86, 3121 (1982).

32. (a) W. J. Hehre, R. Ditchfield, L. Radom, and J. A. Pople, *J. Am. Chem. Soc.,* 92, 4796 (1970); (b) L. Radom, W. J. Hehre, and J. A. Pople, *ibid.,* 93, 289 (1971).

33. J. S. Binkley, unpublished calculations.

34. (a) For an early discussion of the use of bond separation reactions for the calculation of heats of formation, see: L. Radom, W. J. Hehre, and J. A. Pople, *J. Chem. Soc. A,* 2299 (1971); (b) K. B. Wiberg, *J. Comput. Chem.,* 5, 197 (1984); (c) S. W. Benson, *Thermochemical Kinetics,* 2nd ed., Wiley-Interscience, New York, 1976.

35. See reference 4 of Chapter 7 for a comprehensive listing of electrophile-transfer reactions that have been investigated in the gas phase.

36. For a discussion, see: J. F. Wolf, R. H. Staley, I. Koppel, M. Taagepera, R. T. McIver, Jr., J. L. Beauchamp, and R. W. Taft, *J. Am. Chem. Soc.,* 99, 5417 (1977).

37. (a) R. H. Nobes, W. J. Bouma and L. Radom, *Chem. Phys. Lett.,* 89, 497 (1982); (b) M. L. McKee and W. N. Lipscomb, *J. Am. Chem. Soc.,* 103, 4673 (1981); (c) J. V. Ortiz and W. N. Lipscomb, *Chem. Phys. Lett.,* 103, 59 (1983).

38. For a recent review, see: S. Fliszar, *Charge Distributions and Chemical Effects,* Springer-Verlag, New York, 1983.

39. For recent discussions, see: (a) P. Coppens and E. D. Stevens, *Adv. Quantum Chem.,* 10, 1 (1977); (b) P. Coppens, *Angew. Chem. Int. Ed. Engl.,* 16, 32 (1977); (c) A. Streitwieser, *Tetrahedron,* 37, 345 (1981).

40. D. R. Lide, *J. Chem. Phys.,* 33, 1519 (1960).

41. J. S. Muenter and V. W. Laurie, *J. Chem. Phys.,* 45, 855 (1966).

42. (a) R. C. Benson and W. H. Flygare, *J. Chem. Phys.,* 51, 3087 (1969); (b) M. K. Kemp and W. H. Flygare, *J. Am. Chem. Soc.,* 91, 3163 (1969); (c) W. J. Hehre and J. A. Pople,

ibid., **97,** 6941 (1975); (d) K. B. Wiberg and J. J. Wendoloski, *J. Phys. Chem.,* **83,** 497 (1979).

43. R. Krishnan and J. A. Pople, *Int. J. Quantum Chem.,* **20,** 1067 (1981).

44. (a) R. F. W. Bader and T. T. Nguyen-Dang, *Adv. Quantum Chem.,* **14,** 63 (1981), and references therein; (b) R. F. W. Bader, P. J. MacDougall, and C. D. H. Lau, *J. Am. Chem. Soc.,* **106,** 1594 (1984).

45. D. L. Grier and A. Streitwieser, *J. Am. Chem. Soc.,* **104,** 3556 (1982), and references therein.

46. K. B. Wiberg and J. J. Wendoloski, *J. Phys. Chem.,* **88,** 586 (1984).

47. (a) R. F. Hout, Jr. and W. J. Hehre, *J. Am. Chem. Soc.,* **105,** 3728 (1983); (b) M. M. Francl, R. F. Hout, Jr. and W. J. Hehre, *ibid.,* **106,** 563 (1984); (c) W. J. Hehre, C. F. Pau, R. F. Hout, Jr., and M. M. Francl, *Molecular Modeling. Computer Assisted Descriptions of Molecular Structure and Reactivity,* Wiley, New York, to appear in 1986.

7

APPLICATION
OF THE THEORY

7.1. INTRODUCTION

The number of diverse problems to which *ab initio* molecular orbital calculations have been directed is very large. It is impractical to present an all-encompassing treatment here, and our selection of topics is necessarily representative rather than comprehensive. Our goal has been to illustrate what *can be done* with the theory, rather than what *has been done*. Most of the selections closely relate to the authors' own research, and many worthy contributions of others have not been touched upon.

This chapter is divided into a number of separate and largely independent sections. It begins (Section 7.2) with a discussion of intramolecular interactions, in particular, interactions between directly-bonded groups and between those connected to a common center. The concept of aromatic stabilization is explored.

The properties of reactive intermediates are dealt with in Section 7.3. Carbocations, the first class of compounds considered, have received a great deal of attention from theorists, and have often served as an important testing ground for new computational models. Our present treatment includes a systematic survey of high-level theoretical work on small carbocations, as well as discussion of the best available calculations on a few larger systems. The structures and stabilities of carbodications are addressed briefly. The properties of alkali and alkaline earth metal compounds are considered next, and their relationship to free carbanions explored.

The section concludes with a discussion of simple carbenoids and metal carbenes, and of their relationship to free singlet carbenes.

Molecules that violate conventional structure rules are discussed in Section 7.4. Considered first are compounds incorporating square-planar (as opposed to tetrahedral) four-coordinate carbon and heavier main-group and transition-metal analogues. The structures and stabilities of perpendicular olefins and bridged acetylenes, and their relationship to normal planar and linear systems, respectively, are addressed briefly. The section concludes with a discussion of the fascinating structures of transition-metal carbenes and transition-metal alkyl complexes.

Section 7.5 concerns the structures and stabilities of molecules containing several (main-group) metal atoms. The structures and binding energies of small beryllium clusters are discussed in the context of the bulk metal. The surprising geometries of polylithiated hydrocarbons are also addressed. Comparisons with analogous hydrocarbons demonstrate consequences of extreme electron deficiency.

The chapter concludes (Section 7.6) with a discussion of reaction potential surfaces and, in particular, a survey of reaction transition structures for simple chemical transformations. This is a recent area of application of theory, and a major direction for future research. While the examples selected are necessarily simple, they often reveal interesting (and unexpected) features.

This chapter is largely concerned with situations that are experimentally uncharacterized or, at best, incompletely characterized. As such, it complements Chapter 6 where theory is compared with well-established experimental data.

7.2.   INTRAMOLECULAR INTERACTIONS

Theoretical procedures enable systematic studies of intramolecular interactions to be carried out. Extensive and complete sets of data can often be easily gathered computationally in areas where experimental information is lacking or is difficult to obtain. These data may then be used to evaluate the energies of formal reactions that measure a particular interaction of interest. Theory may also be employed to predict changes in molecular structure which reflect differences in intramolecular interactions. Alternatively, calculations may be carried out on structures artificially constrained to allow individual interactions to be assessed.

This section is divided into four parts. The first considers interactions between directly-bonded groups, in particular, the effects of substituents within the more common classes of organic molecules. The second part addresses the interaction of substituents bonded to a common carbon center, and the third, substitution in a benzene ring. The section concludes with a discussion of aromatic stabilization and an attempt to generalize familiar concepts to three-dimensional systems.

7.2.1.   Interactions Between Directly-Bonded Groups

Stabilization reactions (7.1) are often used to analyze energy data from theoretical calculations.

$$RX + R'H \rightarrow RH + R'X. \tag{7.1}$$

Here, the effects of a substituent X in RX and in R'X are compared. For example, if R' is H, reaction (7.2),

$$RX + H_2 \rightarrow RH + HX, \tag{7.2}$$

compares the interaction of R with X, with the sum of the separate interactions of R with H and X with H. A molecule of hydrogen is needed to balance the equation. When R' is methyl, reaction (7.3),

$$RX + CH_4 \rightarrow RH + CH_3X, \tag{7.3}$$

compares the effect of a substituent X in RX with its effect in methane. We shall frequently make use of the energies of reactions (7.3), that is, methyl stabilization energies, in this section.

The energies of many other processes, for example, (7.4),

$$XNH_3^+ + NH_3 \rightarrow XNH_2 + NH_4^+, \tag{7.4}$$

may be obtained by subtracting appropriate methyl stabilization energies, for example, $\Delta E(7.4) = \Delta E(7.6) - \Delta E(7.5)$.

$$XNH_2 + CH_4 \rightarrow NH_3 + CH_3X \tag{7.5}$$

$$XNH_3^+ + CH_4 \rightarrow NH_4^+ + CH_3X \tag{7.6}$$

Thus, a relatively small set of tabulated data reveals the interactions within many diverse systems.

Stabilization reactions find immediate application in the analysis of the interactions of directly-bonded groups X and R in systems X–R. Examples of stabilization energies calculated for XCH_2^+, $XCH_2^\cdot$, XCH_2^-, XCH (singlet), and XCH (triplet) are found in Table 7.1. The effects of a variety of substituents (replacing hydrogen) in ethane, ethylene, acetylene, benzene, formaldehyde, formic acid, and hydrogen cyanide are examined in Table 7.2. Finally, effects of substituents in systems containing directly-bonded nitrogen and oxygen centers are surveyed in Table 7.3.

In the discussion which follows, we do not attempt to detail the more subtle aspects of substituent effects in these systems; instead, we mention briefly the manner in which data of this type enable assessment of the importance of conjugative, hyperconjugative, and inductive effects in various situations [1].

It is convenient to characterize qualitatively the electronic properties of common substituents. Note, however, that substituent effects depend in part on the substrate to which the substituents are attached. The electropositive Li, BeH, and BH_2 groups are π-electron acceptors, with an ordering

$$BH_2 > BeH > Li.$$

TABLE 7.1. Calculated Methyl Stabilization Energies for Carbocations, Carbon-Centered Radicals, Carbanions, and Singlet and Triplet Carbenes (kcal mol^{-1})[a,b]

X	XCH_2^+	$XCH_2^{\cdot c}$	XCH_2^{-d}	XCH Singlet	XCH^c Triplet
H	0	0	0 (0)	0	0
Li	81	9	3 (25)	19	30
BeH	19	9^e	31 (47)	9	22
BH_2	24	11	55 (72)	36	13
CH_3	29	2	−5 (3)	10	5
NH_2	93	10	0 (6)	60	13
OH	53	6	10 (13)	46	8
F	9	3	17 (20)	28	3
$CH{=}CH_2$	54	21	28 (37)	17	
$C{\equiv}CH$	33	15	(45)	14	
$C{\equiv}N$	−9	11	48 (58)	9	21
$CH{=}O$	−4	4	60 (65)	−1	17
Na	93	8^e	6 (18)	17	
MgH	40	8^e	20 (38)		
AlH_2	28	9^e	43 (62)	16	
SiH_3	16	5^e	27 (43)	2	
PH_2	44	4^e	26 (39)	17	
SH	47	5^e	21 (34)	32	
Cl	12	3^e	27 (46)	16	

[a] 3-21G//3-21G (3-21G$^{(*)}$//3-21G$^{(*)}$ for molecules incorporating second-row elements) unless otherwise noted.

[b] Stabilization energies defined by reaction (7.3) with appropriate X.

[c] UHF.

[d] 3-21+G//3-21+G (3-21+G$^{(*)}$//3-21+G$^{(*)}$ for molecules incorporating second-row elements); 3-21G//3-21G (3-21G$^{(*)}$//3-21G$^{(*)}$) values given in parentheses.

[e] UHF. 6-31G*//6-31G*. C. Schade, unpublished results.

They are also σ-electron donors,

$$Li > BeH > BH_2 .$$

The electronegative substituents NH_2, OH, and F have the opposite properties, that is, they are π-electron donors,

$$NH_2 > OH > F,$$

and σ-electron acceptors,

$$F > OH > NH_2 .$$

The methyl group can act as a hyperconjugative electron donor or acceptor, that is,

TABLE 7.2. Calculated Methyl Stabilization Energies for Common Organic Systems (kcal mol^{-1})^{a,b}

X	C$_2$H$_5$X	C$_2$H$_3$X	HCCX	PhX	XCHO	XCOOH	XCN
Li	-4	6	35	6	34	83	49
CH$_2^+$	10	29	11		-23	-19	-25
CH$_3$	1	4	4	2	10	11	13
C≡CH	2	4	6		5	0	7
CH=CH$_2$	1	7	8		13	14	12
CH$_2$CH$_3$	1	3	8		12	13	13
C$_6$H$_5$					-3		13
C≡N	2	3		2	-3	-8	13
CH=O	3	7	2		4	0	0
COOH	9	7	-2		0	-5	-5
CF$_3$	4	1	-7		-7	-11	-13
NH$_2$	3	13	15	12	37	35	16
NH$_3^+$	6	1	-19		-5	-8	-42
NO$_2$	5	6	-12		3	-8	-26
O$^-$	8	46	69		85	89	82
OH	5	12	6		36	28	3
F	6	9	-5	12	27	20	-13
Na	-4	6	37		31	84	53
SiH$_3$	-1	3	14	1	-3	1	15
SH	2	4	5		11	6	5
Cl	3	3	-4	2	9	1	-6

^{a} 3-21G//3-21G (3-21G$^{(*)}$//3-21G$^{(*)}$ for molecules incorporating second-row elements).

^{b} Stabilization energies defined by reaction (7.3) with appropriate R.

TABLE 7.3. Calculated Methyl Stabilization Energies for Nitrogen- and Oxygen-Based Systems (kcal mol^{-1})^{a,b}

X	XNH$_2$	XNH$_3^+$	XNH^{-c}	XOH	XOH$_2^+$	XO^{-c}
H	0	0	0 (0)	0	0	0 (0)
Li	30	87	9 (46)	58	124	2 (52)
BeH	36	29	55 (88)	59	59	61 (113)
BH$_2$	44	11	57 (92)	47	42	60 (115)
CH$_3$	-5	5	-9 (8)	-3	10	-7 (25)
NH$_2$	-17	-18	-14 (1)	-22	-22	-25 (2)
OH	-23	-40	-12 (2)	-38	-53	-26 (3)
F	-36	-69	-4 (10)	-57	-103	-20 (10)

^{a} 3-21G//3-21G unless otherwise noted.

^{b} Stabilization energies defined by reaction (7.3) with appropriate R.

^{c} 3-21+G//3-21G. G. W. Spitznagel, unpublished calculations. Numbers in parentheses from 3-21G//3-21G calculations.

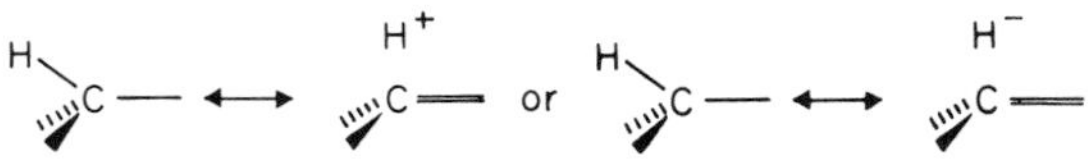

The electron-donating and -accepting properties of second-row substituents generally parallel those of their first-row analogues, although the second-row elements are more electropositive. AlH_2 and MgH are π-electron acceptors,

$$AlH_2 > MgH,$$

and the right-hand groups are π donors,

$$PH_2 > SH > Cl.$$

Both types of π effects appear to be less important than those of the corresponding first-row groups. The silyl substituent is a hyperconjugative (π-electron) acceptor. Na, MgH, AlH_2, SiH_3, and even PH_2 substituents are σ-electron donors (relative to carbon),

$$Na > MgH > AlH_2 > SiH_3 > PH_2,$$

while only Cl is a distinct σ acceptor.

Among the larger substituents, the vinyl and phenyl groups are (approximately) equally effective either as π acceptors or donors. The ethynyl group is a much better π acceptor than π donor. All of these groups are generally classed as σ-electron acceptors relative to methyl and other alkyl substituents. The formyl, cyano, nitro, and trifluoromethyl substituents are all strong σ and π acceptors.

a.   Stabilization of Carbocations.   Methyl cation is best stabilized by strong π donors, for example, NH_2, and by strong σ donors, for example, Li and Na (Table 7.1).

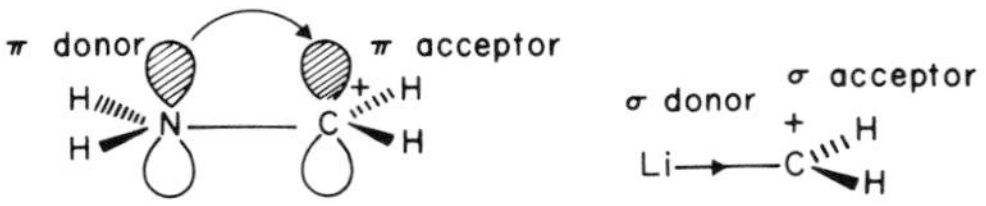

The order of stabilizing abilities of common first-row substituents, $NH_2 > OH > F$, parallels both the decrease in their π-donor ability, and the increase in their σ-acceptor ability. The corresponding second-row groups, PH_2, SH and Cl, are not as effective in stabilizing a carbocation center, and also do not follow the same pattern; the SH group is more effective than either PH_2 or Cl substituents.

The anomalously large stabilization for BH_2 may in part be due to hyperconjugation in the preferred perpendicular conformation, that is,

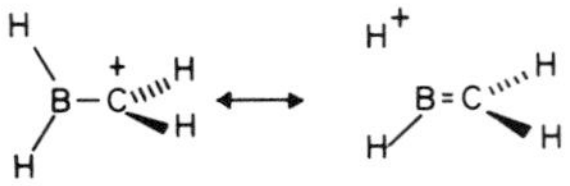

Constrained to a planar geometry, the group is still an effective σ donor, although the stabilization it affords a carbocation center is much smaller (4 kcal mol^{-1}). The calculated stabilization energies for the corresponding second-row substituents (Na, MgH, AlH$_2$) are consistently larger than those for the first-row groups. However, the stabilization afforded a carbocation center by a silyl group is less than that for methyl [2a], under the competing influences of the σ-donor property of SiH$_3$ (leading to carbocation stabilization) and its π-acceptor ability (leading to destabilization).

Large stabilization energies are also found for methyl, vinyl, and ethynyl substituents, all of which act as π-electron donors in this instance. Small destabilization results from both formyl and cyano groups.

b. Stabilization of Radicals. Stabilization energies for radicals measure the change in CH bond dissociation energy (in methane) as a result of substitution (Table 7.1). For the most part, the effects are small, although large changes do occur for groups that can act as strong π-electron acceptors such as BH$_2$, strong π donors such as NH$_2$, or for substituents that facilitate delocalization of the odd electron, for example, CH=CH$_2$. In the latter case, the observed bond-energy lowering provides one measure of the resonance energy of the resultant free radical, for example, 21 kcal mol^{-1} (3-21G//3-21G) for the allyl radical.

c. Stabilization of Carbanions. Stabilization energies for methyl anions are dominated by conjugative effects (Table 7.1). The BH$_2$ group, the best π acceptor, is the most effective (one-heavy-atom) first-row substituent and is rivaled only by the nitrile and formyl groups.

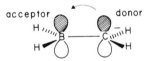

The ordering of stabilizing effects, Li < BeH < BH$_2$ and NH$_2$ < OH < F, is consistent with the directions of both conjugative and inductive contributions. Stabilization afforded methyl anion by fluorine, the best of the latter group, is less than half that provided by BH$_2$, the best of the former group. The 3-21+G//3-21+G (Table 7.1) and higher-level calculations show that a methyl substituent slightly destabilizes the methyl anion [2b, c], in stark contrast to its strong stabilizing effect in the methyl cation.

The second-row substituents MgH and AlH$_2$ show slightly decreased stabilization compared with the analogous first-row groups. The SH and Cl substituents exhibit slightly increased stabilization, and the SiH$_3$ and PH$_2$ groups greatly increased stabilization (relative to first-row substituents). These trends are the reverse of those noted for carbocations. The range of substituent effects for second-row groups Na to Cl (44 kcal mol^{-1} at 3-21G$^{(*)}$, 37 kcal mol^{-1} at 3-21+G$^{(*)}$) is somewhat smaller than that found for the analogous first-row substituents (69 kcal mol^{-1} at 3-21G, 60 kcal mol^{-1} at 3-21+G), consistent with the decreased range of electronegativities.

Strong stabilization results from the vinyl, ethynyl, cyano, and formyl substituents, all of which are not only good π-electron acceptors but are also capable of delocalizing excess electronic charge over their heavy-atom skeleton, for example,

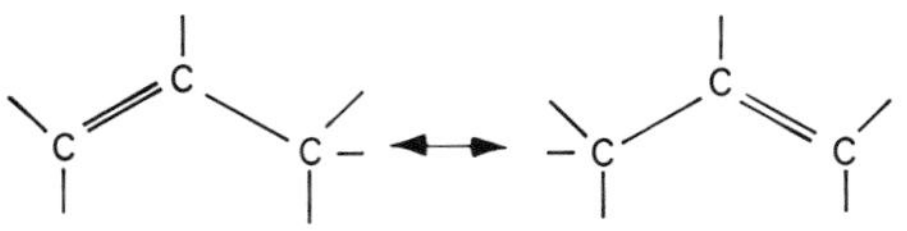

The last two groups are the most effective; the nitrogen and oxygen centers that they incorporate are better able than carbon to accommodate the negative charge, that is,

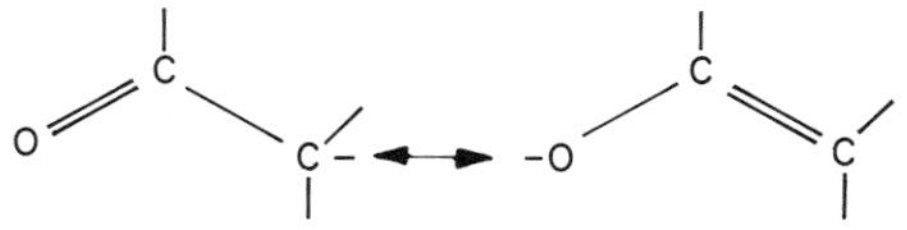

d. Stabilization of Singlet and Triplet Carbenes. Stabilization energies for singlet carbenes parallel those for carbocations, for substituents $X = CH_3$, NH_2, OH, F, and their second-row analogues (Table 7.1). While both are stabilized by π-donor groups, the carbenes are somewhat less sensitive to substitution. Lithium and sodium, the strongest σ-donor substituents, which are highly effective in stabilizing carbocations by delocalizing the positive charge, are not nearly as potent for carbenes.

Triplet carbenes show stabilization energies that are broadly similar (although generally larger) than those for free radicals. Very large stabilization is afforded triplet propargylene, $HC\equiv C\ddot{C}H$, the optimum geometry of which closely resembles that of allene [3] from which it can be (formally) derived by removal of two hydrogens.

The BH_2, NH_2, OH, and F substituents (and their second-row analogues) stabilize singlet methylene much more than the corresponding triplet form. The effects appear to be large enough to reverse the energetic preference of the parent compound for a triplet electronic ground state. On the other hand, the Li and BeH substituents preferentially stabilize triplet methylene. $Li\ddot{C}H$ and $HBe\ddot{C}H$, were they ever to be synthesized, would certainly be ground-state triplets. A methyl group is predicted to be slightly more effective in stabilizing singlet methylene than the triplet form. Ethylidene (like the parent carbene) would be expected to prefer a triplet ground state (see also Section 7.6.3c). Better (MP4SDQ/6-31G*//3-21G) calculations (Table 7.4) support these basic conclusions. Both the individual singlet and triplet stabilization energies and the difference in stabilization energies, that is, energies for reaction (7.7),

$$XCH(triplet) + CH_2\,(singlet) \rightarrow CH_2\,(triplet) + XCH(singlet), \qquad (7.7)$$

are nearly identical to those obtained from the lower-level treatment.

TABLE 7.4. Calculated Stabilization Energies for Singlet and Triplet Carbenes, XCH, and Effect of Substituents on the Singlet/Triplet Energy Difference in Methylene (MP4SDQ/6-31G*//3-21G, kcal mol^{-1})

X	Stabilization Energy[a]		Substituent Effect on Singlet/Triplet Energy Difference[b]
	Singlet	Triplet	
H	0	0	0
Li	9	27	+18
BeH	−2	19	+21
BH_2	37	16	−21
CH_3	12	5	−7
NH_2	62	13	−49
OH	51	11	−40
F	35	6	−29

[a]Stabilization energies defined by reaction (7.3) with appropriate X. Data from: B. T. Luke, J. A. Pople, M.-B. Krogh-Jespersen, M. Karni, J. Chandrasekhar, and P. v. R. Schleyer, *J. Am. Chem. Soc.*, in press.

[b]Energies of reaction (7.7) with appropriate X. The best theoretical estimate of the singlet-triplet energy difference in methylene is 10 ± 2 kcal mol^{-1} (triplet ground state). For a discussion, see: P. Saxe, H. F. Schaefer, III, and N. C. Handy, *J. Phys. Chem.*, **85**, 745 (1981). The most recent experimental splitting is 9.05 kcal mol^{-1}, see: A. R. W. McKellar, P. R. Bunker, T. J. Sears, K. M. Evenson, R. J. Saykally, and S. R. Langhoff, *J. Chem. Phys.*, **79**, 5251 (1983), and references therein.

e. **Stabilization of Larger Systems.** Methyl stabilization energies, that is, energies of reaction (7.3), for substituents in a variety of simple organic systems are given in Table 7.2. In effect, these comparisons relate substituents to hydrogen as a standard, and substrates to methyl as a standard.

Calculated stabilization energies for substituted ethanes, that is, energies for the reaction

$$CH_3CH_2X + CH_4 \rightarrow CH_3CH_3 + CH_3X \qquad (7.8)$$

are small and, except for X = Li, SiH_3 and Na, are positive, indicating stabilization. The largest effects are noted for charged substituents, indicative of the increased polarizability of ethane relative to methane. The destabilizing effects are consistent with the notion that the α carbons in ethyl lithium, ethylsilane, and ethyl sodium bear negative charge (see also Sections 7.3.2a and 7.3.2b), and that the methyl substituent destabilizes carbanions (Table 7.1).

Stabilization energies for substituted ethylenes and benzenes (relative to those for substituted methanes), that is, energies of reactions (7.9) and (7.10),

$$CH_2=CHX + CH_4 \rightarrow CH_2=CH_2 + CH_3X, \qquad (7.9)$$

$$C_6H_5X + CH_4 \rightarrow C_6H_6 + CH_3X, \qquad (7.10)$$

are all positive. These data closely parallel one another (for a given substituent) and seldom differ by more than 2 kcal mol^{-1}. The largest effects are noted for CH_2^+ and O^- substituents where charge may be extensively delocalized.

Energies of reaction (7.11), relating substituent effects in acetylenes to those in methanes, show much wider variation.

$$HC \equiv CX + CH_4 \rightarrow HC \equiv CH + CH_3 X. \qquad (7.11)$$

The O^- substituent is best stabilized, followed by the alkali metals, lithium and sodium. Results for the latter pair reflect the inherent stability of the acetylide anion (relative to methyl anion); molecules such as lithium acetylide involve significant contributions from ionic resonance structures, that is,

$$Li-C \equiv CH \quad \longleftrightarrow \quad Li^+ \quad {}^-C \equiv CH$$

Data relating substitution in formaldehyde, formic acid, and hydrogen cyanide, reactions (7.12), (7.13) and (7.14), respectively, to that in methane closely (but not entirely) parallel those for substituted acetylenes.

$$XCHO + CH_4 \rightarrow H_2 CO + CH_3 X \qquad (7.12)$$

$$XCOOH + CH_4 \rightarrow HCOOH + CH_3 X \qquad (7.13)$$

$$XCN + CH_4 \rightarrow HCN + CH_3 X \qquad (7.14)$$

The most striking difference occurs for the CH_2^+ substituent which is stabilizing (due to π-electron and charge delocalization) in HCCX, but is destabilizing (due to unfavorable π-electron and σ-electron effects) in XCHO and XCOOH.

The effects of electronegative substituents are a composite of their π-donor ability $(NH_2 > OH > F)$, which is stabilizing in unsaturated systems, and their σ effects, which may either be stabilizing, for example, in $C_2 H_5 X$, or destabilizing, for example, in HCCX and NCX.

f. Methyl Stabilization Energies for Amines, Alcohols, and Their Protonated and Deprotonated Forms: Effects of Substituents on the Acid and Base Strengths of Nitrogen and Oxygen Compounds. Methyl stabilization energies for amines are positive for π-acceptor and σ-donor substituents. The ordering of effects, Li $<$ BeH $<$ BH$_2$, suggests that π-accepting ability is the more important factor. Negative methyl stabilization energies are exhibited by the remaining substituents, CH$_3$ $<$ NH$_2$ $<$ OH $<$ F, and are due mainly to σ effects. Methyl stabilization energies for protonated amines also arise mainly from σ effects; the lone pair on nitrogen (in free amines) is no longer available for π donation. The most positive stabilization energies are found for substitution by lithium (the best σ donor) and the most negative for fluorine (the best σ acceptor).

The ordering of methyl stabilization energies in alcohols, BH$_2$ $<$ BeH $<$ Li, differs from that for amines. Due to the poorer π-donor and better σ-acceptor abilities of OH compared with NH$_2$, σ effects now dominate. σ effects are also important for the remaining substituents, for which methyl stabilization energies increase in the order, $CH_3 < NH_2 < OH < F$. The same trend, that is, greater influence of σ effects for oxygen compared with nitrogen systems, is evident for protonated alcohols, for which the methyl stabilization energies decrease smoothly from lithium to fluorine (Table 7.3).

Large positive methyl stabilization energies for anionic systems, XNH^- and XO^-, that is, deprotonated amines and alcohols, occur only for BeH and BH_2 substituents. The fact that the stabilization energies for lithium are only slightly positive suggests that the strong π-accepting ability of the group has been largely counteracted by its strong σ-donor ability. Small (negative) methyl stabilization energies are noted for the remaining substituents.

More relevant are energies of reactions relating the acid and base strengths of amines and alcohols to those of the parent compounds, that is, ammonia and water, respectively. As already indicated, and illustrated for the case of amine basicities relative to ammonia, such processes may be written in terms of methyl stabilization reactions. A further example illustrates that the acidities of alcohols relative to water, reaction (7.15),

$$XO^- + H_2O \longrightarrow XOH + OH^-, \tag{7.15}$$

may be written in terms of the difference of stabilization energies for deprotonated alcohols, reaction (7.16),

$$XO^- + CH_4 \longrightarrow HO^- + CH_3X, \tag{7.16}$$

and those for their neutral precursors, reaction (7.17),

$$XOH + CH_4 \longrightarrow H_2O + CH_3X. \tag{7.17}$$

Data are presented in Table 7.5. The largest increase in proton affinities for amines and alcohols follows substitution by lithium. The resulting ions, $LiNH_3^+$ and $LiOH_2^+$, are known species in the gas phase [4], formed from lithium cation attachment to ammonia and water, respectively. Of the remaining substituents, only methyl leads to an increase in base strength. Greatest reduction in amine proton affinity results from substitution by BH_2 or F. The former effect arises primarily because of stabilization of the neutral amine, the latter because of destabilization of the positive ion (see Table 7.3). Similar effects occur for protonated alcohols, and may be similarly interpreted.

Substitution by OH and especially F leads to greatly increased acidity of both amines and alcohols, effects which are largely due to destabilization of the respective neutral molecules. The BH_2 group also enhances acidity for both systems; the overall effect here is smaller and arises from greater stabilization of the deprotonated forms than their neutral precursors. Lithium substitution results in an extreme decrease in acidity, primarily reflecting the large stabilizing effect that the electropositive atom has on ammonia and water.

TABLE 7.5.   Calculated Effects of Substituents on the Energies of
Protonation and Deprotonation of Amines and Alcohols (kcal mol^{-1})^a

X	Amines XNH_2		Alcohols XOH	
	Protonation[b]	Deprotonation[c]	Protonation[b]	Deprotonation[c]
H	0	0 (0)	0	0 (0)
Li	57	−21 (16)	66	−56 (−6)
BeH	−7	19 (52)	0	2 (54)
BH_2	−33	13 (48)	−5	13 (68)
CH_3	10	−4 (13)	13	−1 (28)
NH_2	−1	3 (18)	0	−3 (24)
OH	−17	11 (25)	−15	12 (41)
F	−33	32 (46)	−46	37 (67)

^{a}Based on stabilization energies in Table 7.3.
^{b}3-21G//3-21G data.
^{c}3-21+G//3-21+G data. G. W. Spitznagel, unpublished calculations. 3-21G//3-21G data given in parentheses.

7.2.2.   Bond Separation Reactions and Geminal Interactions

Bond separation reactions, formal processes in which molecules comprising three or more heavy atoms are broken up into the smallest (two-heavy-atom) fragments containing the same component linkages, provide a useful means of viewing interactions between groups that are not directly bonded [5]. For example, the bond separation energy for methylene fluoride, reaction (7.18),

$$CH_2F_2 + CH_4 \longrightarrow 2\ CH_3F, \qquad (7.18)$$

provides a measure of the interaction of two fluorine atoms bonded to a single carbon. Closely related are reactions such as (7.19),

$$CH_2(CN)_2 + CH_4 \longrightarrow 2\ CH_3CN, \qquad (7.19)$$

the energies of which indicate the extent of interaction between two polyatomic groups attached to a common center. Bond separation reactions are isodesmic; that is, the number of bonds of each formal type is conserved. In the first of the above examples, both sides of the reaction involve two CF bonds and six CH bonds; in the second reaction (related to a formal bond separation process), both reactants and products contain two CC single bonds, two CN triple bonds, and six CH linkages.

Bond separation energies, that is, the energies of bond separation reactions, indicate deviations from strict bond-energy additivity. The positive energy for the methylene fluoride reaction indicates a stabilizing interaction between the fluorine atoms. The experimental enthalpy for the bond separation process in malononitrile is negative, indicating the unfavorable nature of the interaction between two cyano groups. As shown in Section 6.5.6, the energies of bond separation and related isodesmic reactions are moderately well described using simple levels of theory. The examples provided in Table 6.71 suggest that, even at the HF/STO-3G level, calculated bond separation energies are in reasonable accord with the experimental data; larger basis sets usually produce even better results.

a. Interactions in $X-CH_2-Y$. An example of the application of bond separation reactions is provided by disubstituted methanes, $X-CH_2-Y$. Here, energies of reaction (7.20),

$$X-CH_2-Y + CH_4 \longrightarrow CH_3-X + CH_3-Y, \qquad (7.20)$$

measure the extent to which groups X and Y, both bonded to the same tetravalent carbon, interact. A selection of results is provided in Table 7.6. The substituents considered might roughly be classified as σ donor, π donor (CH_3, PH_2), σ donor, π acceptor (Li, SiH_3), σ acceptor, π donor (NH_2, OH, F, SH, Cl), and σ acceptor, π acceptor (CN); see also Section 7.2.1.

TABLE 7.6. Calculated Bond Separation Energies for Molecules X-CH$_2$-Y (kcal mol^{-1})[a,b]

X/Y	Li	CH$_3$	NH$_2$	OH	F	CN	SiH$_3$	PH$_2$	SH	Cl
Li	-2									
CH$_3$	-4	1								
NH$_2$	16[d]	3	11							
OH	24[d]	5	13	17						
F	24[d]	6	18	16	14					
CN	26[e]	2	2[c]	2[c]	-4	-8				
SiH$_3$	10[d]	-1[c]	-1[c]	-1	-2	0	2			
PH$_2$	6[d]	2	1	1	0	0[c]	1[c]	1		
SH	16[d]	2	4	5	2	-4[c]	0[c]	0	0	
Cl	28[d]	5	11	6	1	-6[c]	2	0	-1	-4

[a]Energies for reaction (7.20) with appropriate X and Y.
[b]3-21G//3-21G (3-21G[(*)]//3-21G[(*)] for molecules incorporating second-row elements) except where otherwise noted.
[c]3-21G//3-21G.
[d]P. v. R. Schleyer, T. Clark, A. J. Kos, G. W. Spitznagel, C. Rohde, D. Arad, K. N. Houk and N. G. Rondan, *J. Am. Chem. Soc.*, **106**, 6467 (1984).
[e]Linear $H_2C=C=NLi$ form most stable at 3-21G (but not at 6-31G*, where a bridged structure is preferred). The 3-21G bond separation energy for the bridged structure is 22 kcal mol^{-1}.

A large stabilizing interaction is noted for systems CH_2XY, where both X and Y are first-row σ-acceptor, π-donor substituents (NH_2, OH, and F). This is related to the anomeric effect [6] and, as depicted below,

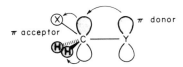

can be viewed as resulting from the mixing of σ and π systems at the tetrahedral center, that is, electrons donated from a π-type lone pair on one substituent are drawn into the σ system of the other, and vice versa. An equivalent explanation is that the energy of the π-acceptor orbital of the XCH_2 group is lowered in response to substitution by an electronegative element, leading to increased interaction with a π-donor substituent Y.

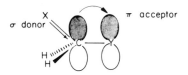

No corresponding stabilization is found where X and Y are both second-row substituents. The destabilization indicated for methylene chloride is an artifact of the small basis set. As previously discussed (Section 6.5.6a), higher-level calculations yield a stabilization energy of nearly zero, consistent with the experimental value. The more electronegative second-row substituents, SH and Cl, function effectively as σ acceptors and do show stabilizing effects when the other group bonded to carbon is a good π donor from the first row, for example, NH_2 or OH. However, none of the second-row substituents appear to function well as a π donor in this context. Not even FCH_2PH_2, which involves the best first-row σ acceptor, exhibits appreciable stabilization.

The same arguments would suggest stabilizing interaction between two σ-donor, π-acceptor groups, for example, Li and SiH_3. That is, electrons donated by X into the CX σ bond could be dispersed into the π system of Y.

The theoretical results are mixed. Geminal interaction of SiH_3 and Li groups is large and stabilizing. Interaction between two silyl substituents is stabilizing, but only slightly so; that between two lithiums is marginally destabilizing. Other factors, for example, electrostatic interactions, are apparently important in these cases.

The fact that small stabilizing interactions are noted between methyl and both fluorine (a typical σ-acceptor, π-donor group) and cyano (a typical σ-acceptor, π-acceptor group) suggests that the methyl group is capable of acting either as a π donor or σ donor, depending on its immediate environment. A strong destabilizing interaction is found for malononitrile (X = Y = CN). Here, both substituents seek to withdraw electron density from the connecting carbon via both the σ and π systems. Significant destabilization also results from attachment of fluoro and cyano substituents to the same sp^3 center, despite the stabilizing interaction that would be expected on the basis that fluorine is a (weak) π donor and cyano is a (strong) σ acceptor. This is probably due to the fact that both groups are relatively strong σ acceptors (and hence both make significant demands on the central carbon).

b. Structural Consequences of Geminal Interactions. Equilibrium bond lengths for a small selection of geminally-disubstituted methanes [7], X—CH$_2$—Y, are presented in Table 7.7. Bond lengths of the corresponding monosubstituted compounds, CH$_3$X and CH$_3$Y, are included for comparison.

A general observation from these data (in relation to the stabilization energies discussed in the previous section) is that favorable geminal interactions generally lead to shortening of the CX and CY linkages while unfavorable interactions usually result in bond lengthening. This is not always the case. A detailed analysis involves consideration of electrostatic effects as well as the particular nature (bonding or antibonding) of the interacting orbitals, and has been presented elsewhere [7i, j]. As mentioned previously (Section 7.2.2a), σ donors are best paired with π acceptors and vice versa. Thus, the geminal interaction of two fluorines (σ acceptors and π donors) leads to substantial shortening of CF linkages (compare the CF bond lengths in methylene fluoride with that in fluoromethane). Interaction of two lithiums (σ donors and π acceptors) has the same effect. Pairing of σ donors with π donors or σ acceptors with π acceptors is usually unfavorable energetically, and leads to bond lengthening. The former situation is illustrated by the structure of F—CH$_2$—SiH$_3$, and the latter by that of malononitrile, CH$_2$(CN)$_2$.

Some experimental data are available. The CC bond distances in malononitrile are slightly longer than that in acetonitrile. Of greater significance, the CF bond

TABLE 7.7. Calculated C—X Bond Lengths in Molecules X—CH$_2$—Y (Å)^a

Y \ X	Li	CH$_3$	F	CN	SiH$_3$
H	2.001	1.542	1.404	1.457	1.883
Li	1.984	1.563	1.593	1.398^b	1.839
CH$_3$	2.017	1.541	1.410	1.461	1.886
F	1.964	1.521	1.372	1.460	1.889
CN	2.179^b	1.547	1.397	1.461	1.908
SiH$_3$	2.012	1.550	1.430	1.455	1.885

^{a}3-21G//3-21G (3-21G$^{(*)}$//3-21G$^{(*)}$ for molecules incorporating second-row elements).

^{b}See Section 7.3.3 and, in particular, Figure 7.13 for a description of the structure of LiCH$_2$CN.

length in methylene fluoride is known to be significantly shorter than that in methyl fluoride (see Section 6.2.4a); the corresponding bond strengths are greater (see Section 6.5.6a). The calculated results and qualitative rationale are also consistent with other experimental observations, and have implications of a more general nature. For example, the anomeric effect in carbohydrate chemistry may be explained in the same context as bond strengthening in methylene fluoride [6]. Stereoelectronic effects in ester and amide hydrolysis reactions under kinetic control may be dictated by which C—O or C—N bond is the most likely to cleave [6c, 8]. This in turn may very well be the linkage that is most weakened by geminal interactions.

7.2.3. Interactions in Aromatic Systems

The treatment of substituent interactions in aromatic systems provided much of the basis on which the first electronic structure theories were founded [9]. Even today, resonance theory continues to play a significant role in chemistry; it influences the way the subject is taught and the manner in which it is applied to new problems. In this section, we examine a number of classical and contemporary problems involving aromatic compounds. We probe the limits of resonance arguments, and suggest alternative (and often equivalent) rationalizations based on qualitative molecular orbital theory and supported by quantitative calculations.

*a. **Disubstituted Benzenes.***   Theoretical calculations may be used to examine systematically interactions between groups in disubstituted benzenes [10]. These may conveniently be measured as energy changes for formal reactions of the type

$$\underset{Y}{\overset{X}{\bigcirc}} + \bigcirc \longrightarrow \overset{X}{\bigcirc} + \overset{Y}{\bigcirc} \qquad (7.21)$$

where a positive energy indicates favorable interaction between substituents X and Y. Representative results for *para*- and *meta*-disubstituted benzenes are presented in Tables 7.8 and 7.9, respectively. Many of the results here would easily have been anticipated on the basis of classical resonance arguments. For example, the interaction energy for *para*-nitroaniline is positive, reflecting reinforcing conjugative interaction, while that for *para*-phenylenediamine is negative, indicative of a saturation effect.

TABLE 7.8. Calculated Interaction Energies in *para*-Disubstituted Benzenes (kcal mol^{-1})[a,b]

X/Y	Li	CH₃	NH₂	OH	F	CN	NO₂
Li	-3.0	CH₃					
CH₃	-0.3	0.0	NH₂				
NH₂	-0.6	-0.5	-1.6	OH			
OH	0.0	-0.4	-1.5	-1.4	F		
F	0.7	-0.2	-0.8	-0.9	-0.7	CN	
CN	3.1	0.5	1.4	0.8	-0.1	-0.2	NO₂
NO₂	4.3	0.8	2.2	1.3	0.0	-2.8	-4.0
O⁻	-5.0	-1.4	-6.4	-4.4	0.7	22.2	30.4
NH₃⁺	15.9	2.0	4.5	1.4	-2.0	-9.0	-8.9

[a]Interaction energies defined by reaction (7.21) with appropriate X and Y.
[b]STO-3G//Std.

TABLE 7.9. Calculated Interaction Energies in *meta*-Disubstituted Benzenes (kcal mol^{-1})[a,b]

X/Y	Li	CH₃	NH₂	OH	F	CN	NO₂
Li	-3.2	CH₃					
CH₃	-0.2	0.0	NH₂				
NH₂	-0.3	0.2	0.8	OH			
OH	0.5	0.3	0.8	0.8	F		
F	1.2	0.3	0.6	0.4	0.0	CN	
CN	3.2	0.2	0.1	-0.4	-0.9	-2.0	NO₂
NO₂	4.4	0.3	0.0	-0.5	-1.3	-2.7	-3.8
O⁻	-13.5	-0.1	0.7	3.8	5.8	14.1	17.6
NH₃⁺	16.1	1.5	2.0	0.0	-3.2	-8.9	-12.0

[a]Interaction energies defined by reaction (7.21) with appropriate X and Y.
[b]STO-3G//Std.

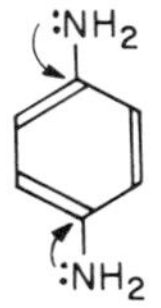

Note, however, that several of the theoretical results do not readily lend themselves to a classical interpretation. For example, why is there a favorable interaction between a *para* σ-accepting NH_2 substituent and the NH_3^+ group? Why is the interaction between the π-accepting Li substituent and a *para* π-donating NH_2 group unfavorable? These and other observations can be rationalized in terms of simple orbital interaction arguments. Thus, the NH_3^+ group lowers the π and π^* levels of benzene; this leads to enhanced interaction with the π-donating NH_2 group, and decreased interaction with π-accepting substituents such as NO_2 and CN. In a similar manner, the strongly electropositive Li substituent raises the π and π^* levels of the ring through a shielding mechanism, leading to poorer interaction with NH_2 and a more favorable interaction with NO_2 and CN.

These same arguments may also be employed to rationalize the interactions in *meta* disubstituted benzenes. Why is there a favorable interaction between the substituents in *meta*-fluoroaniline, and an unfavorable interaction in *meta*-fluoronitrobenzene? Classical resonance theory suggests that there should be no significant interactions at all! In these cases, substitution by electronegative fluorine lowers the energy of the benzene π and π^* orbitals, leading to favorable interaction with π donors, for example, NH_2, and unfavorable interaction with π acceptors, for example NO_2. The electropositive Li substituent behaves in an opposite manner. σ interactions also contribute to the overall stabilization or destabilization: σ donors interact favorably with σ acceptors and vice versa.

b. Analysis of Gas-Phase Acidities and Basicities. Relative basicities of *para*-substituted anilines, many of which have been established experimentally to protonate on nitrogen [11], are measured as energy changes for the proton-transfer reaction (7.22).

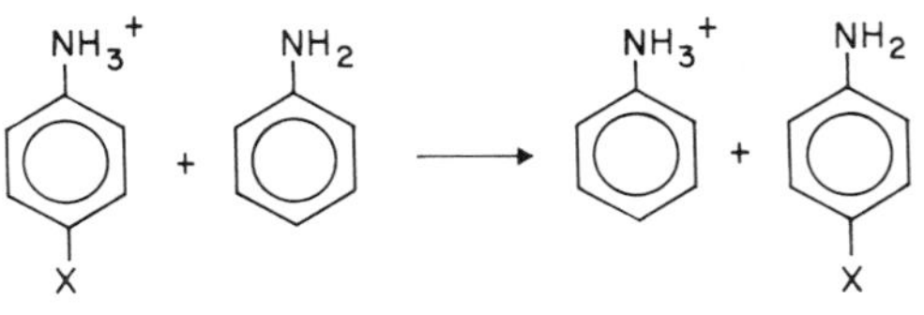

$$(7.22)$$

The total effect of a substituent on basicity may be broken down into its effect in the neutral aniline on the one hand, reaction (7.23),

$$(7.23)$$

and the effect in the corresponding anilinium ion on the other, reaction (7.24),

$$(7.24)$$

(see also Section 7.2.1f and reference 12). Calculated interaction energies for a selection of neutral and protonated anilines are given in Table 7.10, along with theoretical and (where available) experimental relative basicities.

Interaction energies for neutral anilines span only a very small range (from -1.6 to $+2.2$ kcal mol^{-1}). In general, π-electron-donor substituents, for example, NH_2,

TABLE 7.10. Effect of Substituents on Stabilities of _para_-Substituted Anilines and Anilinum Ions and on Calculated and Experimental Relative Gas-Phase Basicities (kcal mol^{-1})

Substituent X	ΔE(aniline)[a,b]	ΔE(anilinium ion)[a,c]	Relative Basicity[d]	
			Theory[a]	Expt.[e]
NO_2	2.2	-8.9	-11.1	f
CN	1.4	-9.0	-10.4	f
CF_3	0.8	-5.1	-5.9	f
CHO	1.0	-3.2	-4.2	f
F	-0.8	-2.0	-1.1	
H	0.0	0.0	0.0	0.0
CH_3	-0.5	2.0	2.5	2.8
OH	-1.5	1.4	2.9	
OCH_3	-1.5	2.8	4.2	3.5
NH_2	-1.6	4.5	6.1	
Li	-0.6	15.9	16.5	

[a] STO-3G//Std.

[b] Energy change for reaction (7.23) with appropriate X.

[c] Energy change for reaction (7.24) with appropriate X.

[d] Energy (enthalpy) change for reaction (7.22) with appropriate X.

[e] Experimental data ($\Delta\Delta H^\circ$) summarized in: D. H. Aue and M. T. Bowers, in _Gas Phase Ion Chemistry_, M. T. Bowers, ed., Academic Press, New York, 1979, vol. 2, p.1.

[f] Protonation on the X substituent favored over protonation at nitrogen. For further discussion on related systems, see Section 7.2.3e.

OH, OCH_3, and F, lead to slight destabilization, while π-electron-acceptor groups, for example, NO_2, CN, CHO, and CF_3, result in slight stabilization. Interaction energies in the corresponding anilinium ions span a much broader range, from -9.0 to $+15.9$ kcal mol^{-1}. Maximum stabilization occurs for lithium, a potent σ donor. The σ- and π-accepting cyano group is the most strongly destabilizing substituent. Interaction moves progressively from slight stabilization to slight destabilization for the series of substituents, NH_2, OH, and F, paralleling their decrease in π-donor ability and increase in σ-acceptor ability.

The utility of this form of analysis is well exemplified by *para*-nitroaniline, a molecule known (in solution) to be a much weaker base than aniline. The usual explanation, that it is the strong resonance stabilization of the neutral base (relative to aniline) that makes protonation unfavorable, that is,

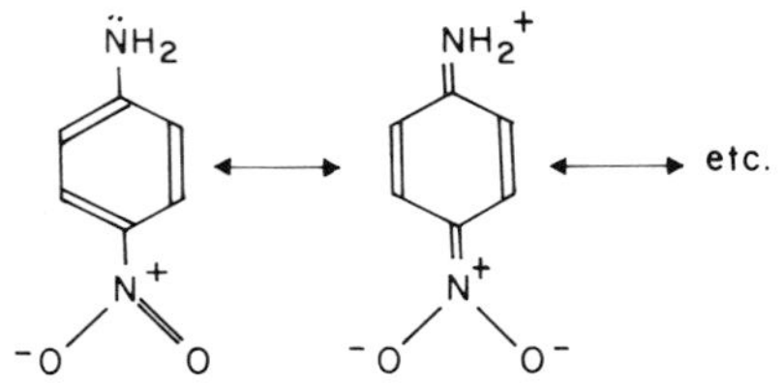

is not in accord with the theoretical data in Table 7.10. In fact, the predicted low nitrogen basicity appears to be due primarily to destabilization of the anilinium ion (-8.9 kcal mol^{-1}) and only marginally to stabilization of the neutral aniline (2.2 kcal mol^{-1}).

Another example illustrating the utility of this type of analysis is provided by the relative acidities of *para*-substituted phenols [13], as given by energy changes of reaction (7.25).

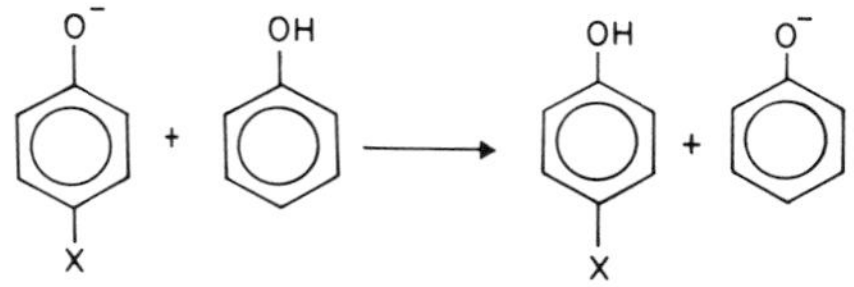

$$(7.25)$$

Here, the overall substituent effect can be partitioned into the effect in the phenoxide ion on the one hand, reaction (7.26),

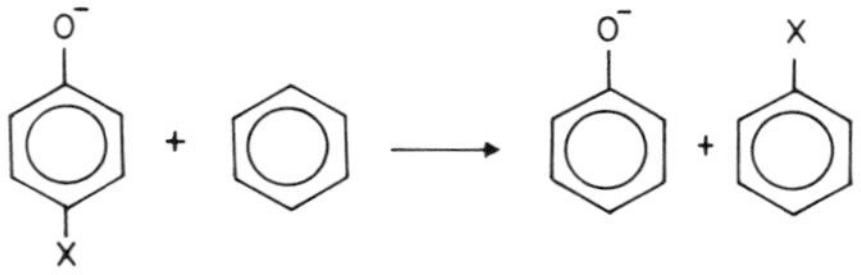

$$(7.26)$$

and the effect in phenol on the other, reaction (7.27).

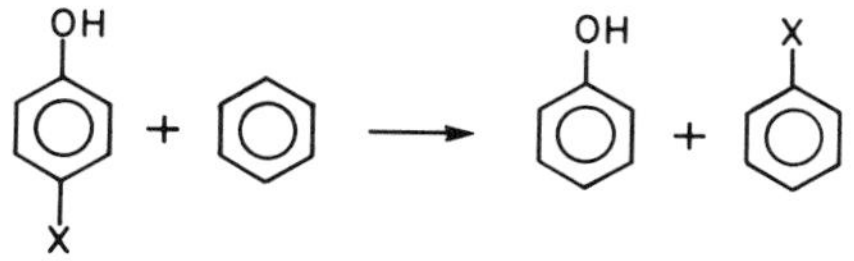

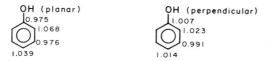

$$(7.27)$$

The theoretical data (Table 7.11) show that the total effect is again dominated by the effect in the charged species, that is, stabilization energies in the neutral phenols span less than 3 kcal mol^{-1} while those in phenoxide ions cover a range that is an order of magnitude greater. The acidity of phenol is increased by substituents that stabilize the phenoxide ion, for example, the σ- and π-electron acceptors, NO$_2$ and CN, and is decreased by substituents that destabilize it, for example, the strong π donor, NH$_2$, or the strong σ donor, Li.

c. ***Torsional Barriers in para-Substituted Phenols.*** Theory and experiment indicate that phenol is planar. As reflected in STO-3G π-electron populations,

OH (planar)
0.975
1.068
0.976
1.039

OH (perpendicular)
1.007
1.023
0.991
1.014

TABLE 7.11.   Effect of Substituents on Stabilities of *para*-Substituted Phenols and Phenoxide Ions and on Calculated and Experimental Relative Gas-Phase Acidities (kcal mol^{-1})

Substitutent X	ΔE(phenol)[a,b]	ΔE(phenoxide ion)[a,c]	Relative Acidity[d]	
			Theory[a]	Expt.[e]
Li	0.0	−5.0	−5.0	
NH$_2$	−1.5	−6.4	−4.9	−4.2
OCH$_3$	−1.3	−4.3	−3.0	−0.8
OH	−1.4	−4.4	−2.9	
CH$_3$	−0.4	−1.4	−1.0	−1.2, −1.3
H	0.0	0.0	0.0	0.0
F	−0.9	0.7	1.6	2.1, 2.6
CF$_3$	0.4	11.9	11.5	
CHO	0.7	14.5	13.8	
CN	0.8	22.2	21.4	17.7
NO$_2$	1.3	30.4	29.2	

[a] STO-3G//Std.

[b] Energy change for reaction (7.27) with appropriate X.

[c] Energy change for reaction (7.26) with appropriate X.

[d] Energy (enthalpy) change for reaction (7.25) with appropriate X.

[e] Experimental data from: (a) R. T. McIver, Jr., and J. H. Silvers, *J. Am. Chem. Soc.*, **95**, 8462 (1973); (b) T. B. McMahon and P. Kebarle, *ibid.*, **99**, 2272 (1972).

delocalization of electron density into the ring from the π lone pair on oxygen is more effective than delocalization from the σ lone pair.

A *para* substituent may either oppose or reinforce the delocalization of the oxygen lone pair, that is,

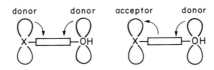

The results of STO-3G calculations (Table 7.12 and reference 14) show that, although the barrier in phenol (5.2 kcal mol^{-1}) is considerably higher than the experimental value (3.6 kcal mol^{-1}), the calculated effects of substituents on the barrier (designated ΔV_2 in Table 7.12) are in excellent agreement with the experimental data. This reflects a cancellation of errors in the isodesmic reaction (7.28) (see also Section 6.4.2b).

TABLE 7.12.   Analysis of Internal Rotation in *para*-Substituted Phenols (XC_6H_4OH)

Substituent X	$\Delta V_2{}^a$ (kcal mol^{-1}) Expt.^{e}	Calc.^{c}	$q_\pi(X)^{b,c}$	$\Delta q_\pi(OH)^{a,c}$	$\Delta\pi_{CO}{}^{a,c}$	Interaction Energy^{c,d} (kcal mol^{-1})
OH	− 0.87	− 0.95	+0.096	− 0.006	− 0.006	− 1.4
F	− 0.60	− 0.53	+0.074	− 0.002	− 0.003	− 0.9
CH$_3$	−0.32	− 0.28	+0.008	− 0.002	− 0.002	− 0.4
H	0	0	0	0	0	0
CHO	+0.87	+0.47	− 0.041	+0.006	+0.004	+0.7
CN	+0.70	+0.66	− 0.029	+0.009	+0.006	+0.8
NO$_2$	+0.98	+1.02	− 0.039	+0.013	+0.009	+1.3

$^a\Delta V_2$, $\Delta q_\pi(OH)$, and $\Delta\pi_{CO}$ are the changes in the twofold component of the rotational barrier, that is, energy (enthalpy) changes for reaction (7.28) with appropriate X, the extent of π-electron donation by OH, and the π-overlap population in the CO bond, respectively, in going from phenol to the substituted phenol.

$^b q_\pi(X)$ is the extent of π-electron donation by X.

^{c} STO-3G//Std.

^{d} Interaction energies defined by reaction (7.29) with appropriate X.

^{e} Experimental data from reference 14.

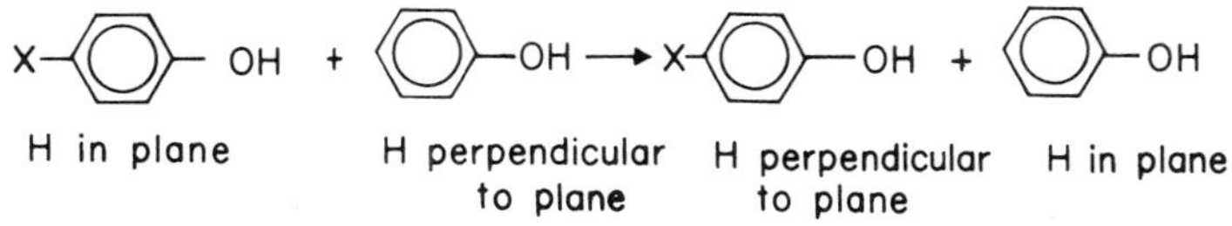

$$\text{X}\!-\!\!\bigcirc\!\!-\text{OH} \;+\; \bigcirc\!\!-\text{OH} \longrightarrow \text{X}\!-\!\!\bigcirc\!\!-\text{OH} \;+\; \bigcirc\!\!-\text{OH} \tag{7.28}$$

H in plane   H perpendicular   H perpendicular   H in plane
                  to plane   to plane

A number of other points are worthy of mention:

1. Twofold barriers decrease with π-electron-donor substituents (positive values of $q_\pi(\text{X})$, the number of π electrons donated by X into the ring), and increase with π-electron-acceptor substituents.

2. π-electron donation by the OH group is suppressed, that is, $\Delta q_\pi(\text{OH})$ is negative, and the degree of double-bond character in the CO linkage (as measured by π-overlap populations π_{CO}) is reduced, that is, $\Delta\pi_{CO}$ is negative, by π-electron donor substituents, and conversely for π-electron-acceptor substituents.

3. Energies of interaction of substituents with the OH group, as measured by energy changes for the isodesmic reaction (7.29),

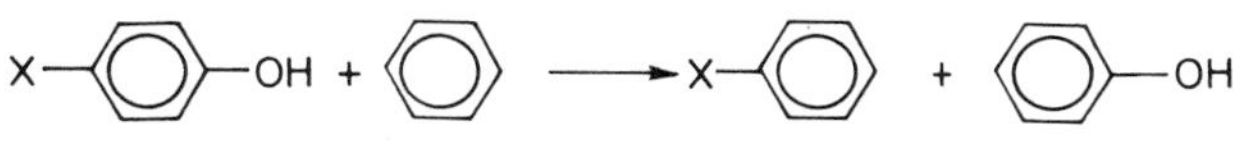

$$\text{X}\!-\!\!\bigcirc\!\!-\text{OH} \;+\; \bigcirc \longrightarrow \text{X}\!-\!\!\bigcirc \;+\; \bigcirc\!\!-\text{OH} \tag{7.29}$$

are favorable for π acceptors and unfavorable for π donors.

d. Proton Affinities of Alkyl Benzenes.   Both experimental and theoretical work have sought to determine the effect of alkyl-group size on proton affinity [15]. Data for substituted benzenes, that is, for reactions (7.30) [15c],

$$\bigcirc_R^+ \;+\; \bigcirc_{CH_3} \longrightarrow \bigcirc_R \;+\; \bigcirc_{CH_3}^+ \tag{7.30}$$

are presented in Table 7.13, and show an ordering of alkyl group effects: *tert*-butyl > *n*-butyl > *iso*-propyl ≈ *n*-propyl > ethyl > methyl. This does not reflect the (Baker–Nathan) sequence obtained calorimetrically in super-acid media [16]. This difference in ordering between gas and solution phases reveals that the enthalpy of solvation of protonated toluene is significantly greater than that of protonated *tert*-butyl benzene, as suggested originally by Schubert and Sweeney [17], but until recently discounted in favor of arguments based on the relative hyperconjugative abilities of CH and CC bonds. The large differential solvation effect is probably due largely to steric factors, that is, the bulkier the alkyl group the better shielded is the cation center from interaction with the solvent [18].

e. Sites of Protonation in Substituted Benzenes.   There are several reasonable alternatives for the site of protonation of a substituted benzene [19]. For example, toluene may protonate *ortho*, *meta*, *para*, or *ipso* to the substituent. The following ordering of stabilities is indicated by STO-3G calculations [20].

TABLE 7.13. Theoretical and Experimental Gas- and Solution-Phase Relative Proton Affinities of Alkyl Benzenes (kcal mol^{-1})

Substituent R	Theory	Experiment	
	ΔE(STO-3G//Std.)^{a,b}	$\delta \Delta H$(gas)^{a}	$\delta \Delta H$(solution)^{c}
Methyl	0.0	0.0	0.0
Ethyl	1.2	0.9	−2.5
n-Propyl	2.3	1.7	−2.0
iso-Propyl	2.3	2.1	−2.9
n-Butyl	2.7	2.1	
tert-Butyl	3.1	2.3 ± 0.5	−3.6

^{a}Energies (enthalpies) for reaction (7.30) with appropriate R. Data from reference 15c.

^{b}Protonation *para* to the substituent assumed.

^{c}Relative enthalpies of transfer from CH_3OH to the "super-acid" (11.5% SbF_5 in HSO_3F) at 233 K. Data from reference 16a.

Other sites, for example, the midpoint of the carbon–carbon bond or the center of the ring, appear to be much less favorable (for a discussion of the structure of protonated benzene, see Section 7.3.1c(v) and reference 21). In super-acid media below 176 K, NMR data reveal only a single isomer, presumed to be the *para* form. At higher temperatures, a single resonance is observed, indicating rapid equilibration among several forms [22].

The proton affinities of many substituted benzenes have been investigated both experimentally and theoretically [12a, 20, 21, 23]. A sampling of calculated affinities for protonation on the ring *ipso*, *ortho*, *meta*, and *para* to the substituent, that is, energies for reaction (7.31),

$$ (7.31) $$

are provided in Table 7.14. Experimental proton affinities (relative to benzene) are also supplied wherever available.

Various traditional patterns of aromatic substitution appear in these data, that is, *ortho–para* orientation with activation relative to benzene, *ortho–para* orientation with deactivation and *meta* orientation with deactivation [24].

Lithium substitution (and to a much lesser extent substitution by BeH) activates all ring positions, the *ortho* and *para* more so than the *meta*. However, *ipso* protonation is strongly preferred, reflecting the stabilizing ability of electropositive elements β to a carbocation center, that is, hyperconjugation.

TABLE 7.14. Theoretical and Experimental Gas-Phase Relative Proton Affinities for Monosubstituted Benzenes (kcal mol^{-1})

Substituent X	Theory (STO-3G//Std.)$^{a, b}$				Experimental^{a}
	ipso	ortho	meta	para	
Li	**62**	32	25	36	
BeH	**14**	2	2	3	
BH$_2$ (planar)	0	-5	**1**	-6	
(perpendicular)	4	6	0	8	
CH$_3$	-1	6	2	8	8
NH$_2$ (planar)	-23	24	-2	**27**	28
(perpendicular)	-9	2	1	**3**	
OH (planar)	-22	**15**	-5	15	13
F	-25	**4**	-8	4	1
CN		**-14**	-15	-15	14
CHO	-16	**-3**	-5	-3	18
NO$_2$	-36	-20	**-18**	-22	11

^{a}Energies (enthalpies) for reaction (7.31) with appropriate X. Experimental data from: Y. K. Lau and P. Kebarle, *J. Am. Chem. Soc.*, **98**, 7452 (1976).

^{b}Values set in bold type designate preferred (ring) site of protonation.

The site of protonation of phenylborane depends on conformation. For a co-planar BH$_2$ substituent, *meta* protonation with (slight) activation is strongly favored relative to *ortho* or *para* protonation, and is slightly better than *ipso* substitution. Oriented perpendicular to the ring, the BH$_2$ substituent directs *ortho-para*.

Note that, while the calculated proton affinities (at the most-favored ring site) for toluene, phenol, and fluorobenzene are in good agreement with the experimental data, those for benzonitrile, benzaldehyde, and nitrobenzene are markedly smaller than the corresponding measured values. In these cases, protonation probably occurs, not on the ring, but on the substituent, that is,

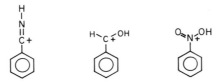

The calculated affinity for protonation *para* to the amino group in aniline is also in close accord with the experimental proton affinity. Here, it has been established that the nitrogen and (*para*) carbon proton affinities are nearly the same, protonation on nitrogen probably being slightly preferred to protonation on carbon [11].

7.2.4. Aromaticity

Bond separation energies provide one measure of the extent of stabilization or de-stabilization resulting from cyclic conjugation, that is, aromaticity. The case of benzene is discussed in Section 6.5.6. Examples for other hydrocarbons and hydro-carbon ions, and heavier main-group and transition-metal analogues, are provided here. We also explore the extension of aromatic stabilization to three-dimensional systems.

*a. **Hydrocarbons and Hydrocarbon Ions.*** Calculated and (where available) ex-perimental bond separation energies for a number of potentially aromatic or anti-aromatic hydrocarbons and hydrocarbon ions are displayed in Table 7.15. The 2π-electron systems, cyclopropenyl cation and cyclopropenylidene, should exhibit considerable stabilization due to delocalization of electrons from the (formal) double bond into the vacant p orbital at the carbocation and carbene centers, re-spectively, that is,

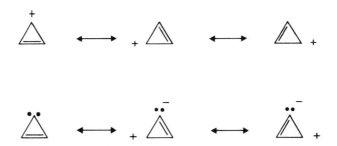

Calculated bond separation energies for both systems are negative, however, in con-trast to the large positive energy found in benzene. This is due to the fact that both of these molecules incorporate a highly-strained three-membered ring. Indeed, sub-traction of the corresponding energy for bond separation in cyclopropene, that is,

$$\triangle + 3CH_4 \rightarrow 2CH_3CH_3 + CH_2{=}CH_2, \qquad (7.32)$$

$$\Delta E(\text{3-21G}//\text{3-21G}) = -60 \text{ kcal mol}^{-1},$$

$$\Delta H(\text{expt.}) = -44 \text{ kcal mol}^{-1},$$

yields positive values for the stabilization energies in these two systems. The signifi-cant difference between calculated and experimental bond separation energies for cyclopropene, reaction (7.32), is due in part to the lack of polarization functions in the 3-21G basis set. 6-31G* calculations yield better results (see Section 6.5.6a).

The calculated bond separation energy for rectangular singlet cyclobutadiene is large and negative. While part of the destabilization is due to ring strain, for ex-ample, the bond separation energy for cyclobutene is large and negative,

TABLE 7.15. Calculated and Experimental Bond Separation Energies for Aromatic and Antiaromatic Hydrocarbons and Hydrocarbon Ions (kcal mol^{-1})

Molecule	Bond Separation Reaction	3-21G//3-21G	Expt.[a]
Cyclopropenyl cation	[structure] $+ 2CH_4 + CH_3^+ \longrightarrow 2CH_3CH_2^+ + CH_2{=}CH_2$	-34	-30^b
Cyclopropenylidene	[structure] $+ 2CH_4 + \ddot{C}H_2 \longrightarrow 2CH_3\ddot{C}H + CH_2{=}CH_2$	-12	
1,3-Cyclobutadiene	[structure] $+ 4CH_4 \longrightarrow 2CH_3CH_3 + 2CH_2{=}CH_2$	-70	
Cyclopentadienyl cation	[structure] $+ 4CH_4 + CH_3^+ \longrightarrow CH_3CH_3 + 2CH_3CH_2^+ + 2CH_2{=}CH_2$	-5	-4^b
Benzene	[structure] $+ 6CH_4 \longrightarrow 3CH_3CH_3 + 3CH_2{=}CH_2$	60	64^c
Cycloheptatrienylidene	[structure] $+ 6CH_4 + \ddot{C}H_2 \longrightarrow 2CH_3CH_3 + 3CH_2{=}CH_2 + 2CH_3\ddot{C}H$	49^d	
Tropylium cation	[structure] $+ 6CH_4 + CH_3^+ \longrightarrow 2CH_3CH_3 + 3CH_2{=}CH_2 + 2CH_3CH_2^+$	80	(73)

[a]Experimental data have been corrected to 0 K and for zero-point vibrational energy.
[b]Data from D. H. Aue and M. T. Bowers, in *Gas Phase Ion Chemistry*, M. T. Bowers, ed., Academic Press, New York, 1970, vol. 2, p. 1.
[c]From Table 6.71.
[d]4-31G//STO-3G.

$$\square + 4CH_4 \rightarrow 3CH_3CH_3 + CH_2\!=\!CH_2, \qquad (7.33)$$

$$\Delta E(\text{3-21G}//\text{3-21G}) = -28 \text{ kcal mol}^{-1},$$

$$\Delta H(\text{expt.}) = -17 \text{ kcal mol}^{-1},$$

the major part of the effect arises because of unfavorable interaction of π electrons. The (singlet) 4π-electron cyclopentadienyl cation also exhibits destabilization due to unfavorable cyclic conjugation. Its modest (negative) bond separation energy may be contrasted with the significant positive value found for cyclopentadiene, that is,

$$\pentagon + 5CH_4 \rightarrow 3CH_3CH_3 + 2CH_2\!=\!CH_2, \qquad (7.34)$$

$$\Delta E(\text{3-21G}//\text{3-21G}) = 11 \text{ kcal mol}^{-1},$$

$$\Delta H(\text{expt.}) = 22 \text{ kcal mol}^{-1}.$$

The calculated (3-21G) equilibrium structures for both singlet cyclobutadiene and cyclopentadienyl cation show localized single and double bonds,

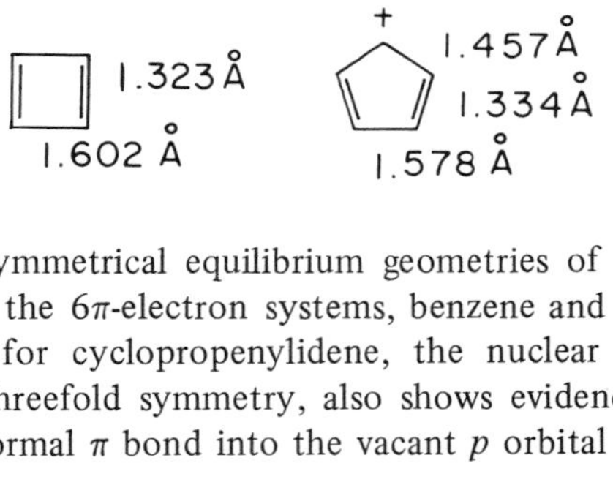

in contrast to the symmetrical equilibrium geometries of the 2π-electron cyclopropenyl cation and the 6π-electron systems, benzene and tropylium cation. The calculated structure for cyclopropenylidene, the nuclear framework of which does not allow for threefold symmetry, also shows evidence of delocalization of electrons from the formal π bond into the vacant p orbital at the carbene center.

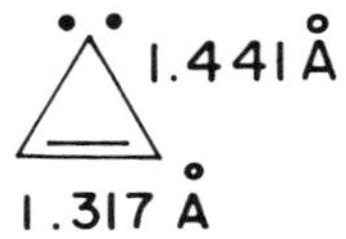

For comparison, the calculated (3-21G) single and double bond lengths in cyclopropene are 1.523 and 1.282 Å, respectively.

Both calculated and experimental bond separation energies for benzene and tropylium cation are, as expected, large and positive, clearly illustrating the energetic stabilization to be gained from cyclic conjugation of six π electrons.

The case of cycloheptatrienylidene is especially interesting [25]. The bond separation energy for the planar molecule (given in Table 7.15) is large and positive, indicating the expected aromatic stabilization. Note, however, that the 3-21G geometry for this molecule incorporates nearly fully localized single and double bonds, that is,

More careful scrutiny reveals that this planar structure is not the lowest-energy form on the C_7H_6 singlet potential surface, and, in fact, is probably not even an energy minimum. Instead, a nonplanar C_2 symmetry form,

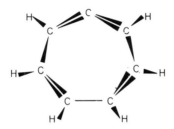

is 16 kcal mol^{-1} lower in energy at 4-31G//STO-3G [25]. *This species is not aromatic!* It incorporates four double bonds including an allenic-type linkage, and is most appropriately termed cycloheptatetraene. The low-temperature matrix infrared spectrum of the molecule is in accord [26]. It would appear, therefore, that the aromatic stabilization afforded planar cycloheptatrienylidene is not sufficient to offset the gain resulting from formation of an additional π bond.

b.   Benzene and Pyridine Analogues.   A number of heavy-atom analogues of benzene and pyridine have now been characterized experimentally. Among the simplest of systems in the former group is silatoluene, the low-temperature matrix infrared spectrum of which has been obtained [27].

Three pyridine analogues, phosphabenzene [28], arsabenzene [29], and stibabenzene [30], have been prepared, and their equilibrium structures determined using microwave spectroscopy.

These show little evidence for bond alternation, and therefore support the notion of π-electron delocalization over the entire ring. Experimental information regarding the delocalization energies (aromatic stabilization) in these compounds is unavailable.

The X-ray crystal structure of at least one compound incorporating a transition metal in a benzene ring has been reported [31].

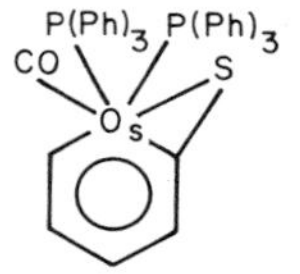

Here too, the ring CC lengths show only slight variation from the value appropriate for benzene.

Bond separation energies (STO-3G//STO-3G and 3-21G//3-21G) for benzene and some of its group IV main-group and transition-metal analogues, that is, the energies of reaction (7.35),

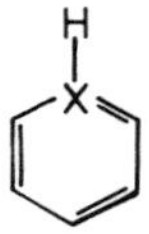

$$+ 5CH_4 + XH_4 \longrightarrow 2CH_2 = CH_2 + 2CH_3CH_3$$
$$+ CH_2 = XH_2 + CH_3XH_3, \tag{7.35}$$

and for pyridine and some of its main-group analogues, reaction (7.36),

$$+ 5CH_4 + XH_3 \longrightarrow 2CH_2 = CH_2 + 2CH_3CH_3$$
$$+ CH_2 = XH + CH_3XH_2, \tag{7.36}$$

are presented in Table 7.16.

Large stabilization energies are found for all main-group systems; the compound incorporating a transition metal exhibits a smaller effect. The limited 3-21G data suggest that the STO-3G calculations overestimate the aromatic stabilization energies by 10–20 kcal mol^{-1}. Calculated stabilization energies for main-group pyridine analogues are closer to that for pyridine than are calculated energies for benzene analogues relative to benzene. This is consistent with the fact that pyridine analogues are thermally stable, whereas the benzene analogues (apparently) are not. It should be noted, however, that the use of energies of reactions such as (7.35) and (7.36) as measures of relative aromatic stabilization does not take into account possible differences in the stabilities of the products. For example, the fact that double bonds between carbon and nitrogen are inherently weaker than ethylenic linkages implies a higher bond separation energy for pyridine than for benzene if all other factors were equal. Furthermore, one likely reason for the low stabi-

TABLE 7.16.   Calculated Bond Separation Energies for
Benzene and its Main-Group and Transition-Metal
Analogues and for Pyridine and its Main-Group
Analogues (kcal mol^{-1})[a]

	STO-3G//STO-3G	3-21G//3-21G
Reaction (7.35)		
X = C	70	60
Si	60	46[b]
Ge	55	
Ti	31	
Reaction (7.36)		
X = N	70	64
P	68	56[b]
As	65	
Sb	76	

[a]Theoretical data from: W. J. Pietro and W. J. Hehre, unpublished
results, except for X=Si which is from J. Chandrasekhar, P. v. R.
Schleyer, R. O. W. Baumgartner, and M. T. Reetz, *J. Org. Chem.*,
48, 3453 (1983).
[b]3-21G(*)//3-21G(*).

lization energy calculated for the transition-metal benzene analogue is the high
thermal stability of molecules incorporating metal-to-carbon double bonds (for
further discussion see Sections 7.4.2 and 7.4.3), which appear as products in
reaction (7.35).

c. Aromaticity in Three Dimensions.   Aromaticity must also be exhibited by
three-dimensional systems. Many capped rings (half sandwiches) or doubly-capped
rings (inverse sandwiches) obey a $4n + 2$ interstitial electron rule, which extends
classical aromatic bonding concepts into three dimensions [32]. Interstitial electrons are like the π electrons of planar annulenes, but bind rings and caps together.

Half-sandwich molecules such as those shown below may conveniently be represented in terms of a carbocyclic ring (annulene) and a capping atom or group X.

The electrons involved in skeletal CC and CH bonding of the annulene ring are
assigned in a conventional manner, that is, two electrons to each single bond. Delocalization of the π electrons of the ring onto the cap or the electrons of the cap
onto the ring serves to bind the two units together. The π symmetry designation is
no longer appropriate, and the term interstitial better describes the electrons which
are shared between ring and cap.

How do $4n + 2$ interstitial electrons lead to aromatic stabilization? The case of
cyclopentadienyllithium is typical [33].

The parent annulene, cyclopentadienyl anion, is aromatic, that is, six π electrons are delocalized over five carbons. The lithium cation, located over the center of the ring, does not contribute to the number of available valence electrons, but acts solely to redistribute these electrons into the interstitial region between the two components. As shown by the interaction diagram in Figure 7.1, the a and e symmetry π orbitals of cyclopentadienyl anion interact with the s and p functions of the cap giving rise to three stabilized molecular orbitals. The familiar *one-below-two-* Hückel pattern of π orbitals, as is found in all annulenes, results.

The $C_6H_6^{2+}$ system further illustrates the $4n + 2$ interstitial electron rule [36]. The thirty valence electrons of benzene comprise six two-electron CC bonds, six two-electron CH bonds and six π electrons. With two fewer (π) electrons, the planar dication is no longer aromatic. It can, however, regain stabilization via interstitial electron delocalization by assuming a half-sandwich pyramidal geometry.

Here, only ten, and not twelve, electrons are tied up in CC σ bonds, leaving six interstitial electrons to bind ring and cap. The molecule is aromatic. A permethylated derivative [37] as well as numerous isoelectronic boron analogues [38] are known. In principle, two additional electrons could be removed yielding the benzene tetracation. Such a species (unlikely as it may seem on purely electrostatic grounds) could achieve interstitial stabilization by yet another structural change to give an inverse sandwich compound, in which the central annulene comprises only four CC σ bonds, and is linked by six interstitial electrons to two caps.

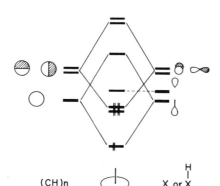

FIGURE 7.1.   Interaction of the valence molecular orbitals of cyclopentadienyl anion with the s and p functions of a capping group.

An interaction diagram (similar to that in Figure 7.1), involving ring and cap orbitals, leads again to three stable delocalized molecular orbitals. While $C_6H_6^{4+}$ is unknown, the isoelectronic borane $B_6H_6^{2-}$ [38] and carborane $C_2B_4H_6$ [38] exist and exhibit octahedral structures. Indeed, similar structural relationships involving differences of two valence electrons are familiar in carborane chemistry [38], for example, between *arachno, nido*, and *closo* geometries.

For any given ring, many caps will provide the proper number of interstitial electrons. Which combinations are the best? One criterion is the extent to which the equilibrium structures of the rings distort due to transfer of π-electron density into the interstitial region. Calculated CC bond distances and ring CH out-of-plane bending angles in a variety of pyramidal half-sandwich compounds (all of which involve six interstitial electrons) are given in Table 7.17; also included are the corresponding distances in the base ring compounds benzene, cyclopentadienyl radical, triplet cyclobutadiene, and cylopropenyl radical. Both STO-3G and 3-21G calculations show, as expected, that CC bond lengths in the complexes are longer than in the free polyenes. The effect is smallest for benzene complexes and largest for cyclopropenyl radical complexes; increases in bond lengths for complexes involving cyclopentadienyl radical and triplet cyclobutadiene are intermediate. The size of the bond length increments is also sensitive to the nature of the capping group. Finally, there is a gradation in the out-of-plane bending angle with ring size and with the nature of the cap. For a given ring system, the bending angle goes from a positive value (bending away from the cap) to a negative value (bending toward the cap) with increasing electronegativity of the capping element. For example, with cyclobutadiene as a base polyene, the bending angle decreases from 12.1° for an Li⁻ cap to -2.7° for CH⁺ as the capping group, according to 3-21G calculations. Also, for a given cap, the bending angle increases with decreasing polyene ring size. Thus, with BeH (or Be) as a cap, the angle increases from -3.1° for a benzene base to 47.1° for cyclopropenyl radical as the base polyene (STO-3G).

The noted geometrical differences between systems may be rationalized in terms of a desire to match best the sizes of the interacting ring and cap orbitals. Large ring sizes and/or tightly held cap orbitals require an upward bending of the polyene hydrogens,

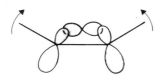

TABLE 7.17. Carbon-Carbon Bond Distances (Å) and Ring CH Out-of-Plane Bending Angles (°) in Half-Sandwich Compounds

Ring System (number of π electrons)	Cap	C—C Distance		Out-of-Plane Bending Angle[a]	
		STO-3G	3-21G	STO-3G	3-21G
Benzene (6)	—	1.387	1.385	0.0	0.0
	Li^+	1.409	1.395	2.5	1.6
	LiH	1.399		1.4	
	BeH^+	1.412		−3.1	
Cyclopentadienyl radical (5)	—	1.393	1.406	0.0	
	Li	1.417	1.416	6.4	2.8
	Be^+	1.436	1.430	−0.1	0.5
	BeH	1.419	1.418	−0.9	−0.5
	B		1.414		−2.6
	BH^+	1.434	1.427	−8.1	−4.4
	C^+	1.427	1.424	−8.6	−4.4
	CH^{2+}	1.458	1.442	−11.7	−8.2
Cyclobutadiene (4)	—	1.431		0.0	
	Li	1.436	1.467	17.2	12.1
	Be	1.463	1.477	7.1	6.6
	BH	1.459	1.467	−1.2	1.0
	C	1.449	1.455	−2.4	−1.0
	CH^+	1.477	1.472	−5.4	−2.7
Cyclopropenyl radical (3)	—	1.421	1.417		
	Be^-	1.509	1.585	47.1	49.6
	BH^-	1.503	1.525	31.8	34.8
	CH	1.473	1.489	19.5	19.5
	N		1.452		12.6
	NH^+	1.490	1.466	10.1	10.6

[a]Positive angles indicate bending of hydrogens away from the cap.

while small base rings and/or diffuse cap orbitals lead to downward bending.

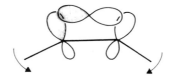

7.3. REACTIVE INTERMEDIATES

Quantitative molecular orbital theory can be used to investigate the properties of short-lived reactive intermediates just as readily as those of more stable molecules. While the scarcity of experimental data greatly enhances the overall value of the calculations, the predictions of theory are difficult to assess. Meaningful results require careful application! The present treatment surveys theoretical results for three classes of organic reactive intermediates: carbocations, carbanions (as well as polar organometallics), and carbenoids (and their singlet carbene analogues). Some of the simpler systems have been examined at very high levels of theory, and the results are not likely to be modified significantly with further refinements. Others have not been treated as well and this field remains a fertile ground for further theoretical exploration.

7.3.1. Carbocations and Carbodications

A large number of carbocations have now been observed and studied both in the gas phase [39] and in solution in super-acid media [40]. Except for a few larger systems where X-ray data are available (Section 6.2.8d), very little is known from experiment about their detailed structure. In contrast, theoretical work on carbocation geometries has been extensive. The literature has been reviewed periodically, and the interested reader is referred elsewhere for thorough documentation [41].

Most of the early *ab initio* calculations were performed with Hartree–Fock wavefunctions using minimal and split-valence basis sets [42]. It was soon recognized, however, that such simple levels often fared poorly in their description of carbocation energies, and that the addition of *d*-type functions to the carbon basis sets preferentially stabilized bridged *(nonclassical)* ions relative to open *(classical)* forms [43]. It was also found that inclusion of electron correlation favored bridged arrangements [44]. The situation is best exemplified by the vinyl and ethyl cations. Early, low-level computational work suggested open, non-bridged structures for both species.

Later calculations, using larger basis sets and including corrections for electron correlation, significantly altered these initial predictions, and the best theoretical results now show hydrogen-bridged structures to be preferred for both cations.

It is likely that the best theoretical results now available for small systems [44h] will stand up to further scrutiny both at even higher theoretical levels and experimentally. As more sophisticated models are brought into practical and widespread use, their performance may well be calibrated in terms of many of the systems discussed in this section.

a. Protonated Alkanes. Even alkanes are susceptible to protonation in the gas phase. Protonated methane, ethane and propane are well characterized experimentally, and the existence of higher protonated alkanes has also been established [45]. While saturated hydrocarbons are not particularly strong bases, that is, relative to alkenes or *n*-donor bases such as pyridine, their absolute (gas-phase) proton affinities are in excess of 127 kcal mol^{-1} (the experimental proton affinity of methane), and they certainly cannot be described as loose complexes between a proton and the alkane. Whatever their structures, they must involve increased coordination of either carbon or hydrogen or both.

Alkanes, once protonated, rapidly lose a hydrogen molecule:

$$RH + H^+ \longrightarrow [RH_2]^+ \longrightarrow R^+ + H_2. \tag{7.37}$$

As shown in Figure 7.2, this tendency increases with increasing stability of the resulting alkyl cations: methyl $<$ ethyl $<$ *iso*-propyl. The entropy of dissociation is typically on the order of 30 cal mol^{-1} deg^{-1} (at room temperature $T\Delta S$ is approximately 10 kcal mol^{-1}), and complexes involving ethyl, *iso*-propyl, or more stable ions, may only be observed at very low temperatures. Even for initial carbon–carbon bond protonation (instead of protonation of a carbon–hydrogen linkage), rapid loss of H_2 occurs, implying that barriers to proton migration in these systems are very low.

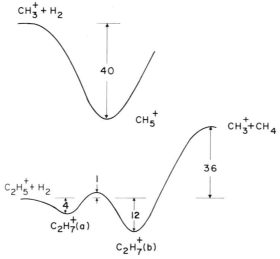

FIGURE 7.2. Experimentally derived enthalpy surfaces for protonated methane (top) and ethane (bottom). The $C_2H_7^+$ isomer labeled (*a*) is believed to correspond to CH protonation and that labeled (*b*) to CC protonation. See reference 45.

The fascinating chemistry of saturated hydrocarbons in strong-acid media has been explored largely through the efforts of Olah and his associates [46]. Hydrogen-deuterium exchange, as well as substitution by other electrophiles, has been extensively documented. Little quantitative information is known, however, either about the structures of protonated alkanes or about their relative stabilities. At present, theoretical calculations afford the best opportunity to obtain such data.

(i) PROTONATED METHANE.   Reasonable structures for protonated methane include the C_s symmetry forms 1 and 2, the C_{2v} symmetry structure, 3, the square pyramid (C_{4v}), 4, and the trigonal bipyramid (D_{3h}), 5 [42a, b, 43a, 44b, h].

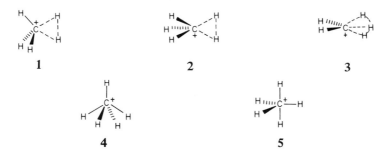

At the HF/6-31G* level, only the first of these is actually a potential energy minimum. Normal-coordinate analyses for the remaining forms indicate that all are unstable with respect to displacements along one or more directions.

According to the HF/6-31G* calculations, the stable structure 1 resembles a complex between methyl cation and hydrogen molecule, rather than between methane and a proton. The calculated H—H bond distance (in the hydrogen molecule fragment) is 0.853 Å, only slightly longer than that found in the free hydrogen molecule at the same level of theory (0.730 Å), and much shorter than that in H_2^+ (1.041 Å). The calculated binding energy for the complex, that is, ΔE for $CH_5^+ \longrightarrow CH_3^+ + H_2$, is 20 kcal mol^{-1}. The MP2/6-31G* equilibrium structure shows a longer H—H linkage (0.949 Å, compared with 0.738 Å in H_2 and 1.041 Å in H_2^+). The binding energy has increased to 46 kcal mol^{-1}.

Other forms of protonated methane provide possible routes to facile hydrogen scrambling. The relative energies displayed in the last column of Table 7.18, obtained at the MP4SDQ/6-311G**//MP2/6-31G* level, again show a preference for structure 1. Structure 2 is very close in energy; rotation of the two fragments that comprise the complex is essentially free. This is to be expected for a sixfold barrier. The energy of structure 3, corresponding to protonation of methane along an HCH bisector, is only 1 kcal mol^{-1} above that of 1. The close proximity of the energies of these three forms was first noted by Dyczmons and Kutzelnigg [44b], and implies a very low activation barrier for complete hydrogen scrambling in the ion. The C_{4v} form, 4, which resembles the solution-phase structures of a number of carbocations [47], is also relatively low in energy. Finally, the energy of the trigonal bipyramid, 5, is significantly above those of all other forms considered.

TABLE 7.18. Relative Energies of CH_5^+ Structures (kcal mol^{-1})

Structure (point group)	3.21G// 3-21G	6-31G*// 6-31G*	HF	MP2	MP3	MP4SDQ	Expt.[a]
			/6-311G**//MP2/6-31G*				
1 (C_s)	0	0	0	0	0	0	0
2 (C_s)	0	0	0	0	0	0	
3 (C_{2v})	5	4	3	1	1	1	
4 (C_{4v})	5	7	8	3	3	4	
5 (D_{3h})	7	13	16	11	11	12	
$CH_3^+ + H_2$	18	20	24	46	43	42	44
$CH_4 + H^+$	115	121	128	132	132	132	133

[a]Experimental data from: D. H. Aue and M. T. Bowers, in *Gas Phase Ion Chemistry*, M. T. Bowers, ed., Academic Press, New York, 1979, vol. 2, p. 1. Corrected for zero-point energies and to 0 K.

Also included in Table 7.18 are calculations performed at the 3-21G//3-21G and 6-31G*//6-31G* levels, as well as results derived with the 6-311G** basis set at the Hartree-Fock and at the second- and third-order Møller–Plesset levels. Comparisons of this sort are especially useful; MP4 calculations may be prohibitively expensive for larger systems. The HF calculations succeed in properly ordering the stabilities of the various CH_5^+ structures. The main effect of the correlation corrections is to lower the relative energies of **3** and **4**. The MP2 calculations mirror the more complete correlation treatments fairly well, although they tend to overestimate slightly the ultimate (MP4) effects. This behavior of the MP2 model has previously been noted, see Section 6.5. The MP3 calculations fare better.

In the gas phase, the addition of molecular hydrogen to the methyl cation is exothermic by 44 kcal mol^{-1} [45] (after corrections for zero-point energy and to 0 K have been made). The calculated (MP4SDQ/6-311G**//MP2/6-31G*) difference is 42 kcal mol^{-1}. The gas-phase proton affinity of methane (133 kcal mol^{-1}) after zero-point energy and 0 K corrections) is also well reproduced by the highest-level calculations (132 kcal mol^{-1}). Both of these reaction energies are underestimated by all three Hartree–Fock models, although even the simplest of the correlation energy schemes successfully reproduces the higher-level values.

(ii) PROTONATED ETHANE. Experimental studies have identified two different forms of $C_2H_7^+$, one, **6**, presumed to correspond to protonation of the carbon-carbon linkage and the other, **7**, analogous to CH_5^+, to protonation of a CH bond [45].

        6   **7**

As previously indicated in Figure 7.2, **6** is believed to be more stable than **7** by 7–8 kcal mol^{-1}. This ordering of stabilities is consistent with the best available theoretical calculations on these systems [42a, b, 44g, h]. At the MP4SDQ/6-311G**// MP2/6-31G* level, structure 6 is 7 kcal mol^{-1} more stable than 7. Hartree-Fock calculations (using 3-21G, 6-31G*, and 6-31G** basis sets) yield the same ordering of stabilities, although energy separations are somewhat larger.

The HF/6-31G* equilibrium geometries for 6 and 7 closely resemble complexes between methyl cation and methane, and between classical ethyl cation and a hydrogen molecule, respectively. The corresponding MP2/6-31G* structures show evidence of significant strengthening of both complexes relative to the Hartree-Fock results. Most striking are the calculated CH distances in 7, which are reduced from 2.7 Å (HF/6-31G*) to 1.3 Å (MP2/6-31G*). This effect has previously been noted by Kohler and Lischka at the CEPA level [44g]. Also of note is the decrease in the central CHC bond angle in 6 from 122° at HF/6-31G* to 106° at MP2/6-31G*.

Hiraoka and Kebarle's measured energy of dissociation (ΔH_0) of protonated ethane into ethyl cation and a hydrogen molecule [45] is 12 kcal mol^{-1} (or ΔH_e = 19 kcal mol^{-1} after the data are corrected for zero-point effects and to 0 K), which compares reasonably well with the MP4SDQ/6-31G**//6-31G* value of ΔH_e = 15 kcal mol^{-1}. The corresponding theoretical energy (ΔH_e) of separation of 6 into methyl cation and methane is 35 kcal mol^{-1}; the experimental value (ΔH_e, corrected for zero-point energy and to 0 K) is 40 kcal mol^{-1}.

Finally, note the similarity of the calculated structure for protonated ethane with theoretical [48a,b] and experimental [48c] geometries for the isoelectronic species $B_2H_7^-$. The success of the theory here lends credence to predictions for $C_2H_7^+$.

b. Protonated Acetylene and Ethylene: Hydrogen Bridging in Simple Carbocations.

As noted earlier, carbocation geometries often show strong tendencies toward bridging by hydrogen, alkyl, or aryl groups. A more compact molecular skeleton affords enhanced opportunity for the limited number of valence electrons to bind the nuclei together. Because it is often difficult to differentiate unambiguously hydrogen-bridged nonclassical from acyclic classical carbocation structures on rather indirect experimental evidence, much controversy has surrounded this area of chemistry [49]. As seen below, the best available calculations favor hydrogen-bridged structures for both vinyl and ethyl cations; the equilibrium geometries of more complicated carbocations, such as cyclopropylcarbinyl and norbornyl, are still controversial. Here, we examine the theoretical predictions (and available experimental results) for vinyl and ethyl cations, and attempt to assess those factors that lead to a favoring or disfavoring of bridged over acyclic ions. Such analyses assist the interpretation of data on larger systems, where high-level theoretical calculations will not always be feasible.

(i) PROTONATED ACETYLENE: THE VINYL CATION. The vinyl cation is the simplest carbocation for which the relative stabilities of acyclic (classical), 8, and hydrogen-bridged (nonclassical), 9, structures may be compared; extensive theoretical studies have been carried out [42a, 43a, 44a,c,d,f,h].

8　　　　　　　　　　**9**

Calculations at the MP4SDQ/6-311G**//MP2/6-31G* level (Table 7.19) indicate the hydrogen-bridged structure, **9**, to be 3 kcal mol^{-1} lower in energy than the acyclic form, **8**. The preference for hydrogen bridging is apparently exaggerated with the lower-order perturbation methods; at the MP2 level, the nonclassical ion is 8 kcal mol^{-1} lower in energy than the open form. Similar behavior has also been noted in comparisons of the IEPA and CEPA methods [44f]. Hartree–Fock treatments generally yield the opposite ordering of stabilities. While this failing is most conspicuous for basis sets lacking d functions, for example, 3-21G, even 6-311G**// 6-31G* calculations show the classical ion, **8**, to be 5 kcal mol^{-1} lower in energy than the nonclassical structure, **9**. The effect of electron correlation observed here, to stabilize the hydrogen-bridged form preferentially, is typical of that noted in other systems, and was first pointed out by Zurawski, Ahlrichs and Kutzelnigg who used the IEPA method [44a].

Little is known experimentally about the structures of vinyl cations. Substituted vinyl cations were established as reactive intermediates only in the past decade, long after aliphatic carbocations were well accepted [50]. The parent vinyl cation and several of its derivatives are stable species in the gas phase. While there is no information about their geometries, the enthalpy of protonation of acetylene, leading to vinyl cation, has been established as 156 kcal mol^{-1} (after zero-point energy and 0 K corrections). The MP3/6-311G**//MP2/6-31G* value is somewhat larger (161 kcal mol^{-1}). Unfortunately, the experimental error limits are too large to allow a conclusive assignment of structure to be made.

(ii) PROTONATED ETHYLENE: THE ETHYL CATION. The ethyl cation (protonated ethylene) has also been the subject of much theoretical work [42a, 43a, 44a,f,h]. Three geometries are likely candidates for its ground-state structure, two

TABLE 7.19. Relative Energies of $C_2H_3^+$ and $C_2H_5^+$ Structures (kcal mol^{-1})^a

Structure (point group)	3-21G// 3-21G	6-31G*// 6-31G*	HF	MP2	MP3	MP4SDQ
			/6-311G**//MP2/6-31G*			
8 Classical vinyl (C_{2v})	−20	−7	−5	8	5	3
9 Bridged vinyl (C_{2v})	0	0	0	0	0	0
10 Classical ethyl (C_s)	−8	−1	b	b	b	b
11 Classical ethyl (C_s)	−7	0	2	8	7	7
12 Bridged ethyl (C_{2v})	0	0	0	0	0	0

^{a}Theoretical data from reference 44h.
^{b}No minimum; collapses to **12**.

acyclic (classical) forms, **10** and **11**, and a hydrogen-bridged (nonclassical) ion, **12**.

10   **11**   **12**

Although protonated ethylene is a familiar species in the gas phase, like protonated acetylene it too has never been directly observed in solution, despite numerous attempts. There is, however, ample evidence for its existence in solution as a reactive intermediate. The facile label scrambling of partially deuterated materials shows that the ion fluctuates between hydrogen-bridged and open forms, but such experiments do not identify which is the more stable [51]. Indirect evidence for a bridged equilibrium structure for ethyl cation derives from the photoelectron spectrum of the corresponding free radical. Houle and Beauchamp [52] have observed that the vertical and adiabatic ionization potentials for this species are separated by approximately 3 kcal mol^{-1}. Assuming that the observed vertical transition corresponds to the instantaneous formation of a classical ethyl cation, that is, with a geometry similar to that of ethyl radical, the observation of a significant energy separation between vertical and adiabatic ionization potentials suggests that (at least partial) rearrangement to a more stable (hydrogen-bridged) species later occurs. Other gas-phase methods, which lead to ethyl cations in energy-relaxed (adiabatic) states, give heats of formation about 4 kcal mol^{-1} lower than that obtained from the vertical ionization potential [53]. Accordingly, the 3-4 kcal mol^{-1} discrepancy may be interpreted to correspond approximately to the difference in energies between the open and the more stable hydrogen-bridged ethyl cations.

The theoretical predictions depend somewhat on the level of calculation (Table 7.19). At HF/3-21G, the classical ion, **10**, is more stable than the bridged structure, **12**, by 8 kcal mol^{-1}. Polarization functions (d on carbon and p on hydrogen) preferentially reduce the energy of the bridged ion. At the highest level (MP4SDQ/6-311G**), **12** is 7 kcal mol^{-1} lower in energy than **11**. The other acyclic structure, **10** (which may be more stable than **11**), does not survive at this level; it collapses without activation to the bridged form, **12**. In fact, neither classical ion represents a true local minimum on the calculated MP2/6-31G* potential surface. Structure **11** does not fall directly, that is, without a barrier, to the hydrogen-bridged ion only because of the imposed C_s symmetry. It might serve as a transition structure for hydrogen scrambling.

c. **Larger Systems.** Here we review the present state of calculations and experiment on selected larger carbocations. For most of these systems, the computational models which have been applied to date are not as sophisticated as the best theoretical levels discussed in the previous sections.

(i) $C_3H_5^+$. Various isomers of $C_3H_5^+$ have been studied at Hartree–Fock and Møller–Plesset levels [43c, 44h]. Among them are the planar and perpendicular allyl cations, **13** and **14**, the 2-propenyl cation, **15**, the l-propenyl cation, **16**, the cyclopropyl cation, **17**, and corner-protonated cyclopropene, **18**.

Calculated relative energies for these systems are found in Table 7.20.

At all levels of theory, the allyl cation, **13**, is the most stable form, in agreement with experiment [54]. Its CC bond length, 1.373 Å at HF/6-31G*, is intermediate between the lengths of normal single and double bonds, and similar to that found experimentally in typical aromatic systems, for example, 1.396 Å in benzene. The perpendicular allyl cation, **14**, which is assumed to model reasonably the rotational transition structure, is 35 kcal mol^{-1} less stable according to the highest-level computations; this value is largely independent of the theoretical model [55]. There has been no direct experimental determination of the rotational barrier in the allyl cation, although an indirect estimate of 34 kcal mol^{-1} has been made [55]. The CC bond lengths in **14**, 1.318 and 1.444 Å, are clearly within the range for normal double and single bonds, respectively.

The 2-propenyl cation, **15**, is found to be the second-lowest-energy structure on

TABLE 7.20. Relative Energies of $C_3H_5^+$ Structures (kcal mol^{-1})^a

Structure (point group)	3-21G// 3-21G	6-31G*// 6-31G*	HF	MP2	MP3	MP4SDQ^b
			|	6-31G**//6-31G*		
13 Allyl (C_{2v})	0	0	0	0	0	0
14 Perpendicular allyl (C_s)	34	34	34	38	35	35
15 2-Propenyl (C_s)	15	17	16	14	13	12
16 1-Propenyl (C_s)	30	33	32	33	32	30
17 Cyclopropyl (C_{2v})	47	38	38	37	36	36
18 Corner-protonated cyclopropene (C_s)	44	43	43	31	33	34

^{a}Theoretical data from reference 44h.

^{b}Values correspond to estimates based on calculated fourth-order effects using the 6-31G* basis set.

the $C_3H_5^+$ potential surface. At the highest theoretical level employed, the energy of **15** is only 12 kcal mol^{-1} above that of the allyl cation, in good agreement with an experimental estimate of 11 kcal mol^{-1} [56]. Lower-level calculations suggest slightly larger separations. A normal-coordinate analysis (at 3-21G) shows that **15** is a local minimum on the $C_3H_5^+$ surface, in accord with experimental gas-phase evidence [57]. The 1-propenyl cation, **16**, is also found to be a local minimum at the 3-21G level. Its energy, about 30 kcal mol^{-1} above allyl, is also moderately insensitive to the level of computation. HF/6-31G* calculations suggest, however, a very small barrier for the 1,2-hydride shift converting the 1-propenyl cation, **16**, to the 2-propenyl cation, **15**. This barrier may disappear altogether at higher (correlated) levels of theory.

Woodward and Hoffmann [58] first aroused interest in the ring-opening of the cyclopropyl cation, **17**; their qualitative predictions were soon verified both by solvolysis [59] and super-acid [60] studies. At the highest level of theory examined, **17** lies 36 kcal mol^{-1} above the planar allyl cation, but only 1 kcal mol^{-1} above its perpendicular conformer, **14**. The cyclopropyl cation is not an energy minimum at 3-21G. It may, however, be a reasonable alternative to the perpendicular allyl cation, **14**, as a transition structure for the stereomutation of the planar allyl cation. This might occur via the symmetry-allowed disrotatory twisting of both methylene groups in concert, that is, **13** to **17** to **13'**.

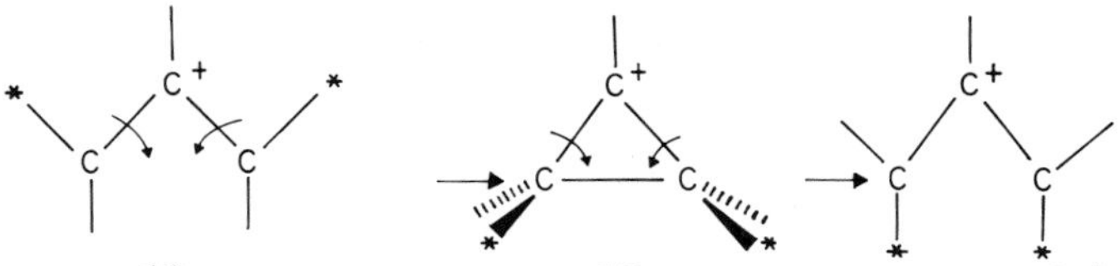

$$(7.38)$$

In the parent system, a stepwise route, in which the perpendicular allyl cation serves as a transition structure, that is, **13** to **14** to **13''** to **14'** to **13'**,

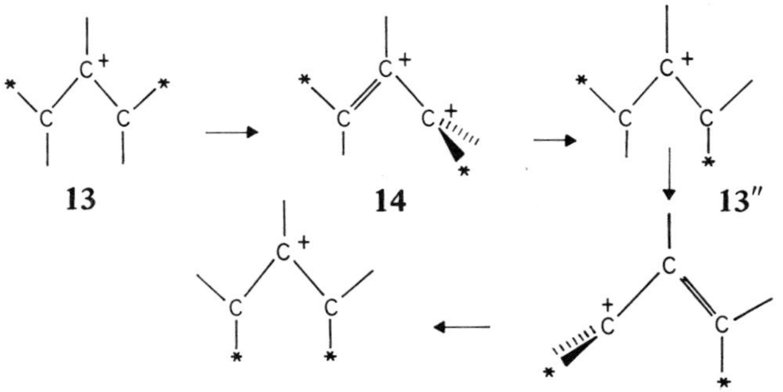

$$(7.39)$$

is favored by the calculations; experimental confirmation is unavailable. As seen by the data in Table 7.21, electron-donor substituents at the 2-position in the allyl

TABLE 7.21. Calculated Energies for Pathways to
Stereomutation of 2-Substituted Allyl Cations (kcal mol^{-1})

Substituent X	Energy[a]	
	Perpendicular Allyl	Cyclopropyl
H	35	39
CH$_3$	33	19
NH$_2$	42	-34
OH	36	-10
F	39	15

[a]Relative to planar allyl cation; STO-3G and 6-31G* data, see reference 61 for details.

cation change this preference [60, 61]. While none of the groups examined significantly alter the relative energies of planar and perpendicular allyl cations, all lead to preferential stabilization of the cyclopropyl (relative to planar allyl) cation. Substitution by methyl or fluorine reduces the (planar) allyl cation-cyclopropyl cation energy difference by about half. Stereomutation in these systems should occur via pathway (7.38). Substitution by NH$_2$ and OH groups is actually predicted to reverse the ordering of stabilities of **13** and **17**.

The 1-propenyl cation, **16**, and corner-protonated cyclopropene, **18**, might be viewed as the classical and nonclassical forms of methyl-substituted vinyl cation. At HF/6-31G**, the classical form, **16**, is 10 kcal mol^{-1} more stable than the bridged ion, **18**. The inclusion of correlation, however, favors the bridged, relative to the open, structure (see also Section 7.3.1b), and at the highest level, the classical form is favored by only 3 kcal mol^{-1}.

Either **16** or **18** would be an acceptable transition structure for a 1, 2-methyl shift resulting in carbon scrambling, that is,

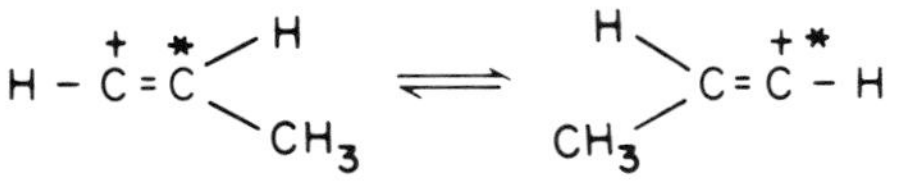

$$H - \overset{+}{C} = \overset{*}{C} \overset{H}{\underset{CH_3}{\diagdown}} \rightleftharpoons \overset{H}{\underset{CH_3}{\diagup}} C = \overset{+\;*}{C} - H \qquad (7.40)$$

Structures **15** and **16** might also be involved in the scrambling of hydrogens in the allyl cation. Conclusive experiments have yet to be reported.

(ii) C$_3$H$_7^+$. A number of C$_3$H$_7^+$ structures have been investigated theoretically using Hartree–Fock [42d, 43b] as well as CEPA [44f] and Møller–Plesset [44h] schemes. These cations include 2-propyl, **19**, 1-propyl, **20** and **21**, and corner- and edge-protonated cyclopropanes, **22** and **23**, respectively.

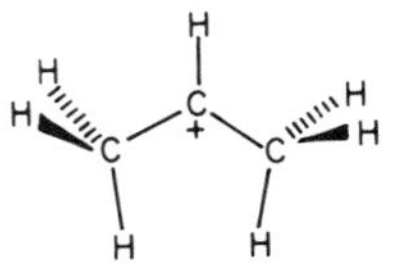

19 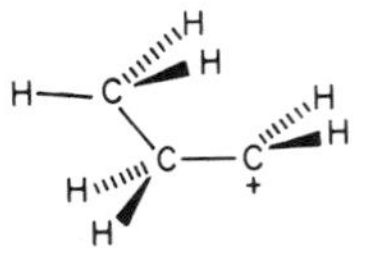   20 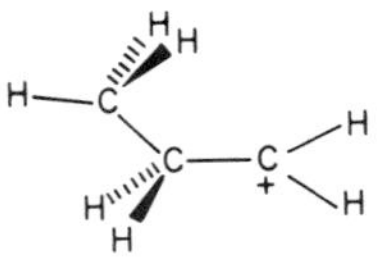   21

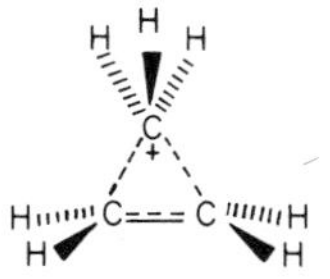

22 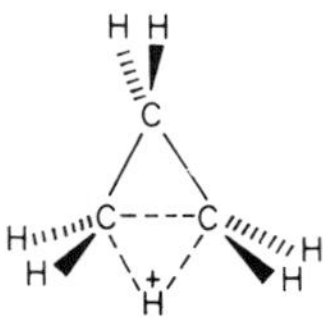   **23**

Calculated relative energies for four of these five forms are summarized in Table 7.22. At the HF/6-31G* level, the 1-propyl cation, **20**, is not a local minimum, and converts without activation to corner-protonated cyclopropane, **22**. Presumably, the alternative conformation of the 1-propyl cation, **21**, is also unstable at this level of theory; it has been artificially constrained from collapsing to **22** by an imposed symmetry plane.

The structure shown for the 2-propyl cation, **19**, may not be the true global minimum on the $C_3H_7^+$ surface. A normal-coordinate analysis performed using the 3-21G basis set reveals an imaginary frequency, corresponding to a distortion from C_{2v} to C_2 symmetry. However, the calculated energy lowering resulting from such a distortion is minute, and the C_{2v} structure, **19**, has been used for all higher-level calculations.

Protonated cyclopropane has long been implicated as an intermediate in electrophilic substitution, and in a variety of carbon scrambling processes [62]. In a classic experiment, Baird and Aboderin [63] passed cyclopropane into D_2SO_4; the recovered reactant was partially deuterated and the 1-propyl product had deuterium incorporated not just at C_3, but at C_1 and C_2 as well. There is also evidence for a second $C_3H_7^+$ species in the gas phase, with an energy approximately 8 kcal mol^{-1} above that of 2-propyl cation [64]. While the structure of this isomer is unknown, it is certain that it must be very short lived.

The calculated relative energies of $C_3H_7^+$ isomers are sensitive to theoretical level. The MP4SDQ/6-31G** estimates indicate corner-protonated cyclopropane, **22**, to be 8 kcal mol^{-1} higher in energy than 2-propyl cation. The corresponding edge-protonated species, **23**, is only slightly less stable at this level (9 kcal mol^{-1} above

TABLE 7.22. Relative Energies of $C_3H_7^+$ Structures (kcal mol^{-1})^a

Structure (point group)	3-21G// 3-21G	6-31G*// 6-31G*	HF	MP2	MP3	MP4SDQ^b
			/6-31G**//6-31G*			
19 2-Propyl (C_{2v})	0	0	0	0	0	0
20 1-Propyl (C_s)	18	19	19	20	20	20
22 Corner-protonated cyclopropane (C_{2v})	13	14	14	5	7	8
23 Edge-protonated cyclopropane (C_{2v})	27	19	17	5	8	9

^{a}Theoretical data from reference 44h.

^{b}Values correspond to estimates based on calculated fourth-order effects using the 6-31G* basis set.

19), although it does not appear to be an energy minimum on the 3-21G surface; a normal-coordinate analysis carried out at the 3-21G level reveals an imaginary frequency corresponding to displacement toward the corner-protonated form, **22**. This ordering of stabilities of edge- and corner-protonated cyclopropanes is different from that obtained by Lischka and Kohler [44f], who used an approximate CEPA method, a double-ζ + polarization basis set (roughly comparable to 6-31G**) but only STO-3G equilibrium geometries. We conclude that a rather flat surface connects these two isomers; in agreement with experiment, their interconversion is facile.

The energy of the (unstable) 1-propyl cation, **20**, is 20 kcal mol^{-1} above 2-propyl at MP4SDQ/6-31G**. Two experimental determinations of this difference exist [39b, 65]. The more recent, due to Beauchamp and Houle [65], is 24 kcal mol^{-1}, and is based on the photoelectron spectrum of 1-propyl radical; the older value [39b] is 16 kcal mol^{-1}.

(iii) $C_4H_7^+$, THE CYCLOPROPYLCARBINYL AND CYCLOBUTYL CATIONS. $C_5H_9^+$, THE METHYLCYCLOPROPYLCARBINYL AND METHYLCYCLOBUTYL CATIONS. The cyclopropylcarbinyl cation, **24**, is remarkable both for its unusual stability and for its ability to rearrange rapidly. While little is known about the gas-phase species, cyclopropylcarbinyl is one of the most thoroughly studied carbocations in solution, both as a reactive intermediate in nucleophilic solvents [66], and as a long-lived entity in super acids [67]. Several theoretical investigations have been carried out [68]. Many derivatives are known, for example, **25**, but it is the parent ion which offers both the experimentalist and the theoretician the most challenging interpretive problems.

24   **25**

Consider, for example, the unusual structure of the bisected cyclopropylcarbinyl cation, **24**. Two of the ring carbon-carbon linkages are predicted to be abnormally long (1.673 vs. 1.513 Å in cyclopropane at the 3-21G level) while the third bond is found to be very short (1.419 Å). These distortions arise as a result of stabilizing interaction of the vacant p-type function at the carbocation center with the antisymmetric component of the highest-occupied pair of valence (Walsh) cyclopropane orbitals [69].

"Antibonding", density removed, bond shortens.

"Bonding", density removed, bond lengthens.

Note, however, that such interaction can occur only in the bisected conformation of the cation, that is, structure **24**. The difference in energy between this form and the corresponding perpendicular structure, **26**, is 31 kcal mol^{-1} at the 6-31G*//3-21G level.

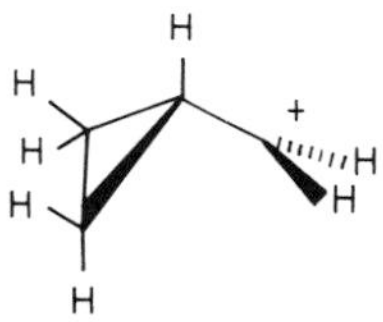

26

This provides an estimate of the stabilization of the carbocation center due to its interaction with the small ring. A rotational barrier of 13.7 kcal mol^{-1} has been determined by NMR spectroscopy for the tertiary dimethylcyclopropylcarbinyl cation, **25**, in super-acid media [70]. The corresponding calculated barrier, 18 kcal mol^{-1} at 3-21G, while larger than the experimental quantity, is much smaller than the barrier found in the parent compound, reflecting the reduced demand placed on the cyclopropane ring by the substituted cation center.

What is the structure of $C_4H_7^+$ in super-acid media, where NMR data show apparent threefold symmetry? No static structure seems likely! The tricyclobutonium ion, **27**, first considered (and ultimately rejected) by Roberts [71], is found to be very high in energy.

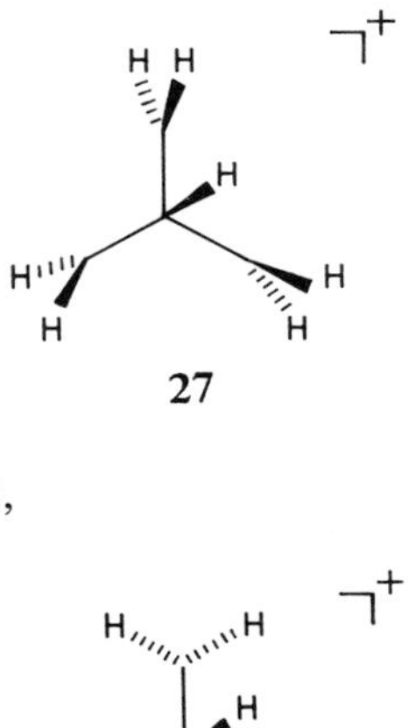

27

An alternative C_{3v} structure, **28**,

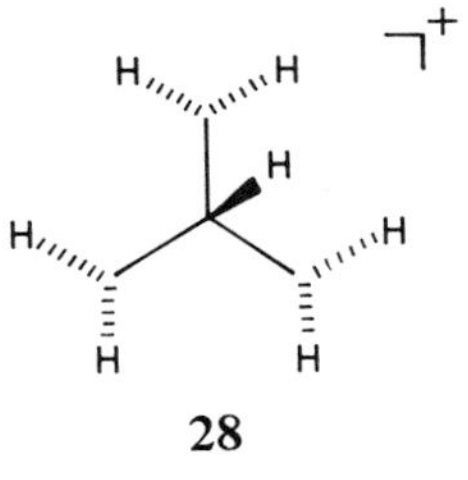

28

is not only unsatisfactory energetically (presumably because it forces the methylene hydrogens into close proximity) but, in addition, fails to account for the observed inequivalence of the methylene hydrogens H_a and H_b in **24**.

Rapid (on the NMR time scale) equilibration of lower-symmetry structures must be invoked in order to account for the apparent threefold symmetry of $C_4H_7^+$. What possibilities exist? In addition to three equivalent cyclopropylcarbinyl cations, **24**, equilibration might involve puckered cyclobutyl cations, **29**, or even structures intermediate between cyclopropylcarbinyl and cyclobutyl cations, completely lacking symmetry.

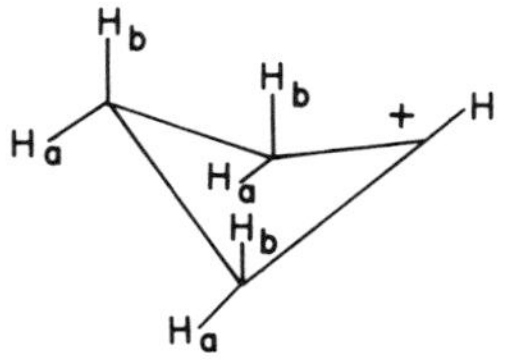

29

Equilibration involving *planar* cyclobutyl cations, **30**,

30

is clearly not acceptable; this possibility fails to account for the observed non-equivalence of the methylene hydrogens.

Note, however, that the methylene hydrogens in systems substituted at the 1-position by methyl or by phenyl are found to be equivalent in the NMR spectra [72]. Such substituents stabilize the localized positive charge in the classical planar cyclobutyl cation more than the delocalized charge in the alternative nonclassical ions.

According to both 3-21G and 6-31G* calculations, the $C_4H_7^+$ potential surface contains at least two minima, either or both of which might be consistent with experimental observations on the parent ion. The first of these is bisected cyclopropylcarbinyl, **24**, and the second, **31**, an ion of C_1 symmetry, the important structural parameters for which are shown below.

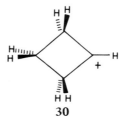

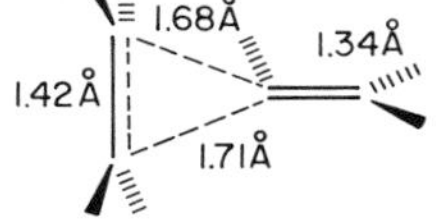

31

This resembles an *intramolecular complex* between vinyl cation and a double bond.

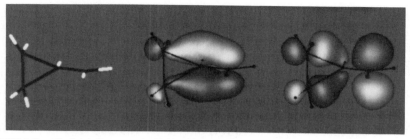

FIGURE 7.3. Skeletal framework (left), highest-occupied (middle) and lowest-unoccupied (right) molecular orbitals for C_1 symmetry form of $C_4H_7^+$. 6-31G*//6-31G*.

The highest-occupied and lowest-unoccupied molecular orbitals of this species (Figure 7.3) support such a description.

The two $C_4H_7^+$ minima are very close in energy at both levels of theory considered (31 is favored over 24 by 0.3 kcal mol^{-1} at 3-21G//3-21G, and by 0.1 kcal mol^{-1} at 6-31G*//6-31G*). Neither planar nor puckered cyclobutyl cations, 30 and 29, are minima on the 3-21G potential energy surface; their characterization on the 6-31G* surface remains to be established. These assignments are consistent with the available experimental data, which may be interpreted in terms of a rapid equilibration process among three equivalent forms of 24, or of 31, or both.

Methyl substitution in $C_4H_7^+$ yields a related series of interesting ions [67a, 68c, 72]. At the 3-21G level, both 1-methylcyclopropylcarbinyl cation, 32, and 1-methylcyclobutyl cation, 33, are found to be minima on the $C_5H_9^+$ potential surface.

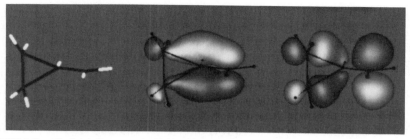

32   33

33 is indicated to prefer a planar carbon skeleton, and is 2.8 kcal mol^{-1} lower in energy than 32. Comparisons involving related systems suggest that this difference will decrease (perhaps leading to a reversal of stabilities) upon improvement of the basis set (see also Section 6.5.5). For example, at the 3-21G level, cyclobutane is 4.6 kcal mol^{-1} lower in energy than methylcyclopropane, while at 6-31G* methylcyclopropane is more stable by 0.5 kcal mol^{-1}. The latter result is in much better agreement with the experimental enthalpy difference (1.1 kcal mol^{-1} favoring methylcyclopropane). Similarly, the 3-21G calculations show an energy separation between cyclobutene and methylenecyclopropane of 17.9 kcal mol^{-1} (the latter is favored) compared with 12.5 kcal mol^{-1} at 6-31G* and 10.3 kcal mol^{-1} experimentally. The two potential minima are probably very close in energy, a result

which is consistent with the available proton and ^{13}C NMR data, which have been interpreted in terms of a rapid equilibration between the 1-methylcyclobutyl cation, **33**, and the three equivalent 1-methylcyclopropylcarbinyl cations, **32**; the equilibrium concentrations of all species are approximately the same [72b].

No minima corresponding to methyl-substituted **31** have been located. All forms explored collapse either to **32** or to **33**.

(iv) $C_4H_9^+$, THE SECONDARY BUTYL CATION. The secondary butyl cation is among the simplest systems on which extensive experimental [73] and theoretical [74] work has been carried out in order to establish the structure. While the issue is still controversial, it now appears that the available data may best be interpreted in terms of a picture in which both open (classical), **34**, and hydrogen-bridged (non-classical), **35**, structures are energy minima (of nearly equal stability) separated by a very small energy barrier.

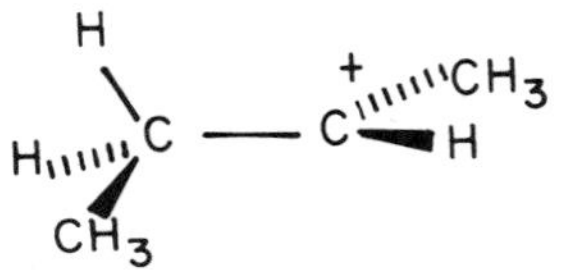

                            34 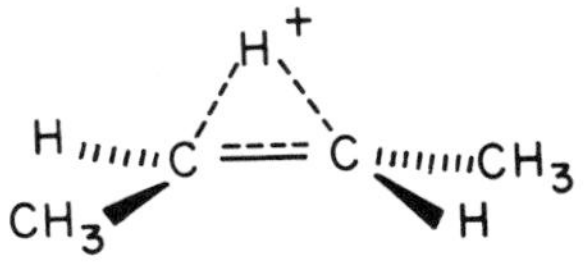   **35**

Table 7.23 lists relative energies for open and hydrogen-bridged forms of ethyl cation, **36** and **37**, secondary butyl cation, **34** and **35**, cyclopentyl cation, **38** and **39**, and 2,3-dimethylbutyl cation, **40** and **41**. These have been obtained from 3-21G// 3-21G calculations and, where available, from higher levels of theory and from experiment.

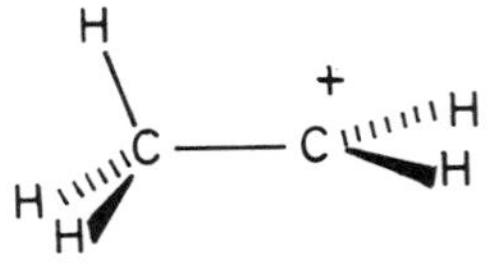

                            36 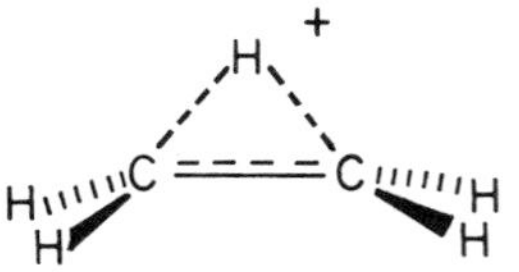   **37**

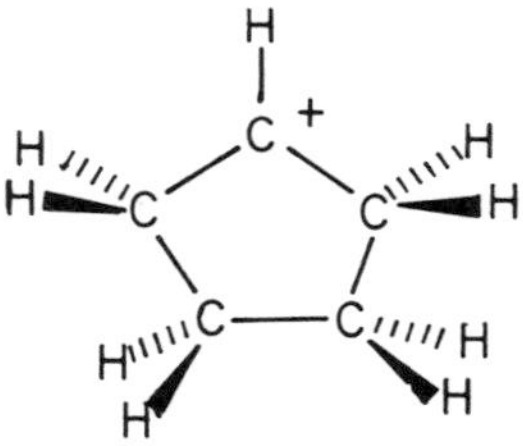

                            38 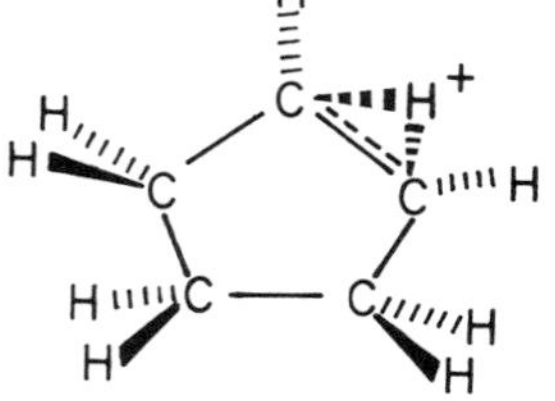   **39**

TABLE 7.23. Relative Stabilities of Open and Hydrogen-Bridged
Cations (kcal mol^{-1})^a

Cation	STO-3G// STO-3G	3-21G// 3-21G	6-31G*// 6-31G*	Best Theory	Expt.
Ethyl	11	8	1	-7^b	-3^c, -3 to -4^d
sec-Butyl	14	12	4	-1^e	
Cyclopentyl	15	12	6^f	0^g	$<2.8^h$
2, 3-Dimethylbutyl		14			3.4^i

^{a}A positive value indicates preference for an open structure over the hydrogen-bridged form. For a review of the experimental and theoretical literature, see: M. Saunders, J. Chandrasekhar and P. v. R. Schleyer, in *Rearrangements in Ground and Excited States*, vol. 1, P. de Mayo, ed., Academic Press, New York, 1980, p. 41 ff.
^{b}MP4SDQ//6-311G**//MP2/6-31G*. Constrained to structure **11**. See Section 7.3.1b.
^{c}Reference 52.
^{d}Reference 53.
^{e}MP3/6-31G**//6-31G*, K. Raghavachari and P. v. R. Schleyer, unpublished results.
^{f}6-31G*//3-21G.
^{g}MP3/6-31G**//3-21G, K. Raghavachari and P. v. R. Schleyer, unpublished results.
^{h}Structure (open or hydrogen-bridged) unknown. Rearrangement barrier <2.8 kcal mol^{-1}. Reference 75.
^{i}Reference 73e.

40   41

At the 3-21G level, all systems are indicated to prefer open geometries. The theoretical result (at this level) for ethyl cation is probably incorrect. As discussed in Section 7.3.1b, the best available calculations indicate the hydrogen-bridged geometry for this species to be favored over the open structure by 7 kcal mol^{-1}. The available (gas-phase) experimental evidence is also consistent with a bridged ground-state geometry and a classical ion some 3–4 kcal mol^{-1} higher in energy [52, 53]. The cyclopentyl cation [75] and 2,3-dimethylbutyl cation [73b,d,e, 76] have been established to prefer open structures in super-acid media and to equilibrate (presumably via hydrogen-bridged ions) with barriers of 3.4 and <2.8 kcal mol^{-1}, respectively. Given that the experimental barriers are lower than the 3-21G values by approximately 10 kcal mol^{-1}, and assuming that the calculated barrier in the secondary butyl cation is also too high by this amount, we are led to the interesting conclusion that, in this particular system, the open and bridged structures are of comparable stability. Higher-level calculations provide support. At 6-31G*//6-31G*, the hydrogen-bridged ion is only 4 kcal mol^{-1} above the open structure; MP3/6-31G**//6-31G* calculations suggest the opposite ordering of stabilities, but an energy separation of only 1 kcal mol^{-1}. Thus, the theory is in accord with experimental observation, although the required normal-coordinate analyses remain to be performed before definite comment can be made on the current interpretation of the experimental data in terms of a three-minimum potential surface.

(v) $C_6H_7^+$, PROTONATED BENZENE. Early work on the structure of protonated benzene [21] with the STO-3G minimal basis set has now been extended to calculations at the 3-21G level. The basic conclusions remain unaltered: (1) the preferred geometry of protonated benzene is the classical benzenium structure, **42**; (2) the calculated barrier to a 1,2-hydrogen shift, via the hydrogen-bridged structure, **43**, is significantly larger than that found in typical secondary or tertiary carbocations at the same level of theory, for example, 24 vs. 12 kcal mol^{-1} in secondary butyl cation and 14 kcal mol^{-1} in 2,3-dimethylbutyl cation; (3) an alternative face-protonated structure, **44**, is very high in energy (86 kcal mol^{-1} above **42** at the 3-21G//3-21G level).

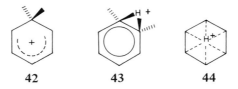

These conclusions are consistent with available experimental information for the parent ion and for simple derivatives [19]. As mentioned previously (Section 6.2.8d and Table 6.28), the crystal structure of the tetrachloroaluminate salt of the heptamethylbenzenium ion is nearly identical to the calculated (3-21G//3-21G) geometry of the parent (classical) ion. The calculated rearrangement barrier (24 kcal mol^{-1}) is significantly higher than that deduced from the temperature dependence of the NMR spectrum of the parent ion in super-acid media (8–10 kcal mol^{-1}) [77]. This is consistent with the tendency for 3-21G to overestimate hydrogen-shift barriers.

d. Carbodications. Doubly-charged cations are known experimentally, both in the gas phase [78] and in solution [79]. While no structural information is available from experiment. energy data for a number of small dications have recently been provided by the development of mass spectrometric charge-stripping techniques [80]. Because of strong coulomb repulsion, dications are prone to fragment into two singly-charged ions. This is observed. Note, however, that the existence of dications implies that these fragmentation processes occur only with substantial

barriers. Here we survey the present theoretical results on a number of small carbodications, in particular, the dications of methane [81], ethane [82], and cyclobutadiene [83], and investigate the dissociation of these species into singly-charged
species. The structure and unusually-high stability of the pyramidal form of the
benzene dication has already been addressed in Section 7.2.4c (see ref. 36 for isomers), and theoretical work on other small carbodications has appeared in the recent literature [84]. The subject of carbodications as *an emerging class of remarkable molecules* has been briefly reviewed [84f].

(i)   THE METHANE DICATION.   The methane dication [81], known experimentally in the gas phase [78c,d], is found to prefer a square-planar structure, **45**,
at HF/6-31G*, making it the simplest example of a molecule incorporating planar
tetracoordinate carbon (see also Section 7.4.1).

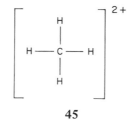

45

Like the isoelectronic methyl cation, it has no π electrons; its six σ electrons are
used to construct four CH bonds. Although **45** is an energy minimum, the deprotonation reaction (7.41),

$$CH_4^{2+} \longrightarrow CH_3^+ + H^+, \tag{7.41}$$

is found to be exothermic by 106 kcal mol^{-1} at the highest (MP4/6-311G**//6-
31G*) level (after corrections for zero-point energy). The kinetic existence of such
a thermodynamically unstable species depends on a substantial activation barrier
for the deprotonation process. This in turn results from balance between two op-
posing factors: the energy gained in formation of a new CH bond, and the coulombic repulsion of a proton and methyl cation. The 6-31G* potential surface yields a
transition structure, **46**, in which one bond is elongated.

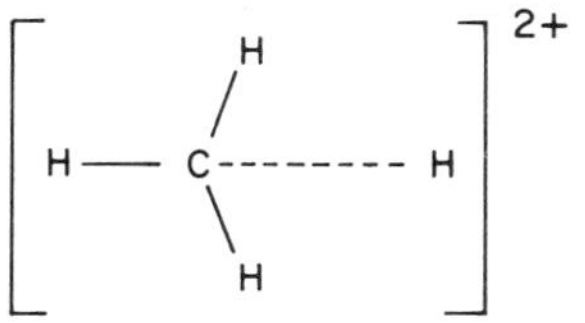

$$\left[\begin{array}{c} H \\ / \\ H \longrightarrow C \text{-------} H \\ \backslash \\ H \end{array} \right]^{2+}$$

46

A barrier to dissociation of 17 kcal mol^{-1} is found at the MP4/6-311G**//6-31G*
level.

(ii) THE ETHANE DICATION. The ethane dication [82], recently postulated in gas-phase experiments [78e, f], is also predicted to possess an unusual equilibrium structure. Although the doubly-bridged form, 47, analogous to the isoelectronic diborane molecule, is a local minimum on the 6-31G* potential surface, a lower energy is found for the open structure 48.

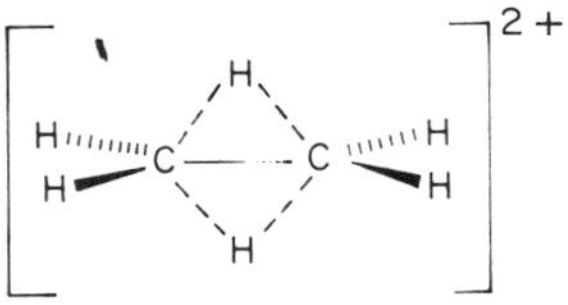

47 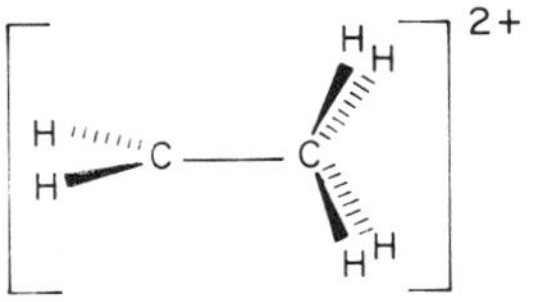   48

The latter may be regarded as a methyl cation substituted by a protonated methyl group, that is, a tricoordinate carbon (a *carbenium ion*) at one end and a pentacoordinate carbon (a *carbonium ion*) at the other end. According to 6-31G* calculations, 47 and 48 are separated by 16 kcal mol^{-1}; this gap is reduced to 9 kcal mol^{-1} at the MP4SDQ/6-31G**//6-31G* level. Another (and apparently more stable) geometry for ethane dication has been suggested on the basis of calculations by Lammertsma, Olah, Barzaghi, and Simonetta [82c]. This ion, 49, can be regarded as a complex between a hydrogen molecule and the perpendicular ethylene dication.

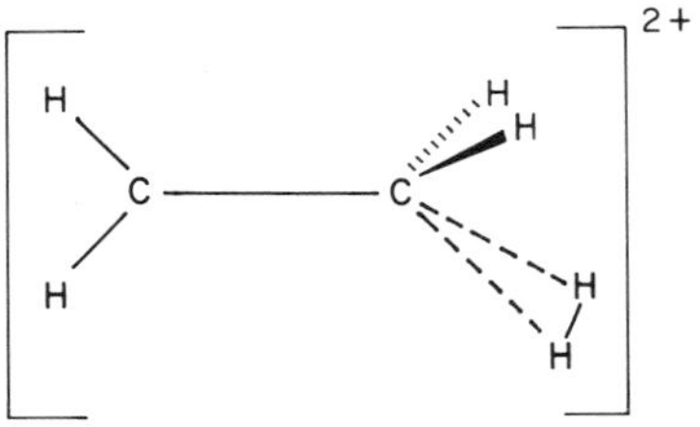

49

Complexation of two H_2 molecules to give $C_2H_8^{2+}$ (diprotonated ethane) has also been described by this group [82c].

Dissociation of 48 into two methyl cations is predicted to be exothermic by 105 kcal mol^{-1} at MP4SDQ/6-31G**//6-31G*, and appears first to involve the aforementioned bridged dication, 47, and then the transition structure 50.

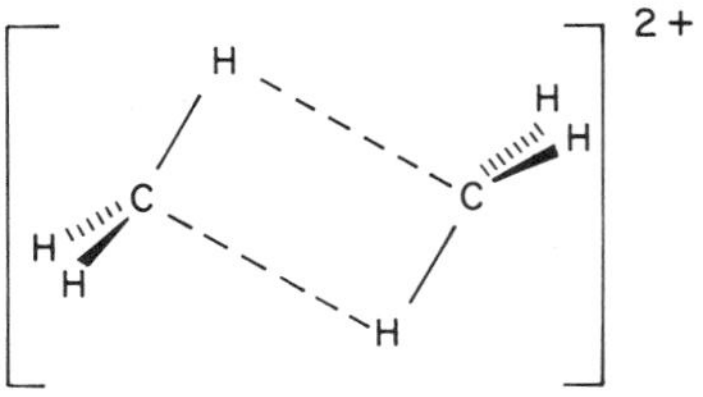

50

The latter closely resembles the transition structure calculated for dissociation of diborane into two boranes. The calculated activation energy for dissociation (27 kcal mol^{-1} at MP4SDQ//6-31G**//6-31G*) is sufficiently high that **48** or **49**, once formed, should be fairly long lived.

(iii)   THE CYCLOBUTADIENE DICATION.   According to Hückel theory, annulene systems with $4n + 2$ π electrons are aromatic; in order to maximize favorable π-orbital overlap, they should be planar. Thus, the cyclobutadiene dication [83] has long been *assumed* to be planar, that is, **51**.

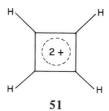

51

A planar skeleton also minimizes angle strain; puckering of a four-membered ring leads to a further decrease in bond angles below ideal tetrahedral values. Finally, transannular repulsions are minimized in a planar geometry; significant puckering might bring opposite carbons (separated by about 2 Å in the planar ring) close enough to interact unfavorably.

Surprisingly, these clear-cut predictions of the qualitative model are not supported by quantitative calculations. The puckered cyclobutadiene dication, **52**, with hydrogens pointing in axial directions, is found to be more stable than the planar form, **51**.

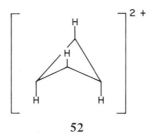

52

Calculations also indicate that alkyl-substituted cyclobutadiene dications prefer puckered structures. Since a number of species of this type are accessible in solution [79], experimental verification of this surprising prediction is possible, at least in principle.

What is the reason for the unexpected preference for **52**? Hückel theory is inappropriate for small doubly-charged species [85]. Chemists tend to emphasize the importance of π effects; when σ and π effects act in opposite directions, it is usually taken for granted that the latter will dominate. The situation in this case may be the reverse. Two factors contribute to the unimportance of π effects in dictating the geometry of cyclobutadiene dication. First, the energy of the π system changes only slightly with (modest) skeletal puckering; the twisting of p orbitals is largely compensated by decreased CC bond distance and π overlap remains essentially constant. More importantly, the resonance stabilization of the cyclobutadiene dication appears to be quite small. σ effects dictate the preferred geometry of this species. Repulsion of the transannular carbons in **51** arises because the valence (Walsh) orbitals of a four-membered ring are 1,3-antibonding [69]. Puckering relieves the unfavorable interaction by skewing the orbitals such that the components of opposite sign overlap less. In short, charge repulsion effects dominate; aromaticity is less important.

(iv) DIHYDRODIBORETE AND OTHER ANALOGUES OF THE CYCLOBUTADIENE DICATION. The discovery of dihydrodiborete followed the observation that the cyclobutadiene dication, which was always assumed to be planar, in fact adopts a nonplanar geometry (see discussion in preceding section). Calculations on the isoelectronic dihydrodiboretes also reveal interesting structures [86]. At the 3-21G level, a puckered geometry, **53a**, with an unusual central CC distance of 1.88 Å (midway between normal bonding and nonbonding contact distances), is favored over the classical structure, **53b** (like bicyclo[1.1.0]butane in which the central CC bond length is only 1.50 Å), the planar Hückel aromatic, **53c**, and all reasonable 1,2-isomers.

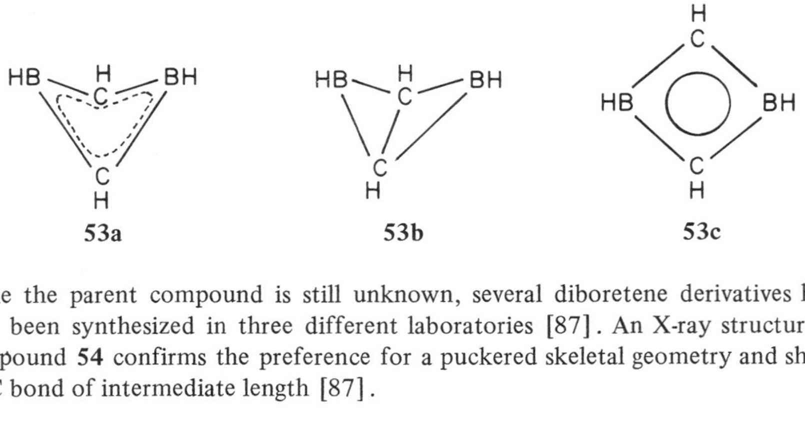

53a	**53b**	**53c**

While the parent compound is still unknown, several diboretene derivatives have now been synthesized in three different laboratories [87]. An X-ray structure of compound **54** confirms the preference for a puckered skeletal geometry and shows a CC bond of intermediate length [87].

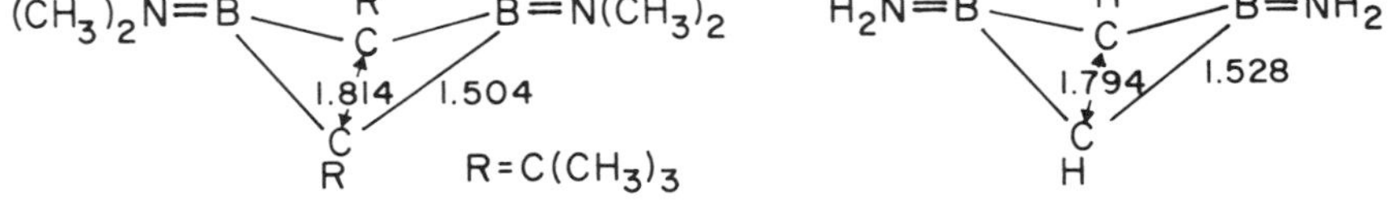

54	**55**

The calculated (3-21G) geometry of a close analogue, **55** (important parameters for which are also given above) is very similar [88]. Both the intermediate CC contact distance and the short BN bond are well reproduced by the theory, as is the puckering angle of the four-membered ring.

Bicyclobutane-2,4-dione, isoelectronic with **55**, is also predicted by the 3-21G calculations to adopt a folded structure, **56a**, in preference both to a classical puckered geometry, **56b**, and a planar Hückel arrangement, **56c**.

| **56a** | **56b** | **56c** |

The central CC length in **56a** is predicted to be 1.76 Å, midway between what would normally be considered appropriate for a single bond and that for a nonbonded contact. The relative instability of **56c** (13 kcal mol^{-1} above **56a** at the 3-21G level) warrants further comment. This appears to be a general phenomenon. Calculated energy differences between planar and puckered forms of cyclobutadiene dication [83a,c] and dihydrodiborete [86] and also significant, and the barrier to inversion in **55** through a planar structure requires 18 kcal mol^{-1} [88]. While these energy differences may need to be refined by higher-level calculations, there seems little doubt that the basic conclusion relating to the inherent instability of *planar* 2π-electron, four-membered rings will remain. The puckered systems are aromatic molecules. While moderate folding does little to affect the delocalization of π electrons, it does significantly reduce the unfavorable 1,3 interactions · found in planar four-membered rings.

The theoretical results on **56** are pertinent for derivatives of squaric acid (the squaraines [89]), some of which may also have nonplanar structures provided good π-donor substituents are not present. When groups with C=N or C=C double bonds [90a] are substituted at C_2 and C_4 in bicyclobutane, open structures with elongated central CC bonds may again result [90b,c].

7.3.2.   Carbanions and Polar Organometallics

Carbanions, perhaps synthetically the most useful of reactive intermediates [91], are seldom free ·of counterion influences. Nevertheless, it is common to equate organolithium, Grignard and other metal-containing reagents, RM, with their equivalent anions, R$^-$. To what extent is this association valid? How does the metal influence the properties of anionic systems? Answers to questions such as these may be difficult to obtain experimentally. The situation is often clouded by the instability of the free carbanion or by solvation and aggregation of the organometallics. Theoretical calculations provide an alternative and viable approach.

Here, we examine in detail both the structural and energetic similarities and differences between free carbanions and a variety of organometallic equivalents. We first deal with methyllithium and other alkali, alkaline-earth and transition-metal analogues of methyl anion. A more limited set of organometallic equivalents, specifically, RLi, RNa and RMgCl, are discussed for larger anions R$^-$.

Experimental structures for a number of simple anions are available from X-ray data. Comparisons with theoretical geometries have already been made (Section 6.2.8c). The available experimental structures of polar organometallics and, in particular, lithium compounds are often fascinating [92]. Alkyllithiums are tetrameric or hexameric, and often incorporate hexacoordinate or even heptacoordinate carbon [93]. Often the lithium atoms are not located where expected. The lithium atom in benzyllithium, for example, does not occupy the center of the π cloud of the benzene ring, but rather bridges the *ortho* and *alpha* positions [94]. Some of these structures will be discussed in this section, while others will be dealt with later in our treatment of the violation of conventional structure rules (Section 7.4) and of molecules with several metals (Section 7.5).

a. Methyllithium and Other Analogues of Methyl Anion. Monomeric methyllithium has not been investigated extensively since its initial observation under matrix-isolation conditions [95]. Calculations clearly provide the best method at present for the systematic study of this species. We provide here a current survey of theoretical results for methyllithium and its dimer, trimer and tetramer, as well as a brief discussion of data for related organometallic anion equivalents.

Table 7.24 summarizes calculated structures, dipole moments and Mulliken charges for methyllithium at various theoretical levels. Data for methyllithium dimer, trimer and tetramer are also provided.

There is little variation in calculated structural parameters with theoretical level. All models predict that the HCH bond angles in methyllithium are smaller than those in other first-row CH_3X compounds. As shown in Figure 7.4 (6-31G* data), the average HCH bond angle in these systems increases monotonically across the first row, and (roughly) parallels the electronegativity of the heavy atom bonded to methyl. Simply interpreted, methyl anion (the limiting structure for methyllithium) prefers to be highly puckered, while methyl cation (the limiting structure for methyl fluoride) prefers to be planar.

The dipole moments at higher levels of theory are in good agreement with the experimental estimate of 6 debyes [93]. The calculated Mulliken populations vary significantly with theoretical level. Both 3-21G and 6-31G* calculations depict much greater charge separation than suggested by the STO-3G model.

What is the nature of carbon–lithium bonding? This central question is still under active debate. An unambiguous and generally acceptable definition of *covalent* and *ionic* is hard to achieve. While the large dipole moment of methyllithium is an indication of its ionic character, many lithium compounds do not exhibit salt-like behavior. For example, alkyllithium aggregates are soluble in hydrocarbons; in their pure form, many are liquids or are low-melting solids (methyllithium is an exception) [93]. The existence of polylithiated compounds such as CLi_4 also argues against completely ionic formulations. That is, $C^{4-} \cdots 4Li^+$ seems intu-

TABLE 7.24. Structures, Mulliken Charges, and Dipole Moments for Methyllithium, and Its Dimer, Trimer, and Tetramer.

Level	$r(CLi)^a$ (Å)	$r(CH)^a$ (Å)	$\langle HCH \rangle^a$ (degrees)	Mulliken Charges (electrons)			Dipole Moments (debyes)
				C	Li	H	
Methyllithium (C_{3v})							
STO-3G	2.008	1.083	106.2	−0.242	+0.158	+0.028	4.27
3-21G	2.001	1.094	107.0	−0.905	+0.520	+0.142	5.50
6-31G*	2.001	1.093	106.2	−0.781	+0.574	+0.119	5.72
MP2/6-31G*	2.003	1.099	106.8				
Methyllithium Dimer (C_{2h})							
STO-3G	2.127	1.090	105.7				0
3-21G	2.150	1.099	106.5	−1.027	+0.540	+0.162	0
Methyllithium Trimer (C_{3h})							
STO-3G	2.090	1.092	105.6	−0.301	+0.179	+0.041	0
3-21G	2.116	1.099	106.7	−1.035	+0.495	+0.180	0
Methyllithium Tetramer (T_d)							
STO-3G	2.189	1.097	102.2	−0.311	+0.193	+0.039	0
3-21G	2.236	1.102	103.9				0

^{a} Average values for dimer and trimer structures.

403

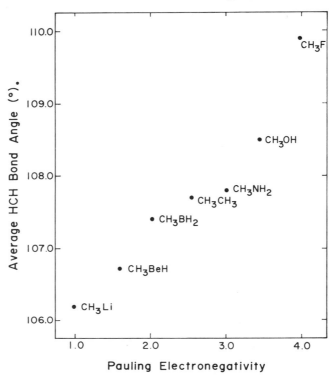

FIGURE 7.4. Correlation of average HCH bond angles in molecules CH_3X vs. Pauling electronegativity. 6-31G* data.

itively unlikely. X-ray electron-density difference maps, obtained by Amstutz, Dunitz, and Seebach for alkyllithiums, do not provide definitive evidence for the description of these compounds in terms of either covalent or ionic models [96].

The available theoretical evidence is also mixed. Mulliken-type analyses suggest significant charge on lithium in simple compounds, although they do not support description in terms of a completely ionic picture. As previously mentioned, however, the extent of charge separation indicated by the Mulliken procedure is sensitive to the theoretical level (see Section 6.6.2 and Table 7.24). Other methods for assigning charges to atoms have often led to different conclusions [97–100]. According to Hinchliffe and Saunders [97], "the C–Li bond in methyllithium is covalent in character, since the overlap density is fairly equally shared by C and Li. The atoms, however, carry substantial charges corresponding to C, −0.215; H, −0.141; Li + 0.638." In contrast, Streitwieser and Collins [98], who employed an electron projection function to evaluate charge density and covalency, concluded that methyllithium "by our criteria is wholly ionic and has large charge transfer (ca. 0.7 electron)." Graham, Marynick and Lipscomb [99] have examined the nature of the C–Li bond by means of electron-density difference maps at various theoretical levels. They concluded, "methyllithium appears to be about 60% ionic, but it clearly is far from the limit of complete ionicity expected for molecules such as LiF." Work by Francl, Hout, and Hehre [100] further suggests that the bonding in methyllithium is largely ionic, although not to quite the extent of that in inorganic lithium compounds, e.g., LiF, LiOH, and LiCl. These authors have fitted calculated electron-density surfaces (6-31G*//6-31G* level) for a variety of lithium compounds to a collection of nuclear-centered spheres. The resulting best-sphere radii for lithium in its compounds (Table 7.25) are nearly identical to that for free Li$^+$. Similarly, calculated radii for the remaining atoms, for example, the carbon and hydrogens in methyllithium, quite closely approach those in the corresponding free anions, for example, methyl anion.

Lithium compounds exhibit a strong tendency to aggregate. Indeed, it is difficult to observe monomeric species even in strongly solvating media. The calculated dimerization energy of methyllithium is 46 kcal mol^{-1} (3-21G//3-21G); the corresponding trimerization and tetramerization energies are 90 and 138 kcal mol^{-1} respectively. Thus, the average carbon-lithium bond dissociation energy in methyllithium tetramer is approximately 35 kcal mol^{-1} greater than that in the monomeric species (45 kcal mol^{-1} according to the highest-level calculations

TABLE 7.25. Comparison of Best-Sphere-Fit Radii for Atoms in Lithium Compounds, LiX, with those in Li$^+$ and in the Corresponding Anions, X$^-$ (6-31G*//6-31G*)^a

Molecule LiX	Atom	Radii in LiX	Radii in X$^-$
Li$^+$	Li	0.862	
Li$^{\cdot}$	Li	1.759	
LiH	Li	0.962	
	H	1.711	1.570
LiCH$_3$	Li	0.957	
	C	2.015	1.944
	H	1.274	1.351
LiNH$_2$	Li	0.927	
	N	1.918	1.843
	H	1.184	1.302
LiOH	Li	0.928	
	O	1.756	1.701
	H	1.097	1.265
LiF	Li	0.929	
	F	1.588	1.533
LiCl	Li	0.948	
	Cl	2.024	1.992

^{a}Theoretical data from reference 100 and W. J. Hehre, unpublished calculations.

TABLE 7.26. Structures and Mulliken Charges of Incorporated Methyl Group in Organometallic Analogues of Methyl Anion[a]

Molecule	STO-3G//STO-3G			3-21G//3-21G[b]		
	r(CH)	<(HCH)	Q(CH$_3$)[c]	r(CH)	<(HCH)	Q(CH$_3$)[c]
CH$_3$Li	1.083	106.2	−0.16	1.094	107.0	−0.52
CH$_3$Na	1.091	104.3	−0.67	1.091	108.3	−0.48
CH$_3$K	1.085	107.1	−0.35			
CH$_3$Rb	1.090	104.9	−0.60			
CH$_3$Cu	1.083	107.0	−0.24			
CH$_3$BeCl	1.084	107.3	−0.11	1.090	107.4	−0.34
CH$_3$MgCl	1.084	106.3	−0.43	1.089	107.8	−0.38
CH$_3$ZnCl	1.083	105.9	−0.40			
CH$_3^-$	1.117	99.8	−1.00	1.121	103.2	−1.00

[a] Bond lengths in angstroms, bond angles in degrees. Theoretical data from W. J. Hehre and K. D. Dobbs, unpublished calculations.
[b] 3-21G$^{(*)}$//3-21G$^{(*)}$ for molecules incorporating second-row elements.
[c] Mulliken charge on methyl group (electrons).

carried out to date [101]). This additional stabilization provides the incentive for aggregation.

Calculated equilibrium geometries and Mulliken atomic charges for other organo-metallic analogues of methyl anion are provided in Table 7.26. Data for methyllithium and for methyl anion are also included for comparison. The local geometry of the methyl group in all the organometallics considered is nearly the same. STO-3G CH bond lengths in methyl sodium and methyl rubidium are marginally longer than those in the remaining compounds, and HCH bond angles slightly smaller; these effects are indicative of increased anionic character. Calculated (STO-3G) Mulliken charges on methyl in CH$_3$Na and CH$_3$Rb are also somewhat larger than those in the other molecules. Note, however, that STO-3G Mulliken populations generally provide an unrealistic account of charge distributions in highly polar compounds, see Table 7.24. Indeed, methyl group (Mulliken) charges from 3-21G (3-21G$^{(*)}$) calculations for methyllithium and methyl sodium are nearly identical, as are those for CH$_3$BeCl and CH$_3$MgCl. The theory at this level indicates less ionic character for CH$_3$BeCl and CH$_3$MgCl than for the alkali-metal systems. This situation is examined in greater detail in the next section.

b. Alkali Metal and Alkaline Earth Analogues of Larger Carbanions. Here we examine the similarities and differences between free carbanions and alkali metal and alkaline earth equivalents. A measure of the influence of structure on anion stability is given by the energies of reaction (7.42).

$$R^- + CH_4 \rightarrow RH + CH_3^-. \tag{7.42}$$

The energies for the analogous reaction for lithium compounds, (7.43),

$$RLi + CH_4 \longrightarrow RH + CH_3Li, \tag{7.43}$$

provides a basis for comparison. Theoretical data (both 3-21G//3-21G and 3-21+G//3-21G) for the two series of reactions are provided in Table 7.27 for a variety of inorganic and organic R groups. As noted in Section 6.5.8, diffuse functions have a significant effect on the properties of anions, and good agreement with (gas-phase) experimental data, for example, for relative acid strengths, is realized only after they are included. Note, however, that the calculated energies for reactions (7.43) involving lithium compounds are generally unaffected by the addition of diffuse functions to the basis set. This difference does not obviously follow from comparisons of total electron density surfaces of free anions and of their organolithium equivalents. For example, those for methyl anion and methyllithium (Figure 7.5) are nearly identical. Only for lithium amide are the theoretical energies altered significantly by addition of diffuse functions (in the same direction and by about the same amount as for NH_2^-). The amino group is the best π donor, and evidently its stabilizing interaction with an empty p orbital on lithium is increased by the availability of diffuse functions. Noting this exception, the small effect of diffuse functions in these systems is a further indication that calculations at this level do not point to a limiting description of lithium compounds as separated ion pairs (see also the discussion in Section 7.3.2a).

Also included in Table 7.27 are stabilization energies for the corresponding sodium and magnesium chloride (Grignard) systems.

$$RNa + CH_4 \longrightarrow RH + CH_3Na \tag{7.44}$$

$$RMgCl + CH_4 \longrightarrow RH + CH_3MgCl \tag{7.45}$$

Again, both $3\text{-}21G^{(*)}//3\text{-}21G^{(*)}$ and $3\text{-}21+G^{(*)}//3\text{-}21G^{(*)}$ data have been tabulated. Significant differences between the two theoretical levels are found for sodium amide, sodium hydroxide and sodium fluoride as well as for fluoromagnesium chloride. Here, as in lithium amide, the availability of diffuse functions apparently increases the extent of π donation by the heteroatom lone pairs into vacant p functions of the metal. The large effect noted in sodium cyanide must be of a different origin; the cyano group is a π acceptor and not a donor. The cyanide anion is also the most stable of the systems considered. It is not unreasonable that its organometallic analogues will resemble more closely a separated ion pair than will analogues of less-stable anions.

As shown in Figure 7.6, the calculated (3-21+G//3-21G) stabilities of organolithium compounds closely parallel those of the corresponding anions. Lithium chloride and lithium azide also fit the correlation line, although the theoretical data for lithium amide, lithium hydroxide, lithium fluoride and lithium formate do not [102]. The stability of these systems (relative to methyllithium) is greater

TABLE 7.27. Stabilization Energies for Anions and their Lithium, Sodium and Magnesium Chloride Analogues[a]

R	$R^- + CH_4 \rightarrow RH + CH_3^-$		$RLi + CH_4 \rightarrow RH + CH_3\,Li$		$RNa + CH_4 \rightarrow RH + CH_3\,Na$		$RMgCl + CH_4 \rightarrow RH + CH_3\,MgCl$	
	3-21G	3-21+G	3-21G	3-21+G	3-21G(*)	3-21+G(*)	3-21G(*)	3-21+G(*)
CH_2CH_3	2	−4	−4	−5	−4	−4	−4	−4
CH_3	0	0	0	0	0	0	0	0
▽	14	3	−1	−1	−1	0	−1	
$CH{=}CH_2$	21	14	6	6	6		3	2
NH_2	1	15	29	41	20	26	27	26
(phenyl)	32		6					
$CH_2CH{=}CH_2$	37	30	18	16	16	16	3	
OH	13	42	58	57	47	58	54	51
$C{\equiv}CCH_3$	53	45	33					
$C{\equiv}CH$	54	52	35	35	37		26	
F	31	75	74	74	68	75	74	63
(cyclopentadienyl)	81		65		62			
$C{\equiv}N$	84	83	49	50	53	61	33	34
N_3		89		64				
$COOH$		90	83	80	84	88		
Cl	123	109	79	78	83	82	59	57

[a] kcal mol^{-1}. Data from W. J. Hehre, unpublished calculations, and C. Rohde, Diplomarbeit, Erlangen, 1982.

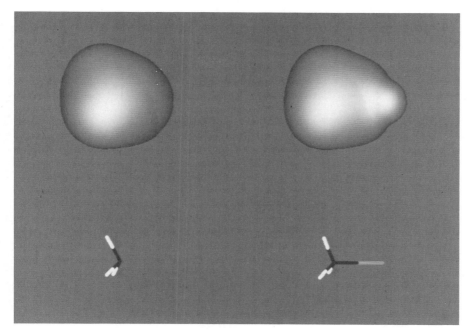

FIGURE 7.5. Total electron density surfaces for methyl anion (left) and methyllithium (right). Skeletal frameworks are underneath. 3-21+G//3-21G.

than parallels with corresponding anions would indicate. We suspect that this additional stabilization is due to the delocalization of the π-type lone pairs on the heteroatoms into the low-lying vacant p orbitals on lithium. More generally, it might be stated that the noted deviations are a result of inherent differences in LiX bond energies, that is, the relative ion-pair energies are attenuated from those of the free anions because of the presence of the (lithium) counterion. This suggests that the set of carbon-based compounds should give one correlation line, the set of oxygen systems another line, and so forth. This is apparently the case. The relative stabilities for the carbon-based lithium compounds shown in Figure 7.6 clearly fit a single line; data for a series of oxygen-based anions (including formate) are shown in Figure 7.7, and also correlate linearly with those for the corresponding lithium compounds [103]. The slopes of the two correlation lines are similar, but the origins have shifted. The fact that the data for N_3^- and Cl^- lie near the correlation line for carbon systems in Figure 7.6 is probably fortuitous.

The slope of the least-squares line for the carbon-based systems shown in Figure 7.6 is greater than unity, indicating that the anions are generally more sensitive to structural variations than are corresponding lithium analogues. This is consistent with the notion that the organolithium compounds are not adequately represented as fully separated ion pairs.

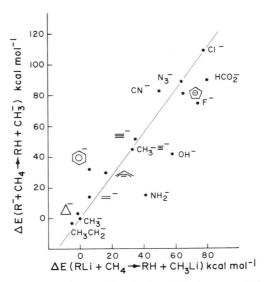

FIGURE 7.6. Correlation of stabilities of carbanions (relative to methyl anion) with those of the analogous lithium compounds (relative to methyllithium). 3-21+G//3-21G data.

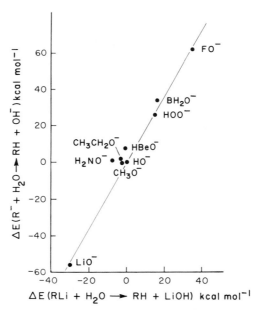

FIGURE 7.7. Correlation of stabilities of oxygen-based anions (relative to hydroxide anion) with those of the analogous lithium compounds (relative to lithium hydroxide). 3-21+G//3-21G data.

410

Relative energies for sodium and magnesium chloride (Grignard) compounds, from reactions (7.44) and (7.45), are plotted against the corresponding anion stabilities, from reaction (7.42), in Figures 7.8 and 7.9, respectively. Structural effects on the sodium compounds very closely parallel those already noted for the lithium systems, that is, the slope of the least-squares line (again excluding data for non-carbon-centered systems) is nearly identical to that for lithium compounds. Data for the Grignard reagents also parallel those for the corresponding lithium (and sodium) systems. The least-squares line is steeper, indicating somewhat less sensitivity to structural variations.

As previously detailed (Section 6.5.8), the calculations are moderately successful in reproducing experimental gas-phase relative acidities once diffuse-type functions have been included in the basis set. Do the theoretical relative acidities (as well as the relative stabilities of the corresponding alkali metal and Grignard compounds) correlate with acidity data in solution, where carbanions are commonly employed synthetically? Table 7.28 lists aqueous-phase pK_a values obtained from the MSAD (McEwen, Streitwieser, Applequist, Dessey) scale [104] for a series of compounds on which calculations have been performed. Also listed are the corresponding free energies (2.3 RT ΔpK_a at T = 293 K) relative to that in methane. The latter quantities are plotted against the corresponding gas-phase acidities, reactions (7.42),

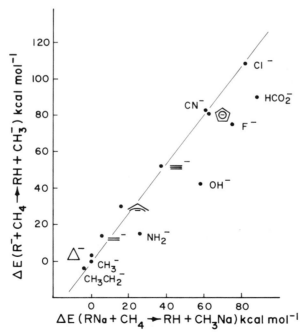

FIGURE 7.8. Correlation of stabilities of carbanions (relative to methyl anion) with those of the analogous sodium compounds (relative to methyl sodium). 3-21+G//3-21G (3-21+G(*)//3-21G(*)) data.

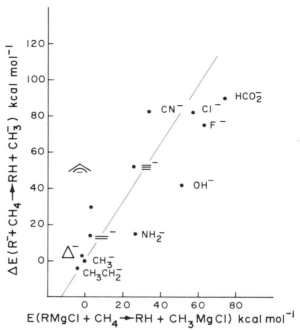

FIGURE 7.9. Correlation of stabilities of carbanions (relative to methyl anion) with those of the analogous Grignard (MgCl) compounds (relative to methyl magnesium chloride). 3-21+G// 3-21G (3-21+G(*)//3-21G(*)) data.

using 3-21+G//3-21G data, in Figure 7.10, and against the corresponding lithium stabilization energies, reactions (7.43), in Figure 7.11. Comparisons for sodium compounds and for Grignard reagents are not plotted, but would show broadly similar features. Both sets of theoretical data show rough correlation with the solution-phase acidities. The slope of the best correlation line (not drawn) for the lithium compounds is closer to unity than that for the free anions. These systems (as well as the corresponding sodium and Grignard compounds) may in fact provide better models for the properties of carbanions in solution than data for free anions!

The structures of alkali metal and alkaline earth compounds are often unusual (from a conventional bonding point of view), and will be discussed more generally in Sections 7.4 and 7.5. The preferred geometry of the lithium (cation) allyl (anion) system, like those of lithium benzene and lithium cyclopentadienyl (already discussed in Section 7.2.4c), places the metal above the π system.

TABLE 7.28. Experimental Solution-Phase
Stabilization Energies for Anions

Anion	$pK_a{}^a$	$\dfrac{2.3\,RT}{\Delta pK_a{}^b}$
$CH_3CH_2^-$	42	2.7
CH_3^-	40	0.0
$\overline{CH_2CH_2CH}^-$	39	−1.3
$C_6H_5^-$	37	−4.0
$CH_2{=}CH^-$	36.5	−4.7
$CH_2{=}CH{-}CH_2^-$	35.5	−6.0
NH_2^-	34	−8.0
HCC^-	25	−20.1
$C_5H_5^-$	16	−32.2
OH^-	15.7	−32.6
F^-	3.2	−49.0
Cl^-	−7	−63.0

[a] From: J. March, *Advanced Organic Chemistry, Reactions, Mechanisms and Structure*, 2nd ed., McGraw Hill, New York, 1977, p. 227–230.
[b] Relative to methane, kcal mol^{-1}.

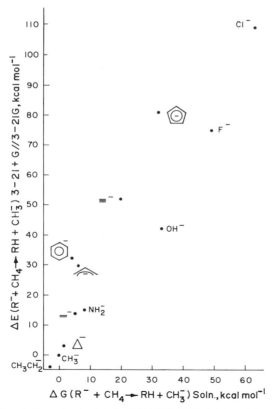

FIGURE 7.10. Correlation of calculated (3-21+G//3-21G data) gas-phase and experimental solution-phase acidities (relative to the acidity of methane). Experimental data from reference 104.

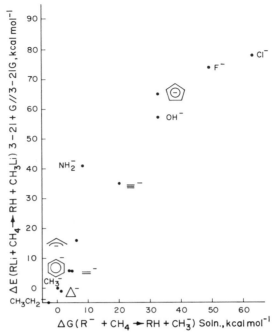

FIGURE 7.11. Correlation of calculated (3-21+G//3-21G data) stabilities of lithium compounds (relative to methyllithium) and experimental solution-phase acidities (relative to the acidity of methane). Experimental data from reference 104.

Calculated geometries for the analogous sodium and magnesium chloride systems are similar. The optimum structure for lithium formate (and its sodium and magnesium chloride analogues) has the metal in plane and bridging both oxygens.

The geometries for the remaining systems considered here are broadly similar to those exhibited by normal compounds, although some differences are evident, for example, lithium hydroxide is linear whereas water is bent.

7.3.3. Carbenoids

Carbenoid compounds [105], R_2CMX (M = metal, X = halogen), may often undergo α-elimination to yield singlet carbenes, R_2C: [106]. Nevertheless, carbenoids are distinct entities, which show their own distinct reactivity patterns. ^{13}C NMR investigations at low temperature sometimes demonstrate the existence of isomers, for example, for CBr_3Li [107], even though only a single (tetrahedral) geometry is to be expected from conventional bonding considerations.

What are the structures of carbenoids? A number of systems have now been investigated theoretically [108], including CH_2MX for all combinations of the alkali metals lithium and sodium, and the halogens fluorine and chlorine, as well as CHF_2Li. A silicon analogue, SiH_2LiF, has also been considered [109]. Computational results for CH_2LiF [108i] are typical.

Consider the interaction of the filled lone pair and vacant p orbitals on singlet methylene (the HOMO and LUMO, respectively) with Li^+F^-. Lithium will seek the filled and fluorine the empty methylene orbitals. This gives rise to the three (6-31G*) geometries displayed in Figure 7.12. In the first, 57, the lithium end of LiF engages the methylene lone pair. The calculated geometry shows a linear arrangement of the two fragments. The species can be represented as the donor-acceptor complex $H_2C \longrightarrow LiF$.

The second alternative, 58, involves interaction of the lone-pair electrons on fluorine with the vacant p orbital at the carbene center. Ideally, both the angle that the methylene plane makes with the C—F bond, as well as the LiFC bond angle, should be 90°. The 6-31G* structure shows small distortions from these values. In particular, the methylene plane is bent back slightly, while the Li—F linkage leans forward slightly. The latter distortion is probably due to a weak bonding interaction between lithium and the methylene hydrogens. Isomer 58 resembles a complex $LiF \longrightarrow CH_2$.

While 57 and 58 are minima on the rather flat CH_2LiF potential surface at all theoretical levels examined, the highest-level calculations (MP4/6-31G*//6-31G*) show that they are separated from the apparent global minimum, 59, by energy barriers of only 2 and 6 kcal mol^{-1}, respectively (Table 7.29). Structure 59 benefits

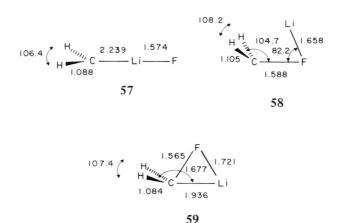

FIGURE 7.12.   Calculated (6-31G*) equilibrium geometries for CH_2LiF isomers.

TABLE 7.29. Relative Energies and Energy Barriers for $CH_2 LiF$ Isomers (kcal mol^{-1})[a]

Structure	3-21G//3-21G	6-31G*//6-31G*	MP4/6-31G*//6-31G*[b]
59	0	0	0
58	30	26	22
57	28	20	24
CH_2 + LiF	56	43	49
57 $\longrightarrow$ **59**[c]	1[d]	1	2
58 $\longrightarrow$ **59**[c]	3[d]	7	6

[a]Data from reference 108i.

[b]Corrected for zero-point energies.

[c]Values listed are *barriers* for these processes.

[d]4-31G//4-31G values.

from the simultaneous interaction of the nonbonded electrons on fluorine with the vacant p orbital on methylene, and of the electropositive lithium with the methylene lone pair. Hence, **59** is predicted at all theoretical levels to be significantly more stable than all alternatives (Table 7.29); the calculated binding energy for **59**, that is, ΔE for $CH_2 LiF \longrightarrow CH_2$ + LiF, is quite large (49 kcal mol^{-1} at MP4/6-31G*//6-31G*, following corrections for zero-point energy). The corresponding binding energies for the two alternative $CH_2 LiF$ structures, **57** and **58**, are much smaller, 25 and 27 kcal mol^{-1}, respectively. The theoretical equilibrium geometry of **59** is itself unusual; *all four ligands lie in a single hemisphere!* Relative to the structures of the other isomers, the geometry of **59** also shows evidence for the complementary and supportive nature of the two interactions. Both C—F and C—Li bond distances in **59** are shorter than they are in either of the alternatives, and the Li $\cdots$ F separation is larger. Overall, the best conventional representation for **59** is probably the ion-pair arrangement, $LiCH_2^+F^-$, that is, the C—Li linkage is relatively short (1.936 Å compared with 2.001 Å in methyllithium at 6-31G*) and both C—F and Li—F bonds are relatively long (1.564 and 1.721 Å compared with 1.365 and 1.555 Å in methyl fluoride and lithium fluoride, respectively). Why is this limiting description favored over the alternative (and entirely reasonable) ion-pair arrangement, FCH_2^- Li^+? The calculated energies for reactions (7.46) and (7.47) reveal that lithium stabilizes a methyl cation [110] to a much greater extent than fluorine stabilizes a methyl anion [1f, 108f, 111] (see also Section 7.2.1).

$$LiCH_2^+ + CH_4 \longrightarrow LiCH_3 + CH_3^+ \qquad (7.46)$$
$$78 \text{ kcal mol}^{-1} \text{ (6-31G*//6-31G*)}$$

$$CH_2 F^- + CH_4 \longrightarrow CH_3 F + CH_3^- \qquad (7.47)$$
$$9 \text{ kcal mol}^{-1} \text{ (MP2/6-31+G*//4-31+G)}$$

The basic carbenoid structural types are widespread and exhibited by a large

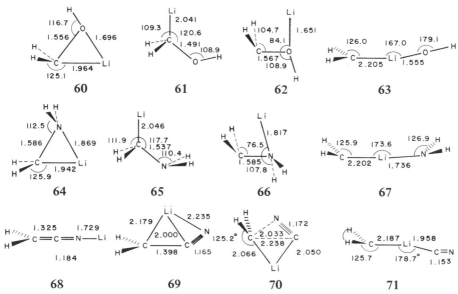

FIGURE 7.13. Calculated (3-21G) equilibrium geometries for isomers of $CH_2 LiOH$, $CH_2 LiNH_2$, and $CH_2 LiCN$. From refs. 108d, g, j.

number of species. Important geometrical parameters for isomers of $CH_2 LiOH$, $CH_2 LiNH_2$, and $CH_2 LiCN$, as determined from 3-21G calculations, are summarized in Figure 7.13. Relative energies (3-21G//3-21G, 6-31G*//3-21G and MP2/6-31G*// 3-21G levels) are provided in Table 7.30. Isomers of the first two of these are qualitatively similar to those for $CH_2 LiF$, except that in both cases conventional (tetrahedral carbon) structures also appear. Lithiated acetonitrile is different. From their calculated structures, both low-energy isomers, **68** and **69**, would seem to be well represented in terms of ion-pair complexes between lithium cation and different resonance structures of deprotonated acetonitrile, that is,

$$ \underset{H}{\overset{H_{\prime\prime\prime}}{\diagdown}} C = C = N^- \quad \longleftrightarrow \quad \underset{H}{\overset{H_{\prime\prime\prime}}{\diagdown}} C \!\!-\!\!-\!\! C \equiv N $$

The contrast with $CH_2 LiF$ is striking and is due primarily to differences in the abilities of fluoro and cyano substituents to stabilize a carbanion center. Compare the energy of reaction (7.47) with that of (7.48).

$$ CH_2 CN^- + CH_4 \longrightarrow CH_3 CN + CH_3^- \qquad (7.48) $$
$$ 48 \text{ kcal mol}^{-1} \text{ (3-21+G//3-21+G)} $$

The interesting carbenoid $CCl_3 Li$ offers new possibilities. The infrared spectrum of the matrix-isolated species suggests the presence of (at least) two isomers [112]; the principal form has been assigned to possess at least C_3 and possibly C_{3v} sym-

TABLE 7.30.   Relative Energies of Isomers of CH_2LiOH, CH_2LiNH_2, and CH_2LiCN (kcal mol^{-1})

Structure	3-21G//3-21G	6-31G*//3-21G	MP2/6-31G*//3-21G
CH_2LiOH^a			
60	0	0	0
61	27	15	14
62	23	22	20
63	42	27	50
$CH_2LiNH_2{}^a$			
64	0	0	0
65	24	15	14
66	24	21	18
67	60	56	71
CH_2LiCN^b			
68	0	6	
69	5	0	
70	54	52	
71	63	61	

[a]T. Clark, P. v. R. Schleyer, K. N. Houk and N. G. Rondan, *J. Chem. Soc., Chem. Commun.*, 579 (1981).

[b]T. Clark, unpublished calculations.

metry. In addition, frequencies corresponding to a (weakened) C—Cl bond and to a Li—Cl linkage have been assigned. The ^{13}C NMR of what appears to be a single isomer of the species in tetrahydrofuran has been reported [107]. Five structures for CCl_3Li have been examined at the 3-21G$^{(*)}$//3-21G$^{(*)}$ and 6-31G*//3-21G$^{(*)}$ levels [113]. Geometries for four of these are given in Figure 7.14; relative energies are provided in Table 7.31. The remaining form, **76**, was found to collapse without activation to structure **72**.

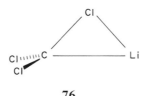

76

Some of these structures closely parallel those already found in other systems, for example, CH_2LiF. A conventional form, **73**, appears, as well as the remarkable inside-out structure, **75**. The latter may be thought of as a $CCl_3^- Li^+$ ion pair, in which the pyramidal trichloromethyl anion is stabilized by delocalization of negative charge onto the lithium. Structure **74** is also unique; two chlorine atoms bridge the C—Li bond.

As mentioned earlier (Section 7.3.2), lithium compounds show considerable ten-

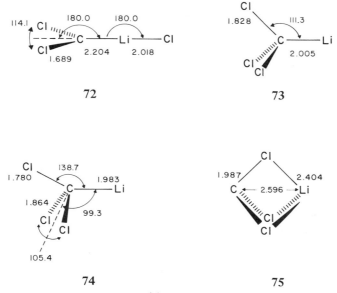

FIGURE 7.14. Calculated $(3\text{-}21\text{G}^{(*)})$ equilibrium geometries for CCl_3Li isomers.

dency to aggregate. Carbenoids are no exception; NMR evidence supports the existence of dimers in solution [107]. What effects might association have on carbenoid structures and charge distributions? Will aggregation facilitate or inhibit the principal reaction of carbenoids, to act as methylene transfer agents? The calculated (3-21G) structure of CH_2LiF dimer, **77**, [108g] (the heavy-atom bond lengths for which are given in Figure 7.15) is qualitatively similar to that for the monomer, **59**, at the same level of theory. So too are Mulliken charges. Both C—F and C—Li linkages in the dimer are (slightly) longer than in the monomer; this suggests that methylene transfer should be facilitated. Calculated (3-21G) reaction energies provide further support. Loss of singlet methylene from monomeric CH_2LiF, reaction (7.49), is endothermic by 56 kcal mol^{-1} at this level, while transfer from the dimer, reaction (7.50), resulting in formation of a $CH_2LiF:LiF$ complex, **78** (also shown in Figure 7.15), has a reduced endothermicity of 34 kcal mol^{-1}.

TABLE 7.31. Relative Energies of CCl_3Li Isomers (kcal mol^{-1})^a

Structure	$3\text{-}21\text{G}^{(*)}//3\text{-}21\text{G}^{(*)}$	$6\text{-}31\text{G}^*//3\text{-}21\text{G}^{(*)}$
72	0	0
73	6	12
74	6	12
75	13	22

^{a}Data from reference 113.

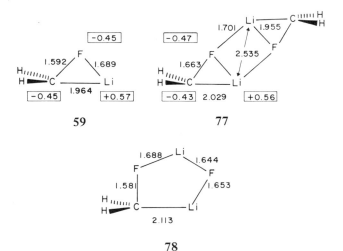

78

FIGURE 7.15. Calculated (3-21G) heavy-atom bond lengths and (enclosed in rectangles) Mulliken atomic charges for CH_2LiF, **59**, and its dimer, **77**. Calculated (3-21G) bond lengths for complex of CH_2LiF and LiF, **78**, the product of loss for the methylene from CH_2LiF dimer.

$$CH_2LiF \longrightarrow CH_2 + LiF \qquad\qquad (7.49)$$

$$(CH_2LiF)_2 \longrightarrow CH_2 + CH_2LiF:LiF \qquad\qquad (7.50)$$

a. Reactions of Carbenoids with Olefins. Methylene transfer to alkenes is the principal reaction of carbenoids. Cyclopropanes are formed [114].

For the Simmons-Smith reaction [115-117], in which the carbenoid is $IZnCH_2I$ (or a solvated and/or aggregated analogue), a *butterfly* transition structure, **79**, has been proposed [115c,d].

79

Kobrich's transition structure for CCl_2 transfer from $LiCCl_3$ is similar, but he has suggested that the reaction is preceded by complexation of lithium to the olefinic π bond [116]. Hoeg *et al.* [117] also have provided a similar formulation, but represent the carbenoid as Li^+ and Cl^- coordinated on the backside of $:CCl_2$. Closs [105a,b] proposed the transition structure **80** for reactions of arylcarbenoids with alkenes. Here, the π bond acts as a nucleophile, effecting an S_N2-like displacement of halogen from the carbenoid.

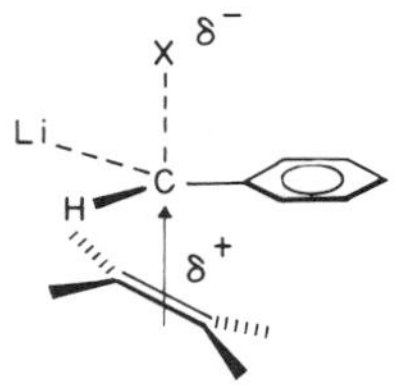

80

Moser [118] has suggested the importance of coordination with the metal in alkene addition reactions of copper carbenoids, for example, **81**.

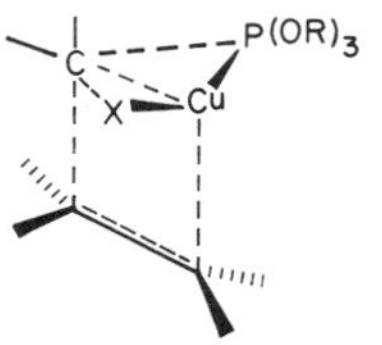

81

Comparison of the frontier molecular orbitals of the ground-state cyclic form of CH_2LiF with those of singlet methylene reveals similarities and differences [119] (Figure 7.16). The carbenoid HOMO, essentially a representation of the highly-polar carbon-lithium σ bond, is also more-or-less a lone pair at carbon. It is responsible for the carbanion-like character of halo-organometallics [121]. The similarity of this orbital to the HOMO of singlet methylene is striking. On the other hand, the LUMO of the model carbenoid, localized primarily on the metal, is quite different from that of singlet methylene, which is concentrated on carbon.

Calculations (3-21G//3-21G) suggest [122] that, prior to their reaction, CH_2LiF and ethylene form a complex, **82**, the important geometrical features of which are displayed in Figure 7.17. The calculated binding energy (12 kcal mol^{-1} at 3-21G// 3-21G) is large enough to guarantee stability in the gas phase. Both solvation and aggregation [107] of the carbenoid would, however, be expected to reduce the binding energy of the complex, and it is unlikely that it will be stable in solution. The calculated transition structure for cyclopropane formation, **83** in Figure 7.17, lies 16 kcal mol^{-1} above reactants. It shows that the LiF moiety is substantially decomplexed, freeing the carbene character of methylene. The two new carbon-carbon linkages are just beginning to form, while the double bond in ethylene has

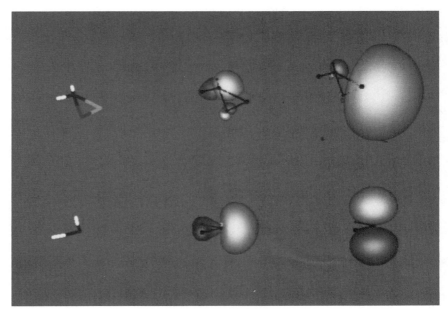

FIGURE 7.16. Skeletal frameworks (left) highest-occupied (middle) and lowest-unoccupied (right) molecular orbitals of CH_2LiF (top) and singlet methylene (bottom). 3-21G//3-21G.

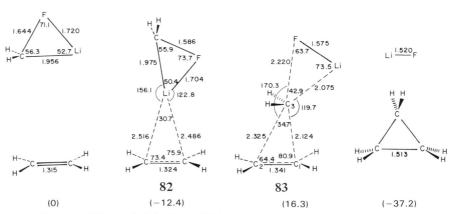

FIGURE 7.17. Calculated (3-21G) equilibrium geometries for the complex of CH_2LiF and ethylene, **82**, and the transition structure, **83**, on the reaction pathway for cyclopropane formation. Energies (kcal mol^{-1}, relative to CH_2LiF and ethylene) are provided.

only slightly weakened. Thus, carbenoid addition occurs with an early transition structure. Overall, the geometry of **83** is quite similar to transition structures found at the same level of theory for olefin cycloadditions involving halocarbenes [123].

b. Metal Carbenes as Methylene Transfer Agents: The Role of the Tebbe Reagent in Olefin Metathesis. Transition-metal carbene complexes act as methylene transfer agents [124] and catalyze olefin metathesis [125] and oligomerization [126]. The favored mechanism for metathesis involves the intermediacy of an alkene-complexed metal carbene in equilibrium with a metallacyclobutane [127].

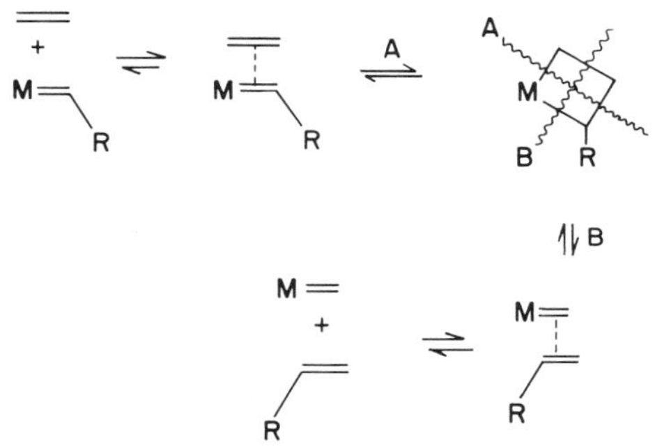

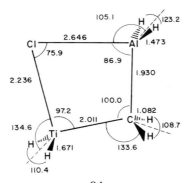

The Tebbe reagent [128], $Cp_2Ti{=}CH_2/ClAlMe_2$, displays a strikingly similar pattern of reactivity [129]. This encourages speculation that it may be a weak *intermolecular complex* between the carbene and a Lewis acid, that is,

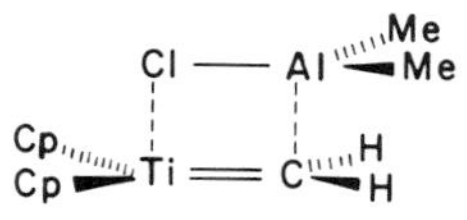

The STO-3G equilibrium structure for a model compound, $H_2Ti{=}CH_2/ClAlH_2$, **84** (shown below),

84

does not support such a notion [130a]. The Ti—C bond length is calculated to be nearly 0.2 Å longer than that in the free carbene, and approaches a normal single-bond value, for example, 2.095 Å in TiMe$_4$. The Ti—Cl and C—Al distances are only slightly longer than normal single-bond lengths, for example, 2.167 Å in TiCl$_4$ and 1.899 Å in AlMe$_3$. The Al—Cl bond is far longer than that found in the free Lewis acid, for example, 2.050 Å in AlCl$_3$. STO-3G structures for a number of methyl- and chloro-substituted compounds show similar features [130b]. The theoretical data suggest representation of the model reagent as an intramolecular complex in which chlorine acts as the electron donor and aluminum as the acceptor, that is,

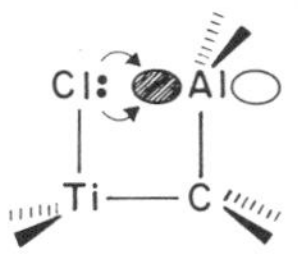

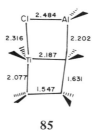

This is also in accord with proton NMR data [128] that have previously been interpreted in terms of a zwitterionic metallacycle, that is,

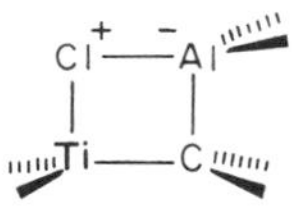

How can such a species behave chemically like a free carbene? The calculated structures of the model complex show the two fragments, H$_2$TiCH$_2$ and ClAlH$_2$, to be strongly associated (the calculated complexation energy is 125 kcal mol^{-1} at the STO-3G level), and it seems likely that the equilibrium abundance of free carbene is miniscule. The STO-3G structure of the model reagent complexed to ethylene, **85**, shows some evidence of a weakened interaction with the Lewis acid.

$$Cl \overset{2.484}{\rule{2cm}{0.4pt}} Al$$

2.316   2.202

Ti   2.187

2.077   1.631

1.547

85

The calculated complexation energy of **85**, relative to the dissociation products titanacyclobutane and ClAlH$_2$, is smaller than the corresponding energy for the uncomplexed model Tebbe reagent, although it is still too large to permit easy loss of the Lewis acid.

Is it possible that solvent aids in the displacement of the Lewis acid [131]? Attack of a Lewis base on the Tebbe reagent would (presumably) occur at aluminum, from one of two sides. STO-3G structures for the two ammonia adducts, 86 and 87,

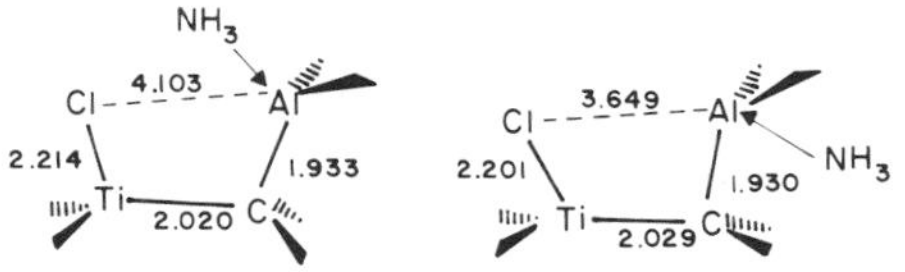

86 **87**

are essentially identical to 84, insofar as the local geometry of the metal carbene is concerned. Prior complexation of 84 to a Lewis base is, therefore, not expected to affect greatly the reactivity of the incorporated metal carbene toward olefins. On the other hand, simultaneous complexation of both an olefin and base appears to have the desired effect. Compare the STO-3G geometry of 88 with that of isolated titanacyclobutane.

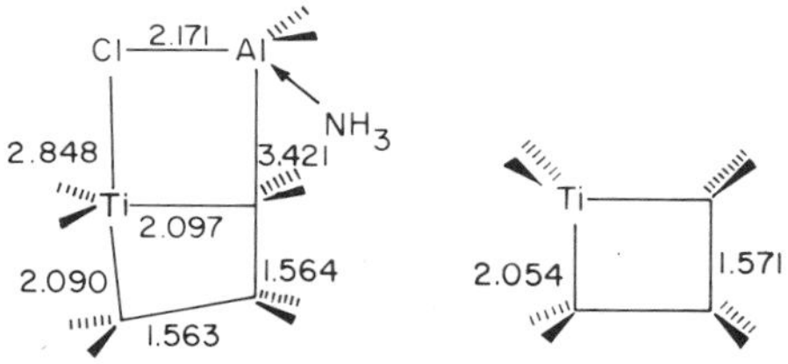

88

Here the Lewis acid has been almost completely removed (in the form of a donor–acceptor adduct with the Lewis base) leaving behind a titanacyclobutane moiety which is nearly identical to the free metallacycle, and which is the first intermediate isolated in the Tebbe-catalyzed metathesis reaction.

7.4. VIOLATION OF CONVENTIONAL STRUCTURE RULES

The theory of tetrahedral carbon (van't Hoff and Le Bel, 1874 [132]) predicts the shapes of methane as tetrahedral, ethylene as trigonal planar (two tetrahedra edge-to-edge) and acetylene as linear (two tetrahedra face-to-face). The best available calculations for the energies of alternative singlet forms of methane, ethane, ethylene and acetylene (Figure 7.18) show qualitatively how good the van't Hoff–Le Bel hypothesis really is, and reveals the essential rigidity which characterizes classical organic molecules. Thus, both square-planar and square-pyramidal methane geometries are less stable than the tetrahedral structure by more than the C—H bond dissociation energy (104 kcal mol^{-1}). These two forms are probably incapable of

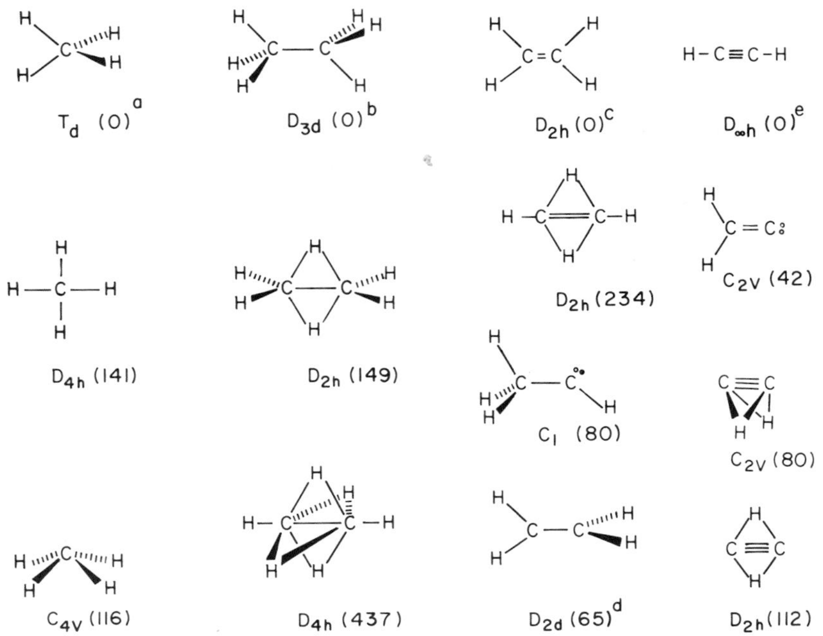

FIGURE 7.18. Calculated experimental relative energies of alternative forms of methane, ethane, ethylene and acetylene (kcal mol^{-1}): (a) MP2/6-311+G**//6-311+G**, see Table 7.32; (b) MP2/6-31G*//6-31G*, see ref. 184; (c) MP4/6-31G*//6-31G* ref. 220a, also: K. Ragha-vachari, M. J. Frisch, J. A. Pople, and P. v. R. Schleyer, *Chem. Phys. Lett.*, **85**, 145 (1982); (d) value for *cis-trans* isomerization in 1,2-dideuteroethylene, J. E. Douglas, B. S. Rabinovitch, and F. S. Looney, *J. Chem. Phys.*, **23**, 315 (1955); (e) MP3/6-31G**//6-31G*, see ref. 153.

existence, and can only be examined by calculations. Similarly, hydrogen-bridged structures for ethane, ethylene and acetylene appear to be hopeless alternatives energetically.

Singlet structures which might possibly be accessible experimentally include perpendicular ethylene (the transition structure for *cis-trans* isomerization of ethylene), vinylidene (H_2CC, a likely transition structure for scrambling of doubly-labeled acetylenes) and ethylidene (H_3CCH, a possible transition structure for 1, 2-hydrogen exchange in ethylene). None of these appears to represent a potential energy minimum, and all are much higher in energy than their respective lowest-energy forms [220].

From 1885, when Baeyer announced his strain theory [133], chemists have probed the limits of conventional bonding concepts by conceiving and investigating strained structures [134]. Geometries can be forced to be abnormal by the constraints of rigid molecular frameworks. For example, the unknown molecules *windowpane* [135, 136], **89**, and the carbon-centered *annulene* [136], **90**, may be constrained to incorporate a more-or-less planar tetracoordinate carbon, and *adamantene* [137], **91**, a nearly perpendicular double bond.

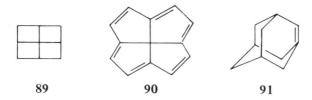

| 89 | 90 | 91 |

Actually, *anti-Bredt* [134, 138] twisted ethylene systems like **91** would have considerable singlet diradical character, and probably incorporate what is essentially a CC single bond [139]. Furthermore, molecular orbital calculations show **89** and **90** actually to favor distorted tetrahedral arrangements at the central carbons rather than the idealized planar geometries [135, 136].

Were it ever possible to achieve the synthesis and experimental characterization of molecules with planar tetracoordinate carbons or perpendicular double bonds, these would not be expected to pose fundamental violations of classical bonding rules. If, on the other hand, such species proved to be thermodynamically stable, unreactive molecules, new and unexpected principles would be operating. The isolation of [1.1.1] propellane, **92**, is a case in point [140].

$$H_2C \quad CH_2 \quad CH_2$$

92

This molecule is as unstable and as highly reactive as its distorted and strained structure implies; it owes its existence only to the lack of any low-energy decomposition or oligomerization pathways.

How might fundamental violations of accepted bonding rules be achieved more easily? One possibility would be to find substituents capable of stabilizing a planar carbon center over a tetrahedral one, a twisted double bond over a planar structure and a doubly-bridged acetylene over a linear arrangement. Suitable substituents might make doubly- or even quadruply-bridged ethanes the most stable geometries. What kind of substituents are needed? The large number of different groups which have already been studied experimentally, for example, alkyl, aryl, and halogen, and have failed to produce basic structural alterations, are all *electron sufficient or electron excessive*. That is, they possess sufficient electrons to allow electron-pair (covalent) bonds to be formed; any electrons left over are relegated to lone pairs. Perhaps *electron-deficient substituents* might induce quite different geometries. The known structures of electron-deficient carbon compounds, for example, carbocations and transition-metal organometallics, are often quite startling from the usual organic chemist's viewpoint [134, 141]. Many show carbon associated with far more than the usual maximum number of neighbors, and many depict bonding arrangements which differ quite markedly from those found in normal organic molecules, where there are sufficient electrons to form two-electron, two-center linkages. These observations encourage more careful scrutiny.

Here, we examine possible compounds which incorporate square-planar carbon or heavier group IV elements. Our goals are to uncover species that might be accessible experimentally either as low-energy transition structures for processes involving normal tetrahedral systems or which themselves are stable structures. While no such systems have been established to exist at present, several reasonable possibilities arise from the calculations. We shall also examine the possibility of stable or low-energy ethylene analogues with perpendicular structures and of doubly-bridged acetylenes.

7.4.1. Planar Tetracoordinate Carbon

Eight valence electrons are available for bonding in methane. In the tetrahedral structure, all participate in the construction of the four CH linkages, two electrons for each bond. The valence molecular orbitals of planar methane are shown in Figure

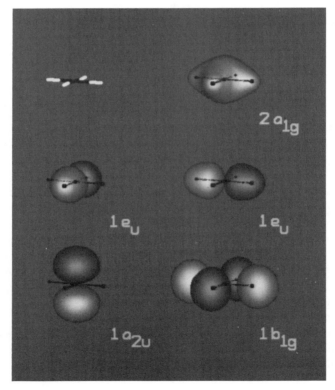

FIGURE 7.19.   Valence molecular orbitals of square-planar methane from lowest to highest energy (top to bottom, left to right). $2a_{1g}$, $1e_u$ and $1a_{2u}$ occupied and $1b_{1g}$ unoccupied. STO-3G//STO-3G.

TABLE 7.32. Calculated Energies of Square-Pyramidal and Square-Planar Singlet Methane Structures Relative to the Lowest-Energy Tetrahedral Structure (kcal mol^{-1})[a]

Theoretical Level	Pyramidal (C_{4v})	Square Planar (D_{4h})
STO-3G//STO-3G	206	240
3-21G//3-21G	166	170
3-21+G//3-21+G	157	159
6-31G*//6-31G*	146	171
6-311+G**//6-311+G**	130	155
MP2/6-31G*//6-31G*	137	160
MP2/6-311+G**//6-311+G**	116	141

[a] Data from G. W. Spitznagel and P. v. R. Schleyer, unpublished calculations.

7.19. Only six valence electrons contribute to CH bonding. These fill orbitals labeled $2a_{1g}$ and $1e_u$. The remaining two electrons are allocated to a nonbonding p orbital (labeled $1a_{2u}$) perpendicular to the molecular plane. As the data in Table 7.32 attest, the energetic consequences are severe [142]. The best calculations tabulated (MP2/6-311+G**//6-311+G**) show that square-planar methane lies 141 kcal mol^{-1} above the lowest-energy tetrahedral structure. The alternative (square) pyramidal geometry is predicted to be not quite as unfavorable, but its energy (116 kcal mol^{-1} above tetrahedral methane) is still larger than a typical C—H bond dissociation energy. Note that both improvements in the basis set and treatment of electron correlation preferentially stabilize the square-planar and square-pyramidal structures over the tetrahedral form. Some of the energy differences presented in this and in the following sections are likely to be similarly modified with improvements in the theoretical models.

How might substituents preferentially stabilize planar methane over the normal tetrahedral form [143]? Both π-acceptor and σ-donor groups should be effective. The former will act to delocalize electron density from the nonbonding lone pair on carbon (the highest occupied molecular orbital); the latter will function by alleviating the electron deficiency of the σ framework. The data in Table 7.33 concur. Electronegativity effects are most important, although substitution by π acceptors also favors the planar form. Indeed, planar lithiomethane is only 34 kcal mol^{-1} less stable than the tetrahedral form according to the best available calculations. The planar-tetrahedral energy difference in the corresponding sodium system is 32 kcal mol^{-1} at the 3-21+G//3-21+G level. These energies are low enough, at least in principle, to be established experimentally, were it possible to investigate the stereomutation of monomeric compounds.

The influence of two lithium atoms is spectacular! According to 3-21+G//3-21+G calculations (Table 7.34), the energy difference between *cis*-planar and tetrahedral dilithiomethane is only 5 kcal mol^{-1} (the latter is still preferred). Surprisingly, the corresponding *trans*-planar compound is much less favorable (17 kcal mol^{-1} above the *cis* at 3-21G//3-21G). *Cis*-planar dilithiomethane is a 2π-electron Hückel aromatic; the lone pair on carbon in planar methane (labeled $1a_{2u}$ in Figure 7.19) has

TABLE 7.33. Calculated Planar-Tetrahedral Energy
Differences for Substituted Methanes (kcal mol^{-1})

Molecule	STO-3G//STO-3G	3-21+G//3-21+G
CH_4	240	159
CH_3Li	52	34
CH_3BeH	100	70
CH_3BH_2	155	102
CH_3CH_3	243	155
CH_3NH_2	248	
CH_3OH	249	184
CH_3F	265	211
CH_3Na	76	32
CH_3MgH		62
CH_3AlH_2		79
CH_3SiH_3		131
CH_3PH_2		136
CH_3SH		164
CH_3Cl		87^a
CH_3CN	217	160^b

[a]Complex of CH_3^+ and Cl^- formed in the planar structure.
[b]3-21G//3-21G.

delocalized into empty p orbitals on the adjacent lithiums. The molecular orbital that results (the HOMO in *cis*-planar dilithiomethane) clearly shows the delocalization over the three centers.

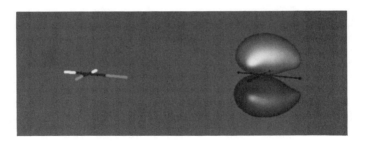

The calculated planar-tetrahedral energy difference in trilithiomethane (5 kcal mol^{-1} at 3-21G//3-21G) is only slightly smaller than that found in the disubstituted compound (7 kcal mol^{-1} at the same level of theory); a larger (14 kcal mol^{-1}) separation is predicted for tetralithiomethane. Evidently, other factors are at work

TABLE 7.34. Calculated Planar-Tetrahedral Energy Differences for Singlet Lithium- and Sodium-Substituted Methanes (kcal mol^{-1})

Molecule	STO-3G//STO-3G[a]	3-21G//3-21G[a,b]	3-21+G//3-21+G[b]
CH_3Li	52	37	34
cis-CH_2Li_2	17	7	5
trans-CH_2Li_2		24	
$CHLi_3$	10	5	
CLi_4	22	14	
CH_3Na	76	35	32
cis-CH_2Na_2	59	9	13

[a]Reference 143a.
[b]Reference 143d.

to reduce the beneficial effects of charge delocalization. Substitution by sodium leads to similar results: cis-planar CH_2Na_2 is only 9 kcal mol^{-1} higher in energy than the tetrahedral form according to 3-21G//3-21G calculations.

Ring strain may also be employed to lower the planar-tetrahedral energy difference [143]. A square-planar carbon (with idealized 90° bond angles) is more easily incorporated into a small ring than is a tetrahedral center. Incorporation of a planar carbon center into cyclopropane, for example, requires 130 kcal mol^{-1} (3-21G//3-21G), a difference which, while still very large, is considerably less than the 170 kcal mol^{-1} separation calculated for methane at the same level of theory. Strain arguments alone would suggest an even larger effect in cyclopropene. In fact, the planar-tetrahedral energy gap here is actually larger than that found in cyclopropane (168 kcal mol^{-1} vs. 130 kcal mol^{-1}). The π system now contains four electrons (two from the carbon lone pair and two from the CC double bond) and the molecule is antiaromatic.

1,1-Dilithiocyclopropane, which draws on the σ-donor, π-acceptor properties of two lithium substitutents, and allows for the possibility of aromaticity and for reduction of strain, is predicted by the 3-21G//3-21G calculations to prefer a planar geometry, **93**. The conventional tetrahedral structure, **94**, is 7 kcal mol^{-1} higher in energy.

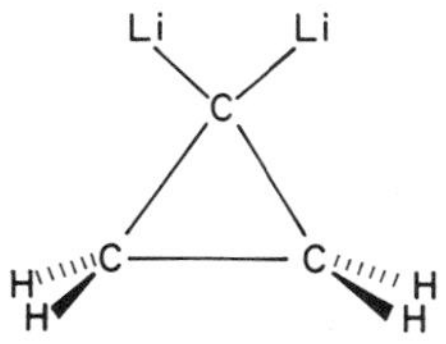

93 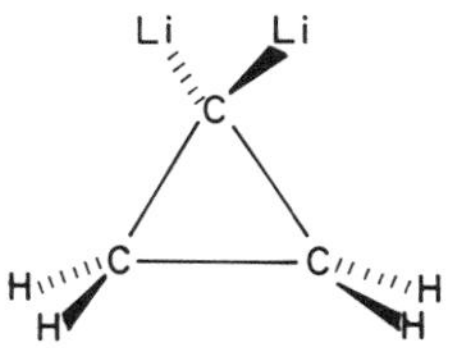   **94**

Anti-van't Hoff geometries are likely to be widespread. One need only search for systems with insufficient electrons to form normal two-center, two-electron bonds.

a. *Stereochemistry of Methane Analogues.* The tetracoordinate compounds of group IVa elements beneath carbon, that is, silicon, germanium, tin, and lead [144], as well as those of the group IVb elements, that is, titanium, zirconium, and hafnium [145], prefer tetrahedral geometries. Theoretical calculations [146] (Table 7.35) suggest, however, that square-planar structures for these systems may be more accessible energetically than are those for compounds of carbon. What is the cause of this increase in relative stability of square-planar geometries for these systems? There are only three stabilized molecular orbitals in square-planar methane (Section 7.4.1). These are the bonding combinations of the carbon $2s$ atomic orbital with the completely symmetric arrangement of the $1s$ functions on the four hydrogens, and of the carbon $2p_x$ and $2p_y$ orbitals with the degenerate pair of single-node hydrogen $1s$ combinations (see Figure 7.19). The remaining arrangement of hydrogen $1s$ functions is completely antibonding, and is the lowest unoccupied molecular orbital in planar methane (labeled $1b_{1g}$ in Figure 7.19). Note, however, that the corresponding orbital in planar silane is the highest occupied of the system; the π-type lone pair of a_{2u} symmetry (the HOMO in planar methane) has become the lowest unfilled molecular orbital (Figure 7.20). The same ordering holds for planar germane and stannane. Evidently, the b_{1g} symmetry orbital has been stabilized by decreased overlap between the hydrogens, that is, Si—H (Ge—H, Sn—H) bonds are longer than C—H linkages, and by the availability of functions of d symmetry on the central atom. Bonding in the transition-metal hydride TiH_4 (the valence-orbital manifold of which is displayed in Figure 7.21) involves functions of d symmetry; four sets of TiH bonding orbitals may be constructed for both tetrahedral and square-planar geometries. Any difference in stabilities of these two arrangements should be small, and might arise primarily because of increased repulsion between the TiH bonds in the planar geometry.

How might preferential stabilization of the square-planar forms of main-group and transition-metal methane analogues be achieved [146]? Unlike the situation in planar methane, σ-donor substituents should not be particularly effective [147]:

TABLE 7.35. Calculated Planar-Tetrahedral Energy Differences for Methane Analogues (kcal mol^{-1})^a

Molecule	STO-3G//STO-3G	3-21G//3-21G	6-31G*//6-31G*
CH_4	240	170	171
SiH_4	152	95^b	93
GeH_4	166		
SnH_4	127		
TiH_4	43		

^{a}Except for methane, the highest occupied molecular orbital in the planar structure is of b_{1g} symmetry and corresponds to the antibonding combination of hydrogen $1s$ functions (see Figure 7.20). The a_{2u} methane HOMO is a nonbonding out-of-plane p orbital on carbon (see Figure 7.19).

^{b}3-21G$^{(*)}$//3-21G$^{(*)}$. The corresponding 3-21G//3-21G difference is slightly larger (105 kcal mol^{-1}), which shows the relative importance of d-orbital effects.

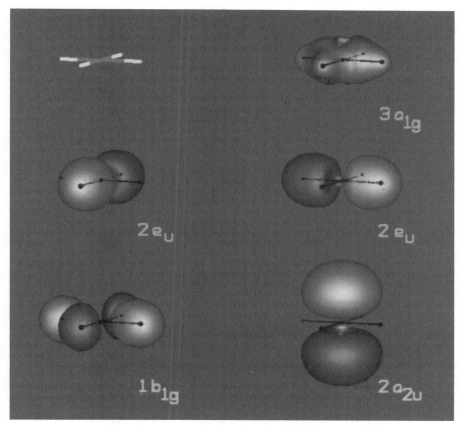

FIGURE 7.20. Valence molecular orbitals of square-planar silane from lowest to highest energy (top to bottom, left to right). $3a_{1g}$, $2e_u$ and $1b_{1g}$ occupied and $2a_{2u}$ unoccupied. STO-3G//STO-3G.

eight electrons may already be accommodated in the stabilized molecular orbitals of both tetrahedral and square-planar arrangements. On the other hand, π donors should (and do) lead to a favoring of the square-planar geometries of the main-group hydrides (Table 7.36). For example, an amino group is predicted by STO-3G calculations to lower the planar-tetrahedral energy difference in silane by 47 kcal mol^{-1}, to 105 kcal mol^{-1}. 3-21G$^{(*)}$//3-21G$^{(*)}$ calculations predict an even smaller energy separation (64 kcal mol^{-1}). Large effects are also found for fluorine and chlorine substitution in silane. The observed ordering of substituent effects:

$$NH_2 > Cl > F$$

does not parallel the usual sequence of π-donor abilities, see Section 7.2.1. Additional factors are evidently at work. Sizable effects, in all cases favoring planar struc-

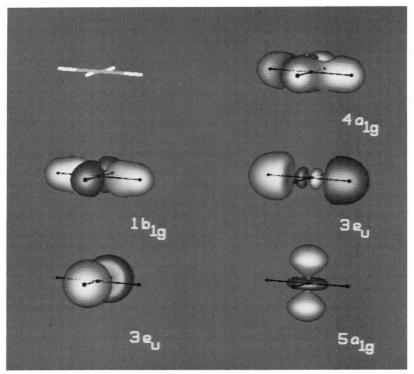

FIGURE 7.21. Valence molecular orbitals of square-planar tetrahydridotitanium from lowest to highest energy (top to bottom, left to right). $4a_{1g}$, $1b_{1g}$ and $3e_u$ occupied and $5a_{1g}$ unoccupied. STO-3G//STO-3G.

tures, are also found for substituted germanes and stannanes. The same ordering of substituent effects calculated for the silanes is also noted here.

The predicted ordering of effects for a given substituent:

$$SiH_3 X < GeH_3 X > SnH_3 X \qquad X = NH_2, F, Cl$$

warrants some comment. Shown in Figure 7.22 are the highest occupied molecular

TABLE 7.36. Calculated Planar-Tetrahedral Energy Differences for Substituted Methane Analogues (kcal mol^{-1})[a]

X	H	F	Cl	Br	NH_2
$SiH_3 X$	152 (95)	115 (77)	112 (77)	120	105 (64)
$GeH_3 X$	166	123	116	124	113
$SnH_3 X$	127	97	91	97	88
SiX_4	152	70 (75)	80		63 (53)
GeX_4	166	64	74		62
SnX_4	127	48	55		42

[a]STO-3G//STO-3G. 3-21G$^{(*)}$//3-21G$^{(*)}$ data for silicon compounds given in parentheses.

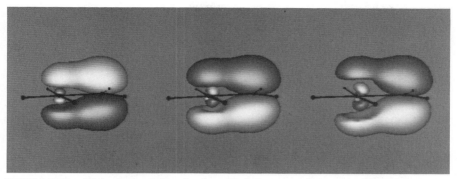

FIGURE 7.22. Highest-occupied molecular orbitals in square-planar aminosilane (left), amino-germane (middle), and aminostannane (right). STO-3G//STO-3G.

orbitals in aminosilane, aminogermane and aminostannane. These functions result from interaction of the lone pair on the amino substituent with the vacant p function on the group IVa element. Note that the relative sizes of the atomic components are most nearly identical for the case of germanium; the nitrogen lone pair appears to be larger than the $3p$ function on silicon and smaller than the $5p$ orbital on tin. Note also that the observed substituent effects parallel the electronegativities of the group IVa elements, Ge > Si > Sn, suggesting that σ effects might also play a role in dictating structural preference.

Ring strain may also be employed to stabilize preferentially square-planar over tetrahedral geometries (Table 7.37). Incorporation of silicon into a saturated three-membered ring reduces the planar-tetrahedral energy difference by 17 kcal mol^{-1} according to 3-21G$^{(*)}$//3-21G$^{(*)}$ calculations. Comparable effects are noted for the analogous germanium and tin systems, and for the transition-metal compound. Further advantage may accrue by incorporation of the planar center into an unsaturated ring. Relative to their normal (tetrahedral) structures, the silicon, germanium, and tin analogues of planar cyclopropene benefit not only from increased ring strain but also from aromaticity. The two π electrons associated with the carbon-carbon double bond may be delocalized into the vacant p function at the planar center. The results of the combination of effects are substantial. The separation of planar and tetrahedral forms of silacyclopropene is reduced to 49 kcal mol^{-1} ac-

TABLE 7.37. Calculated Planar-Tetrahedral Energy Differences for Main-Group and Transition-Metal Analogues of Cyclopropane and Cyclopropene (kcal mol^{-1})[a]

X	XH$_4$	XH$_2$ /‾\ H$_2$C—CH$_2$	XH$_2$ /‾\ HC=CH
C	240	190	196
Si	152 (95)	127 (78)	80 (49)
Ge	166	141	88
Sn	127	108	68
Ti	43	32	23

[a] STO-3G//STO-3G. 3-21G$^{(*)}$//3-21G$^{(*)}$ data for silicon compounds given in parentheses.

cording to the $3\text{-}21G^{(*)}//3\text{-}21G^{(*)}$ calculations. Large effects also occur for the analogous germanium- and tin-containing ring systems. The lowering noted for the transition-metal compound is smaller and must arise only because of increased ring strain. The $C{=}C$ π bond is of incorrect symmetry to interact with the lowest unoccupied function on the planar metal center (see Figure 7.21).

Finally, square-planar geometries may be preferentially stabilized by coordination with external nucleophiles. Hexacoordinate silicon, germanium and tin compounds in which local octahedral symmetry is attained about the central atom, that is, $MX_4 : Nu_2$, have been known for many decades, and numerous crystal structures are available [148]. Typical are those for MCl_4 : pyridine$_2$ (M=Si, Ge) compounds, which exhibit square-planar geometries for the central MCl_4 moiety [149]. The existence (and known structures) of such compounds can be rationalized on the basis of donation of lone-pair electrons from the nucleophile into the unoccupied out-of-plane p orbital on the central atom. This means that MX_4 systems in which the LUMO has been lowered in energy by strong σ-withdrawing X groups will be stabilized by coordination to a greater extent than those where such lowering has not taken place. For example, $SiF_4 : (NH_3)_2$ should be a stronger complex (toward dissociation into tetrahedral SiF_4 and two ammonia molecules) than $SiH_4 : (NH_3)_2$. $3\text{-}21G^{(*)}//3\text{-}21G^{(*)}$ calculations concur; dissociation of the hydride is exothermic by 7 kcal mol^{-1} while the analogous process for the fluoride complex is endothermic by 48 kcal mol^{-1}.

Because of the ability of nucleophilic solvents to act as strong Lewis acids, it should be difficult to prepare in such solvents compounds incorporating a chiral center at silicon, germanium, tin or lead. Racemization would occur following the production of a local square-planar structure through solvent coordination.

b. ***Structure of the Tebbe Reagent.*** The Tebbe reagent [128], $Cp_2Ti{=}CH_2$: $ClAlMe_2$ (presented here for convenience as an intermolecular complex between a metal carbene and a Lewis acid), provides an example of a stable molecule in which a tetracoordinate carbon is associated with two electropositive metals. While the details of its geometry remain unknown (the crystal structure of a related compound, $Cp_2Ti{=}CH_2 : ClAl(CH_2CMe_3)_2$, is disordered [150]), proton NMR data [128] support interpretation, not in terms of an intermolecular complex, but rather as a metallacycle, that is,

The calculated structure [130] of the model compound $H_2Ti{=}CH_2 : ClAlH_2$ (already described in Section 7.3.3b) is suggestive either of a metallacycle (as above), or even better of an intramolecular complex, chlorine acting as an electron-pair donor and aluminum as an acceptor.

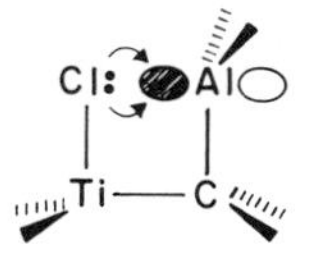

In either case, the carbon does, in fact, appear to be directly bonded to both metals, and the opportunity arises for it to assume a planar geometry. Disappointingly, as indicated in the structure above, the model Tebbe reagent is found to prefer a tetrahedral geometry at carbon. Titanium and aluminum are evidently not as effective as lithium in stabilizing a planar center. Furthermore, the strain inherent in the four-membered metallacycle is not as great as would be associated with a three-membered ring. The corresponding planar structure is 54 kcal mol^{-1} higher in energy at the STO-3G//STO-3G level. If previous experience is any guide (see, for example, Table 7.32), additional improvements in the basis set and treatment of electron correlation will serve to reduce the difference in energy between the tetrahedral and planar forms. It is not likely, however, that the calculated preference for a tetrahedral arrangement at carbon will be reversed.

Compounds of the type

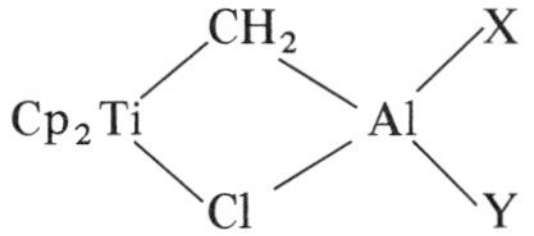

$$X = Me, \ Y = Cl, \ CH_2 CMe_3$$

have been synthesized, and the methylene hydrogens observed by NMR to be magnetically inequivalent at room temperature [151]. If the hydrogens were to become equivalent at some higher temperature, this would imply that one of several possible dynamic processes was taking place. Among the most reasonable are dissociation-recombination of the metal carbene/Lewis acid adduct, sequence (7.51),

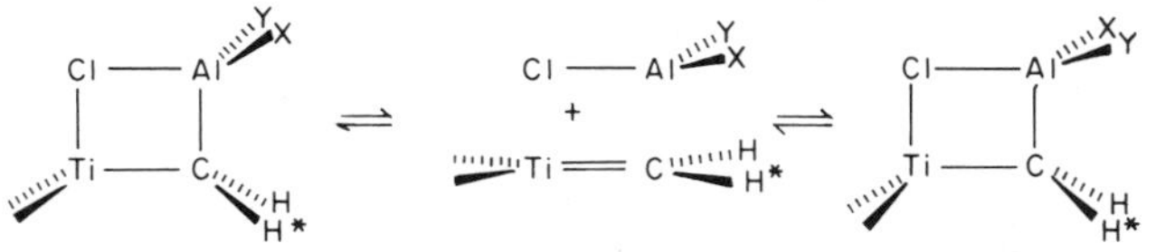

(7.51)

cleavage of the intramolecular aluminum–chlorine coordinate bond, followed by rotation about the carbon–aluminum linkage, sequence (7.52),

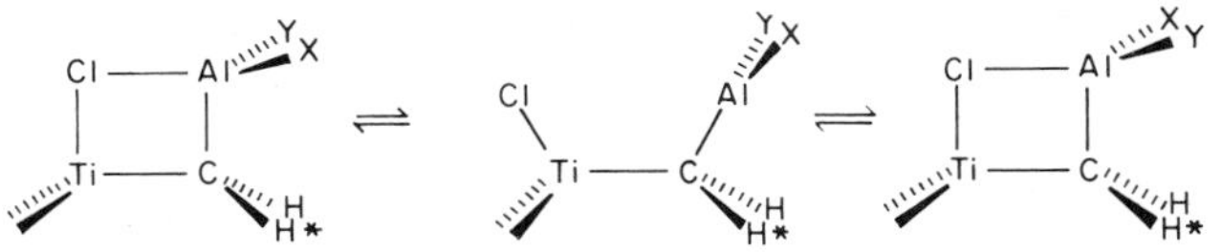

(7.52)

or passage through a planar carbon species (the desired process), sequence (7.53).

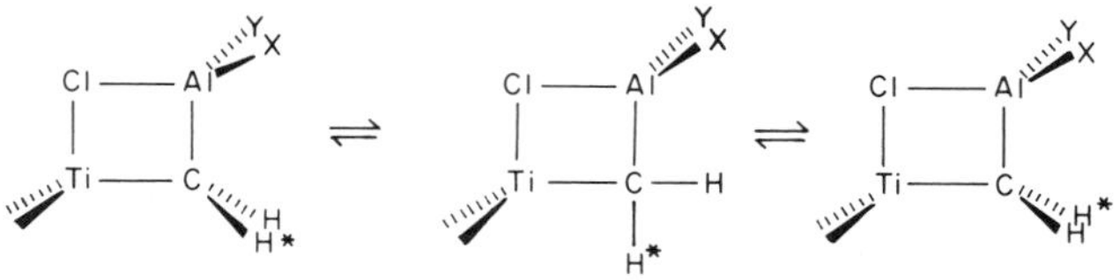

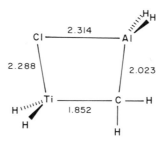

$$(7.53)$$

The calculations indicate that the energy of the model Tebbe reagent is significantly lower than that of its separated metal carbene and Lewis acid components (by 125 kcal mol^{-1} at STO-3G). The dissociation-recombination process, (7.51), is, therefore, not as favorable a pathway for equilibration of the methylene hydrogens as is rearrangement involving a planar carbon structure. Sequence (7.52), involving rupture of the coordinate linkage between aluminum and chlorine, is found to be somewhat more competitive, although better calculations, or calculations on analogues closer to experimentally accessible Tebbe reagents [152], are needed to assess which of the two competing pathways would actually be followed were the equilibrium of methylene hydrogens to be observed.

Calculated STO-3G heavy atom bond distances in the model Tebbe reagent incorporating square-planar carbon are shown below.

Compared with the analogous lengths in the tetrahedral carbon complex, **84**, this structure is more consistent with representation in terms of an intramolecular adduct between a (perpendicular) metal carbene and a Lewis acid. In particular, the incorporated titanium-carbon bond length is somewhat shorter than that in the normal complex (1.852 Å vs. 2.011 Å), and is nearly identical to that in the free (perpendicular) metal carbene (1.837 Å; see Section 7.4.2, following). It would appear that this remarkable structure incorporates not only a planar tetracoordinate carbon but a perpendicular double bond as well!

7.4.2. Other Violations of Van't Hoff Stereochemistry. Doubly-Bridged Acetylenes and Perpendicular Ethylenes

Doubly-bridged dilithioacetylene, **95**, is predicted by MP3/6-31G**//3-21G calculations to be 10 kcal mol^{-1} more stable than the conventional linear form, **96** [153a,b].

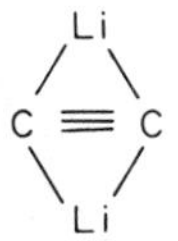

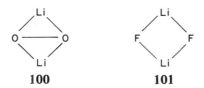

95    **96**

Both forms are predicted to be minima on the 3-21G potential surface. The contrast with acetylene itself is striking; here the doubly-bridged structure, **97**, is 107 kcal mol^{-1} higher in energy than the linear molecular, **98** (MP4/6-31++G**// 6-31G**). A nonplanar bridged structure, **99**, which utilizes both acetylene π orbitals, is only slightly less unfavorable; it lies 72 kcal mol^{-1} above linear acetylene at MP4/6-31++G**//6-31G**.

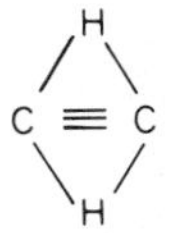

97    **98**    **99**

Both the conventional linear form, **98**, and the bridged structure, **99**, are minima on the 6-31G** surface, but **97** is not.

The situation is completely altered if carbon is replaced by silicon. According to high-level *ab initio* calculations [153c, 154], the global minimum for Si_2H_2 is a nonplanar doubly-bridged structure analogous to **99**.

What are the reasons for the change in gross structure between acetylene and dilithioacetylene? Lithium has much greater ionic character than hydrogen; electrostatic factors, more effective in stabilizing the compact bridged species, **95**, than the open structure, **96**, may be important [155]. If so, then their importance would increase, as would the favoring of bridged structures, with increasing electronegativity of the atoms associated with lithium. Structures **100** [156] and **101** are known examples [155b,c, 157].

100   **101**

The theoretical results suggest that the availability of p and d functions on lithium (unoccupied in the free atom) has little effect on the selection of doubly-bridged lithioacetylene, **95**, over the open form, **96**. Thus, the calculated difference in energies at the 6-31G*//6-31G* level (7 kcal mol^{-1} favoring **95**) is lowered by only 1 kcal mol^{-1} upon removal of the p and d functions from the lithium basis set. Removal of the valence s functions as well from the lithium basis set is of greater consequence, and results in the doubly-bridged and open forms becoming roughly

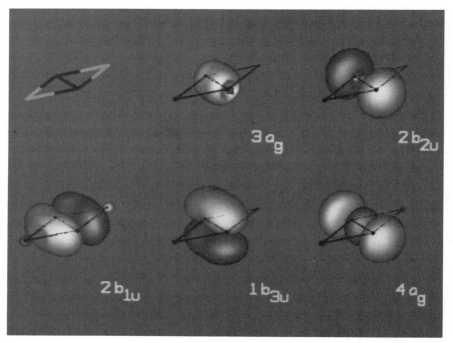

FIGURE 7.23. Occupied valence molecular orbitals of doubly-bridged dilithioacetylene from lowest to highest energy (top to bottom, left to right). 6-31G*//6-31G*.

equal in energy [158]. It would appear, therefore, that the possibility of multi-center covalent bonding involving lithium contributes to the preference for **95** over **96**, but that the overall effect is modest. Electrostatic effects probably dominate [153, 155].

Calculated CC bond lengths in doubly-bridged and linear forms of dilithioacetylene are similar (1.250 and 1.235 Å, respectively, according to 6-31G* calculations). While both are longer than the central linkage in the parent hydrocarbon (1.185 Å at 6-31G*), they are shorter than the double bond in ethylene (1.317 Å at the same level of theory). As is typical for bridging lithium structures, the C—Li distance in **95** (2.027 Å) is significantly longer than that in the linear alternative, **96** (1.904 Å). The valence manifold of doubly-bridged dilithioacetylene (Figure 7.23) contains two distinguishable π-type orbitals, one an out-of-plane function (labeled $1b_{3u}$) and the second an orbital (labeled $2b_{1u}$) which involves in-plane p functions on both carbons and lithiums.

Calculated energies for planar and perpendicular 1,1-dilithioethylene structures, **102** and **103**, respectively, are nearly identical [153a, 159].

102   **103**

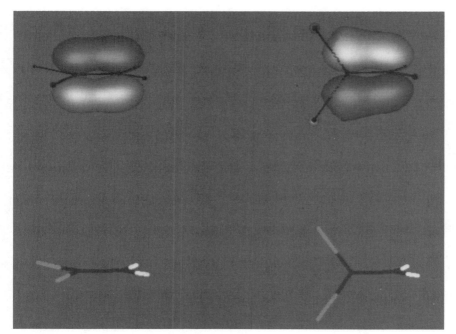

FIGURE 7.24.   Highest-occupied (π-type) molecular orbitals in planar 1,1-dilithioethylene (left) and perpendicular 1,1-dilithioethylene (right). STO-3G//STO-3G.

The perpendicular form is actually preferred (by 2 kcal mol^{-1}) according to both 6-31G*//4-31G and 3-21G//3-21G calculations [159c]. The same ordering and energy difference has been reported by Laidig and Schaefer using a polarization basis set (comparable to 6-31G*) and limited configuration interaction [159b]. Note also that the calculated (4-31G) C=C bond length in the planar structure (1.356 Å) is actually slightly longer than that in the perpendicular form (1.334 Å). While both lengths are greater than that found in the parent hydrocarbon at the same level of theory (1.330 Å), the differences are not large. A comparison of the highest occupied molecular orbitals of planar and perpendicular 1,1-dilithioethylene (Figure 7.24) confirms the expectation (based on calculated C=C lengths) that both species incorporate fully-formed π bonds. The orbital in the perpendicular structure involves a component on the CLi$_2$ fragment, which, while lithium–carbon σ bonding, is primarily localized on carbon, and is of the correct symmetry to interact hyperconjugatively with the out-of-plane p function on the adjacent methylene group. It is evident that the usual notion relating *cis-trans* isomerization barriers in olefins to π-bond strengths is not applicable here (see also discussion following on transition-metal carbene complexes).

The ground state of 1,1-dilithioethylene appears to be a triplet; as with the corresponding singlets, both planar and perpendicular structures are of comparable stability [159]. Calculated geometries for the two forms are also nearly identical, but differ in some respects from the corresponding singlet structures. Important parameters for the perpendicular singlet and triplet (obtained from 4-31G calculations) are compared below.

Li
1.866
1.334 ,,,,,H
104.1 C ═══ C
Li   H

singlet

Li
2.064
1.323 ,,,,,H
75.5 C ═══ C
Li   H

triplet

Note especially the small LiCLi bond angle in the triplet structure (75.5° vs. 104.1° in the singlet). This results in a separation between the lithiums of only 2.53 Å, compared with 2.94 Å in the corresponding singlet and 2.67 Å in diatomic lithium at the same level of theory. This distortion may be attributed to the benefits of three-center π bonding involving the two lithiums and the carbon to which they are attached.

In 1875, van't Hoff explored further consequences of his tetrahedral carbon model. He reasoned that three tetrahedra placed edge-to-edge would give perpendicular allene, while four tetrahedra similarly arranged would yield planar butatriene. Van't Hoff's cumulene rule (cumulenes with an odd number of double bonds assume planar structures, while those with an even number adopt perpendicular geometries) may, however, also be violated. Consider the cumulene series:

$$HB \atop HB \Big\rangle C(=C)_n =CH_2 \qquad n = 0, 1, 2$$

All involve the BBC ring, which functions similarly to a pair of lithium substituents [160]. Calculations (Figure 7.25) suggest that anti-van't Hoff structures are favored for all of these systems [161]! Structural data further show that the CC double-bond lengths are largely unaffected by conformation, and are, in fact, quite similar to those found in the analogous hydrocarbons. On the other hand, the local geometries of the BBC rings undergo significant changes as a result of twisting. Inspection of the occupied valence molecular orbitals in these compounds reveals why. While both Walsh cyclopropane-type orbitals [69], one of a_1 symmetry and the other of b_2 symmetry, are occupied in the van't Hoff geometries, in the anti-van't Hoff structures the latter orbital has been replaced in the valence manifold by a function of b_1 symmetry.

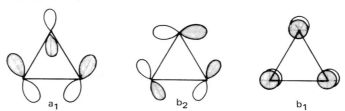

a_1 b_2 b_1

This orbital, analogous to the occupied π-type function in the aromatic cyclopropenyl cation (see Section 7.2.4a), is the basis for the enhanced stability of the anti-van't Hoff forms. Loss of the b_2 symmetry function (which is BB antibonding) accounts for the observed shortening of the boron-boron bond in the anti-van't Hoff structures relative to the normal forms.

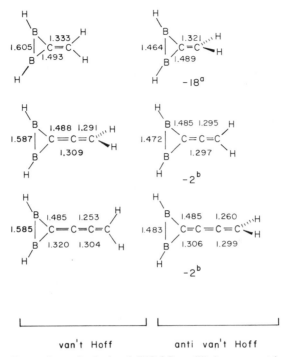

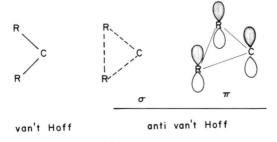

FIGURE 7.25. Comparison of calculated STO-3G equilibrium geometries for planar and perpendicular cumulenes $(HB)_2C(=C)_n=CH_2$, $n = 0$, 1, 2. Calculated energies of anti van't Hoff cumulenes (relative to van't Hoff forms) are tabulated underneath. (a) MP3/6-31G**//3-21G, reference 161b; (b) 4-31G//STO-3G, reference 161a.

In general, for molecules which obey van't Hoff stereochemistry, for example, tetrahedral methane, planar ethylene, and perpendicular allene, the terminal CR_2 group is best described in terms of a pair of two-electron, two-center bonds, while compounds preferring anti-van't Hoff geometries will utilize two three-center, two-electron bonds, one of σ and one of π type.

Whenever such multicenter bonding can be achieved easily, unconventional structures may result.

A similar situation is found for transition-metal carbene complexes, at least when the metal is coordinatively unsaturated [162]. The STO-3G rotational barriers in dihydridomethylidenetitanium, **104**, and in the analogous zirconium compound, **105**, are only 13 and 19 kcal mol^{-1}, respectively. These are clearly not representative of the actual π-bond strengths in these 8-electron complexes. As was the case with 1,1-dilithioethylene and related systems (see previous discussion), both planar and perpendicular forms of both compounds exhibit nearly equal central double-bond lengths.

104

105

The barrier hindering torsion about the double bond in the analogous 16-electron dicyclopentadienylmethylidenetitanium complex, **106**, is much larger (52 kcal mol^{-1} according to the STO-3G calculations). Interestingly, the metal–carbon bond in the higher-energy perpendicular conformer is actually slightly shorter than that in the lower-energy planar form, the same situation as was previously noted for the corresponding structures of 1,1-dilithioethylene.

106

The probable reason for the small barriers in the highly-electron-deficient titanium and zirconium carbenes, and for the significant increase in barrier upon increased saturation of the metal, is easily understood by inspection of the high-lying valence orbitals. The π orbitals for both planar and perpendicular forms of dihydrido-methylidenetitanium (Figure 7.26) are virtually identical; those for the analogous zirconium complexes (not shown) reveal only subtle differences. The reason is clear. The contribution of the metal to the π bond is a hybrid, which comprises mainly a d_{xz}-type atomic orbital, and only small contributions from p_x-type functions. This combines with the $2p_x$ atomic orbital on carbon. The corresponding representation for the perpendicular conformer utilizes the $2p_y$ function on carbon instead of $2p_x$; the contribution of the metal now arises primarily from a d_{yz} function. The important point is that it is the d-type functions on the metal, and not the valence p-orbitals, which contribute to the π bond [163]. In the 8-electron complexes, where the d orbitals are not completely tied up in ligand binding, the d_{xz} component is available to interact with the $2p_x$ on carbon in the planar methylidene complex,

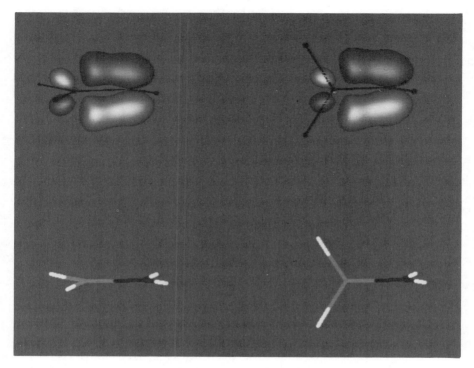

FIGURE 7.26. Highest-occupied (π) molecular orbitals of planar (left) and perpendicular (right) structures of dihydridomethylidenetitanium. Skeletal frameworks are underneath. STO-3G//STO-3G.

and the d_{yz} component available to interact with $2p_y$ in the perpendicular conformation. The increased barrier in the 16-electron complex, **106**, probably reflects the increased involvement of the metal d orbitals in ligand binding: only the d_{xz} component remains available.

7.4.3. Structures of Transition-Metal Carbenes and Transition-Metal Alkyls. The Role of Hyperconjugation in Carbon–Metal Bonding

Equilibrium geometries for a number of metal carbenes have been determined by X-ray and neutron diffraction. That for the 14-electron tantalum carbene, **107** [165],

107

shows a remarkable TaCC bond angle of 165°, much larger than the 120° angle which would be expected for an sp^2-hybridized carbon center. Related compounds in which the metal center is formally associated with 16 or fewer electrons show comparable distortions [166]. The available neutron diffraction structures for metal carbenes show significant lengthening of the alkylidene CH bonds, and MCH bond angles that are well below normal 120° values [166c]. Long alkylidene CH bonds in these compounds are also suggested by low CH-stretching frequencies and NMR coupling constants [166c].

The geometrical distortions in metal alkylidenes may be due to hyperconjugation [167], that is,

$$M=C\underset{R}{\overset{H}{\big\langle}} \longleftrightarrow \bar{M}\equiv C-R \quad \overset{+}{H}$$

In particular, the longer-than-normal alkylidene CH bonds, the smaller-than-normal MCH angles and the large MCR bond angles are all readily accommodated. Hyperconjugation in the plane normal to the π system leads to $^-M=C^+$ polarization, that is, the reverse of $^+M=C^-$ polarization of the π bond. This is supported by the results of STO-3G calculations on a series of titanium carbenes [167b], the principle structural features of which are shown in Table 7.38.

Both the 8-electron and 16-electron methylidene complexes, **108** and **109**, exhibit symmetrical, nonbridged structures, as does the cyano adduct, **110**. A partially-bridged geometry is indicated for the ethylidene complex, **111**. The amino derivatives, **112** and especially **113**, in which the NH_2 group is twisted perpendicular to the plane of the double bond, also show distortions to nonsymmetrical, bridged structures. Electron donation from the nitrogen lone pair into the CH linkage, which acts to stabilize hyperconjugative structures and promote bridging, is possible only in the perpendicular arrangement, **113**. Although the structure of the 8-electron fluoromethylidene complex, **114**, is severely distorted, that of the corresponding 16-electron system, **115**, is approximately symmetrical. Thus, both substitution at carbon and at the metal center appear to be important in determining the structure of the complexes. The preferential stabilization of partially-bridged forms by methyl, fluoro, and amino substituents at the methylidene carbon, and the associated destabilization by nitrile, is easily rationalized by the hyperconjugative model, that is, π-electron donors stabilize the $^-M=C^+$ valence structure while acceptors destabilize it. The distortion seems to depend on the electron deficiency of the central metal atom. Coordinatively saturated systems, for example, **115**, do not exhibit extreme distortions even when the substituents at the methylidene carbon would stabilize a partially-bridged structure.

The metal–carbon linkages in the distorted structures are shortened relative to those with nonbridged geometries. This is also apparent in the experimental structures of tantalum carbenes [164b]. In addition, only the distorted complexes display abnormally long CH bonds. Again, the available experimental data are in accord [166c].

TABLE 7.38. Selected Geometrical Parameters for Titanium Carbenes[a]

Structure		Geometrical Parameter			
		$r(\text{Ti}{=}\text{C})$	$r(\text{CH})$	$<(\text{TiCH})$	$<(\text{TiCX})$
H₂Ti=CH₂	**108**	1.833	1.087	125.4	125.4
(C₅H₅)₂Ti=CH₂	**109**	1.876	1.087	127.2	127.2
H₂Ti=C(H)C≡N	**110**	1.852	1.088	111.5	124.1
H₂Ti=C(H)CH₃	**111**	1.790	1.134	83.9	165.0
H₂Ti=C(H)NH₂	**112**	1.806	1.122	88.6	161.9
H₂Ti=C(H)NH₂	**113**	1.780	1.178	76.3	169.4
H₂Ti=C(H)F	**114**	1.779	1.164	79.3	169.9
(C₅H₅)₂Ti=C(H)F	**115**	1.899	1.100	125.0	130.6

[a] STO-3G; bond lengths in Å, bond angles in degrees; data from reference 167b.

Hyperconjugative structures of simple metal–alkyl complexes are also reasonable, that is,

and suggest the possibility of unusual geometries [168]. Indeed, the known structures of the tetrabenzyl complexes of group IV transition metals (Ti, Zr, and Hf) are nothing less than extraordinary [169]! Each incorporates four different MCC bond angles, ranging from close to 90° to greater than 120°. The geometry of tetrabenzyltitanium, **116,** is typical.

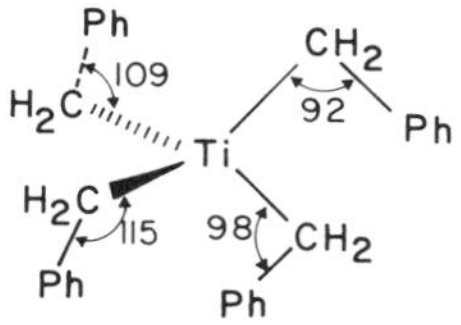

116

In contrast, the structures of the analogous main-group compounds exhibit near-tetrahedral bond angles [169c]. Distortions of this type are not limited to complexes of the early transition metals: ThCC bond angles in η^5 $(C_5 Me_5)Th(CH_2 Ph)_3$, **117,** are all approximately 90° [170].

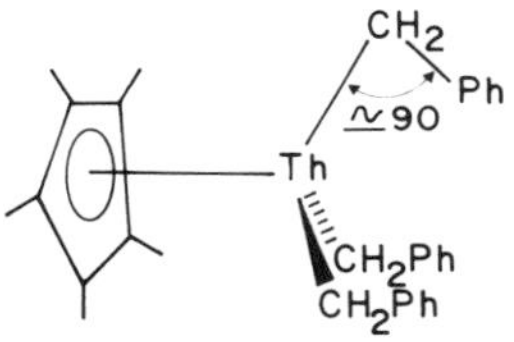

117

Neither are they exclusive to coordinated benzyl groups. Simple alkyl groups in electron-deficient complexes have also been shown to exhibit highly distorted geometries, for example, the TiCC bond angle in $Cl_3 Ti$ (diphos)$CH_2 CH_3$, **118,** is 86° [171].

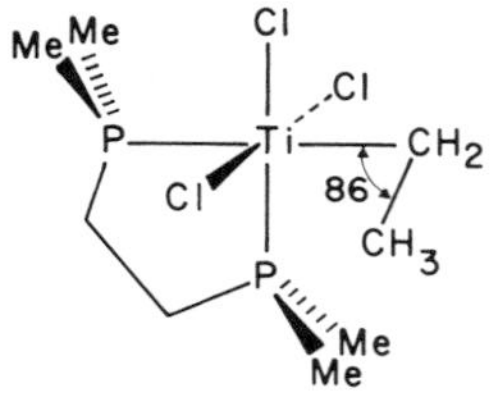

118

TABLE 7.39. Calculated Equilibrium TiCR Bond Angles and Distortion Energies for Selected Titanium Complexes

Molecule $H_3 Ti—CH_2—R$	Equilibrium TiCR Bond Angle (degrees)	Distortion Energy (kcal mol^{-1})^a
$H_3 TiCH_3$	111.7	5
$H_3 TiCH_2 CH_3$	112.7	3
$H_3 TiCH_2 Ph$	110.5	2
$Cp_2 (H)TiCH_3$	113.3	7

^{a}Energy required for a distortion of TiCR bond angle from equilibrium value to a (constrained) value of 90°.

Calculated (STO-3G) TiCR bond angles for a selection of titanium methyl, ethyl and benzyl complexes, presented in Table 7.39, are all well inside the normal range for sp^3-hybridized carbon. Note, however, that the required energy for closure of the TiCR bond angle from its equilibrium value to 90° in each of the highly-electron-deficient complexes is very small; bent structures are energetically more accessible for ethyl and benzyl complexes than for the methyl systems. Distortion of $H_3 TiCH_2 Ph$ is probably assisted by the ability of the aromatic π system to donate to the electron-deficient metal center, that is,

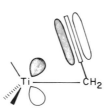

The corresponding distortion energy for the 16-electron complex, $Cp_2 Ti(H)CH_3$, is slightly higher than that for the 8-electron complex, $H_3 TiCH_3$, although it is still small relative to typical values of 15–25 kcal mol^{-1} for distortion in main-group compounds. The titanium center in the 16-electron complex, while not severely electron deficient, still has available an empty d orbital to accept electrons from the alkyl group. Crystal structures for the 16-electron complexes, $Cp_2 Zr(CHPh_2)_2$ [172] and $Cp_2 Hf(CHPh_2)_2$ [173], exhibit MCC bond angles that fall in the range of 114–121°. It seems likely that related complexes, in which the central metal is not severely electron deficient, will assume nondistorted geometries in the solid state.

It is apparent that hyperconjugative resonance structures are less important for electron-deficient metal-alkyl complexes than they are for metal alkylidenes. Although model structures for distorted titanium complexes are predicted to be less stable than normal (nondistorted) structures, the calculated energy differences are very slight. Sizable geometric distortions to accommodate better the packing of molecules in the solid state may occur with relative ease.

7.5. MOLECULES WITH SEVERAL METALS

We examine now the structures and stabilities of additional molecules containing more than one metal atom. A number of compounds incorporating two electropositive metals have already been considered, for example, the Tebbe reagent (Sections 7.3.3b and 7.4.1b) and 1,1-dilithioethylene (Section 7.4.2). Their fascinating structures, in part consequences of extreme electron deficiency and high ionic character, suggest that similar, and perhaps even more remarkable, properties will be exhibited by molecules with even more metals, for example, metal clusters, exemplified by small beryllium aggregates, polylithiated hydrocarbons, and lithiocarbons.

7.5.1. Lithiocarbons and Lithiated Hydrocarbons

This section is devoted to *lithiated hydrocarbons*, $C_xH_yLi_z$, with more than one lithium, and to *lithiocarbons*, C_xLi_z. Since organolithium compounds are frequently equated with *carbanions* (see Section 7.3.2), the presence of more than one alkali metal in a molecule implies more than one negative charge. Although *polyanions* are commonly discussed in the chemical literature [174], this is an oversimplified description. Coulomb repulsion destabilizes such multiply-charged species, and none appear to be known in the gas phase. The naphthalene dianion, **119**,

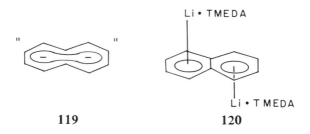

119   **120**

is illustrative. Naphthalene cannot even bind a single electron in the gas phase (its electron affinity is negative [175]), let alone two electrons. The observed chemical intermediate is stabilized electrostatically by interaction with two metallic counterions, as exemplified by the X-ray structure of the dilithium compound, **120** [176]. The cations are associated with the opposite faces of different rings.

Polylithium compounds are common synthetic intermediates; some are easy to obtain [174]. For example, pyrolysis of methyllithium gives dilithiomethane [177]. Reaction with excess butyllithium converts propyne to C_3Li_4 [178], and 1,3-pentadiyne, $CH_3C\equiv C-C\equiv CH$, to C_5Li_4 [179].

Lagow has prepared CLi_4, C_2Li_6, and a variety of polylithiated hydrocarbons by high-temperature gas-phase reactions of lithium atoms with various organic substrates [180]. Wu has detected CLi_3, CLi_4 and even CLi_5 and CLi_6 mass spectroscopically, as products resulting from the diffusion of lithium vapor through graphite membranes [181]. The same research has also led to lithiated carbocations such as CLi_3^+ and CLi_5^+ [180-182].

a. 1,2-Dilithioethane and Higher α,ω-Dilithioalkanes. 1,2-Dilithioethane is inherently interesting as the simplest possible ethane vicinally substituted by two metals; it also serves as a model for double-lithiated stilbene and acenaphthene, for both of which X-ray crystal structures are available [183]. In the known structures, the two lithium atoms occupy nearly-symmetrical or symmetrical *trans* positions, that is, on opposite faces of the molecule [184].

The calculated (3-21G) equilibrium geometry of *trans* 1,2-dilithioethane is the unsymmetrical double-bridged structure, **121**, which incorporates a CCLi bond angle of only 72°. The symmetrically-bridged structure, **122**, is only slightly higher in energy (2 kcal mol^{-1} at the MP2/6-31G*//3-21G level, Table 7.40), but is not an energy minimum. Instead, it represents a transition structure for the exchange of the two vicinal lithiums, that is, a *dyotropic rearrangement* [185].

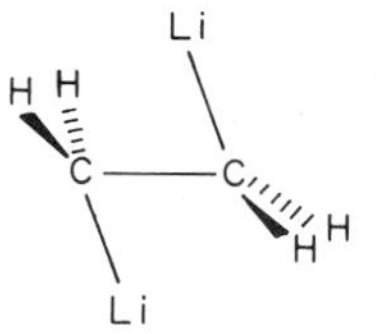

121 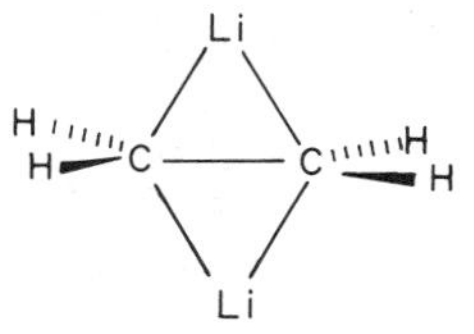   **122**

Calculations indicate that *cis*-(eclipsed) 1,2-dilithioethane, **123**, is much less stable than the *trans* compound (by 29 kcal mol^{-1} at the MP2/6-31G*//3-21G level), and is not a minimum on the 3-21G potential surface, but that a *gauche* form, **124** (LiCCLi dihedral angle of 84°), which is a minimum, is only 8 kcal mol^{-1} higher in energy than **121**. It too is bridged unsymmetrically. The corresponding symmetrical form, **125**, a possible transition structure for dyotropic rearrangement in *gauche*-1,2-dilithioethane, lies 2 kcal mol^{-1} higher in energy [184].

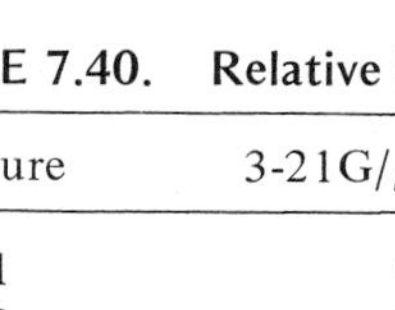

123 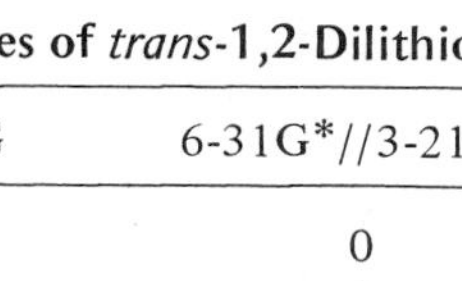   **124** 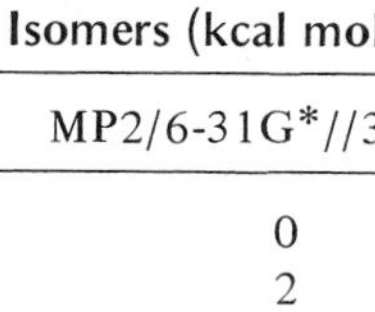   **125**

TABLE 7.40.   Relative Energies of *trans*-1,2-Dilithioethane Isomers (kcal mol^{-1})^a

Structure	3-21G//3-21G	6-31G*//3-21G	MP2/6-31G*//3-21G
121	0	0	0
122	2	4	2
123	29	28	29
124	8	7	8
125	9	9	10

^{a}Data from reference 184.

In principle, ethane itself might also undergo dyotropic rearrangement via the doubly-bridged diborane-like structure, **126**.

126

As discussed in Section 7.4, this is highly unlikely; the energy of **126** is 149 kcal mol^{-1} above that of normal, staggered ethane at the MP2/6-31G*//6-31G* level (higher than the CH bond energy). The greater ionic character of lithium, as well as the availability of vacant p orbitals to participate in multicenter bonding involving both carbons (and to some extent the hydrogens), contribute to the pronounced difference in bridging proclivities of lithium and hydrogen.

Only indirect evidence supports the existence of 1,2-dilithioethane. Elimination of LiH to give vinyllithium is very rapid, and complicates attempts to detect the species in solution. While the gas-phase elimination reaction (7.54) is endothermic (by 21 kcal mol^{-1} at 3-21G//3-21G), another process (7.55), involving elimination followed by recombination to give a complex, **127**, between vinyllithium and lithium hydride, is highly exothermic (by 30 kcal mol^{-1} at 3-21G//3-21G).

$$LiCH_2CH_2Li \rightarrow H_2C=CHLi + LiH \qquad (7.54)$$

$$LiCH_2CH_2Li \rightarrow H_2C=CH\underset{Li}{\overset{Li}{\diamond}}H \qquad (7.55)$$

127

Thus, **127** appears to be the global minimum on the $C_2H_4Li_2$ potential surface (3-21G), even though 1,1-dilithioethane is 18 kcal mol^{-1} less stable than the best *trans*-1,2-disubstituted compound, **121**, at the same level of theory [186].

Other mixed dimers involving LiH and LiR are likely to be important intermediates in lithium chemistry. For example, 1,3-dilithiopropane, **128**, is also much less stable in ether solution than primary alkyllithiums; elimination to yield allyllithium and LiH occurs rapidly [187]. If elimination is blocked, as it is in 1,3-dilithio-2,2-dimethylpropane, unusual chemical stability in solution is observed [187]. Again, the gas-phase elimination reaction of **128**, reaction (7.56), to allyllithium, **129**, and lithium hydride, is predicted to be endothermic (by 17 kcal mol^{-1} at 3-21G//3-21G), while the elimination, followed by formation of the complex **130**, reaction (7.57), is found to be exothermic by 29 kcal mol^{-1} at this level.

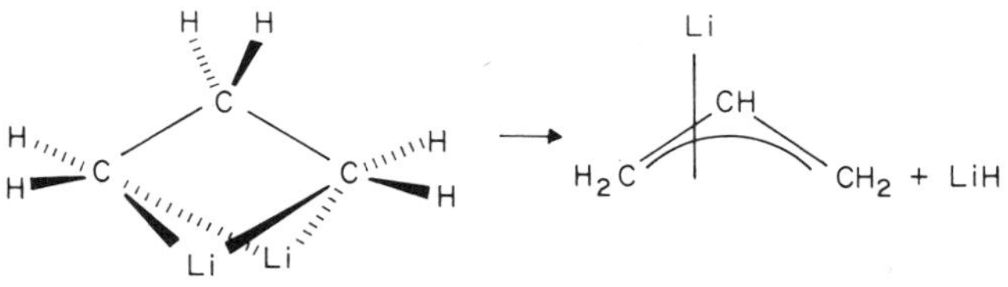

(7.56)

128 **129**

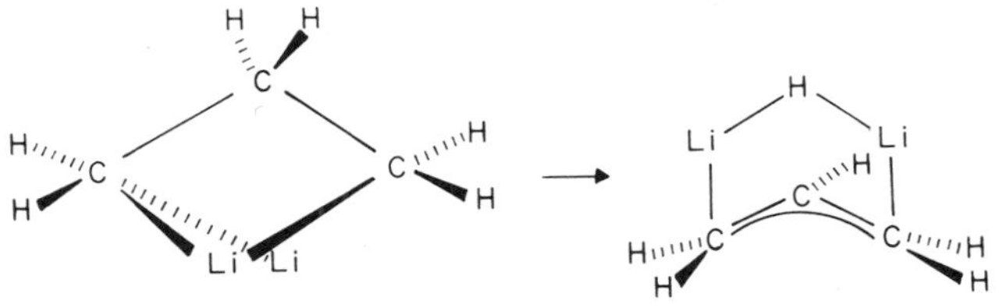

(7.57)

128 **130**

Complex **130** benefits from the interaction of the empty p functions on the two lithiums with the highest occupied molecular orbital on allyl anion.

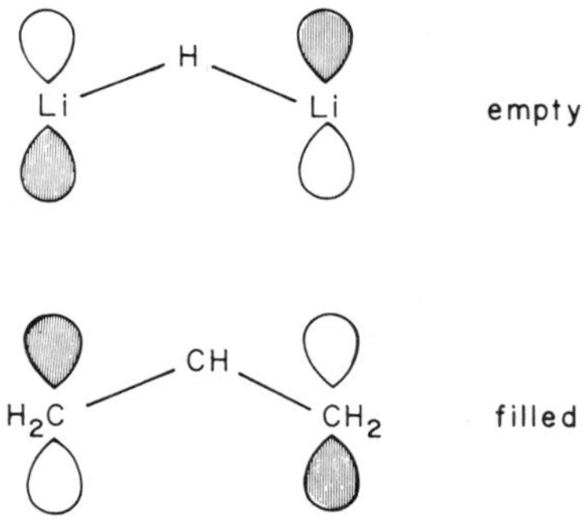

empty

filled

The most stable geometry of 1,3-dilithiopropane is the doubly-bridged structure, **128**. This form is predicted to be 25 kcal mol^{-1} lower in energy (3-21G//3-21G) than an alternative W-shaped structure (not shown) in which each lithium is associated with only a single carbon. The difference reflects the electrostatic benefits resulting from intramolecular interactions between electron-deficient and electron-rich centers.

1,4-Dilithiobutane (and higher α,ω-dilithioalkanes) are easily prepared, in contrast to the lower homologs. The preferred structures again involve double bridging by lithium, for example, 131.

131

Extended linear forms are much higher in energy. For example, the fully extended form of 1,4-dilithiobutane is 35 kcal mol^{-1} above **131** at 3-21G//3-21G [188]. This approaches the dimerization energy of methyllithium (46 kcal mol^{-1}, see Section 7.3.2a), a system that does not suffer from the geometry constraints of a polymethylene chain. Consistent with the experimental observation of **131**, its isomerization to a complex between 3-butenyllithium and lithium hydride, **132**, is predicted to be endothermic.

132

b.   Dilithioethylene.   A number of possible structures for dilithioethylene have been examined at the 3-21G level (Table 7.41) [159c]. Two other forms, perpendicular and planar 1,1-dilithioethylene, have already been discussed in Section 7.4.2. They are not as stable as the best of the structures dealt with here, in contrast to the behavior of more conventional substituents which generally favor 1,1 over 1,2-substitution [189, 190].

The global minimum on the C$_2$H$_2$Li$_2$ potential surface at this level appears to be the *cis*-doubly-bridged structure, **133**. The more conventional *cis* structure, **137**, is not an energy minimum; normal-mode analysis yields a single imaginary frequency. The best *trans* form located is **134**, the geometry of which incorporates acute LiCC bond angles (86° at 3-21G). A similar structure was also encountered on the dilithioethane potential surface (structure **121** in Section 7.5.1a). Unlike the latter system, however, the corresponding symmetrical doubly-bridged *trans* form, **138**, is not a transition structure for dyotropic rearrangement in 1,2-dilithioethylene. Not only is it very much higher in energy than the alternative dilithioethylene structures explored, but a normal-mode analysis reveals two imaginary frequencies. Movement along the normal coordinates corresponding to these frequencies leads to two other minima, an unusual (and low-energy) form **135**, and the alternative *trans*-doubly-bridged structure **136**. The reason for the instability of **138** may be rationalized in terms of a limiting description of C$_2$H$_2$Li$_2$ as C$_2$H$_2^{--}$ and 2Li$^+$.

TABLE 7.41. Relative Energies of Dilithioethylene
Structures (kcal mol^{-1})

Structure		3-21G//3-21G
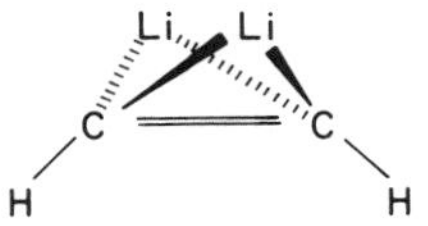	133	0
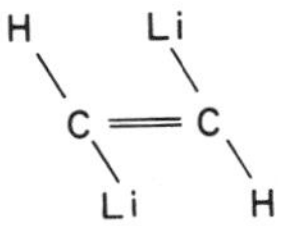	134	2
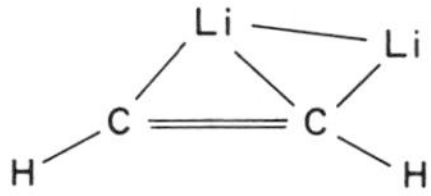	135	5
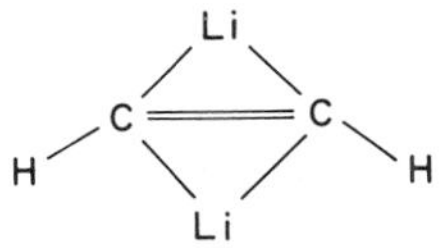	136	16
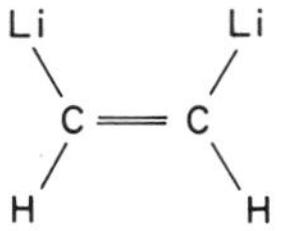	137	15
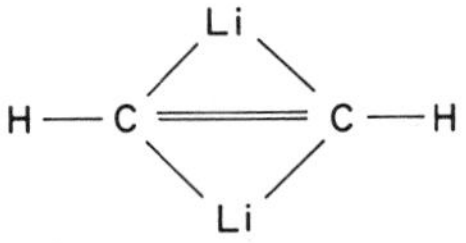	138	51

Here the preferred geometry of the dianion would be expected to be strongly bent, *cis* or *trans*, like the isoelectronic diimide structures. Thus, **138**, with its linear hydrocarbon arrangement, would not be very satisfactory. Dyotropic exchange of the lithiums in **134** might occur in a stepwise manner, process (7.58), involving inversion of hydrogen at one end of the molecule at a time and the intermediacy of a structure such as **136**.

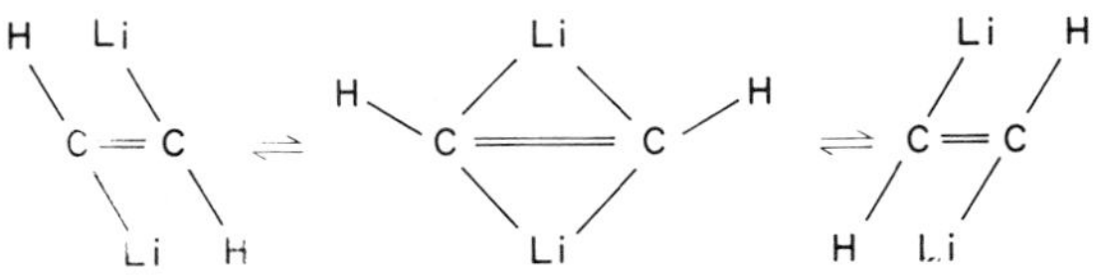

$$\text{134} \quad \rightleftharpoons \quad \text{136} \quad \rightleftharpoons \quad \text{134} \tag{7.58}$$

Although polylithioalkenes appear to be formed in gas-phase reactions [180 b, d], none of the species discussed here have as yet been prepared in solution. Moreover, even the lowest-energy structures, **133–135**, are calculated to be, at best, only marginally stable thermodynamically toward dissociation into C_2Li_2 and H_2; this may well contribute to the synthetic difficulties. It should be noted, however, that solvation and aggregation may enhance the thermochemical stabilities of dilithioalkenes relative to their dissociation products.

c. Lithiocarbons.   While binary compounds of carbon and lithium may seem exotic, 30 or more examples have now been observed. As yet, most of these are not well characterized in the usual chemical sense. The exception is lithium carbide, a readily available material which is inert toward both air and water [191]. The two principal ways of preparing lithiocarbons are H/Li exchange in solution, and high-temperature reactions of lithium vapor either with organic molecules or with graphite [180, 181]. Flash vaporization mass spectrometry [180e] has confirmed the formation not only of species such as CLi_4 and C_3Li_4, but also of dimers and trimers of these molecules, as well as compounds derived by loss or gain of one or more lithium atoms.

What is the nature of the bonding in lithiocarbons? Are they to be regarded simply as polyanions? Calculations suggest not. Bond lengths, Mulliken atomic charges and dipole moments for both tetrahedral and planar forms of CH_4, CH_3Li, *cis*-CH_2Li_2, $CHLi_3$, and CLi_4, obtained from 3-21G calculations, are provided in Table 7.42. In their tetrahedral forms, C–Li bonds shorten with increasing lithium substitution.

Mulliken charges on lithium decrease with increasing number of metal atoms. While this is not surprising, what is, is the fact that the negative charge on carbon is predicted to *decrease* from methyllithium to tetralithiomethane. There is no indication from these data that lithiocarbons are highly ionic species. As to be expected, positive charge is better situated on lithium than on carbon, but the latter does not appear to become excessively negative, at least according to the Mulliken analysis.

It may be useful to regard lithiocarbons as essentially alkali–metal–atom clusters associated with a carbon impurity. Lithium–lithium bonding appears to play an increasingly important role as the number of lithium atoms increases. As a consequence, the normal valency of carbon in these compounds may be exceeded, sometimes greatly so. Species such as CLi_5, CLi_6, and even CLi_8 appear to be stable molecules, and (according to calculations) assume geometries in which carbon is associated with more than its normal complement of valence electrons.

We survey here the structures and stabilities of three series of lithiocarbons. (Previous discussions of anti-van't Hoff geometries, Sections 7.4.1a and 7.4.2, have already touched upon the properties of CLi_4 and C_2Li_2.) The first are *hypermetalated* compounds as exemplified by CLi_5 and CLi_6. The factors responsible for their stability, as well as their remarkably normal geometries, will be examined. Isomers of C_3Li_4 and C_4Li_4 will also be considered, and compared and contrasted with the analogous hydrocarbons. Particularly interesting will be structures analogous to allene in the C_3Li_4 manifold and to cyclobutadiene and tetrahedrane among the isomers of C_4Li_4.

TABLE 7.42. Equilibrium Bond Lengths, Mulliken Charges and Dipole Moments in Singlet Lithiated Methanes (3-21G//3-21G)

Compound	Bond Lengths (Å)		Mulliken Charges (electrons)			Dipole Moment (debyes)
	$r(C–Li)$	$r(C–H)$	C	Li	H	
CH$_4$ (tetrahedral)		1.083	−0.80		+0.20	0.00
(planar)		1.084				0.00
CH$_3$Li (tetrahedral)	2.001	1.094	−0.94	+0.52	+0.14	5.50
(planar)	1.910	1.066 (*trans*)				5.09
		1.081 (*cis*)				
CH$_2$Li$_2$ (tetrahedral)	1.984	1.101	−0.92	+0.32	+0.14	5.60
(*cis* planar)	1.856	1.104	−0.92	+0.34	+0.12	4.27
CHLi$_3$ (tetrahedral)	1.952	1.112	−0.87	+0.24	+0.15	3.96
(planar)	1.839 (*trans*)	1.127	−0.81	+0.24 (*trans*)	+0.08	1.17
	1.766 (*cis*)			+0.25 (*cis*)		
CLi$_4$ (tetrahedral)	1.929		−0.80	+0.20		0.00
(planar)	1.982		−0.72	+0.18		0.00

(i) CLi_5 and CLi_6. While CH_5 and CH_6 are likely to exist only as weak complexes between methane and H or H_2, in accord with the octet rule, the situation is entirely different when lithium is involved instead of hydrogen. Both trigonal-bipyramidal CLi_5, **139**, and octahedral CLi_6, **140**, are predicted by 3-21G calculations to be stable toward all possible dissociation reactions [192].

139   **140**

For example, loss of lithium atom from CLi_5 is calculated to be endothermic by 54 kcal mol^{-1}, and loss of Li_2 from CLi_6 endothermic by 65 kcal mol^{-1}. The related series of molecules, $CLi_{5-n}H_n$ (n = 0-4) and $CLi_{6-n}H_n$ (n = 0-5), behave similarly [192].

How can it be that these species flagrantly exceed the maximum valence dictated by the octet rule? The calculated structures of CLi_5 and CLi_6 (shown above) are unspectacular; they incorporate C—Li bonds which are only slightly longer than those found in normal-valent compounds at the same level of theory, for example, 2.00 Å in CH_3Li and 1.92 Å in CLi_4. Normal bond lengths are also found in other hypermetalated species, for example, OLi_3 and OLi_4 [193]. A clue to the noted high stabilities and symmetrical equilibrium geometries of these compounds is to be found by examination of the valence-orbital manifold of a typical member, CLi_6 (Figure 7.27). In octahedral symmetry, the first eight electrons occupy orbitals that are analogous to those found in normal tetrahedral- or octahedral-symmetry species. These are clearly C—Li bonding but Li $\cdots$ Li antibonding. The last pair of electrons goes into a totally symmetrical orbital which, while it is carbon–lithium antibonding, possessing a spherical node, is nevertheless strongly lithium–lithium bonding. It is the 12 pairwise bonding contacts between adjacent lithium atoms which contribute to the high stability of CLi_6.

Further evidence for the mode of binding in these species comes from the calculated Mulliken charges at carbon, − 0.81 and − 0.93 electrons for CLi_5 and CLi_6, respectively, compared with − 0.80 electrons in tetrahedral CLi_4 (see Table 7.42) [194]. Thus, the extra electrons in the hypermetalated compounds *are not* primarily associated with carbon (which maintains what is essentially its original octet), but rather are free to contribute to Li $\cdots$ Li bonding, that is, to build a *metallic cage* around the carbon. One is tempted to suggest that the exact identity of the central atom is secondary, and that *hyperlithiation* should be a general phenomenon. Calculations [192, 193, 195] and experiments [181, 196] concur.

(ii) C_3Li_4. Other than C_2Li_2, C_3Li_4 is the most commonly encountered lithiocarbon. It is readily prepared in solution by lithiation of propyne or allene with butyllithium [178, 197]. C_3Li_4 is a major product obtained from the reaction

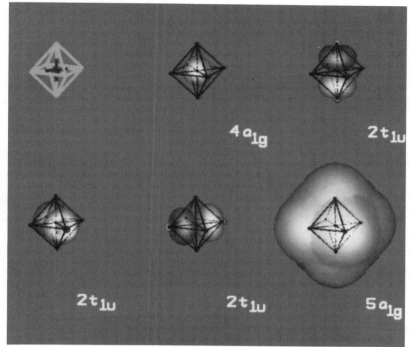

FIGURE 7.27. Valence molecular orbitals of CLi$_6$ from lowest to highest energy (top to bottom, left to right). STO-3G//STO-3G.

of lithium and carbon vapors [198], and is also formed by reaction of lithium atoms with a variety of organic substrates [180b,d,f]. The structure of C$_3$Li$_4$ has not been determined experimentally, although the infrared spectrum of the species has been interpreted in terms of an allene-like skeleton, rather than one incorporating a CC triple bond [178].

A number of possible C$_3$Li$_4$ structures have been investigated computationally [199]. Important geometrical parameters (STO-3G) are summarized below, and relative stabilities (STO-3G//STO-3G and 4-31G//STO-3G) are provided in Table 7.43.

TABLE 7.43.   Relative Energies of C$_3$Li$_4$ Isomers (kcal mol^{-1})[a]

Structure	STO-3G//STO-3G	4-31G//STO-3G
141	0	0
142	14	12
143	34	19
144	36	19
145	16	32

[a] Data from reference 199.

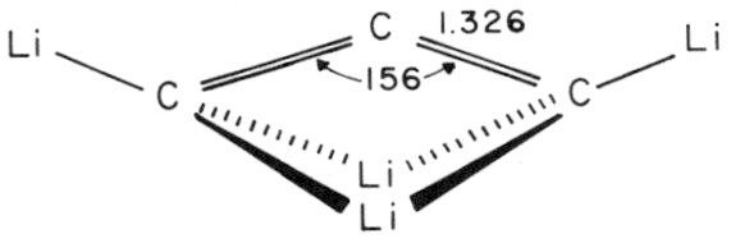

141

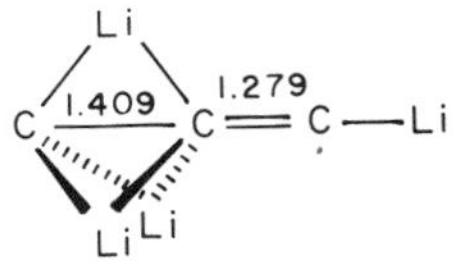

142

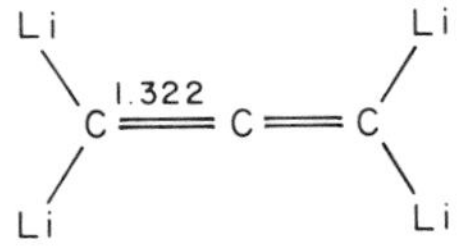

143

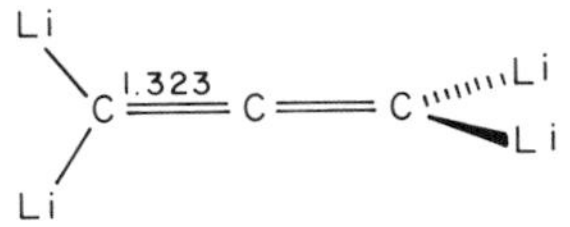

144

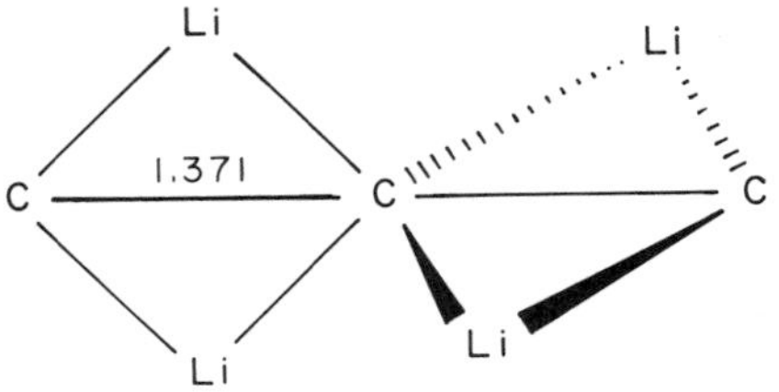

145

Most classical C_3Li_4 structures, that is, those based on the corresponding hydrocarbon geometries, are not favorable, especially those incorporating 3-membered rings. The exception is the allene analogue, **144**. Note, however, that the corresponding planar structure, **143**, involves CC bond lengths nearly the same as in **144** and is of comparable stability. This contrasts with the hydrocarbon system, where twisting away from a perpendicular geometry is energetically costly and leads to rupture of one of the π bonds. The similarity of **143** and **144** may be rationalized in a manner similar to that already discussed for planar and perpendicular 1,2-dilithioethylene (Section 7.4.2).

While the quadruply-bridged form, **145**, which might be thought of as a nonclassical allene, is more stable than either classical allene structure at the STO-3G//STO-3G level, this result is not supported by the 4-31G//STO-3G calculations. The search for both classical and nonclassical propyne analogues has been without success; imposition of a C_{3v} symmetry constraint leads instead to the triply-bridged structure, **142**, which is the second-lowest-energy minimum on the C_3Li_4 potential surface. The best structure found, **141**, has a curious geometry with two terminal and two bridging lithiums, as well as a W-shaped LiCCCLi backbone. All atoms bend away from the central carbon into a single hemisphere! That is to say, the central carbon actually prefers a pyramidal over a tetrahedral arrangement. The peculiar geometry of **141** is easy to rationalize in terms of its limiting description as C_3^{4-} (isoelectronic with CO_2) interacting with four lithium cations. Two of the negative charges reside in sp lone pairs at the ends of the C_3 chain; these are each coordinated by a single lithium cation. The other two negative charges are delocalized into two degenerate, four-electron, three-center π orbitals at right angles to one another. A lithium (cation) is associated with each. Even though slight bending of the carbon

skeleton occurs (to facilitate π overlap), the resulting LiCCLi dihedral angle, involving the two terminal carbons and the two out-of-plane lithiums (93° according to
the STO-3G calculations), is nearly the idealized value.

(iii) C_4Li_4. CH bonds attached to small strained rings have enhanced s character (because the corresponding CC bonds incorporated into the rings have enhanced p character). Carbanions resulting from deprotonation at strained centers are
more stable than their acyclic counterparts, for example, the cyclopropyl anion is
4 kcal mol^{-1} more stable than the 2-propyl anion (see Section 7.3.2b). The effect
increases with decreasing ring size and with involvement of more than one ring.
The best available calculations indicate that the acidity of bicyclobutane (deprotonation from the bridgehead positions) is 16 kcal mol^{-1} greater than that of cyclopropane, and the acidity of non-existent tetrahedrane is 29 kcal mol^{-1} above cyclopropane [158]. Organolithium compounds and related systems with high ionic
character show similar effects to carbanions; logic leads one to anticipate that
highly-strained hydrocarbons should be stabilized to an unusually high degree by
electropositive metallic substituents.

It was Dill et al. [200] who first pointed out that tetralithiotetrahedrane, **146**,
might be more likely to exist than the parent hydrocarbon, **147** [201]; theoretical
calculations supported their expectations [202].

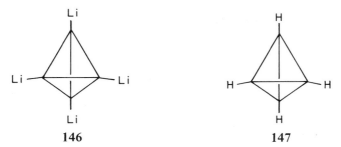

146   **147**

At the STO-3G//STO-3G level, the energy of the isodesmic reaction (7.59), which
provides a rough measure of the strain, is -196 kcal mol^{-1} for the hydrocarbon
(X = H) [202] but only -67 kcal mol^{-1} for the tetralithio compound (X = Li).

$$C_4X_4 + 6CH_3CH_3 \longrightarrow 4\,(CH_3)_3CX \qquad (7.59)$$

Even more remarkable is the reaction energy of -2 kcal mol^{-1} for an alternative
form of C_4Li_4, **148**, in which the metals are associated with the faces of the tetrahedra of carbon atoms, rather than with the corners [202].

148

The strain in tetrahedrane has vanished!

Note, however, that acetylene, another likely product of tetrahedrane dissociation, is also greatly stabilized by lithium substitution. MP2/6-31G*//6-31G calculations show a bond separation energy for dilithioacetylene, that is, ΔE for reaction (7.60), of 114 kcal mol^{-1} [203, 204].

$$C_2 Li_2 + 2CH_4 \rightarrow C_2 H_2 + 2CH_3 Li \qquad (7.60)$$

This means that dimerization of $C_2 Li_2$ to give tetralithiotetrahedrane (endothermic by 20 kcal mol^{-1} at MP2/6-31G*) is actually slightly more unfavorable than dimerization of acetylene to give tetrahedrane (14 kcal mol^{-1} at the same level of theory). Even so, **148** must be regarded as a remarkable molecule, and well illustrates how an inherently unfavorable carbon skeleton may be stabilized through substitution with electropositive metals.

Structure **148** is not a minimum on either the STO-3G or 3-21G potential energy surfaces (normal-mode analyses reveal two imaginary frequencies [204]). It does, however, appear to be a local energy minimum at higher theoretical levels [203].

Note the striking resemblance of **148**, with a tetrahedron of carbons on the inside and lithiums on the outside on the tetrahedral faces, to the methyllithium tetramer (Section 7.3.2a), which has a tetrahedron of lithium atoms on the inside and methyl groups outside on the faces. There are direct precedents for **148** in *Zintl compounds*, $Si_4 M_4$ and $Ge_4 M_4$ (M = Na, K, Rb, and Cs), which also exhibit face-centered tetrahedral geometries [205, 206].

Lithium substitution is also highly effective in stabilizing singlet cyclobutadiene. The $C_4 Li_4$ isomer, tetralithiocyclobutadiene **149**, is slightly more stable than **148** at lower levels of theory, but is slightly less stable at higher levels, for example, by 4 kcal mol^{-1} at 6-31G* (6-31G basis set on lithium).

149

The ordering of stabilities of the parent hydrocarbons is the reverse: at 6-31G*// 6-31G*, tetrahedrane is 27 kcal mol^{-1} higher in energy than singlet cyclobutadiene.

Various dilithioacetylene dimers are predicted to have considerably lower energies than the best $C_4 Li_4$ forms, that is, **148** and **149**. Structure **150**, a favorable possibility [204], benefits from a large number of metal–carbon interactions.

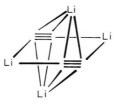

150

The crystal structure of dilithioacetylene comprises just such units [206].

There is some evidence for the existence of C_4Li_4 in solution. Photolysis of C_2Li_2 in liquid ammonia produces a new compound, distinct from the starting material [202]. Field-desorption mass spectrometry reveals a C_4Li_4 radical cation, although its structure, as well as that of the putative C_4Li_4 photolysis product, has yet to be established.

7.5.2. Small Beryllium Clusters

The structures and other properties of small metal clusters are of interest for a number of reasons [207]. Theoretical calculations [208] have dwelt on their role, both as models for studies of bulk crystals and crystal surfaces and on the interaction of atoms and small molecules with those surfaces. Small clusters are also important in their own right, and aggregates of several elements have been observed both in the gas phase and trapped in rare-gas matrices [209]. Of special interest is the role that small silver clusters play in the chemistry of photography [207c, 210].

Perhaps the simplest and most fundamental questions to ask about small clusters are their preferred structures (and the relationship of these structures to that of the bulk material), and the nature and strength of the binding of the individual atoms. It is also of interest to explore at what point these properties approach those of the bulk material. These questions have been addressed in a recent study of small beryllium clusters [211, 212]. Optimum (6-31G) geometries for the lowest singlet states of species Be_n ($n = 2, 3, 4, 5$) are presented in Figure 7.28. Binding energies, relative to complete dissociation into separated beryllium atoms, obtained at several different theoretical levels, are given in Table 7.44. The following points are worthy of specific mention.

While diatomic beryllium, 151, is only weakly bound, a conclusion that is supported by numerous other theoretical studies [212], the calculated binding energies in larger aggregates, 152–157, are significant. Strong binding appears to be more a function of having a collection of atoms than of the binding strengths of the individual contacts. According to the highest-level calculations (MP4SDQ/6-31G*//6-31G), binding energies (per beryllium atom) are 4 kcal mol^{-1} for Be_3, 152, 17 kcal mol^{-1} for tetrahedral Be_4, 153, and 18 kcal mol^{-1} for trigonal bipyramidal Be_5, 155. The calculated binding energy for square-planar Be_4, 154, is less than a third of that for the tetrahedral structure. The square-based pyramidal and pentagonal-planar alternatives of Be_5, 156 and 157, respectively, are also less strongly bound but not to as great an extent. The experimental heat of sublimation of beryllium metal (77 kcal mol^{-1} [213]) is not directly comparable, as each beryllium atom in the crystal has 12 nearest neighbors compared with 3–4 neighbors in the best pentamer structure considered here. Perhaps when one considers that 3–4 times the number of bonds need to be broken for loss of an atom from the bulk metal rather than from the model clusters, the level of agreement between calculated and experimental results is reasonable. Note that the calculated binding energies are very sensitive to theoretical level, seemingly to a greater extent than previous comparisons of relative energies would suggest (see especially Section 6.5.2). Hartree–Fock models underestimate the stabilities of the three-dimensional structures, whereas for Be_4 and

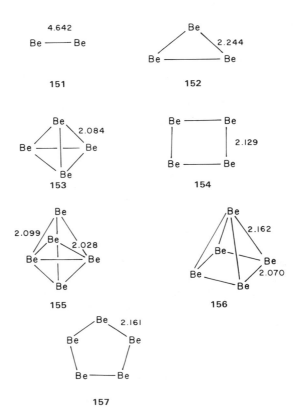

FIGURE 7.28. Calculated (6-31G) equilibrium geometries for small beryllium clusters.

TABLE 7.44. Binding Energies for Beryllium Clusters (kcal mol^{-1})^a

Cluster			6-31G	6-31G*	MP2/6-31G*	MP3/6-31G*	MP4SDQ/6-31G*
					//6-31G		
Be$_2$		**151**	0	1	1	1	0
Be$_3$		**152**	-1	2	25	20	13
Be$_4$	(T_d)	**153**	40	46	98	81	68
	(D_{4h})	**154**	20	23	35	26	18
Be$_5$	(D_{3h})	**155**	41	52	143	106	88
	(C_{4v})	**156**	45	53	101	78	62
	(D_{5h})	**157**	62	66	81	63	50

^{a} Data from reference 211.

464

Be_5, they overestimate the stabilities of planar (two-dimensional) forms. On the other hand, low-level correlation treatments overestimate the stabilities of all clusters relative to their separated atoms. The perturbation series has not yet converged at the MP4SDQ level, and it is not clear what effects additional terms or improvements in the underlying basis set will have. Clearly, the description of the relative energies of small metallic clusters, such as those dealt with here, cannot be given properly without adequate treatment of electron correlation. The sensitivity of cluster structure to the level of treatment of correlation, and to the properties of molecules chemisorbed onto cluster surfaces, remains to be studied.

Compact three-dimensional structures are greatly favored for the larger clusters over less crowded two-dimensional forms. Binding appears to be best achieved in those situations where each beryllium is associated with the maximum possible number of neighbors. Metallic beryllium assumes a hexagonal-close-packed (HCP) lattice, whereby sets of four adjacent atoms are arranged approximately tetrahedrally, and sets of five atoms in trigonal bipyramidal structures. The experimental Be—Be distances in the crystal (2.286 Å between atoms in a layer and 2.226 Å between atoms in different layers [214]) are somewhat larger than the corresponding lengths calculated for the trigonal bipyramidal structure (2.028 and 2.099 Å, respectively).

The building up of the larger clusters from the smaller units is easily rationalized in terms of simple frontier molecular orbital arguments. Figure 7.29 shows the result of interaction of the highest occupied and lowest unoccupied molecular orbitals of trigonal Be_3, **152**, with those of Be atom, to yield tetrahedral Be_4, **153**. The degenerate pair of highest occupied orbitals on Be_3 overlap constructively with the pair of lowest unfilled p functions, and are stabilized; in addition, the filled $2s$ atomic orbital on beryllium atom interacts favorably with the empty π-type function on Be_3. A similar diagram constructed for interaction of tetrahedral Be_4, **153**, and beryllium atom (leading to trigonal bipyramidal Be_5, **155**), given in Figure 7.30, leads to the same general conclusions.

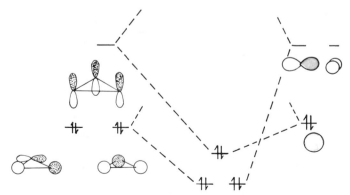

FIGURE 7.29. Representation of favorable interactions between highest-occupied and lowest-unoccupied molecular orbitals of triangular Be_3 and a beryllium atom approaching symmetrically from the top, and leading to tetrahedral Be_4.

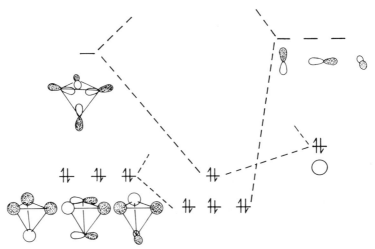

FIGURE 7.30. Representation of favorable interactions between the highest-occupied and lowest-unoccupied molecular orbitals of tetrahedral Be_4 and a beryllium atom approaching a face, and leading to trigonal bipyramidal Be_5.

7.6.   REACTION POTENTIAL SURFACES

The elucidation of reaction mechanisms is a major challenge for chemical theory. Although the geometries of reactants and products may generally be obtained experimentally using a wide variety of spectroscopic methods, these same techniques provide little if any detailed information about connecting pathways. *Transition structures*, local energy maxima along such reaction pathways, are not subject to experimental scrutiny, at least not at present [215]. *Activation energies*, that is, energies of transition structures relative to reactants, may often be obtained from rate measurements using *transition state theory* [216], and qualitative structural information may sometimes be inferred from *kinetic isotope effects* [217] and *activation entropies* [218]. However, the only way that the geometries of reaction transition structures may be obtained is from theory. Computational methods may also be employed to characterize short-lived *reactive intermediates*, which correspond to shallow local minima on potential surfaces. While such species may sometimes be detected experimentally, detailed information regarding their structures is usually difficult to obtain (see also Section 7.3).

Theory can be used to examine any arrangement of nuclei. A reaction potential surface can be explored to any degree of detail desired, using the same models applied to the equilibrium structures of reactants and products. It is possible to establish *a priori* whether or not a given structure corresponds to a local minimum, that is, a stable intermediate, or a saddle point, that is, a transition structure. Favored reaction pathways can then be established as those involving progression from reactants to products over the lowest-energy transition structures.

The application of theory to the description of reaction pathways is fraught with dangers. Having found a transition structure, it is difficult to rule out completely the possibility of alternative routes proceeding through other transition structures of even lower energy. Calculated potential energy surfaces may be very sensitive to the level of theory employed. Structures obtained using low levels, for example, Hartree-Fock methods with minimal or split-valence basis sets, may differ significantly from those that would result from calculations using larger basis sets, or including electron correlation. The success of simple theoretical models in determining the properties of stable molecules may not carry over into the description of reaction pathways. Transition structures are normally characterized by weak "partial" bonds, that is, those being broken or formed. In the extreme situation, bond cleavage is nearly complete, and the transition structure is accurately described only as an admixture of diradical and ionic states. Here Hartree-Fock models may perform poorly. Detailed assessment of a variety of quantum chemical methods is needed.

The characterization of even simple reaction potential surfaces may entail the location of one or more transition structures, and is likely to require many more individual calculations than are necessary to obtain equilibrium geometries for either reactant or product. While there has been significant progress in recent years toward the development both of analytical procedures for the calculation of first and second derivatives of the energy, and of methods for the location of minima and saddle points using this information [219], even the most advanced procedures are generally slow to converge on structures far away from a poorly chosen starting point. It may well be advantageous for a preliminary mapping of the potential surfaces to be carried out at a low level of theory, in advance of efforts made to locate points of interest accurately with higher-level methods. No matter what strategy is eventually employed, the practitioner should expect to expend considerable effort on potential surface explorations.

There still remain major difficulties associated with the characterization of many-dimensional potential surfaces. There is a scarcity of information at present, which makes it difficult to predict the detailed geometry of a particular transition structure. Experience will grow with each new exploration, and in time the geometries of transition structures will be as familiar as those of stable compounds are today. This area is one of the richest fields for chemical theory, for ultimately it will lead to an intimate understanding of the way chemical reactions proceed.

7.6.1. Classification of Reaction Types

Before discussing individual applications of theory, it is useful to classify chemical reactions in terms of the key features of the potential surface.

a. Isomerizations. Isomerizations are transformations from one reactant molecule, A, into an *isomer*, B.

$$A \rightleftharpoons B. \tag{7.61}$$

A section of the potential surface is represented in Figure 7.31. Here, A and B are separate local minima, and each of the possible paths connecting them will have an

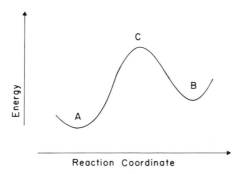

FIGURE 7.31. Reaction profile for endothermic isomerization of a molecule A into an isomer B via a transition structure C.

energy maximum. The *lowest maximum* of all such A → B paths occurs at the saddle point C, which is the preferred transition structure of the reaction. The energy difference $E(C) - E(A)$ is the activation energy for the A → B reaction, that is, the least energy required for the reaction to proceed (ignoring molecular vibration and tunneling corrections). It is possible that the preferred transition structure may correspond to some dissociated form of the molecule, in which case the favored path for the rearrangement would be dissociation followed by recombination.

Isomerizations may be subclassified as *nondegenerate*, where A and B are chemically distinct species, or *degenerate*, where A and B differ only by interchange of equivalent nuclei, so that both have the same energy. The rearrangement of $CH_3 CN$ to $CH_3 NC$ is an example of a nondegenerate process. The inversion of pyramidal ammonia and the internal rotation of ethane are examples of degenerate processes.

b. Addition or Combination Reactions.   Addition reactions occur when two species, A and B, come together to form an aggregate species, AB, that is,

$$A + B \rightarrow AB. \tag{7.62}$$

Such a reaction may occur with (Figure 7.32 (left)) or without (Figure 7.32 (right)) an activation barrier. Note that the product of reaction cannot survive for a long period unless a third body, M, such as a solvent molecule or the walls of the reaction vessel, is present to remove excess energy. Thus, reaction (7.62) is more properly written

$$A + B + M \longrightarrow AB + M^*, \tag{7.63}$$

where the * indicates excitation. The reverse processes,

$$AB \longrightarrow A + B, \tag{7.64}$$

are generally termed *dissociation reactions*, if A or B are radicals, for example,

$$HF \longrightarrow H^{\cdot} + F^{\cdot}, \tag{7.65}$$

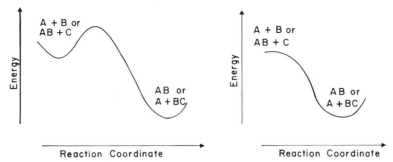

FIGURE 7.32. Reaction profile for addition of a molecule A to a molecule B, leading to the aggregate species AB, or of transfer of an atom (or fragment) B between atoms (or fragments) A and C, with activation (left) or without activation (right) in the forward (exothermic) direction.

and *elimination reactions*, if all species are ground-state singlet molecules. In the latter reaction, the smaller of A and B is said to be *eliminated* from the compound species AB, for example,

$$CH_3-CH_3 \rightarrow CH_2=CH_2 + H_2. \qquad (7.66)$$

c. **Bimolecular Transfer Reactions.** Bimolecular transfer reactions are processes in which two reactants combine and then separate to form two different products, that is,

$$AB + C \rightarrow A + BC. \qquad (7.67)$$

They may occur with an activation energy in both directions (Figure 7.32 (left)), or without activation in the exothermic direction (Figure 7.32 (right)). Bimolecular transfer reactions may be degenerate, as in the isotopic exchange reaction

$$CH_4 + D^{\cdot} \rightarrow CH_3D + H^{\cdot}, \qquad (7.68)$$

or nondegenerate, as in the process

$$CH_4 + F^{\cdot} \rightarrow CH_3F + H^{\cdot}. \qquad (7.69)$$

The reactions may be further subclassified according to the spin multiplicity of the species involved. The above are examples of radical-transfer processes.

d. **Reaction Profiles.** A complete ground-state potential energy surface for a particular species may describe reactions of more than one type. For example, the lowest singlet potential surface of CH_2O will reveal *both* the connection between the isomers formaldehyde, $H_2C=O$, and hydroxycarbene, HCOH, *and* the hydrogen elimination to give H_2 and CO. Each of these distinct reactions on the same potential surface has its own transition structure.

7.6.2.   Computational Procedures

The major objective in a survey of a potential surface $E(R)$ (Section 2.2) is the location of the stationary points. Mathematically, stationary points have an energy gradient equal to zero, that is,

$$\frac{\partial E}{\partial r_i} = 0, \qquad i = 1, 2, \ldots, 3N - 6. \qquad (7.70)$$

Here r_i is some set of $3N - 6$ internal coordinates for an N-atom molecule. Once a stationary point has been located, it may be characterized as a *local energy minimum*, a *local energy maximum*, or a *local energy saddle point*. These can be distinguished from one another by construction and diagonalization of the $(3N - 6) \times (3N - 6)$ symmetrical matrix, the elements of which are the second derivatives of the energy with respect to the internal coordinates, that is,

$$\frac{\partial^2 E}{\partial r_i \partial r_j}, \qquad i, j = 1, 2, \ldots, 3N - 6. \qquad (7.71)$$

If the eigenvalues of this matrix are all positive, the stationary point is an energy minimum, and corresponds to a particular isomeric equilibrium structure. If the eigenvalues are all negative, the energy is a local maximum, a situation of little chemical interest. We reserve the term saddle point (more strictly first-order saddle point) to describe those stationary points with *one and only one* negative eigenvalue. These correspond to situations in which the energy is at a minimum for all but one independent direction; in this one direction, represented by a single linear combination of internal coordinates, it is at a maximum. Saddle points *may* represent transition structures for reactions connecting two minima (or exit channels on the surface). The particular direction corresponding to a negative eigenvalue of the second derivative matrix is termed the *transition vector*.

In order to confirm that a saddle point is a transition structure for a particular reaction, it is necessary to establish a downhill *reaction path* connecting the saddle point with the two minima. Even then, there remains the possibility that other paths exist with lower-energy saddle points.

Many difficulties are encountered in locating saddle points. Simple minimization techniques, which can be used for equilibrium structures, cannot be applied to saddle points, since the energy is neither a minimum nor a maximum. One method is to use quadratic extrapolation if the first and second derivatives have been evaluated at a trial point. Thus, if $\mathbf{a}$ is the one-dimensional array $\partial E / \partial r_i$ ($i = 1, 2, \ldots, 3N - 6$) and $\mathbf{B}$ is the square matrix of second energy derivatives, (7.71), then the displacement from the starting point to the stationary point is given by (7.72),

$$\mathbf{x} = -\mathbf{B}^{-1}\mathbf{a}, \qquad (7.72)$$

(where $\mathbf{B}^{-1}$ is the inverse of the matrix $\mathbf{B}$), *provided that the surface is quadratic.*

This method is useful *if the trial point is close to the saddle point*, but is usually not effective in the early stages of an exploration when the two may be far apart.

Another technique sometimes adopted in searching for a transition structure is the use of a *trial reaction coordinate*, followed by refinement using gradient procedures. For example, the angle $\alpha = <HAB$ might be selected as a trial reaction coordinate for a 1,2-hydrogen-shift reaction

$$H-A-B \longrightarrow \overset{H}{\underset{A \overset{\alpha}{\diagup} B}{\diagup}} \longrightarrow A-B-H \qquad (7.73)$$

Next, a series of structures is obtained for different values of α, varying from the initial value (corresponding to the equilibrium structure of H-A-B) to the final value (corresponding to A-B-H). If the energies of these structures are *continuous*, we then have a section of the potential surface, $E(\alpha)$, for which the maximum corresponds to a saddle point. This procedure is not always effective. Progression in one direction may lead to a surface section that does not meet the section obtained by starting from the other direction. Under these circumstances, a more elaborate multidimensional scan of the surface is required to locate the saddle point.

7.6.3. Reaction Surfaces for Two-Heavy-Atom Systems

Singlet reaction potential surfaces, including those corresponding to C_2H_2, CHN, C_2H_4, CH_3N, CH_2O, N_2H_2, and CH_5N, have been investigated computationally [220]. They allow the establishment of isomers, provide estimates of the relative energies of stable forms, and finally assess the heights of any barriers separating these forms.

The HF/6-31G* model has been used for *locating* and *characterizing* the stationary points. While equilibrium geometries calculated at this level are generally in reasonable accord with experimental structures (see Sections 6.2.2 and 6.2.3), the reliability of HF/6-31G* as a means of describing the geometries of transition structures remains to be documented by comparisons with higher-level models. Single-point calculations for all energy minima and transition structures have been carried out at higher levels. The best of these, MP4SDQ/6-31G**//6-31G*, has been corrected for the effects of triple substitutions E (MP4/6-31G*) - E (MP4SDQ/6-31G*) (see Section 6.5.9). Although this does not correspond precisely to the full MP4/6-31G** theory, we refer to the results as such. Corrections for zero-point vibrational energy are also significant and have been estimated from 6-31G* harmonic frequencies (scaled by 0.9, see Section 6.3.9c). The imaginary frequency arising from the normal reaction coordinate at transition structures has been excluded from the zero-point energy calculation. The complete set of relative energies calculated are listed in Table 7.45. Also presented are relative energies obtained at the HF/3-21G level. They provide some indication of the performance of the simple model relative to more sophisticated treatments, assessment of which will be of value where use of the latter is not practical.

TABLE 7.45. Calculated Relative and Zero-Point Energies for Two-Heavy-Atom Molecules (kcal mol^{-1})^a

Molecule or Transition Structure	3-21G// 3-21G	HF/	MP2/	MP3/	MP4^b	Relative Zero-Point Energy
			6-31G**//6-31G*			
C$_2$H$_2$						
HC≡CH, **158**	0	0	0	0	0	0
H$_2$C=C:, **159**	39	34	49	42	44	−2
TS(HC≡CH → H$_2$C=C:)	64	49	51	49	48	−4
CHN						
HC≡N, **160**	0	0	0	0	0	0
HN≡C, **161**	9	11	19	16	16	−1
TS(HC≡N → HN≡C), **162**	68	51	53	51	49	−4
C$_2$H$_4$						
H$_2$C=CH$_2$, **163**	0	0	0	0	0	0
H$_3$C—C̈H, **164**	73	69	83	79	80	−3
TS(H$_2$C=CH$_2$ → H$_3$C—C̈H)	94	81	81	81	80	−3
TS(H$_2$C=CH$_2$ → H$_2$C=C: + H$_2$), **165**	121	111	106	106	104	−6
TS(H$_3$C—C̈H → HC≡CH + H$_2$), **166**	123	126	123	124	121	−6
CH$_3$N						
H$_2$C=NH, **167**	0	0	0	0	0	0
HC̈—NH$_2$, **168**	27	33	41	39	39	0
H$_3$C—N:, **169**	71	78	98	91	92	−3
TS(H$_2$C=NH → H$_3$C—N:)	88	77	83	82	81	−5
TS(H$_2$C=NH → HC̈—NH$_2$), **171**	102	95	91	92	90	−5
CH$_2$O						
H$_2$C=O, **172**	0	0	0	0	0	0
trans HC̈OH, **173**	47	49	58	54	56	0
cis HC̈OH, **174**	54	55	63	60	61	−1
TS(H$_2$C=O → *trans* HC̈OH), **176**		101	90	94	89	−5
TS(H$_2$CO → H$_2$ + CO), **177**		105	96	98	93	−6
N$_2$H$_2$						
trans HN=NH, **178**	0	0	0	0	0	0
cis HN=NH, **179**	8	7	6	6	6	0
H$_2$N—N:, **180**	9	19	28	26	26	−1
TS(*trans* HN=NH → H$_2$N—N:), **182**	90	89	78	83	78	−6
CH$_5$N						
H$_3$C—NH$_2$, **183**	0	0	0	0	0	0
H$_2$C · · · NH$_3$, **184**	66	72	74	74	73	−1
TS (H$_3$C—NH$_2$ → H$_2$C · · · NH$_3$), **185**	90	96	90	92	90	−5

^{a}Theoretical data from reference 220 and J. A. Pople, unpublished calculations.
^{b}Estimates: E(MP4SDQ/6-31G**) + E(MP4/6-31G*) − E(MP4SDQ/6-31G*).

472

a. C_2H_2.   Two isomers of C_2H_2 are reasonable: acetylene, **158**, and vinylidene, **159**.

$$H-C\equiv C-H \qquad\qquad \begin{matrix} H \\ \diagdown \\ \diagup \\ H \end{matrix} C=C\colon$$

158 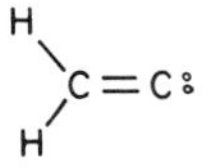   **159**

While only the first of these has been characterized experimentally, the existence of vinylidene has been inferred from chemical trapping experiments [221], for example,

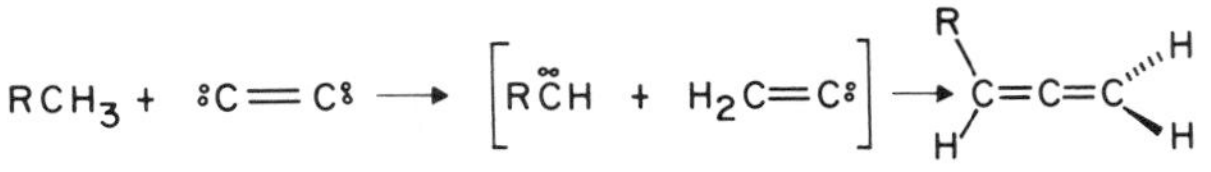

$$RCH_3 + \;\colon\!\!C\!=\!\!C\colon \longrightarrow \left[R\tilde{C}H + H_2C\!=\!\!C\colon\right] \longrightarrow \begin{matrix} R \\ \diagdown \\ \diagup \\ H \end{matrix}C\!=\!\!C\!=\!\!C\overset{H}{\underset{H}{\diagdown}} \qquad (7.74)$$

Efforts to detect the species spectroscopically have thus far been unsuccessful.

Both acetylene and vinylidene are found to be local minima at HF/6-31G*, and the barrier for rearrangement (transition structure not shown) is found to be quite large. With the 6-31G** basis, the interconversion barrier is still 15 kcal mol^{-1}, but this decreases rapidly when correlation energy is included. The barrier at the MP3/6-31G** level, 7 kcal mol^{-1}, is comparable to values obtained using configuration interaction methods by Dykstra and Schaefer [222]. Increasing the size of the basis set (to 6-311G**), taking into account fourth-order Møller-Plesset contributions, particularly those due to triple substitutions (MP4/6-311G**//MP2/6-31G*), and applying a zero-point energy correction lowers the barrier to less than 1 kcal mol^{-1} [222b]. In view of these results, it seems unlikely that vinylidene will have a significant lifetime. Indeed, structure **159** may be a saddle point on the C_2H_2 potential surface, representing a transition structure for the experimentally known degenerate rearrangement in which the two hydrogen atoms of acetylene change places [222f].

The HF/3-21G model provides a reasonable description of the relative stabilities of acetylene and vinylidene; however, it overestimates the relative energy of the transition structure.

b. **CHN.**   Both hydrogen cyanide, **160**, and hydrogen isocyanide, **161**, have been characterized experimentally.

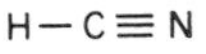

$$H-C\equiv N \qquad\qquad H-N\equiv C\colon$$
160 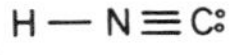   **161**

Several independent experimental determinations of their relative stabilities in the gas phase are available. Based on the intensity of infrared absorption of HNC in a sample of "pure" HCN at 1000 K, Maki and Sams have estimated the difference in energies of the two isomers as 10.3 kcal mol^{-1} [223]. An earlier unsuccessful attempt by Brown and coworkers to detect the $J = 1 \to 0$ HNC microwave line suggested a difference of at least 10.8 kcal mol^{-1} [224]. A somewhat higher separation of 14.8 ± 2 kcal mol^{-1} has been estimated by Pau and Hehre [225], who measured the relative free energies of deprotonation and dedeuteration of DCNH$^+$ by ICR spectroscopy. Finally, work by Ellison and coworkers [226] suggests that HNC lies somewhere between 17.2 and 26.3 kcal mol^{-1} above HCN in enthalpy. While reaction of CN$^-$ with HI leads both to HNC and HCN, proton abstraction from either HBr or HCl results only in HCN, the more stable isomer. The barrier for cyanide to isocyanide rearrangement in this system, that is, **161** → **160**, has not as yet been measured experimentally.

Calculations at the MP4/6-31G**//6-31G* level place HNC 16 kcal mol^{-1} above HCN. The barrier to rearrangement, through the cyclic transition structure, **162**, is predicted to be 33 kcal mol^{-1} at this level of theory, reducing to 30 kcal mol^{-1} after effects of zero-point vibration are taken into account.

162

This is consistent with the long-lived existence of HNC. The calculated (HF/6-31G*) transition structure shows a CN bond length of 1.169 Å, somewhat longer than that calculated in either HCN (1.132 Å) or HNC (1.154 Å) but considerably shorter than a CN double bond, for example, 1.250 Å in CH$_2$NH.

Other high-level quantum chemical calculations reach similar conclusions about the HCN potential surface [227]. In particular, all find HNC to be 10–15 kcal mol^{-1} higher in energy than HCN, and the two isomers to be connected by a significant energy barrier. The latter result is also consistent with the observation that both HCN and HNC are present in interstellar space [228].

The HF/3-21G calculations underestimate the HCN/HNC energy difference, and overestimate the separation between HCN and the rearrangement transition structure.

c. **C_2H_4.** Two potential minima have been located on the HF/6-31G* singlet potential surface for C_2H_4: ethylene, **163**, and ethylidene, **164**.

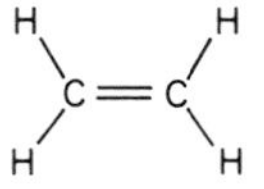

163 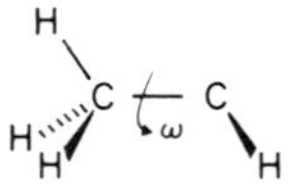   **164**

The latter is predicted to be 77 kcal mol^{-1} higher in energy than ethylene at the MP4/6-31G**//6-31G* level (after zero-point correction). Interestingly, the calculated equilibrium structure of **164** has C_1 symmetry. This preferred conformation, in which the HCCH dihedral angle (ω) is 165°, facilitates overlap of a CH bond of the methyl group with the formally vacant orbital at the carbene center.

The transition structure separating the two forms is also of C_1 symmetry; however, the highest-level calculations show that the energy of this structure is essentially the same as that for ethylidene. In view of experience with the C_2H_2 surface, that is, preferential stabilization of the cyclic transition structure over the acyclic forms with increase in theoretical level, it is highly likely that *ethylidene is not a minimum on the C_2H_4 potential surface* [229]. Rather, it is a possible transition structure for 1,2-hydrogen scrambling in ethylene. This would leave but a single identified singlet C_2H_4 isomer.

Other parts of the C_2H_4 singlet potential surface have also been explored [229b]. Of particular interest is the lowest-energy path for elimination of H_2 from ethylene to give acetylene. Experimentally, it is not known whether 1,1- or 1,2-elimination is more facile. Nor has it been established whether elimination of H_2 can occur at lower activation energies than that required for the radical dissociation process, $C_2H_4 \rightarrow C_2H_3^{\cdot} + H^{\cdot}$. Transition structures, **165** and **166**, have been located on the HF/6-31G* potential surface.

$$ \text{165} \qquad\qquad \text{166} $$

Structure **165**, corresponding to 1,1-elimination from ethylene, leads initially to vinylidene. As the latter does not appear to be stable, see Section 7.6.3a, further rearrangement to acetylene would be expected. Transition structure **166** corresponds to 2,2-elimination from ethylidene. MP4SDQ/6-31G** calculations suggest that **165** is the lowest-energy transition structure for the elimination reaction, $C_2H_4 \rightarrow C_2H_2 + H_2$. At this level, the barrier for the process is predicted to be about 93 kcal mol^{-1}, significantly below the energy required for radical dissociation.

The energy of ethylidene relative to ethylene appears to be underestimated by approximately 7 kcal mol^{-1}, and the energy of the isomerization transition structure appears to be overestimated by 14 kcal mol^{-1} at the HF/3-21G level. This leads to a major overestimation (by 21 kcal mol^{-1}) of the barrier for the ethylidene $\rightarrow$ ethylene transformation at this level.

d.   CH$_3$N.   Among the reasonable singlet structures of formula CH$_3$N, the one with maximum valence is methyleneimine, **167**.

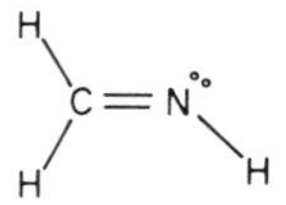

H
\
C = N
/   \
H   H

167

A single hydrogen shift could occur from carbon to nitrogen, leading to amino-methylene, **168**, or from nitrogen to carbon, leading to methylnitrene, **169**.

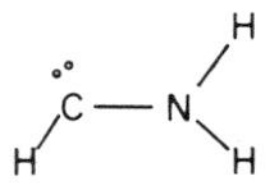

168 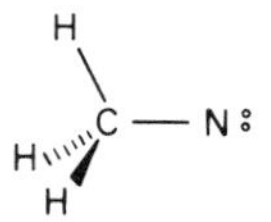   **169**

While only methyleneimine is known experimentally (as a highly reactive short-lived species [230]), all three forms of CH$_3$N are local minima on the HF/6-31G* potential surface. MP4/6-31G** calculations confirm that methyleneimine is the most stable of the three.

Interchange of the *syn* and *anti* hydrogens on carbon is predicted to occur by inversion at nitrogen, that is, via the C_{2v} symmetry transition structure, **170**, rather than by rotation about the carbon-nitrogen double bond. This result was noted previously by Lehn and Munsch [231].

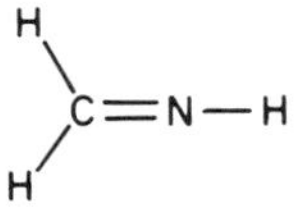

H
\
C = N — H
/
H

170

The MP4/6-31G** barrier is 31 kcal mol^{-1}, somewhat higher than known *syn-anti* interconversion barriers in substituted imines.

Methylnitrene, **169**, might first be considered in C_{3v} symmetry. Here, by analogy with diatomic NH, it would be expected to have two electrons in degenerate (*e*-type) molecular orbitals, leading to a triplet ($^3A_1$) ground state [232]. This is indeed found at the MP4/6-31G** level. The corresponding singlet form will have a ^1E state, and would be subject to Jahn–Teller distortion. The local minimum found has C_s symmetry and is 89 kcal mol^{-1} above methyleneimine after zero-point correction. However, MP4/6-31G*//6-31G* calculations yield an energy for the transition structure (not shown) for rearrangement of **169** to **167** which is actually lower than that for methylnitrene itself, and thus do not support the existence of a stable singlet for methylnitrene.

Aminomethylene, **168**, appears to be a stable molecule, due in part to a high degree of π delocalization from the nitrogen lone pair into the formally vacant p-type orbital on the carbene center. This is reflected in the exceptionally short CN bond length (1.309 Å) found by the HF/6-31G* calculations (compared with 1.250 Å in methyleneimine and 1.453 Å in methylamine at the same level of theory). The energy of aminomethylene is calculated at MP4/6-31G** to be only 39 kcal mol^{-1} above that of methyleneimine, consistent with an experimental lower bound of 34 kcal mol^{-1} for this difference from ion cyclotron resonance spectroscopy [233]. The transition structure connecting methyleneimine and aminomethylene is, according to HF/6-31G*, a planar form, **171**.

171

At the MP4/6-31G** level (after zero-point correction), the energy of **171** is nearly 86 kcal mol^{-1} above methyleneimine, again quite close to the dissociation limit, so that isomeric interconversion may occur via hydrogen dissociation followed by recombination.

The HF/3-21G calculations underestimate the energy separations between methyleneimine and the remaining CH_3N isomers but overestimate the relative energies of the two transition structures.

e. CH_2O. Three stable forms have been located on the singlet CH_2O potential surface: formaldehyde, **172**, and *trans-* and *cis-*hydroxymethylene, **173** and **174**, respectively [234].

172   173   174

At MP4/6-31G**, the *trans* carbene, **173**, is 56 kcal mol^{-1} above formaldehyde; the *cis* compound, **174**, is 5 kcal mol^{-1} higher in energy. Experimentally, the enthalpy difference between formaldehyde and *trans* hydroxymethylene has been determined by ion cyclotron double-resonance spectroscopy to be 54 kcal mol^{-1} [235].

As with singlet aminomethylene (Section 7.6.3d), the HF/6-31G* equilibrium geometries for **173** and **174** show evidence for delocalization of the nonbonding electrons on oxygen into the vacant orbital at the carbene center. The CO bond length in **173** is 1.300 Å (1.298 Å in **174**) which, although longer than the full double bond in formaldehyde (1.184 Å at the same level of theory), is far shorter than the calculated single-bond length in methanol (1.399 Å).

Interconversion of **173** and **174** occurs by internal rotation (as opposed to inversion) via the transition structure **175** and with a calculated barrier of 24 kcal mol^{-1} (MP4/6-31G**) from the *cis* form. The transition structure for conversion of *trans* hydroxymethylene to formaldehyde is characterized by the planar geometry **176**.

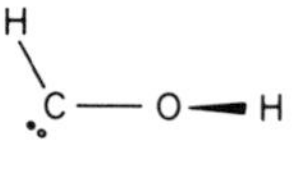

175 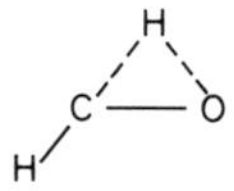   **176**

The barrier via **176**, 29 kcal mol^{-1} above **173**, suggests that singlet hydroxymethylene, once formed, will be relatively long lived. This is in accord with the results of the ion cyclotron resonance experiments [235].

The elimination reaction

$$H_2CO \longrightarrow H_2 + CO \tag{7.75}$$

has also been investigated [234 e–j]. Least-motion removal of H_2 from formaldehyde, while retaining C_{2v} symmetry, is forbidden by orbital symmetry considerations; the true transition structure, **177**, distorts to C_s symmetry.

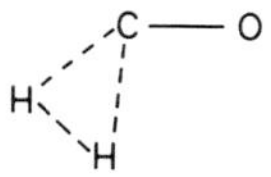

177

MP4/6-311G**//MP2/6-31G* calculations yield a barrier of 80 kcal mol^{-1} for this process.

The 3-21G calculations again underestimate the energy separations between the three stable CH_2O structures.

f. N_2H_2.   Diazene, the high-valence isomer of N_2H_2, can exist in either *trans* or *cis* forms, **178** and **179**, respectively. A 1,2-hydrogen shift leads to the alternative aminonitrene structure, **180**.

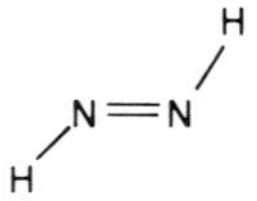

178 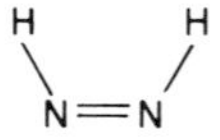   **179** 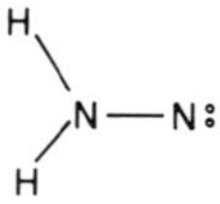   **180**

All three forms are energy minima on the singlet HF/6-31G* potential surface. At MP4/6-31G**, *trans*-diazene, **178**, is 6 kcal mol^{-1} lower in energy than the *cis* isomer, **179** [236]. Interconversion of the two is by way of nitrogen inversion via the planar transition structure **181**, in which one HNN bond angle is nearly linear while the other angle is little changed from the value in either *cis*- or *trans*-diazenes.

181

The calculated inversion barrier (52 kcal mol^{-1}) is noticeably higher than that for the analogous process in methyleneimine.

Aminonitrene, **180**, is predicted to lie only 25 kcal mol^{-1} above *trans*-diazene (at MP4/6-31G**//6-31G* after zero-point correction), in accord with other high-level quantum mechanical calculations [236]. The relatively high stability of **180** is due in part to electron delocalization of the lone pair of the NH$_2$ group into the vacant π-symmetry orbital at the nitrene center. This is also evidenced by the calculated geometry. At the HF/6-31G* level, the NN bond length in aminonitrene is 1.215 Å, essentially identical to the formal double bonds in *cis*- and *trans*-diazenes (1.215 and 1.216 Å, respectively) and far shorter than a normal NN single linkage, for example, 1.413 Å for hydrazine.

The transition structure, **182**, connecting aminonitrene to *trans*-diazene has a nonplanar geometry, that is, C_1 symmetry, at the HF/6-31G* level.

182

However, MP4/6-31G** calculations carried out using this geometry yield an energy higher than that of an alternative planar form. The nonplanarity of **182** is likely an artifact of the calculation. The calculated activation energy for conversion of **180** to **178** (via a planar transition structure) is 72 kcal mol^{-1}, and is nearly identical to the energy estimated for NH bond cleavage (71 kcal mol^{-1}), suggesting that the concerted rearrangement and the dissociative routes to isomerization are probably competitive [236].

3-21G calculations provide an accurate account of the energy difference between *cis*- and *trans*-diazene. However, they underestimate the energy of H$_2$NN (relative to *trans*-diazene) by nearly 20 kcal mol^{-1}, and overestimate the energy of the corresponding isomerization transition structure by 12 kcal mol^{-1}.

g. *CH$_5$N.* The most stable structure for CH$_5$N is known to be the staggered form of methylamine, **183**.

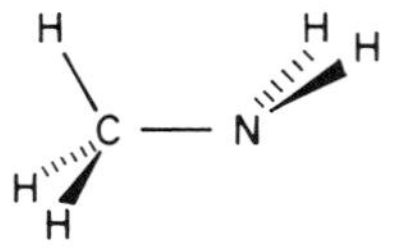

183 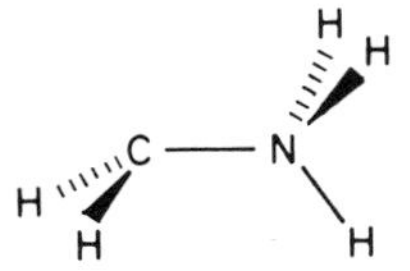   184

A hydrogen shift from carbon to nitrogen leads to an ylide-type structure, $H_2C^--NH_3^+$, which may also be regarded as a complex between singlet methylene and ammonia, $H_2C \cdots NH_3$. The staggered structure **184** is found to be the most stable form of this species; it is confirmed to be a local minimum on the HF/6-31G* surface [237]. The transition structure connecting **183** and **184** is a bridged C_s structure, **185**.

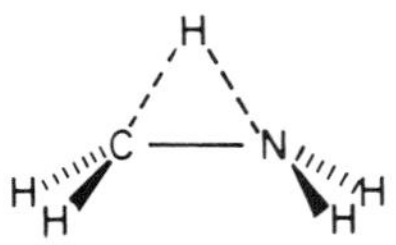

185

The $H_2C \cdots NH_3$ complex, **184**, is quite strongly bound relative to singlet methylene and ammonia, by 27 kcal mol^{-1} according to the highest-level (MP4/6-31G**, including zero-point correction) calculations. Its rearrangement to methylamine, **183**, requires an activation energy of 13 kcal mol^{-1} at this level. It should also be noted that the energy of the rearrangement transition structure, **185**, is below that of the separated products (by 14 kcal mol^{-1} at MP4/6-31G**). Thus, insertion of singlet methylene into an NH bond in ammonia (to give methylamine) proceeds without activation.

Corresponding studies have been made on methanol and methyl fluoride [220] to search for possible metastable complexes, $H_2C \cdots OH_2$ and $H_2C \cdots FH$. For both of these systems, rearrangement and/or dissociaton barriers are zero or negligibly small.

7.6.4.   Reaction Surfaces for Small Cationic Systems

As noted in Section 7.3.1, cations provide a particularly attractive target for theoretical study. Very few gas-phase structures for ions are known [238]; information about ion geometries available from experiment derives mainly from X-ray data on solid-state species (see Section 6.2.8d). Mass spectrometry and related techniques provide a source of thermochemical data from which detailed potential energy surfaces can be (and have been) constructed [239, 240]. In many cases, however, the experimental information is indirect, and its interpretation ambiguous. The elucidation of reaction surfaces for ions by theoretical means can remove some of these ambiguities, and can serve as an important complement to experiment. Theory can also be employed as a powerful predictive tool, with which to lead experimental work. Representative examples of these aspects of computational investigations are presented here.

a.   HOC⁺.   A New Interstellar Molecule.   Interstellar radio emission from hydrogen cyanide was first detected over a decade ago [241]. Shortly afterwards, it was proposed that unidentified microwave lines at 90.66 and 89.19 GHz were due to HNC [242] and to HCO^+ [243], respectively. Support for these assignments came from accurate quantum chemical calculations [227a,b, 244] and final confirmation from laboratory measurements of their microwave spectra [245].

Theoretical calculations [246, 247] played a significant role in the identification of HOC^+, the remaining member of this isoelectronic set of molecules. Of particular value was information from theory regarding the reaction profile for rearrangement of HOC^+ to its more stable isomer, HCO^+, the thermodynamics of possible modes of formation and of destruction of HOC^+, and (most importantly) the equilibrium geometry of HOC^+. The latter quantity leads directly to a prediction of the $J = 0 \rightarrow 1$ rotational transition frequency as a possible fingerprint for both laboratory and interstellar observation.

The reaction profile for the rearrangement of HOC^+ to HCO^+, calculated at the MP3/6-311G***//MP3/6-31G** level [248] and corrected for zero-point energy, is displayed in Figure 7.33. At this level of calculation, HOC^+ lies 38 kcal mol^{-1} above HCO^+. Note, however, that the calculated barrier to rearrangement is also substantial (36 kcal mol^{-1}), suggesting that rearrangement is not likely to be facile, that is, HOC^+, once formed, should be a relatively long-lived molecule.

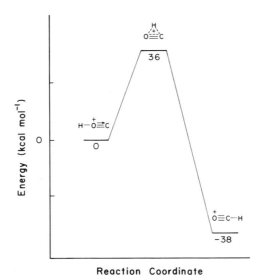

Reaction Coordinate

FIGURE 7.33.   Reaction profile for rearrangement of HOC^+ to HCO^+. MP3/6-311G***// MP3/6-31G**, corrected for differential zero-point energy.

Based on relative abundances of various species in interstellar clouds, Herbst and Klemperer [247a,b] proposed that formation of HCO^+ occurs primarily via reaction of H_3^+ and CO, that is, (7.76)

$$H_3^+ + CO \longrightarrow HCO^+ + H_2 .$$ (7.76)

These authors also suggested that HOC^+ might be produced in a similar manner, (7.77),

$$H_3^+ + OC \longrightarrow HOC^+ + H_2.$$ (7.77)

The reaction profile encompassing both of these processes, as obtained from MP3/6-31G**//4-31G calculations and corrected for zero-point effects, is shown in Figure 7.34 [246]. The data make clear that both HCO^+ and HOC^+ are formed in exothermic processes with no apparent activation barriers. Reaction (7.77) is, as earlier advanced [247a,b], a viable pathway by which HOC^+ might be produced in the interstellar medium. Note that the excess energy resulting from reaction (7.77) (10 kcal mol^{-1}) is substantially less than that required to surmount the barrier to rearrangement to the more stable HCO^+ isomer (36 kcal mol^{-1}). Thus, HOC^+ formed via (7.77) should be stable toward intramolecular rearrangement.

Experimental attempts to form HOC^+ via reaction of H_3^+ with CO in a flowing afterglow apparatus have up till now proven unsuccessful. The calculations suggest why [246]. Reaction of HOC^+ with CO (as would occur under the conditions of the flowing afterglow experiment) is predicted to proceed without activation, to

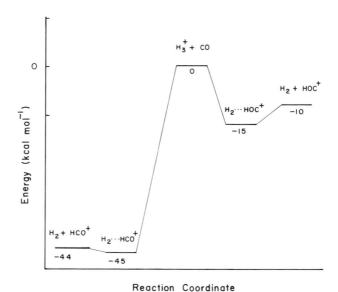

Reaction Coordinate

FIGURE 7.34.   Reaction profile for HCO^+ and HOC^+ formation. MP3/6-31G**//4-31G, corrected for differential zero-point energy.

yield a complex $[C{=}O \cdots H{-}C{=}O]^+$, some 54 kcal mol^{-1} lower in energy than the reactants. In the absence of stabilizing collisions, this complex will rapidly dissociate, either back to the original reactants or (better) to the more stable products, HCO^+ and CO. Thus, CO is an efficient catalyst for conversion of HOC^+ into HCO^+.

An accurate theoretical structure for HOC^+ has been determined by correcting geometries obtained from a number of computational levels for known systematic deficiencies [246] (see also Section 6.2.10). This yields $r_e(CO) = 1.155 \pm 0.003$ Å and $r_e(OH) = 0.988 \pm 0.003$ Å, and a predicted $J = 1 \to 0$ rotational transition frequency of 89.0 ± 0.8 GHz. Subsequent to these predictions, Gudeman and Woods [250] recorded a previously unknown microwave line at 89.487 GHz, following electronic discharge through a mixture of hydrogen, carbon monoxide, and argon. This they assigned to HOC^+. More recently, this same line, attributed also to HOC^+, has been observed in interstellar space [251].

b. *The Nature of the Methoxy Cation.* There has been considerable confusion in the literature as to the identity of the $[COH_3^+]$ ion produced in the mass spectrum of dimethyl ether. The most stable COH_3^+ isomer, formed by α hydrogen cleavage of primary alcohols [252], is the hydroxymethyl cation, **186**, with a heat of formation of 169 kcal mol^{-1} [253]. On the other hand, appearance potential measurements [253] yield a heat of formation of approximately 192 kcal mol^{-1} for the species resulting from dimethyl ether. While in early studies [252, 254] it was assumed that this ion was the methoxy cation, CH_3O^+, **187**, later work [255] suggested that the fragment from dimethyl ether was formed with considerable excess energy and was simply *hot* hydroxymethyl cation. Other workers [253, 256] suggested that the ion was **188**, previously observed as a product of reaction of H_2 and HCO^+ in high-pressure mass spectrometry experiments at low temperature [257]. Note, however, that **188** is a weak complex (only 4 kcal mol^{-1} more stable than separated HCO^+ and H_2) and is not a likely product of a fragmentation process [257b].

186   **187**   **188**

Calculations (MP3/6-31G*//4-31G and MP3/6-31G**//4-31G levels) [258] indicate that the singlet methoxy cation, **187**, collapses without activation to hydroxymethyl cation, **186**. The corresponding triplet form of **187** is, on the other hand, kinetically stable, but is predicted to lie 90 kcal mol^{-1} above **186**. These results would seem to rule out either singlet or triplet methoxy cation as the species observed in the mass spectral fragmentation of dimethyl ether.

How is hydroxymethyl cation produced from dimethyl ether? The theoretical potential surface (Figure 7.35) does not support the proposal that **186** arises from

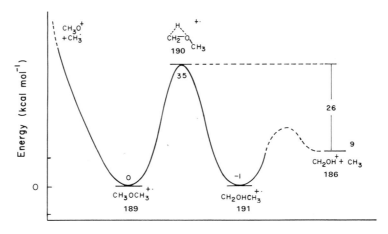

Reaction Coordinate

FIGURE 7.35. Reaction profile for interconversion and dissociation of $CH_3OCH_3^{+\cdot}$ and $CH_2O(H)CH_3^{+\cdot}$. MP3/6-31G**//4-31G, corrected for differential zero-point energy.

a simultaneous 1,2-hydrogen shift and CO bond cleavage following ionization [259], but rather suggests a closely related two-step mechanism. This involves an initial 1,2-hydrogen shift (with a barrier of 35 kcal mol^{-1}) via the transition structure **190**, to yield the stable oxonium radical cation $CH_2O(H)CH_3^{+\cdot}$, **191**, which the calculations find to be marginally (1 kcal mol^{-1}) more stable than the radical cation of dimethyl ether, **189**. Subsequent CO bond cleavage yields the hydroxymethyl cation, **186**, and methyl radical, in a reaction that is found to be endothermic by 9 kcal mol^{-1}. In total, this mechanism shows that hydroxymethyl cation derived from ionized dimethyl ether carries 26 kcal mol^{-1} excess energy, in good agreement with the experimental data [253, 255]. This profile represents a *textbook example* of a fragmentation involving rate-determining isomerization prior to dissociation to yield product ions with considerable excess energy [260].

c.   *The Methyleneoxonium Radical Cation, $CH_2OH_2^{+\cdot}$, and Related Systems.*
Because methanol is the only low-energy isomer on the CH_4O energy surface [220a, 261], it has generally been assumed that the corresponding radical cation would also be best described in terms of a similar structure, **192**. Calculations [262] suggest, however, that an alternative geometry, corresponding to the methyleneoxonium radical cation, $CH_2OH_2^{+\cdot}$, **193**, is the lowest-energy form. In contrast to neutral CH_4O, where the structure $CH_2 \cdots OH_2$ represents a weak complex 85 kcal mol^{-1} above methanol and with a long C $\cdots$ O bond (1.805 Å at MP2/6-31G*) [261], **193** is a tightly-bound species with a CO length that is even shorter than that in the methanol-like structure **192**. 6-31G** geometries for the two isomers are given below.

∠HOH=112.5

0.973 H +•
O 115.2
105.2 | 1.474
1.075 (118.4
H—C᷉ʻʻʻʻʻʻH
1.083 H
∠HCH=111.7
192

H H
0.959 (O +•
144.9
1.454
139.9
C᷉
1.071 H
H
∠HCH=125.4
193

The MP3/6-31G**//6-31G** potential surface (Figure 7.36) shows that $CH_2OH_2^{+\cdot}$ lies in a deep well, 11 kcal mol^{-1} lower in energy than $CH_3OH^{+\cdot}$. The most stable $CH_4O^{+\cdot}$ isomer, **193**, should be experimentally observable. In fact, subsequent to the theoretical predictions, the species was prepared from ionized ethylene glycol and identified mass spectrometrically [263]; an energy of 7 ± 2 kcal mol^{-1} for **192** relative to **193** has been determined [263b].

Rearrangement and dissociation processes in the $CH_4O^{+\cdot}$ system have also been examined theoretically (Figure 7.36). A 1,2-hydrogen shift converting **192** to **193**, via the transition structure **194**, is found to require 27 kcal mol^{-1}. This is 10 kcal mol^{-1} more than the energy required for hydrogen-atom loss to form the hydroxy-methyl cation and, therefore, would not appear to represent a viable means of production of $CH_2OH_2^{+\cdot}$. It is interesting to note that, while loss of a carbon-bound hydrogen atom from **192** (leading to the hydroxymethyl cation) is accompanied by only a small reverse activation energy (2 kcal mol^{-1}), loss of an oxygen-bound hydrogen from **193** (leading to the same products) requires considerable reverse activation (22 kcal mol^{-1}). Dissociations involving CO bond cleavage, that is, **192** → methyl cation + hydroxyl radical and **193** → methylene radical cation +

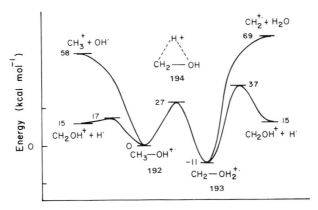

Reaction Coordinate

FIGURE 7.36. Reaction profiles for intramolecular and dissociative rearrangements on the $CH_4O^{+\cdot}$ potential surface. MP3/6-31G**//6-31G*, corrected for differential zero-point energy.

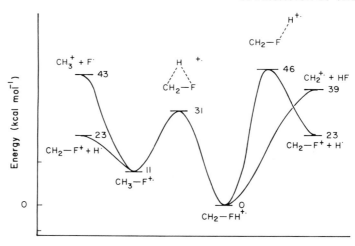

FIGURE 7.37. Reaction profiles for intramolecular and dissociative rearrangements on the $CH_3F^{+\cdot}$ potential surface. MP3/6-31G**//6-31G* and MP3/6-31G**//MP2/6-31G*, corrected for differential zero-point energy.

water, are found to be much higher-energy processes, and not to require activation in the reverse directions.

Similar studies (with similar results) have been carried out for the $CH_3F^{+\cdot}$ and $CH_5N^{+\cdot}$ systems. The methylenefluoronium ion, $CH_2FH^{+\cdot}$, is predicted to lie 11 kcal mol^{-1} lower in energy than the fluoromethane radical cation, $CH_3F^{+\cdot}$ [264] and the methyleneammonium ion, $CH_2NH_3^{+\cdot}$, lies slightly (2 kcal mol^{-1}) lower than the methylamine radical cation, $CH_3NH_2^{+\cdot}$ [262b, 265]. Theoretical potential surfaces connecting different forms for each of these systems are given in Figures 7.37 and 7.38, respectively. Qualitatively, they closely resemble that for the $CH_4O^{+\cdot}$ system (Figure 7.36).

Because species such as $CH_2NH_3^{+\cdot}$, $CH_2OH_2^{+\cdot}$ and $CH_2FH^{+\cdot}$ correspond formally to *ionized ylides,* the name *ylidion* has been introduced to describe this new class of reactive species [266]. The ylidions are specific examples of *distonic* radical cations, species in which the charge and radical sites are separated and which have been found to display special stability [266a,b].

d. $C_2H_3O^+$. Although a large number of reasonable isomers of formula $C_2H_3O^+$ may be drawn (**195** to **202**, below), only the acetyl cation, **195**, and the 1-hydroxy-vinyl cation, **196**, have been observed and characterized [267, 268]. In addition, the results of a recent collisional activation study [269] support the existence of a third stable species, postulated to be either **198** or **199**.

$$CH_3-C\equiv\overset{+}{O} \qquad CH_2=\overset{+}{C}-OH \qquad CH_3-\overset{+}{O}=C\text{:}$$

$$\textbf{195} \qquad\qquad\qquad \textbf{196} \qquad\qquad\qquad \textbf{197}$$

$$\overset{\overset{+}{O}}{\underset{CH_2-CH}{\triangle}}$$

198

$$\overset{+}{C}H_2-CH=O$$

199

$$CH\equiv C-\overset{+}{O}H_2$$

200

$$\overset{+}{C}H=CH-OH$$

201

$$\overset{\overset{+}{O}H}{\underset{CH=CH}{\triangle}}$$

202

Calculated energies for these isomers, as well as a number of alternative geometries, and for several transition structures connecting stable forms are provided in Table 7.46 [270]. A schematic representation of the pathways which have been delineated is given in Figure 7.39.

Consistent with the experimental evidence, the lowest-energy $C_2H_3O^+$ structure is the acetyl cation, **195**. This results from protonation of ketene (the most stable C_2H_2O isomer [271]) on the methylene carbon. Lying 43 kcal mol^{-1} higher in energy (MP3/6-31G**//4-31G) is the 1-hydroxyvinyl cation, **196**, which corresponds to O-protonation of ketene. An isomer of structure CH_3OC^+, **197**, is predicted to be only 9 kcal mol^{-1} less stable than **196** (according to the highest-level calculations), but 52 kcal mol^{-1} above **195**. Structures **195** and **197** may also be viewed as complexes between methyl cation and carbon monoxide (bonded at different ends). Attachment to the carbon end leads to **195**, and a binding energy of 71 kcal mol^{-1} (with respect to dissociation into CH_3^+ and CO), and to the oxygen end to **197**, with a binding energy of only 19 kcal mol^{-1}. Rearrangement of CH_3OC^+, **197**, to CH_3CO^+, **195**, occurs with a barrier of 16 kcal mol^{-1}. This is in contrast to the

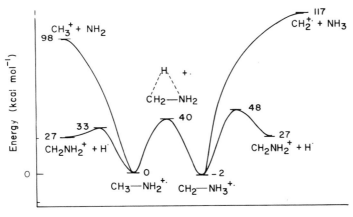

FIGURE 7.38. Reaction profiles for intramolecular and dissociative rearrangements on the $CH_5N^{+\cdot}$ potential surface. MP3/6-31G**//6-31G*, corrected for differential zero-point energy.

TABLE 7.46. Calculated Relative Energies of $C_2H_3O^+$ Isomers and Transition Structures (kcal mol^{-1})

Species	4-31G//4-31G	6-31G**	MP2/6-31G**	MP3/6-31G**	MP3/6-31G**[a]
			//4-31G		
CH_3CO^+, **195**	0	0	0	0	0
CH_2COH^+, **196**	36	43	50	44	43
CH_3OC^+, **197**	34	42	59	53	52
CH_2CHO^+, **198**	71	63	59	59	58
CH_2CHO^+, **199**	67	73	87	80	79
$CHCOH_2^+$, **200**	68	87	89	85	85
$CHCHOH^+$, **201**	69	78	97	87	86
$CHCHOH^+$, **202**	90	93	90	85	84
TS (**199** $\rightarrow$ **195**)	77	78	80	77	74
TS (**199** $\rightarrow$ **196**)	133	132	114	117	112
TS (**199** $\rightarrow$ **198**)	71	69	71	71	69
TS (**197** $\rightarrow$ **195**)	57	56	78	72	68
TS (**198** $\rightarrow$ **197**)	153	136	127	124	120
TS (**201** $\rightarrow$ **196**)	85	81	85	81	76
TS (**201** $\rightarrow$ **200**)	140	149	139	139	133
$CH_3^+ + CO$	57	57	82	76	71
$CH_2CO + H^+$	203	211	210	211	203

[a]Corrected for zero-point vibrational energy.

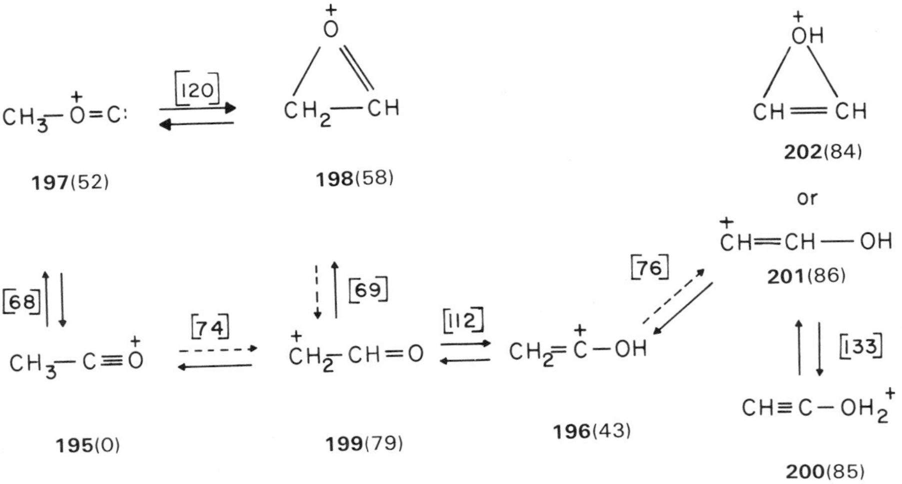

$CH_3-\overset{+}{O}=C:$ $\xrightarrow{\;[120]\;}$ CH_2-CH (with $\overset{+}{O}$ ring)

197(52)   198(58)

$\overset{+}{OH}$ ring with $CH=CH$

202(84)

or

$\overset{+}{C}H=CH-OH$

201(86)

$CH_3-C\equiv\overset{+}{O}$ $\underset{[74]}{\rightleftarrows}$ $\overset{+}{C}H_2-CH=O$ $\underset{[112]}{\rightleftarrows}$ $CH_2=\overset{+}{C}-OH$ [76]

[68]   [69]   [133]

195(0)   199(79)   196(43)

$CH\equiv C-\overset{+}{O}H_2$

200(85)

FIGURE 7.39. Interconversion pathways among $C_2H_3O^+$ isomers. Relative energies (in kcal mol^{-1}) shown in parentheses are for stable structures; those shown in square brackets are for transition structures. MP3/6-31G**//4-31G, corrected for differential zero-point energy.

analogous process $CH_3NC \to CH_3CN$, for which calculations indicate a barrier of 41 kcal mol^{-1} [272] (the experimental barrier is 38 kcal mol^{-1} [273]).

Both the oxiranyl cation, 198, and the formylmethyl cation, 199, are minima on the 4-31G potential surface. Higher-level (MP3/6-31G**) calculations indicate, however, that the formylmethyl cation, 199, is likely to collapse without activation energy to either the oxiranyl (198) or acetyl (195) cations. On the other hand, the oxiranyl cation, 198, is predicted (Table 7.46) to lie 58 kcal mol^{-1} above the acetyl cation, and is competitive with 197 as the third most favorable $C_2H_3O^+$ isomer. The calculated barrier for rearrangement of 198 to 197, involving ring opening via CC bond cleavage with a concomitant 1,2-hydrogen shift, is very high (62 kcal mol^{-1}). Rearrangement of 198 to 195 via 199 requires approximately 21 kcal mol^{-1}. Thus, 198, while relatively high in energy, should be sufficiently long lived to be observable.

Collisional activation mass spectrometry experiments are consistent with the theoretical prediction that the formylmethyl cation, 199, may collapse with little or no activation energy to acetyl cation, 195, or possibly to oxiranyl cation, 198. Precursors such as methyl vinyl ether, ethyl vinyl ether and cyclobutanol, which would have been expected to yield ions with the CH_2CHO^+ structure, 199, instead all give 195. On the other hand, ions produced from 1,3-dichloropropan-2-ol and 2,2-dichloroethanol are found to be distinct from the acetyl cation. In these cases, ionization followed by fragmentation may lead to intermediates of general structure $X-CH_2-CH-OH^+$ or $X-CH-CH_2-OH^+$, respectively, and it has been proposed [270] that 1,3-elimination of HX from such intermediates involves transition structures in which the terminal carbon and oxygen atoms are in close proximity, that is, structures 203 and 204, respectively.

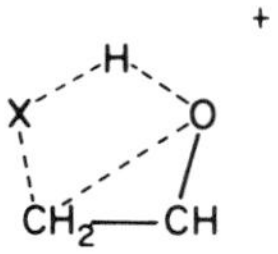

203 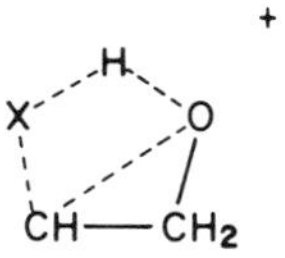   204

The geometries of such structures are favorable for concomitant ring closure and formation of oxiranyl cation, **198**.

The formylmethyl cation, **199**, is also a possible intermediate in the rearrangement of 1-hydroxyvinyl cation, **196**, to acetyl cation, **195**, by means of two consecutive 1,2-hydrogen shifts. The activation energy for the first of these steps is, however, apparently very large (69 kcal mol^{-1} according to the highest-level calculations).

Protonated oxirene, **202**, is not a minimum on the 4-31G potential surface; rather it serves as a transition structure for the degenerate hydroxyl shift in the 2-hydroxyvinyl cation, **201**, that is,

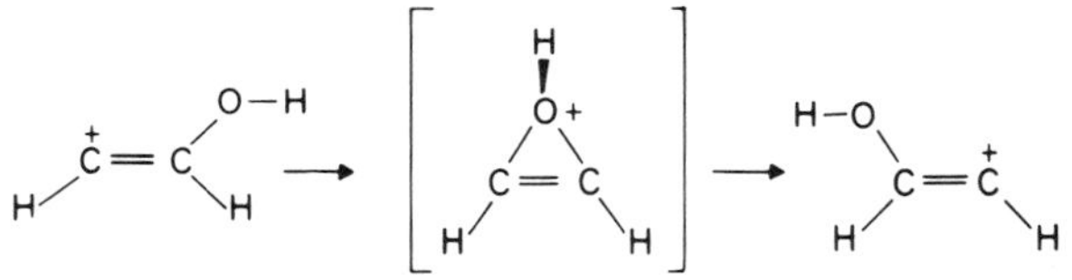

201   202   201

At higher levels of theory, the bridged ion, **202**, is actually more stable than the acyclic form, **201**, although only slightly so.

The 2-hydroxyvinyl cation, **201**, (or protonated oxirene, **202**) may itself rearrange to the more stable 1-hydroxyvinyl cation, **196**, via a 1,2-hydrogen shift across the C=C double bond. Although a transition structure was located at 4-31G, higher-level calculations show its energy to be below that of both **201** and **202**. It seems likely that neither of these species represents a stable form, and that both collapse without activation to **196**.

A high barrier (48 kcal mol^{-1}) is predicted to separate the ethynyloxonium cation, **200**, from **201** or **202** (and hence **196**). This offers some encouragement to the prospect of **200** being observed, despite its relatively high energy (85 kcal mol^{-1} above **195** according to the highest-level calculations).

7.6.5. Reaction Mechanisms

Quantitative molecular orbital calculations can be used to probe reaction mechanisms in far greater detail than is currently possible from experiment. Relationships between the geometries and energies of transition structures and those corresponding to products and reactants can be examined. Old concepts can be tested, and accepted or rejected. New generalizations can be sought. Factors such as steric and electronic effects which influence reaction rates and product distributions and stereochemistry can be explored in detail and their importance assessed. We present here a small selection of examples.

a. ***Mechanistic Aspects of Gas- and Solution-Phase S_N2 Reactions.*** In the absence of solvent, the reaction coordinate for S_N2 displacement, that is,

$$X^- + R\text{-}Y \longrightarrow R\text{-}X + Y^-, \tag{7.78}$$

will take on the form indicated in Figure 7.40 [274]. Two distinct ion-molecule complexes, $X^- \cdots R - Y$ and $Y^- \cdots R - X$, may exist. The energy of the transition structure separating these complexes, $[X \cdots R \cdots Y]^-$, may either be lower or higher than those of reactants or products or both.

Calculations on series of closely-related systems allow scrutiny of relationships between the geometries, energies or other properties of transition structures and overall reaction thermochemistry. For example, the theoretical data displayed in Figure 7.41 [275] indicate a reasonable linear correlation between C—F bond lengths in the transition structures $[X \cdots CH_3 \cdots F]^-$ and the overall thermochemistry of reaction (7.79) for a wide variety of substituents, X.

$$X^- + CH_3F \longrightarrow CH_3X + F^- \tag{7.79}$$

A relationship is also found between (average) FCH bond angles and reaction thermochemistry for the same set of compounds (Figure 7.42), and linear correlations have also been noted between analogous structural parameters and thermochemistry for S_N2 displacements from methane and from methanol [275]. Short C—F bond distances and large angles are characteristic of *early transition structures* and are associated with exothermic reactions (7.79).

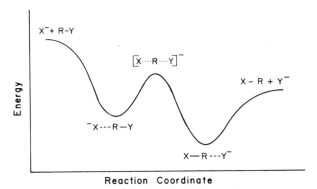

FIGURE 7.40. Reaction coordinate for exothermic gas-phase S_N2 displacement, $X^- + RY \rightarrow RX + Y^-$.

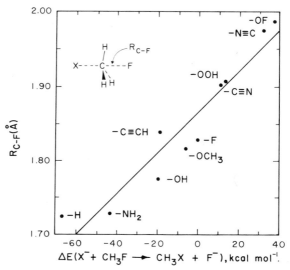

FIGURE 7.41. Correlation of calculated (4-31G) C–F bond lengths (Å) in S_N2 transition structures, $[X\text{-}CH_3\text{-}F]^-$, with the corresponding reaction energies for: $X^- + CH_3F \rightarrow CH_3X + F^-$, kcal mol^{-1}. From reference 275.

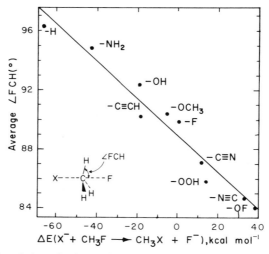

FIGURE 7.42. Correlation of calculated (4-31G) average FCH bond angles (°) in S_N2 transition structures, $[X\text{-}CH_3\text{-}F]^-$, with the corresponding reaction energies for: $X^- + CH_3F \rightarrow CH_3X + F^-$, kcal mol^{-1}. From reference 275.

492

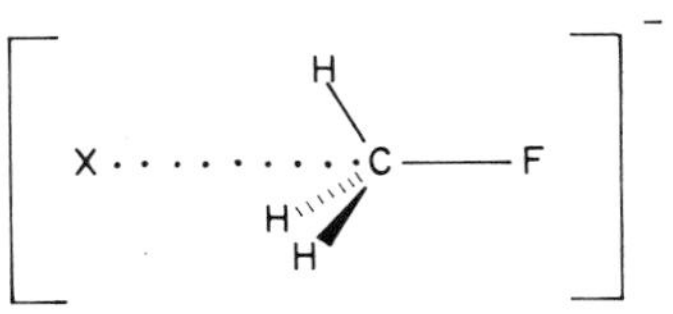

Long C—F distances and small bond angles, characterizing *late transition structures*, are found for endothermic processes.

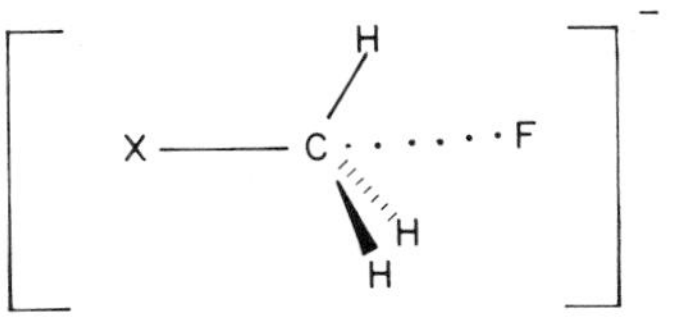

This is, of course, the effect described qualitatively by Bell, Evans, Polanyi, Leffler, and Hammond, among others [276]: "the more exothermic the reaction the more closely will its transition structure resemble the reactants."

S_N2 reactions in solution are now known to be much slower (by up to 20 orders of magnitude) than the corresponding gas-phase processes [274c, 277]. Gas-phase reactions involving partially hydrated nucleophiles appear to display intermediate activity. For example, Bohme and Mackay have shown that the rate constants for the series of reactions

$$OH^-(H_2O)_n + CH_3Br \rightarrow Br^-(H_2O)_n + CH_3OH \qquad (7.80)$$

decrease monotonically with increasing number of water molecules associated with the nucleophile [278]. A qualitative interpretation, which *assumes that the reaction mechanism involving clustered nucleophiles is the same as that for the unhydrated species*, as shown in Figure 7.40, is simply that hydration stabilizes the initially formed complex, $OH^-(H_2O)_n \cdots CH_3Br$, to a greater extent than it does the transition structure, $[OH \cdots CH_3 \cdots Br]^-(H_2O)_n$ [274c, 277b]. This assumption of reaction mechanism and the hypothesis regarding differential solvation effects are subject to test by calculation.

Reaction mechanisms for attack on methyl chloride by chloride anion (leading to methyl chloride and chloride anion), and by mono- and di-hydrated chloride anion, have been explored computationally at the 3-21G level [279]. Two reaction profiles are displayed in Figure 7.43. Both processes involve initial formation of weak complexes between the incoming nucleophile (or hydrated nucleophile) and the substrate, $Cl^- \cdots CH_3Cl$ and $Cl^-(H_2O)_2 \cdots CH_3Cl$, respectively, prior to actual

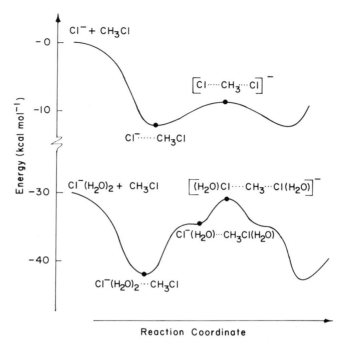

FIGURE 7.43. Reaction profile for $Cl^- + CH_3Cl \rightarrow CH_3Cl + Cl^-$ (top) and for $Cl^-(H_2O)_2 + CH_3Cl \rightarrow CH_3Cl + Cl^-(H_2O)_2$ (bottom). From reference 279.

displacement of the leaving group. This supports the assumption that S_N2 displacements in solution occur by essentially the same mechanisms as the analogous processes in the gas phase. While solvation by only two water molecules hardly alters the relative energies of reactants and the initially formed complexes, these species are better stabilized than is the corresponding reaction transition structure. The rationale is straightforward: the charge in the S_N2 transition structure is more delocalized than that in the reactants or in the initial complex (where the charge resides mainly on the incoming nucleophile rather than on the substrate), and the solvent is more effective in stabilizing charge-localized than delocalized ions. The decrease in rate with increasing hydration [278] appears, as postulated, to be a result of differential solvation rather than to a gross change in reaction mechanism.

Note that the calculated reaction pathway for nucleophilic attack by dihydrated chloride anion contains an additional minimum, corresponding to $Cl^-(H_2O) \cdots CH_3Cl(H_2O)$, and an additional transition structure, corresponding to migration of one of the water molecules from chloride anion to complexed methyl chloride.

b. Stereochemistry of Addition to Alkenes. The origins of *asymmetric induction*, of practical import for stereospecific synthesis [280], have been the subject of several detailed theoretical studies. Houk and coworkers [281] have calculated transition structures for the addition of nucleophiles, free radicals and electrophiles to unsaturated molecules. Their data suggest descriptions of all of these reactions in terms of *staggered* models [283] (shown below), that is, those in which the incoming reagent staggers the alkyl group attached to the reaction center.

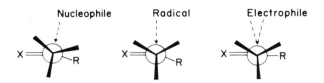

Transition structures (obtained at the 3-21G level) for addition of hydride anion, hydrogen atom and borane to propene are shown in Figure 7.44. These differ from one another primarily in the optimum *angle of approach* of the reagent to the substrate. Approach of borane [285, 286] to propene occurs at an acute angle, while that of H⁻ occurs at an angle greater than 120°; hydrogen-atom attack occurs at an intermediate angle. The differences between reagents noted here appear to carry over into other systems (Table 7.47). All borane additions investigated to date occur with approach angles centering around 75°, free-radical additions around

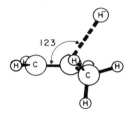

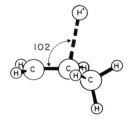

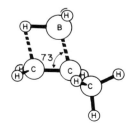

FIGURE 7.44. Calculated transition structures for addition of hydride anion (top), hydrogen atom (middle), and borane (bottom) to propene. 3-21G//3-21G. From reference 281c.

TABLE 7.47.   Calculated Approach Angles for Additions to Multiple Bonds (3-21G//3-21G, degrees)

Reaction	Approach Angle	Reference
Electrophilic Additions		
BH_3 + propene (Markovnikov)	73.1	281b
BH_3 + ethylene	73.4	281c
BH_3 + allyl alcohol[a]	73.7	281c
CH_3BH_2 + ethylene	73.9	281c
BH_3 + propene (anti-Markovnikov)	74.4	281b
BH_3 + isobutene (anti-Markovnikov)	75.4	281d
BH_3 + acetylene	80.4	281d
BH_3 + 2-butyne	81.1	281d
BH_3 + formaldehyde	85.1	281d
Free-Radical Additions		
$H^{\cdot}$ + propene (C_2)	101.5	281b
$F^{\cdot}$ + ethylene	102.2	282e
$H^{\cdot}$ + fluoroethylene (C_1)	102.6	282e
$HO^{\cdot}$ + propene (C_1)	103.3	281d
$HO^{\cdot}$ + ethylene	104.5	281d
$H^{\cdot}$ + ethylene	105.9	281d, 282e
$CH_3^{\cdot}$ + ethylene	107.8	281d
$CH_3O^{\cdot}$ + propene (C_1)	108.2	281d
$CH_3CH_2^{\cdot}$ + ethylene[b]	108.6	281d
Nucleophilic Additions		
H^- + allene (C_2)	98.1	281d
LiH + ethylene	113.9	281d
BH_4^- + formaldehyde	115.0	282f
$CH_3CH_2^-$ + ethylene[b]	119.6	281d
CH_3Li + ethylene	119.8	281d
CH_3^- + ethylene	126.0	281d
H^- + propene (C_2)	123.2	281b
H^- + propene (C_1)	125.1	281b
H^- + ethylene	126.0	281d
H^- + allene (C_1)	128.7	281d

[a] Average of six conformations of allyl alcohol.

[b] Average of *anti* and *gauche* conformations.

105° and nucleophilic additions around 120°. The noted differences in approach geometries may easily be rationalized in terms of simple orbital interaction arguments [282b, 287]. Stabilizing interaction of hydride anion with one end of the lowest unoccupied (π^*) function would be diminished in magnitude by overlap of opposite sign with the other end.

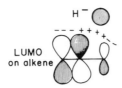

The preferred trajectory of the entering nucleophile will be that which maximizes the primary overlap, namely, that between centers involved in bond formation, while keeping the overlap involving the other end of the double bond to a minimum. On the other hand, overlap of a proton (or other electrophile) approaching one carbon of a double bond will not be reduced by interaction with the other carbon, but would be (slightly) diminished by overlap with an attached alkyl group. The nodal structure of the HOMO in propene illustrates the situation. As a consequence, the entering reagent will lean toward the alkene terminus and away from the alkyl carbon.

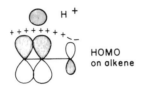

Houk considered the case of the hydroboration reaction in more detail [281c]. The calculated transition structures for addition of borane to ethylene and allyl alcohol, and for methylborane attack on ethylene, are all nearly identical to those presented in Figure 7.44 (for addition of borane to propene). Important bond lengths and calculated activation energies for all of these systems are given in Table 7.48. Except for the unfavorable Markovnikov addition of borane to propene, where steric factors may be involved, the calculated transition-structure geometries parallel the corresponding activation energies, that is, *the lower the activation energy the more closely the reaction transition structure resembles the products*. The effects are small and the correlation is not as good as that between transition-structure geometry and overall reaction thermochemistry previously noted for S_N2 reactions (Section 7.6.5a and especially Figures 7.41 and 7.42).

The similarity of reaction transition structures for closely related processes encouraged model studies of the conformational preferences of alkyl groups attached to borane or to the alkene substrate and the resulting stereochemistry of the chiral adducts [281c]. Activation energies (relative to that for the most stable conformation) for attack of ethylborane on ethylene, and for Markovnikov and anti-Markovnikov addition of borane to 1-butene are given in Figure 7.45. The data yield a clear conclusion: *alkyl groups prefer to be anti periplanar (anti) to the*

TABLE 7.48. Calculated (3-21G//3-21G) Activation Energies (kcal mol^{-1}) and Bond Lengths (Å) for Hydroboration Transition Structures

Reaction	Activation Energy	$r(B \cdots C)$	$r(H \cdots C)$
BH$_3$ + allyl alcohol	6.7	1.742	
BH$_3$ + propene (anti-Markovnikov)	7.1	1.752	1.682
BH$_3$ + ethylene	8.6	1.764	1.688
BH$_3$ + propene (Markovnikov)	10.8	1.753	1.668
CH$_3$BH$_2$ + ethylene	15.7	1.782	1.690

partially formed bonds. These positions, the least sterically crowded, are favored slightly over *outside* sites (away from the alkene double bond), and greatly over positions *inside* the four-centered transition structure, which are clearly the most congested. These results lead to the qualitative model for the stereochemistry of hydroboration shown in Figure 7.46, where the labels S, M, and L designate (ste-

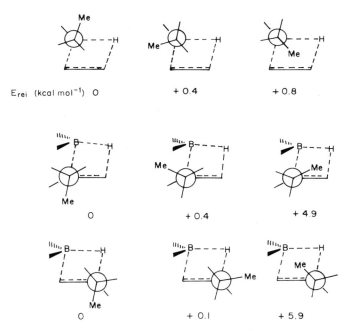

FIGURE 7.45. Calculated activation energies (kcal mol^{-1}, relative to most stable form) for different conformations of: addition of ethylborane to ethylene (top), Markovnikov addition of borane to but-1-ene (middle) and anti-Markovnikov addition of borane to but-1-ene (bottom). From reference 281c.

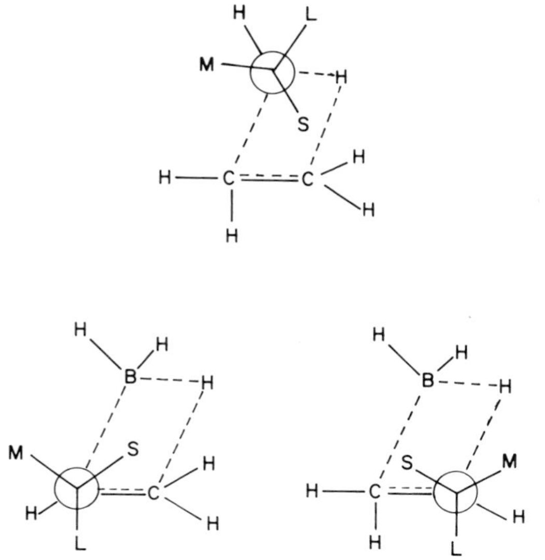

FIGURE 7.46.   Qualitative model describing steric preferences of hydroboration.

rically) small, medium, and large substituents, respectively. It should be applicable in those situations where steric factors are presumed to dominate.

The possible role of electronic effects in directing stereochemical preferences in hydroboration reactions has been investigated by Houk and coworkers in an analogous manner [281c]. For example, the role of a homoallylic oxygen in hydroboration reactions such as (7.81) [288],

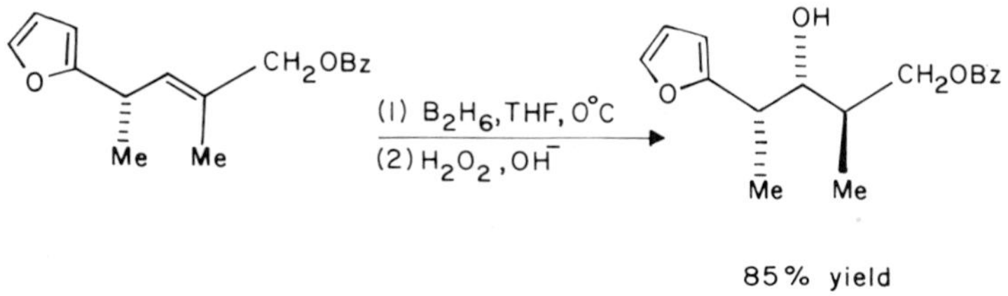

$$(1) \ B_2H_6, THF, 0°C$$
$$(2) \ H_2O_2, OH^-$$

85% yield

(7.81)

has been modeled by attaching a *standard* hydroxymethyl group [289] with a fixed *trans* CCOH dihedral angle to the Markovnikov transition structure for borane addition to propene, and sampling conformations about the two CC linkages at 120° intervals. The most stable conformer has the CH_2OH group *anti* to the partially formed BC bond, and places the CO bond in a *gauche* conformation that allows the oxygen atom to be as near to the alkene as possible.

This allows stabilization of the electron-deficient double bond by optimal interaction with the lone pairs on oxygen. There is some experimental evidence in support of such a picture; rate accelerations for electrophilic additions to alkenes have been noted in rigid systems where the through-space interaction of an ether oxygen and the π bond is possible [290]. The relative energies of the remaining conformers,

TABLE 7.49. Relative Energies (3-21G//3-21G, kcal mol^{-1}) of Model Transition Structures for Addition of Borane to But-1-ene-4-ol

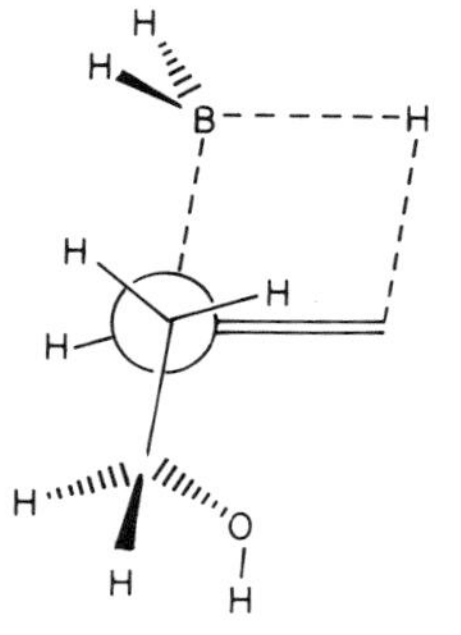

CH$_2$OH-anti
<CCCO=60°

$\omega(B \cdots CCC)$	$\omega(CCCO)$		
	60°	180°	-60°
anti	0.0	1.6	1.1
outside	0.9	1.4	7.1
inside	23.2	7.4	4.3

given in Table 7.49, show the favorability of other arrangements where the oxygen lone pairs and the double bond come into proximity, but (due to the restrictions imposed by the standard model) probably overestimate the instability of the more crowded arrangements.

Hydroboration of a set of conformers of allyl alcohol was investigated in order to model asymmetric syntheses in which an OH or OR group is located on the allylic carbon adjacent to the site where hydride is to be delivered, for example, reactions (7.82) [291].

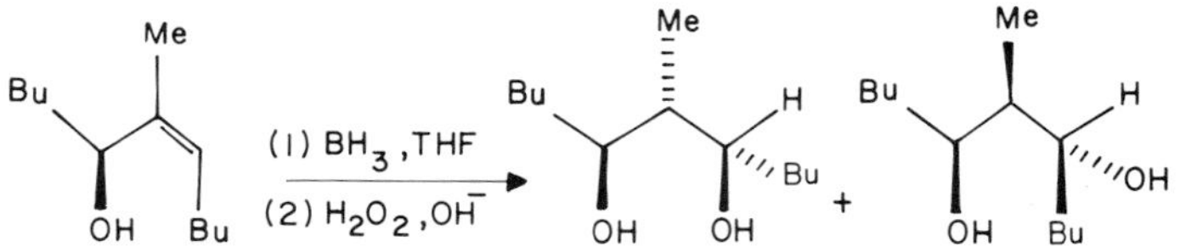

Six arrangements have been considered (Figure 7.47). In the first three (models for allylic alcohols), the *outside* and *inside* conformers, where the OH can hydrogen bond with the alkene, are favored over the *anti* arrangement, in which it is far removed. The same preference is noted for allyl alcohol itself, although the energy differences between conformers is not as great. The remaining transition structures (models for hydroboration of allyl ethers) are all higher in energy. These increase in the order *outside* < *inside* < *anti*, the opposite of the trend noted earlier for borane addition to 1-butene (see Figure 7.46), where the methyl group is preferentially *anti*. Electronic factors apparently dominate and exert a preference for the allylic CO bond to lie nearly in the plane of the alkene. The phenomenon appears widespread: a similar preference has been noted for electrophilic nitrile oxide cycloadditions to allylic ethers [292]. One explanation, that withdrawal of electrons by σ_{CO}^*

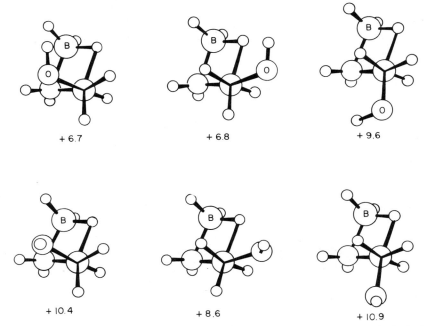

+ 6.7   + 6.8   + 9.6

+ 10.4   + 8.6   + 10.9

FIGURE 7.47. Calculated transition structures and activation energies (kcal mol^{-1}) for hydroboration of allyl alcohol. See text for discussion. From reference 281c.

from the (already electron deficient) alkene double bond is least efficient when the two lie in the same plane, further suggests that the favored conformation of an allylic CO bond in nucleophilic additions will be *anti*. Quantitative molecular orbital calculations concur [284b, c].

c. 1,3- and 1,5-Sigmatropic Rearrangements. Symmetry-Allowed and Symmetry-Forbidden Processes. Concerted migration of an atom or group between the termini of a polyene may occur in one of two topologically distinct ways: *suprafacially*, by remaining on one side of the polyene, or *antarafacially*, by moving from one face to the other.

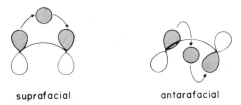

suprafacial   antarafacial

(Migration of an atom or group that is in the polyene plane, for example, the hydroxyl hydrogen in formic acid, may also occur without out-of-plane displacement.) Depending on the number of electrons, the first of these leads either to stabilization of the transition structure by *Hückel aromaticity* or destabilization by *Hückel antiaromaticity*. The second arrangement can also lead to a stabilized or destabilized transition structure, resulting in this case from *Mobius aromaticity* or *antiaromaticity*, respectively. Specifically, systems with $4n + 2$ electrons should proceed via suprafacial migration while those with $4n$ electrons should prefer antarafacial transition structures [58].

Calculated activation energies for a selection of simple 1,3- and 1,5-sigmatropic rearrangements [293] are given in Table 7.50. Only (formally) symmetry-allowed pathways are considered; recent calculations [293e, 294] have shown that a single-configuration treatment is unsatisfactory in describing symmetry-forbidden processes in the 1,3-sigmatropic rearrangement in propene [293e] and in the electro-cyclic isomerization of cyclobutene to *cis*-1,3-butadiene [294]. The theoretical data derive from several different Hartree-Fock and correlated levels. Activation energies calculated with the 3-21G, 4-31G, and 6-31G* basis sets are all similar. While the two correlated levels yield somewhat lower activation energies, in fact these are of the same order of magnitude as the strength of a "typical" CH bond. It is quite possible that rearrangement in propene occurs via dissociation into allyl radical and hydrogen atom followed by recombination [295].

1,3-sigmatropic rearrangement in 2-pentene and 1,3-dicyanopropene (corresponding to substitution of propene on both ends by methyl and cyano groups, respectively) has little effect on the symmetry-allowed antarafacial migration barrier. At the 3-21G//3-21G level, small energy decreases of 3 and 5 kcal mol^{-1}, respectively, are found. No experimental data are available.

A more interesting 1,3-sigmatropic shift leads from vinyl alcohol, the simplest enol, to acetaldehyde, its more stable tautomer. While attempts to prepare vinyl alcohol in solution have generally been unsuccessful [296], the gas-phase species has been observed [297]. This suggests a relatively high barrier connecting the two tautomers, and a mechanism for the solution-phase interconversion involving either intermolecular reactions or the participation of ionic intermediates. No direct experimental measure for the energy separation between keto and enol tautomers is available, although the enthalpy difference between a closely related molecule, acetone, and its enol has been determined by ICR spectroscopy as 7.7 kcal mol^{-1} [298] (see also Section 6.5.5). The barrier hindering interconversion of vinyl alcohol to acetaldehyde has not as yet been measured. The highest-level (CEPA/DZP// 4-31G [293b]) calculations (Table 7.50) show an energy difference between the two tautomers of 13 kcal mol^{-1} (acetaldehyde is the more stable), and a barrier separating them (via a symmetry-allowed antarafacial pathway) of 64 kcal mol^{-1}.

Symmetry-allowed (suprafacial) pathways for 1,5-sigmatropic rearrangement in 1,3-pentadiene and in cyclopentadiene have been examined [293c]. The calculated barriers (53 and 40 kcal mol^{-1}, respectively, at the 3-21G//3-21G level after corrections for zero-point energy) are considerably less than that for the 1,3-shift in propene at the same level of theory, but are larger than the experimental activation energies for these systems (31–37 and 22 kcal mol^{-1} [299, 300], respectively). A 6-31G*//3-21G calculation [293d] yields an even higher barrier (of 59 kcal mol^{-1}) for rearrangement in 1,3-pentadiene. Treatment of electron correlation

TABLE 7.50. Calculated Activation Energies for 1,3- and 1,5-Sigmatropic Rearrangements

System	Mechanism	Theoretical Model	Activation Energy (kcal mol^{-1})	Reference
1,3 Rearrangements[a]				
Propene	Antarafacial	3-21G//3-21G	108	293a
		4-31G//4-31G	108	293b
		6-31G*//3-21G	106	293a
		CEPA/DZP//4-31G	93	293b
		MP2/6-31G*//3-21G	90	293a
2-Pentene	Antarafacial	3-21G//3-21G	105	293a
1,3-Dicyanopropene	Antarafacial	3-21G//3-21G	103	293a
Vinyl alcohol → acetaldehyde	Antarafacial	4-31G//4-31G	83 (10)[c]	293b
		CEPA/DZP//4-31G	64 (13)[c]	293b
1,5-Rearrangements[b]				
1,3-Pentadiene	Suprafacial	3-21G//3-21G	55	293c
		6-31G*//3-21G	59	293d
Cyclopentadiene	Suprafacial	3-21G//3-21G	42	293c

[a]The symmetry-allowed process is antarafacial.
[b]The symmetry-allowed process is suprafacial.
[c]Energy difference between acetaldehyde (the more stable isomer) and vinyl alcohol given in parentheses.

503

would probably lead to lower barriers for both rearrangements, in better agreement with the experimental data. The significant decrease in barrier from 1,3-pentadiene to cyclopentadiene warrants further comment. While purely geometrical factors may be the primary cause, that is, the termini in cyclopentadiene are already in close proximity, it is interesting to speculate that electronic effects may in part be responsible. In the limiting representation, where hydrogen migrates as a cation (as opposed to migration as an anion or free radical), the underlying cyclopentadienyl anion might be viewed as a better aromatic than pentadienyl anion. Calculated sizes [301] for the migrating hydrogen in the two transition structures support such a view; the best-sphere-fit radius of hydrogen in cyclopentadiene is 1.283 Å according to 3-21G//3-21G calculations, a tenth of an angstrom smaller than that in the pentadiene transition structure (1.385 Å). (For comparison, the corresponding radius of hydrogen atom is 1.332 Å.) Thus, the migrating hydrogen in the cyclopentadiene transition structure is more cation-like than that in the analogous process for 1,3-pentadiene.

REFERENCES

1. More detail can be found in published papers on the interaction between directly-bonded groups. See: (a) A. Pross and L. Radom, *Aust J. Chem.*, 33, 241 (1980); (b) A. L. Hinde, A. Pross and L. Radom, *J. Comput. Chem.*, 1, 118 (1980); (c) G. Kemister, A. Pross, L. Radom, and R. W. Taft, *J. Org. Chem.*, 45, 1056 (1980); (d) A. Pross and L. Radom, *Tetrahedron*, 36, 673 (1980); (e) A. Pross and L. Radom, *ibid.*, 36, 1990 (1980); (f) A. Pross, D. J. DeFrees, B. A. Levi, S. K. Pollack, L. Radom, and W. J Hehre, *J. Org. Chem.*, 46, 1693 (1981); (g) J. D. Dill, P. v. R. Schleyer, and J. A Pople, *J. Am. Chem. Soc.*, 98, 1663 (1976); (h) Y. Apeloig, P. v. R. Schleyer, and J. A. Pople, *ibid.*, 99, 1291 (1977); (i) Y. Apeloig, P. v. R. Schleyer, and J. A. Pople, *ibid.*, 99, 5901 (1977); (j) T. Clark and P. v. R. Schleyer, *Tetrahedron Lett.*, 4641 (1979); (k) T. Clark, G. W. Spitznagel, R. Klose, and P. v. R. Schleyer, *J. Am. Chem. Soc.*, 106, 4412 (1984); (l) T. Clark, H. Körner, and P. v. R. Schleyer, *Tetrahedron Lett.*, 21, 743 (1980).

2. (a) J. A. Pople, Y. Apeloig, and P. v. R. Schleyer, *Chem. Phys. Lett.*, 85, 489 (1982); (b) H. Kollmar, *J. Am. Chem. Soc.*, 100, 2665 (1978); (c) G. W. Spitznagel, T. Clark, J. Chandrasekhar, and P. v. R. Schleyer, *J. Comput. Chem.*, 3, 363 (1982).

3. See: W. J. Hehre, W. A. Lathan, J. A. Pople, L. Radom, E. Wasserman and Z. Wasserman, *J. Am. Chem. Soc.*, 98, 4378 (1976), and references therein.

4. (a) R. H. Staley and J. L. Beauchamp, *J. Am. Chem. Soc.*, 97, 5920 (1975); (b) R. L. Woodin and J. L. Beauchamp, *ibid.*, 100, 501 (1978). Quantitative data also exist on the stabilities of other metallic and organometallic cations attached to simple molecules. See: (c) R. R. Corderman and J. L. Beauchamp, *ibid.*, 98, 3998 (1976); (d) R. C. Burnier, T. J. Carlin, W. D. Reents, Jr., R. B. Cody, R. K. Lengel, and B. S. Freiser, *ibid.*, 101, 7127 (1979); (e) R. W. Jones and R. H. Staley, *ibid.*, 102, 3794 (1980); (f) J. S. Uppal and R. H. Staley, *ibid.*, 102, 4144 (1980); (g) R. B. Cody, R. C. Burnier, W. D. Reents, Jr., T. J. Carlin, D. H. McCrey, R. K. Lengel, and B. S. Freiser, *Int. J. Mass Spectrom. Ion Phys.*, 33, 37 (1980); (h) M. Kappes and R. H. Staley, *J. Am. Chem. Soc.*, 103, 1286 (1981); (i) J. S. Uppal and R. H. Staley, *ibid.*, 104, 1229, 1235, 1238 (1982).

5. (a) W. J. Hehre, R. Ditchfield, L. Radom, and J. A. Pople, *J. Am. Chem. Soc.*, 92, 4796 (1970). For an early survey of the use of bond separation energies as a measure of geminal interactions, see: (b) L. Radom, W. J. Hehre, and J. A Pople, *ibid.*, 93, 289 (1971); (c) J. D. Dill, P. v. R. Schleyer, and J. A. Pople, *ibid.*, 98, 1663 (1976).

6. (a) J. T. Edwards, *Chem. Ind. (London)*, 1102 (1955); (b) R. U. Lemieux and N. J. Chiu, *Abstracts*, 133rd. National Meeting of the American Chemical Society, San Francisco, April 1958, no 31N. For general discussions, see: (c) J. F. Stoddart, *Sterochemis-*

try of Carbohydrates, Wiley-Interscience, New York, 1971, p. 72ff; (d) R. U. Lemieux, *Pure Appl. Chem.*, **25**, 527 (1971). The more recent literature is summarized in: (e) A. J. Kirby, *The Anomeric Effect and Related Stereoelectronic Effects at Oxygen*, Springer-Verlag, Berlin, 1983; (f) W. A. Szarek and D. Horton, eds., *Anomeric Effect: Origin and Consequences*, ACS Symposium Series no. 87, American Chemical Society, Washington, D.C., 1979; (g) P. Deslongchamps, *Stereoelectronic Effects in Organic Chemistry*, Pergamon, New York, 1983; (h) P. v. R. Schleyer and A. J. Kos, *Tetrahedron*, **39**, 1141 (1983).

7. For previous *ab initio* studies dealing with the structural consequences of geminal interactions, see: (a) G. A. Jeffrey, J. A. Pople, and L. Radom, *Carbohyd. Res.*, **25**, 117 (1972); (b) G. A. Jeffrey, J. A. Pople, and L. Radom, *ibid.*, **38**, 81 (1974); (c) L. Radom and P. J. Stiles, *Tetrahedron Lett.*, 789 (1975); (d) S. Vishveshwara, *Chem. Phys. Lett.*, **59**, 30 (1978); (e) J. M. Lehn and G. Wipff, *Helv. Chim. Acta*, **61**, 1274 (1978); (f) G. A. Jeffrey, J. A. Pople, J. S. Binkley, and S. Vishveshwara, *J. Am. Chem. Soc.*, **100**, 373 (1978); (g) G. Wipff, *Tetrahedron Lett.*, 3269 (1978); (h) S. Wolfe, M. H. Whangbo and D. J. Mitchell, *Carbohyd. Res.*, **69**, 1 (1979); (i) A. Pross and L. Radom, *J. Comput. Chem.*, **1**, 295 (1980). For a recent review, see: (j) L. Radom, *Prog. Theor. Org. Chem.*, **3**, 1 (1982).

8. Experimental: (a) E. H. Cordes, *Prog. Phys. Org. Chem.*, **4**, 1 (1967); (b) P. Deslongchamps, *Tetrahedron*, **31**, 2463 (1975); (c) P. Deslongchamps, *Heterocycles*, **7**, 1271 (1977). Theoretical: (d) J. M. Lehn and G. Wipff, *J. Am. Chem. Soc.*, **102**, 1347 (1980), and references therein.

9. For a historical review, see: G. W. Wheland, *Resonance in Organic Chemistry*, Wiley, New York, 1955, and references therein.

10. For a recent review, see: (a) A. Pross and L. Radom, *Prog. Phys. Org. Chem.*, **13**, 1 (1981). See also: (b) W. J. Hehre, R. W. Taft, and R. D. Topsom, *ibid.*, **12**, 159 (1976); (c) W. J. Hehre, L. Radom, and J. A. Pople, *J. Am. Chem. Soc.*, **94**, 1496 (1972).

11. (a) S. K. Pollack, J. L. Devlin, III, K. Summerhays, R. W. Taft, and W. J. Hehre, *J. Am. Chem. Soc.*, **99**, 4583 (1977); (b) K. Summerhays, S. K. Pollack, R. W. Taft, and W. J. Hehre, *J. Am. Chem. Soc.*, **99**, 4585 (1977), and references therein. Some systems have been uncovered in which protonation does not occur at nitrogen: (c) S. Kahn and W. J. Hehre, research in progress.

12. (a) M. Taagepera, K. D. Summerhays, W. J. Hehre, R. D. Topsom, A. Pross, L. Radom, and R. W. Taft, *J. Org. Chem.*, **46**, 891 (1981); (b) W. L. Hehre, M. Taagepera, R. W. Taft, and R. D. Topsom, *J. Am. Chem. Soc.*, **103**, 1344 (1981).

13. See: A. Pross, L. Radom, and R. W. Taft, *J. Org. Chem.*, **45**, 818 (1980).

14. L. Radom, W. J. Hehre, J. A. Pople, G. L. Carlson, and W. G. Fateley, *J. Chem. Soc., Chem. Commun.*, 308 (1972).

15. (a) L. Radom, J. A. Pople, and P. v. R. Schleyer, *J. Am. Chem. Soc.*, **94**, 5935 (1972), and references therein to experimental studies; (b) L. Radom, *Aust. J. Chem.*, **27**, 231 (1974); (c) W. J. Hehre, R. T. McIver, J. A. Pople, and P. v. R. Schleyer, *J. Am. Chem. Soc.*, **96**, 7162 (1974). See also: (d) R. W. Taft, M. Taagepera, J. L. M. Abboud, J. F. Wolf, D. J. DeFrees, W. J. Hehre, J. E. Bartmess, and R. T. McIver, Jr., *ibid.*, **100**, 7745 (1975).

16. (a) E. M. Arnett and J. W. Larsen, *J. Am. Chem. Soc.*, **91**, 1438 (1969); see also: (b) D. M. Brouwer and J. A. Van Doorn, *Recl. Trav. Chim. Pays-Bas*, **89**, 88 (1970). For reviews, see: (c) Conference on Hyperconjugation, *Tetrahedron*, **5**, 105 (1959); (d) M. J. S. Dewar, *Hyperconjugation*, Ronald Press, New York, 1962; (e) J. W. Baker, *Hyperconjugation*, Oxford University Press, Oxford, 1952, (f) L. S. Levitt and H. F. Widing, *Prog. Phys. Org. Chem.*, **12**, 119 (1976). Other work is summarized in ref. 15c.

17. W. M. Schubert and W. A. Sweeney, *J. Org. Chem.*, **21**, 119 (1956).

18. For a recent review of solvent effects on gas-phase acidities and basicities, see: R. W. Taft, *Prog. Phys. Org. Chem.*, **14**, 247 (1983), and references therein.

19. For a review, see: D. M. Brouwer, E. L. Mackor, and C. MacLean, in *Carbonium Ions*,

G. A. Olah and P. v. R. Schleyer, eds., Wiley-Interscience, New York, 1970, vol. 1, p. 837ff.

20.   J. L. Devlin III, J. F. Wolf, R. W. Taft, and W. J. Hehre, *J. Am. Chem. Soc.*, **98**, 1990 (1976).

21.   W. J. Hehre and J. A. Pople, *J. Am. Chem. Soc.*, **94**, 6901 (1972), and references therein to earlier theoretical work.

22.   G. A. Olah, R. H. Schlosberg, R. D. Porter, Y. K. Mo, D. P. Kelly, and G. D. Mateescu, *J. Am. Chem. Soc.*, **94**, 2034 (1972).

23.   (a) J. M. McKelvey, S. Alexandratos, A. Streitweiser, Jr., J. L. M. Abboud, and W. J. Hehre, *J. Am. Chem. Soc.*, **98**, 244 (1976); (b) D. J. DeFrees, R. T. McIver, Jr., and W. J. Hehre, *ibid.*, **99**, 3853 (1977).

24.   The *ortho* and *para* substituent effects noted here closely parallel those for substitution directly onto a carbocation center. See reference 15a.

25.   L. Radom, H. F. Schaefer III, and M. A. Vincent, *Nouv. J. Chem.*, **4**, 411 (1980), and references therein to earlier theoretical work.

26.   P. R. West, O. L. Chapman, and J. P. LeRoux, *J. Am. Chem. Soc.*, **104**, 1779 (1982).

27.   T. J. Barton and D. S. Banasiak, *J. Am. Chem. Soc.*, **99**, 5199 (1977).

28.   R. L. Kuczkowski and A. J. Ashe III, *J. Mol. Spectrosc.*, **42**, 457 (1972).

29.   T. C. Wong and L. S. Bartell, *J. Mol. Struct.*, **44**, 169 (1978).

30.   G. D. Fong, R. L. Kuczkowski, and A. J. Ashe III, *J. Mol. Spectrosc.*, **70**, 197 (1978).

31.   G. P. Elliot, W. R. Roper, and J. M. Waters, *J. Chem. Soc., Chem. Commun.*, 811 (1982).

32.   E. D. Jemmis and P. v. R. Schleyer, *J. Am. Chem. Soc.*, **104**, 4781 (1982). This reference contains a thorough summary of previous experimental and theoretical work on the aromatic stabilities of three-dimensional systems.

33.   A symmetrical structure for cyclopentadienyllithium is supported by ^{7}Li NMR [34]. Symmetrical geometries have also been found for a number of simple beryllium analogues by electron and X-ray diffraction [35].

34.   (a) R. H. Cox, H. W. Terry, Jr., and L. W. Harrison, *J. Am. Chem. Soc.*, **93**, 3297 (1971); (b) J. A. Dixon, P. A. Gwinner, and D. C. Lini, *ibid.*, **93**, 3297 (1971); (c) P. Fischer, J. Stadelhofer, and J. J. Weidlein, *J. Organomet. Chem.*, **116**, 65 (1976).

35.   For a review, see: A. Haaland, *Topics Current Chem.*, **53**, 1 (1975).

36.   For a discussion of the isomers of $C_6H_6^{2+}$, see: K. Lammertsma and P. v. R. Schleyer, *J. Am. Chem. Soc.*, **105**, 1049 (1983).

37.   (a) H. Hogeveen and P. W. Kwant, *Acc. Chem. Res.*, **8**, 413 (1975); (b) C. Giordano, R. F. Heldeheg, and H. Hogeveen, *J. Am. Chem. Soc.*, **99**, 5181 (1977).

38.   See: R. N. Grimes, *Carboranes*, Academic Press, New York, 1970.

39.   For reviews, see: (a) D. H. Aue and M. T. Bowers, in *Gas Phase Ion Chemistry*, M. T. Bowers, ed., Academic Press, New York, 1979, vol. 2, p. 91ff; (b) H. M. Rosenstock, K. Draxl, B. W. Steiner, and J. T. Herron, *J. Phys. Chem. Ref. Data, Suppl. 1*, **6**, 1 (1977), and references therein. (c) S. G. Lias, J. F. Liebman, and R. D. Levin, *J. Phys. Chem. Ref. Data*, **13**, 695 (1984). See also: (d) F. Cacace, *Adv. Phys. Org. Chem.*, **8**, 79 (1970).

40.   See, for example: (a) G. A. Olah and C. U. Pittman, Jr., *Adv. Phys. Org. Chem.*, **4**, 305 (1966); (b) D. M. Brouwer and H. Hogeveen, *Prog. Phys. Org. Chem.*, **9**, 179 (1972); (c) G. A. Olah, *Angew. Chem., Int. Ed. Engl.*, **12**, 173 (1973); (d) G. A. Olah, *Topics Current Chem.*, **80**, 19 (1979).

41.   For reviews, see: (a) L. Radom, D. Poppinger and R. C. Haddon, in *Carbonium Ions*, G. A. Olah and P. v. R. Schleyer, eds., Wiley, New York, 1976, vol. 5, p. 2303ff; (b) W. J. Hehre, in *Applications of Electronic Structure Theory*, H. F. Schaefer III, ed., Plenum Press, New York, 1977, vol. 4, p. 277ff.

42.   (a) W. A. Lathan, W. J. Hehre, and J. A. Pople, *J. Am. Chem. Soc.*, **93**, 808 (1971); (b)

W. A. Lathan, W. J. Hehre, L. A. Curtiss, and J. A. Pople, *ibid.*, **93**, 6377 (1971); (c) L. Radom, J. A. Pople, V. Buss, and P. v. R. Schleyer, *ibid.*, **93**, 1318 (1971); (d) L. Radom, J. A. Pople, V. Buss, and P. v. R. Schleyer, *ibid.*, **94**, 311 (1972), and references therein; (e) W. A. Lathan, L. A. Curtiss, W. J. Hehre, J. B. Lisle, and J. A. Pople, *Prog. Phys. Org. Chem.*, **11**, 175 (1974).

43. (a) P. C. Hariharan, W. A. Lathan, and J. A. Pople, *Chem. Phys. Lett.*, **14**, 385 (1972); (b) P. C. Hariharan, L. Radom, J. A. Pople, and P. v. R. Schleyer, *J. Am. Chem. Soc.*, **96**, 599 (1974); (c) L. Radom, P. C. Hariharan, J. A. Pople, and P. v. R. Schleyer, *J. Am. Chem. Soc.*, **95**, 6531 (1973), and references therein; (d) L. Radom, P. C. Hariharan, J. A. Pople, and P. v. R. Schleyer, *ibid.*, **98**, 10 (1976), and references therein.

44. (a) B. Zurawski, R. Ahlrichs, and W. Kutzelnigg, *Chem. Phys. Lett.*, **21**, 309 (1973); (b) V. Dyczmons and W. Kutzelnigg, *Theor. Chim. Acta*, **33**, 239 (1974); (c) J. Weber and A. D. McLean, *J. Am. Chem. Soc.*, **98**, 875 (1976); (d) J. Weber, M. Yoshimine, and A. D. McLean, *J. Chem. Phys.*, **64**, 4159 (1976); (e) H. J. Kohler, D. Heidrich, and H. Lischka, *Z. Chem.*, **17**, 67 (1977); (f) H. Lischka and H. J. Kohler, *J. Am. Chem. Soc.*, **100**, 5297 (1978); (g) H. J. Kohler and H. Lischka, *Chem. Phys. Lett.*, **58**, 175 (1978). A survey of high-level theoretical calculations on C_1 to C_3 carbocations has recently been provided: (h) K. Raghavachari, R. A. Whiteside, J. A. Pople, and P. v. R. Schleyer, *J. Am. Chem. Soc.*, **103**, 5649 (1981), and references therein.

45. K. Hiraoka and P. Kebarle, *J. Am. Chem. Soc.*, **98**, 6119 (1976), and references therein.

46. G. A. Olah, *Carbocations and Electrophilic Reactions*, Wiley-Interscience, New York, 1974.

47. G. A. Olah, G. R. S. Prakash, G. Liang, P. W. Westerman, R. Kunde, J. Chandrasekhar, and P. v. R. Schleyer, *J. Am. Chem. Soc.*, **102**, 4485 (1980).

48. Theory: (a) K. Raghavachari, P. v. R. Schleyer, and G. W. Spitznagel, *J. Am. Chem. Soc.*, **105**, 5917 (1983); (b) A. M. Sapse and L. Osorio, *Inorg. Chem.*, **23**, 627 (1984). Experiment: (c) S. G. Shore, S. H. Lawrence, M. I. Watkins, and R. Bau, *J. Am. Chem. Soc.*, **104**, 7669 (1982), and R. Bau, private communication to P. v. R. Schleyer. These results are in contrast to $Al_2H_7^-$ which prefers a "linear" structure. Theory: (d) J. M. Howell, A. M. Sapse, E. Singman, and G. Snyder, *ibid.*, **104**, 4758 (1982). Experiment: (e) J. L. Atwood, D. C. Hrncir, R. D. Rogers, and J. A. K. Howard, *ibid.*, **103**, 6787 (1981).

49. See, for example, (a) H. C. Brown, *The Nonclassical Ion Problem*, Plenum Press, New York, 1977; (b) C. A. Grob, *Acc. Chem. Res.*, **16**, 426 (1983); (c) H. C. Brown, *ibid.*, **16**, 432 (1983); (d) G. A. Olah, G. K. S. Prakash, and M. Saunders, *ibid.*, **16**, 440 (1983); (e) C. Walling, *ibid.*, **16**, 448 (1983).

50. For reviews, see: (a) G. Modena and U. Tonellato, *Adv. Phys. Org. Chem.*, **9**, 185 (1971); (b) P. J. Stang, Z. Rappoport, M. Hanack, and L. R. Subramanian, *Vinyl Cations*, Academic Press, New York, 1979.

51. (a) P. Ausloos, R. E. Rebbert, L. W. Sieck, and T. O. Tiernan, *J. Am. Chem. Soc.*, **94**, 8939 (1972); (b) J. H. Varachels, G. G. Meisels, R. A. Geanangel, and R. H. Emmel, *ibid.*, **95**, 4078 (1973).

52. F. A. Houle and J. L. Beauchamp, *J. Am. Chem. Soc.*, **101**, 4067 (1979).

53. (a) T. Baer, *J. Am. Chem. Soc.*, **102**, 2482 (1980); (b) D. K. Bohme and G. I. Mackay, *ibid.*, **103**, 2173 (1981).

54. F. P. Lossing, *Can. J. Chem.*, **50**, 3973 (1972).

55. H. Mayr, W. Forner, and P. v. R. Schleyer, *J. Am. Chem. Soc.*, **101**, 6032 (1979).

56. D. H. Aue, W. R. Davidson, and M. T. Bowers, *J. Am. Chem. Soc.*, **98**, 6700 (1976).

57. (a) R. D. Bowen, D. J. Williams, H. Schwarz, and C. Wesdemiotis, *J. Am. Chem. Soc.*, **101**, 4681 (1979); (b) M. T. Bowers, L. Shuying, P. Kemper, R. Stradling, H. Webb, D. H. Aue, J. R. Gilbert, and K. R. Jennings, *ibid.*, **102**, 4830 (1980).

58. R. B. Woodward and R. Hoffman, *The Conservation of Orbital Symmetry*, Verlag Chemie, Weinheim, 1970.

59. See, for example: (a) P. v. R. Schleyer, W. F. Sliwinski, G. W. Van Dine, U. Schöllkopf, J. Praust, and K. Fellenberger, *J. Am. Chem. Soc.*, **94**, 125 (1972); (b) W. F. Sliwinski, T. M. Su, and P. v. R. Schleyer, *ibid.*, **94**, 133 (1972), and references therein.

60. Experimental work: (a) P. v. R. Schleyer, T. M. Su, M. Saunders, and J. C. Rosenfeld, *J. Am. Chem. Soc.*, **91**, 5174 (1969); (b) J. M. Bollinger, J. M. Brinch, and G. A. Olah, *ibid.*, **92**, 4025 (1970); (c) N. C. Deno, R. C. Haddon, and E. N. Nowak, *ibid.*, **92**, 6691 (1970).

61. L. Radom, J. A. Pople, and P. v. R. Schleyer, *J. Am. Chem. Soc.*, **95**, 8193 (1973), and references therein.

62. (a) M. Saunders and E. L. Hagen, *J. Am. Chem. Soc.*, **90**, 6881 (1968); (b) G. A. Olah and A. M. White, *ibid.*, **91**, 5801 (1969); (c) C. C. Lee and D. J. Woodcock, *ibid.*, **92**, 5992 (1970); (d) G. J. Karabatsos, C. Zioudrou, and S. Meyerson, *ibid.*, **92**, 5996 (1970); (e) H. Hogeveen and C. J. Gaasbeck, *Recl. Trav. Chim. Pays-Bas*, **89**, 857 (1970); (f) M. Saunders, P. Vogel, E. L. Hagen, and J. Rosenfeld, *Acc. Chem. Res.*, **6**, 53 (1973); For a review, see: (g) C. C. Lee, *Prog. Phys. Org. Chem.*, **7**, 129 (1970).

63. R. L. Baird and A. A. Aboderin, *J. Am. Chem. Soc.*, **86**, 252 (1964).

64. (a) S. L. Chong and J. L. Franklin, *J. Am. Chem. Soc.*, **94**, 6347 (1972); (b) D. Viviani and J. B. Levy, *Int. J. Chem. Kinet.*, **11**, 1021 (1979); (c) M. Attina, F. Cacace, and P. Giacomello, *J. Am. Chem. Soc.*, **102**, 4786 (1980).

65. F. A. Houle, Ph.D. Thesis, California Institute of Technology (1979).

66. For reviews, see: (a) R. Breslow, in *Molecular Rearrangements*, part 1, P. deMayo, ed., Wiley-Interscience, New York, 1963, chap. 4; (b) M. Hanack and H. J. Schneider, *Angew. Chem. Int. Ed. Engl.*, **6**, 666 (1967); (c) H. G. Richey, Jr., in *Carbonium Ions*, G. A. Olah and P. v. R. Schleyer, eds., Wiley, New York, 1972, vol. 3, p. 1201ff; (d) K. B. Wiberg, B. A. Hess, Jr., and A. J. Ashe III, *ibid.*, p. 1295ff; (e) See ref. 49c, chap. 5; (f) P. Ahlberg, G. Jonsall, and C. Ehndahl, *Adv. Phys. Org. Chem.*, **19**, 223 (1983).

67. (a) G. A. Olah, C. L. Jeuell, D. P. Kelly, and R. D. Porter, *J. Am. Chem. Soc.*, **94**, 146 (1972), and references therein to earlier work; (b) G. A. Olah and R. J. Spear, *ibid.*, **97**, 1539 (1975); (c) J. S. Staral, I. Yavari, J. D. Roberts, G. K. S. Prakash, D. J. Donovan, and G. A. Olah, *ibid.*, **100**, 8016 (1978); (d) J. S. Staral and J. D. Roberts, *ibid.*, **100**, 8018 (1978); (e) M. Saunders and H. U. Siehl, *ibid.*, **102**, 6868 (1980).

68. (a) W. J. Hehre and P. C. Hiberty, *J. Am. Chem. Soc.*, **94**, 5917 (1972), and references therein to earlier theoretical work; (b) W. J. Hehre and P. C. Hiberty, *ibid.*, **96**, 302 (1974); (c) B. A. Levi, E. S. Blurock, and W. J. Hehre, *ibid.*, **101**, 5537 (1979).

69. (a) A. D. Walsh, *Nature*, **159**, 712 (1947); (b) A. D. Walsh, *Trans. Faraday Soc.*, **45**, 179 (1949); (c) L. Salem and J. S. Wright, *J. Am. Chem. Soc.*, **91**, 5947 (1969).

70. D. S. Kabakoff and E. Namanworth, *J. Am. Chem. Soc.*, **92**, 3234 (1970), and references therein to earlier work on the methyl and dimethylcyclopropylcarbinyl cations.

71. (a) J. D. Roberts and R. H. Mazur, *J. Am. Chem. Soc.*, **73**, 3542 (1951). The term non-classical was introduced to the chemical community in this paper. See also: (b) R. H. Mazur, W. N. White, D. A. Semenow, C. C. Lee, M. S. Silver, and J. D. Roberts, *ibid.*, **81**, 4390 (1959).

72. (a) M. Saunders and J. Rosenfeld, *J. Am. Chem. Soc.*, **92**, 2548 (1970); (b) G. A. Olah, R. J. Spear, P. C. Hiberty, and W. J. Hehre, *ibid.*, **98**, 7470 (1976). See also: (c) R. P. Kirchen and T. S. Sorensen, *ibid.*, **99**, 1687 (1977); (d) G. A. Olah, G. K. S. Prakash, D. J. Donovan, and I. Yavari, *ibid.*, **100**, 7085 (1978); (e) P. v. R. Schleyer, D Lenoir, P. Mison, G. Liang, G. K. S. Prakash, and G. A. Olah, *ibid.*, **102**, 683 (1980); (f) K. L. Servis and F. F. Shue, *ibid.*, **102**, 7233 (1980).

73. (a) M. Saunders, E. L. Hagen, and J. Rosenfeld, *J. Am. Chem. Soc.*, **90**, 6882 (1968); (b) J. J. Dannenberg, D. H. Weinberg, K. Dill, and B. J. Goldberg, *Tetrahedron Lett.*, 1241 (1972); (c) G. A. Olah and D. J. Donovan, *J. Am. Chem. Soc.*, **99**, 5026 (1977); (d) M. Saunders and M. R. Kates, *ibid.*, **100**, 7082 (1978); (e) P. H. Myhre and C. S.

Yanoni, *ibid.*, **103**, 230 (1981); (f) M. Saunders, M. R. Kates, G. H. Walker, R. F. Hout, Jr., and W. J. Hehre, research in progress. A comprehensive review of rearrangements in the 2-butyl cation and closely related systems appears in ref. 66f, p. 246 ff.

74. H. J. Kohler and H. Lischka, *J. Am. Chem. Soc.*, **101**, 3479 (1979).

75. M. Saunders, private communication to W. J. Hehre. See also: G. A. Olah and J. Lukas, *J. Am. Chem. Soc.*, **90**, 933 (1968).

76. (a) G. A. Olah and J. Lukas, *J. Am. Chem. Soc.*, **89**, 4739 (1967); (b) G. A. Olah, J. R. DeMember, A. Commeytras, and J. L. Bribes, *ibid.*, **93**, 459 (1971).

77. G. A. Olah, J. S. Staral, G. Asencio, G. Liang, D. A. Forsyth, and G. D. Mateescu, *J. Am. Chem. Soc.*, **100**, 6299 (1978) and references therein to earlier work. A comprehensive review of rearrangements in benzenium ions appears in ref. 66f. p. 313ff.

78. (a) R. G. Cooks, T. Ast, and J. H. Beynon, *Int. J. Mass Spectrom. Ion Phys.*, **11**, 490 (1973); (b) T. Ast, C. J. Porter, C. J. Proctor, and L. H. Beynon, *Bull. Soc. Chim. Beograd.*, **46**, 1935 (1981); (c) T. Ast, C. J. Porter, C. J. Proctor, and J. H. Beynon, *Chem. Phys. Lett.*, **78**, 439 (1981); (d) C. J. Porter, C. J. Proctor, T. Ast, P. D. Bolton, and J. H. Beynon, *Org. Mass Spectrom.*, **16**, 512 (1981); (e) D. Stahl and F. Marquis, *Chimica*, **37**, 87 (1983); (f) M. Rabrénovic and J. H. Beynon, *J. Chem. Soc., Chem. Commun.*, 1043 (1983).

79. Review: (a) G. K. S. Prakash, T. N. Rawdah, and G. A. Olah, *Angew. Chem., Int. Ed. Engl.*, **22**, 390 (1983). See also: (b) C. D. Nenitzescu, in *Carbonium Ions*, G. A. Olah and P. v. R. Schleyer, eds., Wiley-Interscience, New York, 1968, vol. 1, p. 29ff.

80. Review: T. Ast, *Adv. Mass Spectrom.*, **8A**, 555 (1980).

81. (a) J. A. Pople, B. Tidor, and P. v. R. Schleyer, *Chem. Phys. Lett.*, **88**, 533 (1982); see also: (b) P. E. M. Siegbahn, *Chem. Phys.*, **66**, 443 (1982); (c) A. W. Hanner and T. F. Moran, *Org. Mass Spectrom.*, **16**, 512 (1981).

82. (a) G. A. Olah and M. Simonetta, *J. Am. Chem. Soc.*, **104**, 330 (1982); (b) P. v. R. Schleyer, A. J. Kos, J. A. Pople, and A. T. Balaban, *ibid.*, **104**, 3771 (1982); (c) K. Lammertsma, G. A. Olah, M. Barzaghi, and M. Simonetta, *ibid.*, **104**, 6851 (1982).

83. (a) J. Chandrasekhar, P. v. R. Schleyer, and K. Krogh-Jespersen, *J. Comput. Chem.*, **2**, 356 (1981); see also: (b) R. Hoffman, *Angew. Chem., Int. Ed. Engl.*, **21**, 711 (1982); (c) K. Krogh-Jespersen, P. v. R. Schleyer, J. A. Pople, and D. Cremer, *J. Am. Chem. Soc.*, **100**, 4301 (1979).

84. $C_2H_2^{2+}$: (a) J. A. Pople, M. Frisch, K. Raghavachari, and P. v. R. Schleyer, *J. Comput. Chem.*, **3**, 468 (1982); $C_2H_4^{2+}$: (b) K. Lammertsma, M. Barzaghi, G. A. Olah, J. A. Pople, A. J. Kos, and P. v. R. Schleyer, *J. Am. Chem. Soc.*, **105**, 5252 (1983); CH_6^{2+}: (c) K. Lammertsma, M. Barzaghi, G. A. Olah, J. A. Pople, P. v. R. Schleyer, and M. Simonetta, *ibid.*, **105**, 5258 (1983); $C_6H_6^{2+}$: ref. 36. See also: (d) L. Radom and H. F. Schaefer III, *J. Am. Chem. Soc.*, **99**, 7522 (1977); (e) D. P. Craig, L. Radom, and H. F. Schaefer III, *Aust. J. Chem.*, **31**, 261 (1978). For a review, see: (f) P. v. R. Schleyer, *Symposium on Advances in Carbocation Chemistry*, American Chemical Society, Seattle, 1983, p. 413.

85. (a) T. Clark, D. Wilhelm, and P. v. R. Schleyer, *Tetrahedron Lett.*, **23**, 3547 (1982); (b) D. Wilhelm, T. Clark, P. v. R. Schleyer, K. Buckl, and G. Boche, *Chem. Ber.*, **116**, 1669 (1983).

86. (a) K. Krogh-Jespersen, D. Cremer, J. D. Dill, J. A. Pople, and P. v. R. Schleyer, *J. Am. Chem. Soc.*, **103**, 2589 (1981). See also: (b) E. D. Jemmis, J. Chandrasekhar, and P. v. R. Schleyer, *ibid.*, **101**, 2848 (1979).

87. (a) S. M. van der Kerk, P. H. M. Budzelaar, A. van der Kerk-van Hoof, G. J. M. van der Kerk, and P. v. R. Schleyer, *Angew. Chem., Int. Ed. Engl.*, **22**, 48 (1983); (b) R. Wehrmann, C. Pues, H. Klusik, and A. Berndt, *ibid.*, **23**, 372 (1984); (c) M. Hildenbrand, H. Pritzkow, U. Zenneck, and W. Siebert, *ibid.*, **23**, 373 (1984).

88. (a) P. v. R. Schleyer, P. H. M. Budzelaar, D. Cremer, and E. Kraka, *Angew. Chem., Int. Ed. Engl.*, **23**, 374 (1984); (b) P. H. M. Budzelaar, K. Krogh-Jespersen, T. Clark, and P. v. R. Schleyer, *J. Am. Chem. Soc.*, **107**, 2773 (1985).

89. See: R. West, ed., *Oxocarbons*, Academic Press, New York, 1980, especially chap. 10. See also: P. H. M. Budzelaar, H. Dietrich, J. Macheleid, R. Weiss, and P. v. R. Schleyer, *Chem. Ber.*, **118**, 2118 (1985).

90. (a) D. Feller, E. R. Davidson, and W. T. Borden, *J. Am. Chem. Soc.*, **104**, 1216 (1982). Long bonds are found in some bridged annulenes, see: (b) D. Cremer and B. Dick, *Angew. Chem., Int. Ed. Engl.*, **21**, 865 (1982), as well as in some homoaromatic cations, see: (c) D. Cremer, E. Kraka, T. S. Slee, R. F. W. Bader, C. D. H. Lau, T. T. Nguyen-Dang, and P. J. MacDougall, *J. Am. Chem. Soc.*, **105**, 5069 (1983).

91. Reviews of carbanion chemistry: (a) D. J. Cram, *Fundamentals of Carbanion Chemistry*, Academic Press, New York, 1965; (b) E. Buncel and T. Durst, eds., *Comprehensive Carbanion Chemistry*, Elsevier, Amsterdam, 1980; (c) R. Bates and C. A. Ogle, *Carbanion Chemistry*, Springer-Verlag, Berlin, 1983.

92. The Cambridge Crystallographic Database listed 180 X-ray structures for lithium compounds as of Summer 1983. See: (a) S. R. Wilson and J. C. Hoffman, *J. Org. Chem.*, **45**, 560 (1980); (b) W. Setzer and P. v. R. Schleyer, *Adv. Organomet. Chem.*, **24**, 353 (1985).

93. B. J. Wakefield, *The Chemistry of Organolithium Compounds*, Pergamon Press, Oxford, 1974.

94. S. P. Patterson, I. L. Karle, and G. D. Stucky, *J. Am. Chem. Soc.*, **92**, 1150 (1970).

95. (a) L. Andrews, *J. Chem. Phys.*, **47**, 4834 (1967); (b) L. Andrews and T. Carver, *J. Phys. Chem.*, **72**, 1743 (1968); see also: (c) T. L. Brown, *Ann. New York Acad. Sci.*, **136**, 98 (1966).

96. (a) R. Amstutz, J. D. Dunitz, and D. Seebach, *Angew. Chem., Int. Ed. Engl.*, **20**, 465 (1981). (b) R. Amstutz, T. Taube, W. B. Schweizer, D. Seebach, and J. D. Dunitz, *Helv. Chim. Acta*, **67**, 224 (1984).

97. A. Hinchliffe and E. Saunders, *J. Mol. Struct.*, **31**, 283 (1976).

98. (a) A. Streitweiser and J. B. Collins, *J. Comput. Chem.*, **1**, 8 (1980); see also: (b) A. Streitweiser, J. E. Williams, Jr., S. Alexandratos, and J. M. McKelvey, *J. Am. Chem. Soc.*, **98**, 4778 (1976). Similar conclusions have been reached by A. E. Reed, R. B. Weinstock, and F. Weinhold, *J. Chem. Phys.*, **83**, 735 (1985).

99. G. D. Graham, D. S. Marynick, and W. N. Lipscomb, *J. Am. Chem. Soc.*, **102**, 4572 (1980).

100. (a) M. M. Francl, R. F. Hout, Jr., and W. J. Hehre, *J. Am. Chem. Soc.*, **106**, 563 (1984); see also: (b) R. F. Hout, Jr. and W. J. Hehre, *ibid.*, **105**, 3728 (1983); (c) W. J. Hehre, C. F. Pau, R. F. Hout, Jr., and M. M. Francl, *Molecular Modeling. Computer-Aided Design of Molecular Systems*, Wiley-Interscience, New York, in preparation.

101. E. U. Wurthwein, K. D. Sen, J. A. Pople, and P. v. R. Schleyer, *Inorg. Chem.*, **22**, 496 (1983).

102. (a) P. v. R. Schleyer, J. Chandrasekhar, A. J. Kos, T. Clark, and G. W. Spitznagel, *J. Chem. Soc., Chem. Commun.*, 882 (1981). See also: (b) J. E. DelBene, M. J. Frisch, K. Raghavachari, J. A. Pople, and P. v. R. Schleyer, *J. Phys. Chem.*, **87**, 73 (1983).

103. P. v. R. Schleyer and G. W. Spitznagel, unpublished results.

104. For a discussion of the MSAD (McEwen, Streitwieser, Applequist, Dessey) acidity scale, and a tabulation of values, see: J. March, *Advanced Organic Chemistry, Reactions, Mechanisms and Structure*, 2nd ed., McGraw-Hill, New York, 1977, chap. 8.

105. (a) G. L. Closs and L. E. Closs, *Angew. Chem., Int. Ed. Engl.*, **1**, 334 (1962); (b) G. L. Closs and R. A. Moss, *J. Am. Chem. Soc.*, **86**, 4042 (1964). For general discussions of carbenoids as synthetic intermediates, see; (c) H. Siegel, *Topics Current Chem.*, **106**, 55 (1982); (d) K. G. Taylor, *Tetrahedron*, **38**, 2752 (1982); (e) R. A. Moss and M. Jones, Jr., *Reactive Intermediates*, vol. 2, Wiley-Interscience, New York, 1981, p. 79ff; (f) D. Seebach, R. Hassig, and J. Gabriel, *Helv. Chim. Acta*, **66**, 308 (1983).

106. G. Kobrich, *Angew. Chem., Int. Ed. Engl.*, **11**, 473 (1972), and references therein to earlier work.

107. D. Seebach, H. Siegel, J. Gabriel, and R. Hassig, *Helv. Chim. Acta*, **63**, 2046 (1980).

108. (a) T. Clark and P. v. R. Schleyer, *J. Chem. Soc., Chem. Commun.*, 883 (1979); (b) T. Clark and P. v. R. Schleyer, *Tetrahedron Lett.*, **4963** (1979); (c) T. Clark and P. v. R. Schleyer, *J. Am. Chem. Soc.*, **101**, 7747 (1979); (d) J. B. Moffat, *J. Chem. Soc., Chem. Commun.*, 1108 (1980); (e) I. A. Abronin, R. Zahradnik, and G. M. Zhidonomirov, *Zh. Strukt. Khim.*, **21**, 3 (1980); (f) T. Clark, P. v. R. Schleyer, K. N. Houk, and N. G. Rondan, *J. Chem. Soc., Chem. Commun.*, 579 (1981); (g) C. Rohde, T. Clark, E. Kaufmann, and P. v. R. Schleyer, *J. Chem. Soc., Chem. Commun.*, 882 (1982); (h) M. A. Vincent and H. F. Schaefer III, *J. Chem. Phys.*, **77**, 6103 (1982); (i) B. T. Luke, J. A. Pople, P. v. R. Schleyer, and T. Clark, *Chem. Phys. Lett.*, **102**, 148 (1983); (j) P. v. R. Schleyer, T. Clark, A. J. Kos, G. W. Spitznagel, C. Rhode, D. Arad, K. N. Houk, and N. G. Rondan, *J. Am. Chem. Soc.*, **106**, 6467 (1984).

109. T. Clark and P. v. R. Schleyer, *J. Organomet. Chem.*, **191**, 347 (1980).

110. (a) J. Chandrasekhar, J. A. Pople, R. Seeger, U. Seeger, and P. v. R. Schleyer, *J. Am. Chem. Soc.*, **104**, 3651 (1982); (b) Y. Apeloig, P. v. R. Schleyer, and J. A. Pople, *ibid.*, **99**, 1291 (1977).

111. (a) T. Clark, H. Korner, and P. v. R. Schleyer, *Tetrahedron Lett.*, **21**, 743 (1980); (b) A. C. Hopkinson and M. H. Lien, *Int. J. Quantum Chem.*, **18**, 1371 (1980); (c) G. W. Spitznagel, T. Clark, J. Chandrasekhar, and P. v. R. Schleyer, *J. Comput. Chem.*, **4**, 294 (1983).

112. D. A. Hatzenbuhler, L. Andrews, and F. A. Carey, *J. Am. Chem. Soc.*, **97**, 187 (1975).

113. W. J. Hehre, unpublished calculations. See also ref. 108c.

114. (a) W. Kirmse, *Carbene Chemistry*, 2nd ed., Academic Press, New York, 1971, pp. 83–128; (b) A. P. Marchand, in *The Chemistry of Double-Bonded Functional Groups*, supplement A, part 1, S. Patai, ed., Wiley-Interscience, New York, 1977, pp. 553–557, 585–587; (c) T. L. Gilchrist and C. W. Rees, *Carbenes, Nitrenes and Arynes*, Thomas Nelson and Sons Ltd., London, 1969, pp. 85–128.

115. (a) H. E. Simmons and R. D. Smith, *J. Am. Chem. Soc.*, **80**, 5323 (1958); (b) W. v. E. Doering and P. M. LaFlamme, *Tetrahedron*, **2**, 75 (1958); (c) H. E. Simmons and R. D. Smith, *J. Am. Chem. Soc.*, **81**, 4256 (1959); (d) H. E. Simmons, E. P. Blanchard, and R. D. Smith, *ibid.*, **86**, 1347 (1964).

116. G. Köbrich, *Angew. Chem., Int. Ed. Engl.*, **6**, 41 (1967).

117. D. F. Hoeg, D. I. Lusk, and A. L. Crumbliss, *J. Am. Chem. Soc.*, **87**, 4147 (1965).

118. W. R. Moser, *J. Am. Chem. Soc.*, **91**, 1135, 1141 (1969).

119. (a) R. Hoffmann, *J. Am. Chem. Soc.*, **90**, 1475 (1968); (b) R. Hoffmann, D. M. Hayes, and P. S. Skell, *J. Phys. Chem.*, **76**, 664 (1972). Similarities in orbital structure, suggesting similar reactivites, are common among simple molecules. For a discussion, see ref. 120.

120. R. F. Hout, Jr., W. J. Pietro, and W. J. Hehre, *A Pictorial Approach to Molecular Structure and Reactivity*, Wiley-Interscience, New York, 1984.

121. See ref. 116 for typical nucleophilic reactions of carbenoids. See also: (a) G. Köbrich, *Angew. Chem., Int. Ed. Engl.*, **11**, 473 (1972); (b) D. Seebach, R. Hassig, and J. Gabriel, *Helv. Chim. Acta*, **66**, 308 (1983), and references therein.

122. J. Mareda, N. G. Rondan, K. N. Houk, T. Clark, and P. v. R. Schleyer, *J. Am. Chem. Soc.*, **105**, 6997 (1983), and references therein to previous theoretical work.

123. (a) N. G. Rondan, K. N. Houk, and R. A. Moss, *J. Am. Chem. Soc.*, **102**, 1770 (1980); (b) K. N. Houk, N. G. Rondan and J. Mareda, *ibid.*, **106**, 4291 (1984); (c) K. N. Houk and N. G. Rondan, *ibid.*, **106**, 4293 (1984).

124. R. R. Schrock, *J. Am. Chem. Soc.*, **98**, 5399 (1976).

125. For reviews, see: (a) T. J. Katz, *Adv. Organomet. Chem.*, **16**, 283 (1978); (b) N. Calderon, J. P. Lawrence, and E. A. Ofstead, *ibid.*, **17**, 449 (1979).

126. (a) R. R. Schrock, S. McLain, and J. Sancho, *Pure Appl. Chem.*, **52**, 729 (1980); (b) J. Fellman, R. R. Schrock, and G. A. Rupprecht, *J. Am. Chem. Soc.*, **103**, 5752 (1981).

127. For a discussion of the mechanism of olefin metathesis, see: G. W. Parshall, *Homogeneous Catalysis*, Wiley-Interscience, New York, 1980, p. 171ff.

128. F. N. Tebbe, G. W. Parshall, and G. S. Reddy, *J. Am. Chem. Soc.*, **100**, 3611 (1978).

129. (a) F. N. Tebbe, G. W. Parshall, and D. W. Ovenall, *J. Am. Chem. Soc.*, **101**, 5047 (1979); (b) S. H. Pine, R. Zahler, D. A. Evans, and R. H. Grubbs, *ibid.*, **103**, 7358 (1981).

130. (a) M. M. Francl and W. J. Hehre, *Organometallics*, **2**, 457 (1983); (b) M. M. Francl, K. Dobbs, and W. J. Hehre, research in progress.

131. Grubbs has noted that bulky Lewis bases facilitate the formation of titanacyclobutanes. See: J. B. Lee, G. J. Gadja, W. P. Schaefer, T. R. Howard, T. Ikariya, D. A. Strauss, and R. H. Grubbs, *J. Am. Chem. Soc.*, **103**, 7358 (1981).

132. A series of papers marking the 100th anniversary of the van't Hoff-Le Bel hypotheses may be found in: *Van't Hoff-Le Bel Centennial*, O. B. Ramsay, ed., ACS Symposium Series, no. 12, American Chemical Society, Washington, D.C., 1975.

133. A. Baeyer, *Ber.*, 2269, 2277 (1885).

134. For a discussion, see: A. Greenberg and J. F. Liebman, *Strained Organic Molecules*, Academic Press, New York, 1978.

135. (a) J. M. Schulman, M. Sabro, and R. L. Desch, *J. Am. Chem. Soc.*, **105**, 743 (1983), and references therein; (b) K. B. Wiberg and J. J. Wendeloski, *ibid.*, **104**, 5679 (1982).

136. (a) J. Chandrasekhar, E. U. Wurthwein, and P. v. R. Schleyer, *Tetrahedron*, **37**, 921 (1981); (b) E. U. Wurthwein, J. Chandrasekhar, E. D. Jemmis, and P. v. R. Schleyer, *Tetrahedron Lett.*, 843 (1981); (c) E. U. Wurthwein, J. Chandrasekhar, E. D. Jemmis, and P. v. R. Schleyer, *ibid.*, 3306 (1982).

137. See: W. F. Maier and P. v. R. Schleyer, *J. Am. Chem. Soc.*, **103**, 1891 (1981).

138. For reviews, see: (a) G. L. Buchanan, *Chem. Soc. Rev.*, **3**, 41 (1974); (b) G. Kobrich, *Angew. Chem. Int. Ed. Engl.*, **12**, 464 (1973); (c) G. Szeimies, *Reactive Intermediates*, vol. 3, R. A. Abromovitch, ed., Plenum Press, New York, 1983.

139. D. C. Crans and J. P. Snyder, *J. Am. Chem. Soc.*, **100**, 7153 (1980).

140. (a) K. B. Wiberg and F. H. Walker, *J. Am. Chem. Soc.*, **104**, 5239 (1982); (b) K. B. Wiberg, *ibid.*, **105**, 1227 (1983).

141. (a) K. Wade, *Electron Deficient Compounds*, Nelson, London, 1971; (b) P. v. R. Schleyer, in H. C. Brown, ed., *The Nonclassical Ion Problem*, Plenum Press, New York, 1977, p. 13ff.

142. It has already been noted (Section 7.3.1d) that the methane dication with only six valence electrons prefers a square-planar geometry.

143. (a) J. B. Collins, J. D. Dill, E. D. Jemmis, Y. Apeloig, P. v. R. Schleyer, R. Seeger, and J. A. Pople, *J. Am. Chem. Soc.*, **98**, 5419 (1976); (b) M. B. Krogh-Jespersen, J. Chandrasekhar, E. U. Wurthwein, J. B. Collins, and P. v. R. Schleyer, *ibid.*, **102**, 2263 (1980); (c) E. U. Wurthwein, K. D. Sen, J. A. Pople, and P. v. R. Schleyer, *Inorg. Chem.*, **22**, 496 (1983); (d) C. Rhode, unpublished calculations.

144. For a general discussion, see: A. F. Wells, *Structural Inorganic Chemistry*, Fifth ed., Clarendon Press, Oxford, 1984.

145. For a discussion, see: P. C. Wailes, R. S. P. Coutts, and H. Weigold, *Organometallic Chemistry of Titanium, Zirconium and Hafnium*, Academic Press, New York, 1974.

146. (a) E. U. Wurthwein and P. v. R. Schleyer, *Angew. Chem., Int. Ed. Engl.*, **18**, 553 (1979); (b) M. B. Krogh-Jespersen, J. Chandrasekhar, E. U. Wurthwein, J. B. Collins, and P. v. R. Schleyer, *J. Am. Chem. Soc.*, **102**, 2263 (1980); (c) W. J. Pietro and W. J. Hehre, unpublished calculations.

147. Note, however, that strong σ donors, for example, Li, lead to stabilization of square-planar silane by causing it to change states. See ref. 146b.

148. Reviews: (a) I. R. Beattie, *Quart. Rev.*, **17**, 382 (1963); (b) R. S. Walton, *ibid.*, **19**, 126 (1965); (c) J. M. Dumas and M. Gomel, *Bull. Soc. Chim. France*, 9-10, 1885 (1974).

149. (a) H. H. Sisler, H. H. Batey, B. Pfahler, and R. Mattair, *J. Am. Chem. Soc.*, **70**, 3821 (1948); (b) V. V. Udovenko and Y. Y. Fialkov, *Zhur. Neorg. Khim.*, **2**, 434, 868 (1957); (c) Y. N. Nol'nor, *ibid.*, **4**, 2287 (1959); (d) M. F. Lappert, *J. Chem. Soc.*, 542 (1962); (e) I. R. Beattie, G. P. McQuillan, L. Rule, and M. Webster, *ibid.*, 1514 (1963); (f) J. M. Miller and M. Onyszchuk, *J. Chem. Soc.*, *A*, 1132 (1967); (g) G. Vandrish and M. Onyszchuk, *ibid.*, 3327 (1970); (h) P. G. Huett, K. Manning, and K. Wade, *J. Inorg. Nucl. Chem.*, **42**, 655 (1980).

150. U. Klabunde, F. N. Tebbe, G. W. Parshall, and R. L. Harlow, *J. Mol. Catal.*, **8**, 37 (1980).

151. F. N. Tebbe, private communication to W. J. Hehre.

152. K. D. Dobbs and W. J. Hehre, research in progress.

153. (a) Y. Apeloig, P. v. R. Schleyer, J. S. Binkley, J. A. Pople, and W. L. Jorgensen, *Tetrahedron Lett.*, 3923 (1976), and unpublished data. See also ref. 204 and (b) J. P. Richie, *ibid.*, **23**, 4999 (1983); (c) J. S. Binkley, *J. Am. Chem. Soc.*, **106**, 603 (1984).

154. (a) B. T. Luke, J. A. Pople, M. B. Krogh-Jespersen, Y. Apeloig, J. Chandrasekhar, and P. v. R. Schleyer, *J. Am. Chem. Soc.*, submitted; (b) H. Lischka and H. J. Kohler, *ibid.*, **105**, 6646 (1983); (c) J. Kalcher, A. Sax, and G. Olbrich, *Int. J. Quantum Chem.*, **25**, 543 (1984).

155. Streitwieser has recently called attention to the lower Coulombic energies associated with compact ion aggregate arrangements, that is

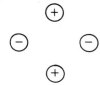

(a) A. Streitwieser, Jr. and J. T. Swanson, *J. Am. Chem. Soc.*, **105**, 2502 (1983). See also: (b) E. Kaufman, T. Clark, and P. v. R. Schleyer, *ibid.*, **106**, 1856 (1984), and references therein; (c) P. v. R. Schleyer, *Pure Appl. Chem.*, **55**, 355 (1983); (d) M. Hodoscek and T. Solmajer, *J. Am. Chem. Soc.*, **106**, 1854 (1984).

156. L. Andrews, *J. Chem. Phys.*, **50**, 4288 (1969).

157. See: B. T. Gowda and S. W. Benson, *J. Chem. Soc.*, *Faraday Trans. 2*, **79**, 663 (1983).

158. C. Rohde, unpublished calculations, Diplomarbeit, Erlangen, 1982.

159. (a) Y. Apeloig, P. v. R. Schleyer, J. S. Binkley, and J. A. Pople, *J. Am. Chem. Soc.*, **98**, 4332 (1976); (b) W. D. Laidig and H. F. Schaefer III, *ibid.*, **101**, 7184 (1979); (c) Y. Apeloig, T. Clark, A. J. Kos, E. D. Jemmis, and P. v. R. Schleyer, *Israel J. Chem.*, **20**, 43 (1980).

160. Indeed, the BBC analogue of dilithiomethane is estimated to prefer an anti-van't Hoff planar geometry over a tetrahedral arrangement by 17 kcal mol^{-1} [161].

This is greater than the best estimate of 9 kcal mol^{-1} for the difference in energies between (the favored) *cis* planar and tetrahedral dilithiomethane structures [143]. The same factors are operative in both systems. See Section 7.4.1a.

161. (a) K. Krogh-Jespersen, D. Cremer, D. Poppinger, J. A. Pople, P. v. R. Schleyer, and J. Chandrasekhar, *J. Am. Chem. Soc.*, **101**, 4843 (1979); (b) P. H. M. Budzelaar, K. Krogh-Jespersen, T. Clark, and P. v. R. Schleyer, *ibid.*, **107**, 2773 (1985).

162. M. M. Francl, W. J. Pietro, R. F. Hout, Jr., and W. J. Hehre, *Organometallics*, **2**, 815 (1983)

163. Minimal basis set calculations used here probably overestimate the importance of *p* functions. Calculations involving split-*d* basis sets show a reduction of the role played by valence *s*- and *p*-type functions. See, for example: J. A. Connor, I. H. Hillier, V. R. Saunders, M. H. Wood, and M. Barber, *Mol. Phys.*, **24**, 497 (1972).

164. For reviews, see: (a) F. J. Brown, *Prog. Inorg. Chem.*, **27**, 1 (1980), and references therein; (b) R. R. Schrock, *Acc. Chem. Res.*, **12**, 98 (1979).

165. (a) G. D. Stucky, unpublished results cited in: (b) C. D. Wood, W. J. McLain, and R. R. Schrock, *J. Am. Chem. Soc.*, **101**, 3210 (1979).

166. (a) R. R. Schrock, L. W. Meserle, C. D. Wood, and L. H. Guggenberger, *J. Am. Chem. Soc.*, **100**, 3793 (1978); (b) M. R. Churchill and F. J. Hollander, *Inorg. Chem.*, **17**, 1957 (1978), and references therein; (c) A. J. Schultz, R. K. Brown, J. W. Williams, and R. R. Schrock, *J. Am. Chem. Soc.*, **103**, 169 (1981); (d) M. R. Churchill and W. J. Young, *Inorg. Chem.*, **18**, 1830 (1979), and references therein; (e) A. J. Schultz, J. M. Williams, R. R. Schrock, G. A. Rupprecht, and J. D. Fellmann, *J. Am. Chem. Soc.*, **101**, 1593 (1979); (f) L. J. Guggenberger and R. R. Schrock, *ibid.*, **97**, 2935 (1975); (g) M. R. Churchill and W. J. Young, *Inorg. Chem.*, **18**, 2454 (1979), and references therein; (h) P. R. Sharp, S. J. Holmes, R. R. Schrock, M. R. Churchill, and H. J. Wasserman, *J. Am. Chem. Soc.*, **103**, 965 (1981); (i) M. R. Churchill, H. J. Wasserman, H. W. Turner, and R. R. Schrock, *ibid.*, **104**, 1710 (1982); (j) J. H. Wengrovius, R. R. Schrock, M. R. Churchill, and H. J. Wasserman, *ibid.*, **104**, 1739 (1982).

167. (a) R. J. Goddard, R. Hoffmann, and E. D. Jemmis, *J. Am. Chem. Soc.*, **102**, 7667 (1980); (b) M. M. Francl, W. J. Pietro, R. F. Hout, Jr., and W. J. Hehre, *Organometallics*, **2**, 281 (1983). For a discussion of the structural consequences of hyperconjugation in carbocations, see: (c) W. J. Hehre, *Acc. Chem. Res.*, **8**, 369 (1975), and ref. 7j.

168. M. M. Francl and W. J. Hehre, unpublished calculations.

169. (a) I. W. Bassi, G. Allegra, R. Scordamaglia, and G. Chioccola, *J. Am. Chem. Soc.*, **93**, 3787 (1971); (b) G. R. Davies, J. A. J. Jarvis, B. T. Kilbourn, and A. J. P. Pioli, *J. Chem. Soc., Chem. Commun.*, 677 (1971); (c) G. R. Davies, J. A. J. Jarvis, and B. T. Kilbourn, *ibid.*, 1511 (1971).

170. E. A. Mintz, K. G. Moloy, T. J. Marks, and V. W. Day, *J. Am. Chem. Soc.*, **104**, 4692 (1982).

171. Z. Dawoodi, M. L. H. Green, V. S. B. Mtetwa, and K. Prout, *J. Chem. Soc., Chem. Commun.*, 802 (1982).

172. J. L. Atwood, G. K. Barker, J. Holton, W. E. Hunter, M. F. Lappert, and R. Pearce, *J. Am. Chem. Soc.*, **99**, 6645 (1977).

173. J. C. Hayes and N. J. Cooper, *J. Am. Chem. Soc.*, **104**, 5570 (1982).

174. (a) B. J. Wakefield, *The Chemistry of Organolithium Compounds*, Pergamon Press, Oxford, 1974; (b) R. B. Bates and C. A. Ogle, *Carbanion Chemistry*, Springer Verlag, Heidelberg, 1983; (c) E. Buncel and T. Durst, eds., *Comprehensive Carbanion Chemistry*, Elsevier, Amsterdam, 1980; (d) J. C. Stowell, *Carbanions in Organic Synthesis*, Wiley, New York, 1979; (e) R. B. Bates, B. A. Hess, Jr., C. A. Ogle, and L. J. Schaad, *J. Am. Chem. Soc.*, **103**, 5052 (1981); (f) J. Klein, *Tetrahedron*, **39**, 2733 (1983).

175. Quoted in : B. K. Janousek and J. I. Brauman, in *Gas Phase Ion Chemistry*, M. T. Bowers, ed., Academic Press, New York, vol. 2, 1979, p. 53ff. This chapter provides a readable account of methods for the determination of gas-phase electron affinities and a survey of available data.

176. J. J. Brooke, W. Rhine, and G. Stucky, *J. Am. Chem. Soc.*, **94**, 7346 (1972).

177. K. Ziegler, K. Nagel, and M. Patheiger, *Z. Anorg. Allgem. Chem.*, **282**, 345 (1955). See also ref. 180i.

178. R. West and P. C. Jones, *J. Am. Chem. Soc.*, **91**, 6156 (1969).

179. T. L. Chwang and R. West, *J. Am. Chem. Soc.*, **95**, 3324 (1973).

180. (a) C. Chung and R. J. Lagow, *J. Chem. Soc., Chem. Commun.*, 1078 (1972); (b) L. G. Sneddon and R. J. Lagow, *ibid.*, 302 (1975); (c) L. A. Shimp and R. J. Lagow, *J. Am. Chem. Soc.*, **95**, 1343 (1973); (d) J. A. Morrison, C. Chung, and R. J. Lagow, *ibid.*, **97**, 5051 (1975); (e) L. A. Shimp and R. J. Lagow, *ibid.*, **101**, 2214 (1979); (f) L. A. Shimp, C. Chung, and R. J. Lagow, *Inorg. Chim. Acta*, **29**, 77 (1978); (g) L. A. Shimp and R. J. Lagow, *J. Org. Chem.*, **44**, 2232 (1979); (h) L. A. Shimp, J. A. Morrison, J. A. Gurak, J. W. Chinn, Jr., and R. J. Lagow, *J. Am. Chem. Soc.*, **103**, 5951 (1981); (i) J. A. Gurak, J. W. Chinn, Jr., and R. J. Lagow, *ibid.*, **104**, 2637 (1982); (j) F. J. Landro, J. A. Gurak, J. W. Chinn, Jr., and R. M. Newman, and R. J. Lagow, *ibid.*, **104**, 7345 (1982); (k) R. J. Lagow, private communication to P. v. R. Schleyer.

181. C. H. Wu and H. R. Ihle, *Chem. Phys. Lett.*, **61**, 54 (1979), and private communication.

182. J. Chandrasekhar, J. A. Pople, R. Seeger, U. Seeger, and P. v. R. Schleyer, *J. Am. Chem. Soc.*, **104**, 3651 (1982); (b) E. D. Jemmis, J. Chandrasekhar, E. U. Wurthwein, P. v. R. Schleyer, J. V. Chinn, Jr., F. J. Landro, R. J. Lagow, and J. A. Pople, *ibid.*, **104**, 4275 (1982); (c) P. v. R. Schleyer, B. Tidor, E. D. Jemmis, J. Chandrasekhar, E. U. Wurthwein, A. J. Kos, B. T. Luke, and J. A. Pople, *ibid.*, **105**, 484 (1983).

183. (a) M. Walczak and G. D. Stucky, *J. Am. Chem. Soc.*, **98**, 5531 (1976); (b) W. E. Rhine, J. N. Davis, and G. D. Stucky, *J. Organomet. Chem.*, **134**, 139 (1977).

184. A. J. Kos, E. D. Jemmis, P. v. R. Schleyer, R. Gleiter, U. Fischbach, and J. A. Pople, *J. Am. Chem. Soc.*, **103**, 4996 (1981).

185. For a discussion of dyotropic rearrangements (involving silicon), see: M. T. Reetz, *Adv. Organomet. Chem.*, **16**, 33 (1977).

186. A. Maercker, M. Theis, A. J. Kos, and P. v. R. Schleyer, *Angew. Chem., Int. Ed. Engl.*, **22**, 733 (1983).

187. J. W. F. L. Seetz, G. Schat, O. S. Akkerman, and F. Bickelhaupt, *J. Am. Chem. Soc.*, **104**, 6848 (1982).

188. P. v. R. Schleyer, A. J. Kos, and E. Kaufmann, *J. Am. Chem. Soc.*, **105**, 7617 (1983).

189. For example, at the 3-21G//3-21G level, 1,1-difluoroethylene is 11.5 kcal mol^{-1} lower in energy than *trans*-1,2-difluoroethylene. Note, however, that calculations at this level show *trans*-1,2-dichloroethylene to be 2.8 kcal mol^{-1} more stable than the corresponding 1,1-disubstituted compound. For a discussion, see Section 6.5.5.

190. For a simple rationalization, see: N. D. Epiotis, J. R. Larson, R. L. Yates, W. R. Cherry, S. Shaik, and F. Bernardi, *J. Am. Chem. Soc.*, **99**, 746 (1977).

191. See: A. T. Dadd and P. Hubberstey, *J. Chem. Soc., Faraday Trans. 1*, **77**, 1865 (1981).

192. P. v. R. Schleyer, E. U. Wurthwein, E. Kaufmann, T. Clark, and J. A. Pople, *J. Am. Chem. Soc.*, **105**, 5930 (1983).

193. (a) P. v. R. Schleyer, E. U. Wurthwein, and J. A. Pople, *J. Am. Chem. Soc.*, **104**, 5839 (1982). See also: (b) P. v. R. Schleyer, in *New Horizons of Quantum Chemistry*, P. O. Lowdin and A. Pullman, eds., Reidel, Dordrecht, 1983, p. 95.

194. Despite the known shortcomings of the Mulliken population analysis (see, for example: (a) D. L. Grier and A. Streitwiser, Jr., *J. Am. Chem. Soc.*, **104**, 3556 (1982)), we expect that the major conclusion reached here, a lack of buildup of charge on the central atom with increased lithium substitution, will stand up to more thorough analysis. See, however, (b) A. E. Ried and F. Weinhold, *ibid.*, **107**, 1919 (1985).

195. E. U. Wurthwein, A. J. Kos, and P. v. R. Schleyer, unpublished calculations.

196. (a) K. I. Peterson, P. D. Dao, and A. W. Castleman, Jr., *J. Chem. Phys.*, **80**, 563 (1985); (b) A. Simon, *Struct. Bonding (Berlin)*, **36**, 81 (1979); (c) E. U. Wurthwein, P. v. R. Schleyer, and J. A. Pople, *J. Am. Chem. Soc.*, **107**, 6973 (1985).

197. F. Jaffe, *J. Organomet. Chem.*, **23**, 53 (1970).

198. L. A. Shimp and R. J. Lagow, *J. Am. Chem. Soc.*, **93**, 1720 (1971).

199. (a) E. D. Jemmis, D. Poppinger, P. v. R. Schleyer, and J. A. Pople, *J. Am. Chem. Soc.*,

99, 5796 (1977); (b) E. D. Jemmis, J. Chandrasekhar, and P. v. R. Schleyer, *ibid.*, 101, 2848 (1979).

200. J. D. Dill, A. Greenberg, and J. F. Liebman, *J. Am. Chem. Soc.*, 101, 6814 (1979).

201. Reviews: (a) W. A. Bennett, *J. Chem. Ed.*, 44, 17 (1967); (b) L. T. Scott and M. Jones, Jr., *Chem. Rev.*, 72, 181 (1972); (c) J. F. Liebman and A. Greenberg, *ibid.*, 76, 312 (1976); (d) M. D. Newton, in *Applications of Electronic Structure Theory*, H. F. Schaeffer III, ed., Plenum Press, New York, 1977, vol. 4, p. 223ff; (e) M. P. Cava and M. J. Mitchell, *Cyclobutadiene and Related Compounds*, Academic Press, New York, 1967. For more recent work, see (f) H. Kollmar, F. Carrion, M. J. S. Dewar, and R. C. Bingham, *J. Am. Chem. Soc.*, 103, 5292 (1981), and references therein.

202. G. Rauscher, T. Clark, D. Poppinger, and P. v. R. Schleyer, *Angew. Chem., Int. Ed. Engl.*, 17, 276 (1978).

203. C. Rohde, K. Raghavachari, T. Clark, and P. v. R. Schleyer, to be published.

204. (a) J. P. Ritchie, *J. Am. Chem. Soc.*, 105, 2083 (1983); (b) R. L Disch, J. M. Schulman, and J. P. Ritchie, *ibid.*, 106, 6246 (1984). We thank Dr. Ritchie for preprints of these papers prior to publication.

205. J. Llanos, R. Nesper, and H. G. von Schnering, *Angew. Chem., Int. Ed. Engl.*, 22, 998 (1983).

206. V. R. Juza, W. Wehle, and H. V. Schuster, *Z. Anorg. Alleg. Chem.*, 352, 252 (1967).

207. (a) E. L. Muetterties, *Science*, 196, 839 (1977); (b) E. L. Muetterties, T. N. Rhodin, E. Brand, C. F. Brucker, and W. R. Pretzer, *Chem. Rev.*, 79, 91 (1979); (c) J. F. Hamilton and R. C. Baetzold, *Science*, 205, 1213 (1979).

208. (a) H. Stoll and H. Preuss, *Phys. Status Solidi B*, 53, 519 (1972); (b) A. B. Kunz, D. J. Mickish, and P. W. Deutch, *Solid State Commun.*, 13, 35 (1973); (c) H. Stoll and H. Preuss, *Phys. Status Solidi B*, 64, 103 (1974); (d) C. W. Bauschlicher, Jr., D. H. Liskow, C. F. Bender, and H. F. Schaefer III, *J. Chem. Phys.*, 62, 4815 (1975); (e) A. L. Companion, *Chem. Phys.*, 14, 7 (1976); (f) C. W. Bauschlicher, Jr., C. F. Bender, H. F. Schaefer III, and P. S. Bagus, *Chem. Phys.*, 15, 227 (1976); (g) W. Kolos, F. Nieves, and O. Novaro, *Chem. Phys. Lett.*, 41, 431 (1976); (h) R. B. Brewington, C. F. Bender, and H. F. Schaefer III, *J. Chem. Phys.*, 64, 905 (1976); (i) K. D. Jordan and J. Simons, *ibid.*, 65, 1601 (1976); (j) S. Goldstein, L. A. Curtiss, and R. N. Euwema, *J. Phys. Chem.*, 9, 4131 (1976); (k) R. F. Marshall, R. J. Blint, and A. B. Kunz, *Phys. Rev. B*, 13, 3333 (1976); (l) C. E. Dykstra, H. F. Schaefer III, and W. Meyer, *J. Chem. Phys.*, 65, 5141 (1977); (m) K. D. Jordan and J. Simons, *ibid.*, 67, 4027 (1977); (n) J. Wood, *Chem. Phys. Lett.*, 50, 129 (1977); (o) W. H. Fink, *J. Chem. Phys.*, 69, 3325 (1978).

209. (a) R. E. Honig, *J. Chem. Phys.*, 22, 126 (1954); (b) W. A. Chupka and M. G. Inghram, *J. Phys. Chem.*, 59, 100 (1955); (c) A. Kant and B. H. Strauss, *J. Chem. Phys.*, 45, 822 (1966); (d) L. Andrews and G. C. Pimentel, *ibid.*, 47, 2905 (1967); (e) K. R. Thompson, R. L. DeKock, and W. Weltner, Jr., *J. Am. Chem. Soc.*, 93, 4688 (1971); (f) H. R. Leider, O. H. Krikorian, and D. A. Young, *Carbon* 11, 555 (1973); (g) E. L. Lee and R. H. Sanborn, *High Temp. Sci.*, 5, 483 (1973); (h) S. L. Bennett, J. L. Margrave, J. L. Franklin, and J. E. Hudson, *J. Chem. Phys.*, 59, 5814 (1973); (i) K. A. Gingrich, A. Desideri, and D. L. Cooke, *ibid.*, 62, 731 (1975); (j) M. Moscovitz and J. E. Hulse, *ibid.*, 66, 3998 (1977); (k) M. Moscovitz and J. E. Hulse, *ibid.*, 67, 4271 (1977).

210. C. E. K. Mees and T. H. James, *The Theory of the Photographic Process*, 3rd. ed., MacMillan, New York, 1966.

211. R. A. Whiteside, R. Krishnan, J. A. Pople, M. B. Krogh-Jespersen, P. v. R. Schleyer, and G. Wenke, *J. Comput. Chem.*, 1, 307 (1980).

212. For previous theoretical work on Be_n species, see: Be_2: ref. 2081 and (a) J. H. Bartlett, Jr. and W. H. Furry, *Phys. Rev.*, 38, 1615 (1931); (b) W. H. Furry and J. H. Bartlett, Jr., *ibid.*, 38, 210 (1932); (c) B. J. Ransil, *Rev. Mod. Phys.*, 32, 239, 245 (1960); (d) S. Fraga and B. J. Ransil, *J. Chem. Phys.*, 36, 1127 (1962); (e) C. F. Bender and E. R. Davidson,

ibid., **47**, 4972 (1967); (f) P. Sutton, P. Bertocini, G. Das, A. C. Wahl, and O. Sinanoglu, *Int. J. Quantum Chem.*, **53**, 479 (1970); (g) G. Malli and J. Oreg, *Chem. Phys. Lett.*, **69**, 313 (1980); (h) B. H. Lengsfield III, A. D. McLean, M. Yoshimine, and B. Liu, *J. Chem. Phys.*, **79**, 1891 (1983); (i) R. J. Harrison and N. C. Handy, *Chem. Phys. Lett.*, **98**, 97 (1983); (j) R. O. Jones, private communication; Be_3: refs. 208g, 208m, and (k) J. F. Harrison and L. C. Allen, *J. Mol. Spectrosc.*, **29**, 432 (1967); (l) M. Bulski, *Mol. Phys.*, **29**, 1171 (1975); (m) S. Godleski, Ph.D. thesis, Princeton University, 1976; Be_4: refs. 208d, h, l, m.

213. W. L. Hildenbrand, L. P. Thread, E. Murad, and F. Ju, NASA Accession No. N65-29905, Rep. No. AFRPL-TR-65-95. A value of 76.5 ± 1.5 is given in L. Brewer, Lawrence Berkeley Lab. Rep. LBL-3720, Feb. 1975.

214. J. Donahue, *Structure of the Elements*, Wiley, New York, 1974.

215. Very-high pressure has been used to lower the enthalpy (by way of the PV term) of reaction transition structures in preference to those of reactants and products. In principle, the transition structure could be made a minimum (on the enthalpy surface) and hence be amenable to detection by spectroscopic means. For discussions of high-pressure chemistry, see: (a) K. E. Weale, *Chemical Reactions at High Pressure*, E. F. N. Spon, Ltd., London, 1967. Also noteworthy is Dunitz's use of X-ray structural data for large sets of molecules to examine reaction trajectories. See: (b) J. D. Dunitz, *X-Ray Analysis and the Structure of Organic Molecules*, Cornell University Press, Ithaca, N.Y., 1979, chap. 7; (c) E. Bye, W. B. Schweizer, and J. D. Dunitz, *J. Am. Chem. Soc.*, **104**, 5893 (1982), and earlier papers in this series.

216. (a) A. A. Frost and R. G. Pearson, *Kinetics and Mechanism*, 2nd. ed., Wiley, New York, 1961; (b) P. J. Robinson and K. A. Holbrook, *Unimolecular Reactions*, Wiley-Interscience, New York, 1972; (c) W. Forst, *Theory of Unimolecular Reactions*, Academic Press, New York, 1973; (d) H. Eyring, S. H. Lin, and S. M. Lin, *Basic Chemical Kinetics*, Wiley-Interscience, New York, 1980.

217. For a discussion and examples of the use of isotope effects in determining reaction mechanisms, see: (a) L. C. S. Melander and W. H. Saunders, Jr., *Reaction Rates of Isotopic Molecules*, Wiley-Interscience, New York, 1980. See also individual reviews in the recent series of monographs, *Isotope Effects in Organic Chemistry*, E. Buncel and C. C. Lee, eds., Elsevier, Amsterdam, in particular, (b) vol. 1, *Isotopes in Molecular Rearrangements*, 1975; (c) vol. 2, *Isotopes in Hydrogen Transfer Processes*, 1976; (d) vol. 5, *Isotopes in Cationic Reactions*, 1980.

218. For a discussion of the use of activation entropies for the elucidation of reaction mechanisms and specific examples, see ref. 216a. Also: S. W. Benson, *Thermochemical Kinetics*, 2nd ed., Wiley, New York, 1976.

219. For a recent review of numerical methods for optimization of equilibrium and transition structure geometries, see: H. B. Schlegel, *J. Comput. Chem.*, **3**, 214 (1982), and references therein.

220. (a) J. A. Pople, K. Raghavachari, M. J. Frisch, J. S. Binkley, and P. v. R. Schleyer, *J. Am. Chem. Soc.*, **105**, 6389 (1983), and references therein. For a review of other theoretical work on many of these same systems, see: (b) H. F. Schaefer III, *Acc. Chem. Res.*, **12**, 288 (1979).

221. See: (a) P. S. Skell, J. J. Havel, and M. J. McGlinchey, *Acc. Chem. Res.*, **6**, 97 (1973); (b) P. J. Stang, *ibid.*, **11**, 107 (1978); (c) P. J. Stang, *Chem. Rev.*, **78**, 383 (1978). Note, however, the failure to detect vinylidene in recent experiments: (d) C. Reiser, F. M. Lussier, C. C. Jensen, and J. I. Steinfeld, *J. Am. Chem. Soc.*, **101**, 350 (1979); (e) C. Reiser and J. I. Steinfeld, *J. Phys. Chem.*, **84**, 680 (1980).

222. (a) C. E. Dykstra and H. F. Schaefer III, *J. Am. Chem. Soc.*, **100**, 378 (1978). See also: (b) R. Krishnan, M. J. Frisch, J. A. Pople, and P. v. R. Schleyer, *Chem. Phys. Lett.*, **79**, 408 (1981). This paper contains a thorough summary of previous theoretical work and an assessment of the effects of basis set and correlation on the relative energies of

C_2H_2 isomers; (c) Y. Osamura, H. F. Schaefer III, S. K. Gray, and W. H. Miller, *J. Am. Chem. Soc.*, **103**, 1904 (1981); (d) L. B. Harding, *ibid.*, **103**, 7469 (1981); (e) J. A. Pople, *Pure Appl. Chem.*, **55**, 343 (1983); (f) R. F. C. Brown, *Pyrolytic Methods in Organic Chemistry*, Academic Press, New York, 1980, p. 125.

223.   A. Maki and R. Sams, *J. Chem. Phys.*, **75**, 4178 (1981).

224.   G. L. Blackman, R. D. Brown, P. D. Godfrey, and H. I. Gunn, *Chem. Phys. Lett.*, **34**, 241 (1975).

225.   C. F. Pau and W. J. Hehre, *J. Phys. Chem.*, **86**, 321 (1982).

226.   M. M. Maricq, M. A. Smith, C. J. S. M. Simpson, and G. B. Ellison, *J. Chem. Phys.*, **74**, 6154 (1981).

227.   (a) P. K. Pearson, G. L. Blackman, H. F. Schaefer III, B. Roos, and U. Wahlgren, *Astrophys. J. Lett.*, **184**, L19 (1973); (b) P. K. Pearson, H. F. Schaefer III, and U. Wahlgren, *J. Chem. Phys.*, **62**, 350 (1975); (c) P. Botschwina, E. Nachbaur, and B. M. Rode, *Chem. Phys. Lett.*, **41**, 486 (1976); (d) J. A. Pople, R. Krishnan, H. B. Schlegel, and J. S. Binkley, *Int. J. Quantum Chem.*, **14**, 545 (1978); (e) P. R. Taylor, G. B. Bacskay, N. S. Hush, and A. C. Hurley, *J. Chem. Phys.*, **69**, 1971 (1978); (f) L. T. Redmon, G. D. Purvis III, and R. J. Bartlett, *ibid.*, **72**, 986 (1980); (g) R. H. Nobes and L. Radom, *Chem. Phys.*, **60**, 1 (1981).

228.   See: G. Winnewisser, *Topics Current Chem.*, **99**, 39 (1981).

229.   This result is consistent with other high-level theoretical work. See: (a) R. H. Nobes, L. Radom, and W. R. Rodwell, *Chem. Phys. Lett.*, **74**, 269 (1980); (b) K. Raghavachari, M. J. Frisch, J. A. Pople, and P. v. R. Schleyer, *ibid.*, **85**, 145 (1982). This reference provides a full account of previous theoretical work. See also: (c) E. M. Evleth and A. Sevin, *J. Am. Chem. Soc.*, **103**, 7414 (1981). For discussions of the experimental evidence with regard to ethylidene, see ref. 220b and: (d) P. P Gaspar and G. S. Hammond, in *Carbenes*, M. Jones, Jr. and R. A. Moss, eds., Wiley, New York, 1975, vol. 2, chap. 6; (e) D. T. T. Su and E. R. Thornton, *J. Am. Chem. Soc.*, **100**, 1872 (1978).

230.   See: D. J. DeFrees and W. J. Hehre, *J. Phys. Chem.*, **82**, 391 (1978), and references cited therein to previous experimental and theoretical work.

231.   J. M. Lehn and B. Munsch, *Theor. Chim. Acta*, **12**, 91 (1968).

232.   See also: J. Demuynck, D. J. Fox, Y. Yamaguchi, and H. F. Schaefer III, *J. Am. Chem. Soc.*, **102**, 6204 (1980).

233.   C. F. Pau and W. J. Hehre, unpublished results.

234.   For other theoretical work, see: (a) J. A. Altmann, I. G. Csizmadia, K. Yates, and P. Yates, *J. Chem. Phys.*, **66**, 981 (1977); (b) R. R. Lucchese and H. F. Schaefer III, *J. Am. Chem. Soc.* **100**, 298 (1978); (c) J. A. Altmann, I. G. Csizmadia, M. A. Robb, K. Yates, and P. Yates, *ibid.*, **100**, 1653 (1978); (d) J. D. Goddard and H. F. Schaefer III, *J. Chem. Phys.*, **70**, 5117 (1979); (e) L. B. Harding, H. B. Schlegel, R. Krishnan, and J. A. Pople, *J. Phys. Chem.*, **84**, 3394 (1980); (f) M. J. Frisch, R. Krishnan, and J. A. Pople, *ibid.*, **85**, 1467 (1981); (g) G. F. Adams, G. D. Bent, R. J. Bartlett, and G. D. Purvis, *J. Chem. Phys.* **75**, 834 (1981); (h) J. D. Goddard, Y. Yamaguchi, and H. F. Schaefer III, *ibid.*, **75**, 3459 (1981); (i) H. B. Schlegel and M. A. Robb, *Chem. Phys. Lett.*, **93**, 43 (1982); (j) M. Dupuis, W. A. Lester, B. H. Lengsfield, and B. Liu, *J. Chem. Phys.*, **79**, 6167 (1983). For a discussion of experimental work, see: (k) W. M. Gelbart, M. L. Elert, and D. F. Heller, *Chem. Rev.*, **80**, 403 (1980).

235.   (a) C. F. Pau and W. J. Hehre, *J. Phys. Chem.*, **86**, 1252 (1982). For other experimental evidence supporting the existence of hydroxymethylene, see: (b) P. L. Houston and C. B. Moore, *J. Chem. Phys.*, **65**, 757 (1976); (c) J. R. Sodeau and E. K. C. Lee, *Chem. Phys. Lett.*, **57**, 71 (1978).

236.   For other theoretical work, see: C. J. Casewit and W. A. Goddard, *J. Am. Chem. Soc.*, **102**, 4057 (1980). This paper also contains references to the experimental literature dealing with N_2H_2 isomers.

237. For other theoretical work, see: (a) F. Bernardi, H. B. Schlegel, M. H. Whangbo, and S. Wolfe, *J. Am. Chem. Soc.*, **99**, 5633 (1977); (b) R. A. Eades, P. G. Gassman, and D. A. Dixon, *ibid.*, **103**, 1066 (1981).

238. (a) R. J. Saykally and R. C. Woods, *Ann. Rev. Phys. Chem.*, **32**, 403 (1981); (b) R. C. Woods, in *Molecular Ions: Geometric and Electronic Structures*, J. Berkowitz and K. O. Groeneveld, eds., Plenum Press, New York, 1983, p. 11; (c) C. S. Gudeman and R. J. Saykally, *Ann. Rev. Phys. Chem.*, **35**, 387 (1984).

239. Reviews: (a) D. H. Williams and I. Howe, *Principles of Organic Mass Spectrometry*, McGraw-Hill, London, 1972; (b) K. Levsen, *Fundamental Aspects of Organic Mass Spectrometry*, Verlag Chemie, New York, 1978; (c) A. L. Burlingame, T. A. Baillie, P. J. Derrick, and O. S. Chizhov, *Anal. Chem.*, **52**, 214 (1980); (d) A. Maccoll, *Org. Mass Spectrom.*, **17**, 1 (1982); (e) J. C. Lorquet, *ibid.*, **16**, 469 (1981); (f) K. Levsen and H. Schwarz, *Mass Spectrom. Rev.*, **2**, 77 (1983); (g) R. G. Cooks, J. H. Beynon, R. M. Caprioli, and G. R. Lester, *Metastable Ions*, Elsevier, Amsterdam, 1973; (h) J. L. Holmes and J. K. Terlouw, *Org. Mass Spectrom.*, **15**, 383 (1980).

240. (a) T. A. Lehman and M. M. Bursey, *Ion Cyclotron Resonance Spectroscopy*, Wiley, New York, 1976; (b) C. H. DePuy and V. M. Bierbahm, *Accounts Chem. Res.*, **14**, 146 (1981); (c) C. H. DePuy, J. J. Grabowski, and V. M. Bierbahm, *Science*, **218**, 955 (1982).

241. L. E. Snyder and D. Buhl, *Astrophys. J.*, **163**, L47 (1971).

242. L. E. Snyder and D. Buhl, *Bull. Am. Astron. Soc.*, **3**, 388 (1971).

243. W. Klemperer, *Nature*, **227**, 1230 (1970).

244. (a) U. Wahlgren, B. Liu, P. K. Pearson, and H. F. Schaefer, III, *Nature*, **246**, 4 (1973); (b) W. P. Kraemer and G. H. F. Diercksen, *Astrophys. J.*, **205**, L97 (1976).

245. (a) R. C. Woods, T. A. Dixon, R. J. Saykally, and P. G. Szanto, *Phys. Rev. Lett.*, **35**, 1269 (1975); (b) R. J. Saykally, P. G. Szanto, T. G. Anderson, and R. C. Woods, *Astrophys. J.*, **204**, L143 (1976); (c) R. A. Creswell, E. F. Pearson, M. Winnewisser, and G. Winnewisser, *Z. Naturforsch.*, **31A**, 221 (1976); (d) G. L. Blackman, R. D. Brown, P. D. Godfrey, and H. I. Gunn, *Nature*, **261**, 395 (1976).

246. R. H. Nobes and L. Radom, *Chem. Phys.*, **60**, 1 (1981).

247. See also: (a) E. Herbst and W. Klemperer, *Astrophys. J.*, **185**, 505 (1973); (b) E. Herbst, J. M. Norbeck, P. R. Certain, and W. Klemperer, *ibid.*, **207**, 110 (1976); (c) D. J. DeFrees, G. H. Loew, and A. D. McLean, *Astrophys. J.*, **257**, 376 (1982); (d) W. P. Kraemer and P. R. Bunker, *J. Mol. Spectrosc.*, **101**, 379 (1983).

248. The 6-311G*** basis set is closely related to 6-311G** (see Section 4.3.3b) but has two sets of d functions on nonhydrogen atoms. See ref. 246 for details.

249. (a) S. D. Tanner, G. I. Mackay, A. C. Hopkinson, and D. K. Bohme, *Int. J. Mass Spectrom. Ion Phys.*, **29**, 153 (1979); (b) S. D. Tanner, G. I. Mackay, and D. K. Bohme, *Can. J. Chem.*, **57**, 2350 (1979). See, however, (c) A. J. Illies, M. F. Jarrold, and M. T. Bowers, *J. Chem. Phys.*, **77**, 5897 (1982).

250. (a) C. S. Gudeman and R. C. Woods, *Phys. Rev. Lett.*, **48**, 1344 (1982). For mass spectrometric observation of HOC^+, see: (b) P. C. Burgers and J. L. Holmes, *Chem. Phys. Lett.*, **97**, 236 (1983), and ref. 249c.

251. R. C. Woods, C. S. Gudeman, R. L. Dickman, P. F. Goldsmith, G. R. Huguenin, W. M. Irvine, A. Hjalmarson, L. A. Nyman, and H. Olofsson, *Astrophys. J.*, **270**, 583 (1983).

252. A. G. Harrison, A. Ivko, and D. Van Raalte, *Can. J. Chem.*, **44**, 1625 (1966).

253. F. P. Lossing, *J. Am. Chem. Soc.*, **99**, 7526 (1977).

254. M. S. B. Munson and J. L. Franklin, *J. Phys. Chem.*, **68**, 3191 (1964).

255. M. A. Haney and J. L. Franklin, *Trans. Faraday Soc.*, **65**, 1794 (1969).

256. (a) R. D. Bowen and D. H. Williams, *J. Chem. Soc., Chem. Commun.*, 378 (1977); (b) P. v. R. Schleyer, E. D. Jemmis, and J. A. Pople, *ibid.*, 190 (1978); (c) M. M. Bursey, J. R. Hass, D. J. Harvan, and C. E. Parker, *J. Am. Chem. Soc.*, **101**, 5485 (1979).

257. (a) F. C. Fehsenfeld, D. B. Dunkin, and E. E. Ferguson, *Astrophys. J.*, **188**, 43 (1974); (b) K. Hiraoka and P. Kebarle, *J. Chem. Phys.*, **63**, 1688 (1975); (c) K. Hiraoka and P. Kebarle, *J. Am. Chem. Soc.*, **99**, 366 (1977).

258. W. J. Bouma, R. H. Nobes, and L. Radom, *Org. Mass Spectrom.*, **17**, 315 (1982).

259. J. D. Dill, C. L. Fischer, and F. W. McLafferty, *J. Am. Chem. Soc.*, **101**, 6531 (1979).

260. D. H. Williams, *Acc. Chem. Res.*, **10**, 280 (1977).

261. L. B. Harding, H. B. Schlegel, R. Krishnan, and J. A. Pople, *J. Phys. Chem.*, **84**, 3394 (1980).

262. (a) W. J. Bouma, R. H. Nobes, and L. Radom, *J. Am. Chem. Soc.*, **104**, 2929 (1982). Similar observations are reported in: (b) M. J Frisch, K. Raghavachari, J. A. Pople, W. J. Bouma, and L. Radom, *Chem. Phys.*, **75**, 323 (1983).

263. (a) W. J. Bouma, J. K. MacLeod, and L. Radom, *J. Am. Chem. Soc.*, **104**, 2930 (1982); (b) J. L. Holmes, F. P. Lossing, J. K. Terlouw, and P. C. Burgers, *ibid.*, **104**, 2931 (1982).

264. W. J. Bouma, B. F. Yates, and L. Radom, *Chem. Phys. Lett.*, **92**, 620 (1982).

265. W. J. Bouma, J. M. Dawes, and L. Radom, *Org. Mass Spectrom.*, **18**, 12 (1983).

266. (a) B. F. Yates, W. J. Bouma, and L. Radom, *J. Am. Chem. Soc.*, **106**, 5805 (1984). For a recent review, see: (b) L. Radom, W. J. Bouma, R. H. Nobes and B. F. Yates, *Pure Appl. Chem.*, **56**, 1831 (1984). An alternative term, *ion/dipole complexes*, has also been suggested. See: (c) H. Schwarz, *Nachr. Chem. Tech. Lab.*, **31**, 451 (1983); see also: (d) F. Marquin, D. Stahl, A. Sawaryn, P. v. R. Schleyer, W. Koch, G. Frenking and H. Schwarz, *J. Chem. Soc., Chem. Commun.*, 504 (1984).

267. J. Vogt, A. D. Williamson, and J. L. Beauchamp, *J. Am. Chem. Soc.*, **100**, 3478 (1978).

268. J. K. Terlouw, W. Heerma, and G. Dijkstra, *Org. Mass Spectrom.*, **15**, 660 (1980).

269. J. K. Terlouw, W. Heerma, and J. L. Holmes, *Org. Mass Spectrom.*, **16**, 306 (1981).

270. (a) R. H. Nobes, W. J. Bouma, and L. Radom, *J. Am. Chem. Soc.*, **105**, 309 (1983). Recent experimental results are consistent with the theoretical predictions: see (b) P. C. Burgers, J. L. Holmes, J. E. Szulejko, A. A. Mommers, and J. K. Terlouw, *Org. Mass Spectrom.*, **18**, 254 (1983); (c) F. Turecek and F. W. McLafferty, *ibid.*, **18**, 608 (1983).

271. See: (a) W. J. Bouma, R. H. Nobes, L. Radom, and C. E. Woodward, *J. Org. Chem.*, **47**, 1869 (1982). Also: (b) C. E. Dykstra, *J. Chem. Phys.*, **68**, 4244 (1978); (c) K. Tanaka and M. Yoshimine, *J. Am. Chem. Soc.*, **102**, 7655 (1980).

272. P. Saxe, Y. Yamaguchi, P. Pulay, and H. F. Schaefer III, *J. Am. Chem. Soc.*, **102**, 3718 (1980), and references therein.

273. F. W. Schneider and B. S. Rabinovitch, *J. Am. Chem. Soc.*, **84**, 4215 (1962).

274. (a) J. I. Brauman, W. N. Olmstead, and W. N. Lieder, *J. Am. Chem. Soc.*, **96**, 4030 (1974); (b) W. E. Farneth and J. I. Brauman, *ibid.*, **98**, 7891 (1976); (c) W. N. Olmstead and J. I. Brauman, *ibid.*, **99**, 4219 (1977). See also: (d) A. P. Wolf, P. Schueler, R. P. Pettijohn, K. C. To, and E. P. Rack, *J. Phys. Chem.*, **83**, 1237 (1979); (e) K. C. To, E. P. Rack, and A. P. Wolf, *J. Chem. Phys.*, **74**, 1499 (1981).

275. S. Wolfe, D. J. Mitchell, and H. B. Schlegel, *J. Am. Chem. Soc.*, **103**, 7692, 7694 (1981), and references therein to earlier theoretical work.

276. For a discussion, see: D. Farcasiu, *J. Chem. Ed.*, **52**, 76 (1975), and references therein. See also reference 2 given in ref. 275 above.

277. (a) K. Tanaka, G. I. Mackay, J. D. Payzant, and D. K. Bohme, *Can. J. Chem.*, **54**, 1643 (1976); (b) M. J. Pellerite and J. I. Brauman, *J. Am. Chem. Soc.*, **102**, 5993 (1980). For a comprehensive bibliography on solvent effects, see S. S. Shaik, *ibid.*, **106**, 1227 (1984).

278. D. K. Bohme and G. I. Mackay, *J. Am. Chem. Soc.*, **103**, 978 (1981).

279. K. Morokuma, *J. Am. Chem. Soc.*, **104**, 3732 (1982).

280. For general reviews, see: (a) J. D. Morrison and H. S. Mosher, *Asymmetric Organic Reactions*, Prentice-Hall, New York, 1971; (b) J. W. ApSimon and R. P. Seguin, *Tetrahedron*, **35**, 2797 (1979); (c) P. A. Bartlett, *ibid.*, **36**, 2 (1980).

281. See: (a) P. Caramella, N. G. Rondan, M. N. Paddon-Row, and K. N. Houk, *J. Am. Chem. Soc.*, **103**, 2438 (1981), and references therein. See also: (b) M. N. Paddon-Row, N. G. Rondan, and K. N. Houk, *ibid.*, **104**, 7162 (1982); (c) K. N. Houk, N. G. Rondan, Y. D. Wu, J. T. Metz, and M. N. Paddon-Row, *Tetrahedron*, **40**, 2257 (1984); (d) K. N. Houk, unpublished calculations. For other calculations of transition structures for olefin addition reactions, see ref. 282.

282. (a) T. Clark and P. v. R. Schleyer, *J. Organomet. Chem.*, **150**, 1 (1978); (b) R. W. Strozier, P. Caramella and K. N. Houk, *J. Am. Chem. Soc.*, **101**, 1340 (1979); (c) S. Nagase, N. K. Ray, and K. Morokuma, *ibid.*, **102**, 4536 (1980); (d) K. N. Houk, R. W. Strozier, M. D. Rozeboom, and S. Nagase, *ibid.*, **104**, 323 (1982); (e) H. B. Schlegel, *J. Phys. Chem.*, **86**, 4878 (1982); (f) O. Eisenstein, H. B. Schlegel, and H. B. Kayser, *J. Org. Chem.*, **47**, 2886 (1982).

283. This deduction was first made by Felkin [284a] and subsequently supported theoretically by Anh [282b, c]. See also refs. 281a, b and: N. G. Rondan, M. N. Paddon-Row, P. Caramella, J. Mareda, P. H. Mueller, and K. N. Houk, *J. Am. Chem. Soc.*, **104**, 4974 (1982).

284. (a) M. Cherest, H. Felkin, and N. Prudent, *Tetrahedron Lett.*, 2199, 2205 (1968); (b) N. T. Anh and O. Eisenstein, *Nouv. J. Chim.*, **1**, 61 (1977); (c) N. T. Anh, *Topics Current Chem.*, **88**, 145 (1980).

285. No barrier is calculated for addition of a proton to propene. Note, however, that the calculated STO-3G equilibrium geometry for hydrogen-bridged propyl cation [42d] incorporates an HCC bond angle of 59°.

286. The calculated transition structure for the addition to ethylene of borane complexed to a single water molecule (as the simplest possible model for the solution-phase reaction) is nearly identical to that for the process involving free BH_3. See: T. Clark, D. Wilhelm, and P. v. R. Schleyer, *J. Chem. Soc., Chem. Commun.*, 606 (1983).

287. (a) H. B. Burgi, J. D. Dunitz, J. M. Lehn, and G. Wipff, *Tetrahedron*, **30**, 1563 (1974); (b) G. Klopman, in *Chemical Reactivity and Reaction Paths*, G. Klopman, ed., Wiley-Interscience, New York, 1974, pp. 55–166; (c) K. N. Houk, in *Frontiers in Free Radical Chemistry*, W. A. Pryor, ed., Academic Press, New York, 1980, pp. 43–72.

288. (a) Y. Kishi, *Aldrichimica Acta* **13**, 23 (1980); (b) H. Nagaota and Y. Kishi, *Tetrahedron*, **37**, 3873 (1981).

289. Standard model bond lengths: $r(CC) = 1.54$ Å, $r(CO) = 1.36$ Å, $r(CH) = 1.09$ Å, $r(OH) = 0.96$ Å. All bond angles are tetrahedral.

290. (a) I. McCay, M. N. Paddon-Row, and R. N. Warrener, *Tetrahedron Lett.*, 1401 (1972); (b) M. N. Paddon-Row and R. N. Warrener, *ibid.*, 1405 (1972); (c) M. N. Paddon-Row, H. K. Patney, and R. N. Warrener, *J. Org. Chem.* **44**, 3908 (1979).

291. (a) M. M. Midland and Y. C. Kwon, *J. Am. Chem. Soc.*, **105**, 3725 (1983); (b) W. C. Still and J. C. Barrish, *ibid.*, **105**, 2487 (1983).

292. S. R. Moses, Y. D. Wu, N. G. Rondan, K. N. Houk, R. Schohe, V. Jager, and F. Fronczek, *J. Am. Chem. Soc.*, **106**, 3880 (1984).

293. (a) W. J. Hehre, unpublished calculations; (b) W. R. Rodwell, W. J. Bouma, and L. Radom, *Int. J. Quantum Chem.*, **18**, 107 (1980); (c) M. G. Rondan and K. N. Houk, *Tetrahedron Lett.*, 2519 (1984); (d) B. A. Hess, Jr. and L. J. Schaad, *J. Am. Chem. Soc.*, **105**, 7185 (1983); (e) F. Bernardi, M. A. Robb, H. B. Schlegel, and G. Tonachini, *ibid.*, **106**, 1198 (1984).

294. J. Breulet and H. F. Schaefer III, *J. Am. Chem. Soc.*, **106**, 1221 (1984).

295. Experimentally, the enthalpy difference between propene and separated allyl radical and hydrogen atom is 88 kcal mol^{-1}; H. M. Rosenstock, K. Draxl, B. W. Steiner, and J. T. Herron, *J. Phys. Chem. Ref. Data, Suppl. 1*, **6**, 1 (1977). Assuming no barrier for dissociation-recombination, this would provide a lower-energy pathway for degenerate hydrogen rearrangement than the concerted processes considered here.

296.   For discussions, see: (a) S. Forsen and M. Nilsson, in *The Chemistry of the Carbonyl Group*, vol. 2, J. Zabicky, ed., Wiley-Interscience, New York, 1970, p. 157ff; (b) E. N. Marvell and W. Whalley, in *The Chemistry of the Hydroxyl Group*, vol. 1, S. Patai, ed., Wiley-Interscience, New York, 1971, p. 719ff. See, however; (c) B. Capon, D. S. Rycroft, T. W. Watson and C. Zucco, *J. Am. Chem. Soc.*, **103**, 1761 (1981); (d) B. Capon and C. Zucco, *ibid.*, **104**, 7567 (1982).

297.   S. Saito, *Chem. Phys. Lett.*, **42**, 399 (1976).

298.   (a) S. K. Pollack and W. J. Hehre, *J. Am. Chem. Soc.*, **99**, 4845 (1977). The keto-enol enthalpy difference given in this paper (13.9 kcal mol^{-1}) was later reduced. See: (b) S. K. Pollack, Ph.D. Thesis, University of California, Irvine, 1980, p. 129ff.

299.   1,3-pentadiene: (a) W. R. Roth and J. Konig, *J. Liebigs Ann. Chem.*, **24**, 699 (1966). It has been shown experimentally that, for a substituted 1,3-pentadiene, hydrogen migration proceeds in a symmetry-allowed suprafacial fashion. See: (b) W. R. Roth and J. Konig, *Chem. Ber.*, **103**, 426 (1970). Cyclopentadiene: (c) W. R. Roth, *Tetrahedron Lett.*, 1009 (1964).

300.   For a recent review of hydrocarbon thermal rearrangements, see: J. J. Gajewski, *Hydrocarbon Thermal Isomerizations*, Academic Press, New York, 1981.

301.   S. D. Kahn, W. J. Hehre, M. D. Rondan and K. N. Houk, *J. Am. Chem. Soc.*, in press.

EPILOGUE

We are perhaps not far removed from the time when we shall be able to submit the bulk of chemical phenomena to calculation [1].
                                                    Joseph Louie Gay-Lussac 1778–1850

Were it not for matters of practicality, the application of quantum mechanics to chemistry would be straightforward and its impact immediate and dramatic. Although both computer and computer program technologies have blossomed in the past decades, and although they will continue to blossom, it is unlikely that *exact* solutions to the Schrödinger differential equation for many-electron systems will become available for decades to come. It is likely that most applications of theory will still need to be carried out at a level dictated in part by practical concerns.

Many properties of small molecules may *now* be accurately calculated using *approximate* quantum mechanical methods. Those of larger systems are also subject to calculation, with greater reservation but often with useful results. As the theoretical models evolve, combining evermore sophisticated treatments of electron correlation with larger and more complete basis sets, they will describe a chemistry which will gradually approach that which characterizes the exact solution of the Schrödinger equation.

Theory is at its best when it precedes rather than follows experiment. The potential of quantitative theoretical methods discussed in this book as *primary exploratory tools* is enormous. Through their application, entirely new and surprising chemistry of novel and unexpected molecules may be unravelled in exquisite detail. It has been said [2] that the true test of any theory is its ability to make risky predictions,

predictions which only that theory and no competitive theories could make. This applies particularly to areas where prior experience provides little or no guide, exactly those areas where truly new knowledge is to be gained.

The science of chemistry as we know it today is in the throes of revolution. Its extension from an experimental to a mathematical science is already well underway.

REFERENCES

1. J. L. Gay-Lussac, *Memoires de la Societe d'Aroueil*, **2**, 207 (1888).
2. K. R. Popper, *Conjectures and Refutations, The Growth of Scientific Knowledge*, Harper and Row, New York, 1963, p. 36.

INDEX

Note: Wherever convenient, chemical compounds have been referred to by name rather than molecular formula. A listing of molecular formulae, ordered:

$$C_mH_nX_oY_pZ_q \text{ (X,Y,Z, heteroatoms in alphabetical order of atomic symbols)}$$

for carbon-containing compounds; and ordered:

$$X_oH_nY_pZ_q \text{ (X,Y,Z, heteroatoms in alphabetical order of atomic symbols)}$$

for systems lacking carbon, appear following the last C or X entry in the index, respectively.

525